INTERFEROMETRY – RESEARCH AND APPLICATIONS IN SCIENCE AND TECHNOLOGY

Edited by **Ivan Padron**

Interferometry – Research and Applications in Science and Technology
Edited by Ivan Padron

Published by InTech
Janeza Trdine 9, 51000 Rijeka, Croatia

Publishing Process Manager Petra Nenadic

Technical Editor Teodora Smiljanic

Cover Designer InTech Design Team

Additional hard copies can be obtained from orders@intechopen.com

Interferometry – Research and Applications in Science and Technology,
Edited by Ivan Padron
p. cm.
ISBN 978-953-51-0403-2

For EU product safety concerns:
IN TECH d.o.o., Prolaz Marije Krucifikse Kozulić 3, 51000 Rijeka, Croatia,
info@intechopen.com or visit our website at intechopen.com.

Meet the editor

Ivan Padron was born in Camajuani, Cuba in 1970. He received his bachelor degree in Physics from the University of Havana in 1993, and his Ph. D in Applied Physics from New Jersey Institute of Technology in 2010. His research has been directed towards diaphragm based Fabry-Perot interferometric sensors. His critical thinking and enterprising personality where acknowledged in two U.S. patents and one patent application, and offered him the opportunity of publish his work in varies peer-reviewed scientific journals, participates in numerous international conferences and to serve as a peer reviewer of the IEEE Sensor Journal. He is a member of the American Physical Society and IEEE Professional Association. His research interests include optical sensors and the application of physics to life sciences and medicine.

Contents

Preface

This book compiles recent studies on interferometry and its applications in science and technology. It is intended as an up-to-date reference of theoretical and experimental aspects of interferometry and their applications.

The book is divided in two sections. The first one is an overview of different interferometry techniques and their general applications. Chapter 1 is concentrated on the classical field interferometry. This chapter describes and explains the structures of different type of fiber interferometers, and the standard methods for signal recovering. Chapter 2 is an overview of the heterodyne interferometry and its applications. It includes some of recent development techniques and a summary of the advantages and disadvantages of the heterodyne interferometer. Chapter 3 describes the Phase-shifting interferometry through an experimental set-up for a polarizing two-window phase-grating common-path interferometer. The system is able to obtain $n = (N+1)$ interferograms with only one shot. Chapter 4 introduces vertical scanning interferometry and shows that the use of phosphor-based LED on vertical scanning interferometry affects the repeatability and accuracy of vertical scanning interferometry, especially repeatability. However, the undesired effects of phosphor-based white LED can be removed by applying a constraint on the input to existing reconstruction algorithm. In Chapter 5 is developed and implemented a similariton based self-referencing method of spectral interferometry for the complete characterization of femtosecond signal. The method is based on the similariton generation from the part of signal and its use as a reference for the interference with the signal in the spectrometer. Chapter 6 offers an overview of Low coherence interferometry including the Basis of Low Coherence Interferometry and Experimental configurations, covering aspects of the spectral domain low coherence Interferometry such as the interference signal, detection system, signal processing and parameter limits. Chapter 7 presents a reliable hybrid method for characterizing stress around the circular hole in a tensile-loaded steel plate. The method utilized only few micro-scale x-displacement data measured by speckle interferometry, in conjunction with phase shifting method using Fourier transform to calculate stress components and eventually stress concentration at an angle of 90 degrees. Chapter 8 is discusses the phenomenon of optical interference for two waves elliptically polarized, and demonstrates with a numerical analysis and a computer simulation the viability of a new method of phase-shifting based on the amplitude variation of two fields considered as the reference beams in a scheme of a three beam interferometer. In Chapter 9, based on the interference of two waves monochromatic and coherences, the phase shifting interferometry is studied and a method to reduce the number of captures needed is proposed on the basis of grating interferometry and modulation of linear polarization. Chapter 10 introduces a new bio-optical method "Path length resolved optical Doppler perfusion monitoring," to determine path length distributions of multiple scattered light in static and dynamic turbid media using phase modulated coherence gated interferometry. Also, is presented the first path length resolved Doppler measurements of multiply scattered light from human skin. Chapter 11 describes two topics related to interferometric measurement in shock tube experiment. Laser Interferometry Computed Tomography technique is applied to the measurement of high-speed, unsteady and 3D flow field induced by discharging shock waves, and Laser Differential Interferometer is applied to velocity and density measurement in micro-scale shock tube.

The second section is devoted to more specific interferometry applications including in chapter12 interferometry for magnetic fusion plasmas, followed by Measurement of Static and Dynamic Phase Objects using Simultaneous Phase Shifting Shearing Interferometry in chapter 13. Chapter 14 offers a biological application of laser interferometry in the Analysis of liposome diffusion. In chapter 15, theoretical and experimental evidences of π phase-shifts in the Fourier spectra of phase gratings and phase grids is presented, which are of considerable relevance when gratings or grids are used for interferometric applications, specifically in Phase shifting interferometry. Chapter 16 proposes a new switching method for in-service optical transmission lines that transfer live optical signals. The method uses optical fibers, instead of using electric apparatus to control the transmission speed. Chapter 17 introduces a novel method for testing large aspherical surfaces by subaperture stitching interferometry. Chapter 18 shows two methods of getting high order momentum states by resonant superradiant scattering and by a sequence of pulsed standing waves, corresponding to traveling wave scattering and standing wave scattering, respectively. Chapter 19 describes the thickness measurement of photoresist thin films using interferometry techniques. Chapter 20 presets a real-time 2-D HI technique based on the use of the correlation image sensor as a two-dimensional array of pixels, each of which simultaneously demodulate the amplitude and phase of incident heterodyne beams at an ordinary frame rate. Chapter 21 is an application of interferometry to analysis of polymer-polymer and polymer-solvent interactions. Chapter 21 describes the baseline approach of radio interferometry based localization and details the many developments that occurred since its introduction; including the mathematical foundation of the localization method and a method for tracking mobile nodes in Wireless Sensor Networks utilizing the phenomena of Doppler effects and radio interferometry.

The book has been made possible as the outcome of the outstanding work done by all authors and by the professional assistance of Publishing Process Manager, Ms. Petra Nenadic, during all phases of editing.

Dr Ivan Padron
New Jersey Institute of Technology,
USA

Part 1

Interferometry Methods and Research

1

Optical Fiber Interferometers and Their Applications

Ali Reza Bahrampour, Sara Tofighi, Marzieh Bathaee and Farnaz Farman
Sharif University of Technology
Iran

1. Introduction

Interference as a wave characteristic of the electromagnetic wave has many applications in science, technology and medicine (Grattan & Meggit, 1997; Wang et al., 2011). The fringe visibility of the first order interference experiments such as the famous double slit Young experiment and Michelson interferometer, is determined by the first order correlation function (Gerry & Knight, 2005). The first order interference is also called the field interference. In Hanbury-Brown and Twiss (HBT) experiment, fringes are due to the intensity interference and visibility is determined by the second order correlation function. (Brown & Twiss, 1956; Scully & Zubairy 2001). In quantum optics, nonlinear Lithography and quantum Lithography, interferometry based on higher order correlation function is of prime importance (Bentley & Boyd, 2004; Boto et al., 2000). However in all of these interferometries, the fringe pattern depends on the optical path difference (OPD) and feature of light source. This chapter is concentrated on the classical field interferometry. The fringe existence is a characteristic of spatial or temporal coherences between the two light beams.

The phenomenon of interference of light is used in many high precision measuring systems and sensors. The optical path can be controlled by optical waveguides and optical fibers. The use of optical fibers allows making such devices extremely compact and economic.

Among the lots of advantages of optical fibers is their ability to reduce the effects of wave front distortion by the atmospheric turbulence and compact beam-splitter and combiner. These abilities made optical fiber as a suitable medium for transportation of light in long baseline interferometers which are used for gravitational wave detection, intruder sensor, structural health monitoring and long length leak detection systems (Sacharov, 2001; Cahill, 2007; Cahill & Stokes, 2008; Jia et al., 2008; Mishra & Soni, 2011, Bahrampour et al., 2012).

Other advantages that make optical fibers become useful elements in sensing technologies are high elongation sensitivity, fast response to internal or external defects such as temperature and tension, electromagnetic noise disturbance immunity, less power consumption and potential for large scale multiplexing (Higuera & Miguel, 2002).

In this chapter the different structures of optical fibers which are important in fiber interferometry are taken into consideration. The structures of different types of fiber interferometers are described. The sensitivity of coherent light optical fiber interferometers is compared with those of the incoherent and white light optical fiber interferometers. The standard methods for signal recovering are explained. A brief discussion on the noise sources appears in this chapter. Due to the immunity of the optical fibers to the lightening and electromagnetic noise, optical fibers are suitable sensors for transient measurement in harsh environments such as current measurement in high voltage transformers (Grattan & Meggit, 1999). The optical fiber hydrophone systems are based on elasto-optic effect in optical fiber coil, which is installed in one arm of an optical fiber interferometer (OFI) (Freitas, 2011). The optical fiber interferometers can be employed as biochemical sensors (Gopel et al., 1991). The cooperation of optical fiber interferometry and Plasmon can improve the sensitivity of biosensors to one molecule detection system (De Vos et al., 2009). The mechanical quantities such as pressure, velocity, acceleration and displacement can be measured by optical fiber interferometers (Shizhuo et al., 2008). Among a lot of applications of optical fiber interferometers, only some applications such as linear and nonlinear photonic circuits and distributed optical fiber sensors are mentioned in this chapter.

2. Optical fibers structures

2.1 Standard fibers

An optical fiber is a cylindrical structure that transports electromagnetic waves in the infrared or visible bands of electromagnetic spectrum. In practice optical fibers are highly flexible and transparent dielectric material. The optical fiber consists of three different layers. Core is the central region which is surrounded by the cladding. These two layers are protected by protective jacket. The core refractive index can be uniform or graded while the cladding index is typically uniform. For light guiding, it is necessary that the core index be greater than the cladding index. Most of the light energy propagates in the core and only a small fraction travels in the cladding. The cladding radius is so large that the jacket has no effect on the light propagation in the optical fiber structure.

Depending on the dimensionless frequency $v = 2\pi a(n_{co}^2 - n_{cl}^2)^{1/2}/\lambda$ where a is the core radius, λ is the wavelength of the light in free space, n_{co} and n_{clad} are the core and clad refractive indices respectively, optical fibers are divided into multimode ($v \gg 1$) and single mode fibers ($0 < v < v_c$), where v_c is cutoff frequency (Agrawal, 2007).

The optical fibers whose core and cladding have very nearly the same refractive index are named weakly guiding fibers. The corresponding eigen value equation is simpler than the exact fiber characteristic equation. The notation $LP_{\nu,\mu}$ introduces the weakly guiding modes. The fundamental mode $HE_{1,1}$is denoted by $LP_{0,1}$(Okamoto, 2006). The normalized propagation constant versus the dimensionless frequency is called the dispersion curve.

Depending on the coupling and optical fiber physical parameters, bounded, radiation and evanescent modes can exist in an optical fiber. The total incident power can be transported by the bounded and radiation modes while evanescent modes store power near the excitation source (Snyder, 1983).

2.2 Polarization maintained optical fibers

Birefringent optical fibers are those fibers that display two distinct refractive indices depending on the polarization direction of the light entering into them. The two principal axes of the birefringent fibers are named the fast and slow axis. For a light beam whose polarization aligned with one of the principal axes of the birefrigent fiber, the light propagates without any disturbance in its polarization state. The birefringence parameter of the fiber is defined by the difference between the two refractive indices corresponding to the two principal axes $B = n_s - n_f$, where n_s and n_f are the refractive indices of the slow and fast axis respectively. Sometimes birefringence is defined in terms of the fiber beat length $L_B = \lambda/B$ that is defined as the length of fiber over which the phase difference between the fast and slow waves becomes 2π. The beat length should be smaller than the perturbation periods introduced in the drawing process as well as the physical bends and twists. Consequently short beat length fibers preserve the polarization direction. This kind of fibers are called Polarization Maintained optical Fibers and denoted by PMF. Several types of PMFs are shown in Fig. 1 (Okamoto, 2006).

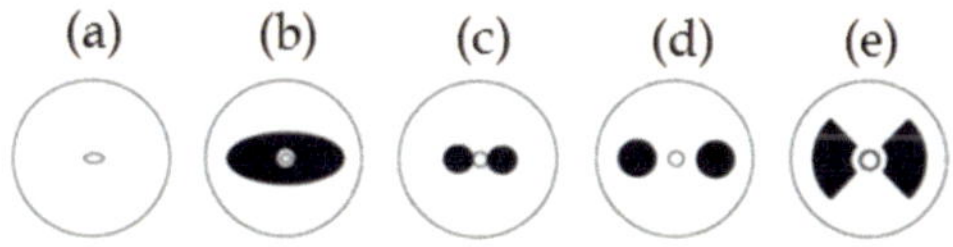

Fig. 1. Cross section of (a) elliptical core fiber (b) elliptical jacket fiber (c) side tunnel fiber (d) PANDA fiber (e) Bow-tie fiber.

2.3 Photonic crystal fibers

Light propagation in standard optical fibers and PMFs is based on the total internal reflection effect. Bragg diffraction effect can also be employed to confine the light in the core of fiber with periodic structure in the cladding. The micro structured fiber which is also called photonic crystal fiber (PCF) as shown in Fig. 2, consists of numerous air holes within a silica host. Usually the air holes are in a periodic arrangement around silica or a hollow core. The silica core PCF is called holey fiber, high delta or cobweb fiber while hollow core is named photonic band gap fiber (PBGF). The simplest structure of the holey fiber is a regular hexagonal lattice of small holes with a defect in the center such that the hole in the center is missed. In the holey fibers, guiding mechanism is also based on the total internal reflection. Air holes in the cladding area cause an effective lowering of the average refractive index (Poli, et al., 2007). In hollow-core fibers, field confinement in the air core is based on the band gap effect.

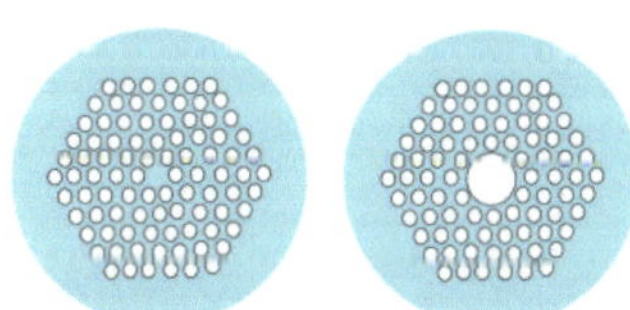

Fig. 2. Cross section of holey fiber (left) and hollow fiber (right).

2.4 Slab optical waveguide

Optical fibers are suitable transmission lines for several applications such as high capacity long-haul communication networks and long length optical interferometry. In many applications such as integrated circuits, the transmission length is less than one millimeter.

Optical waveguides are proper for such applications. A dielectric waveguide consists of a dielectric with refractive index n_1 which is deposited on a substrate with refractive index n_2. The refractive index of the medium above the layer n_1 is indicated by n_3. To achieve true guiding modes, it is necessary that n_1 be larger than n_2 and n_3. The propagating modes of the slab waveguides are TE and TM modes. The mathematical mode analysis of slab waveguides can be found in any standard text book (Adams, 1981). Depending on the propagation constant β, modes of narrow dielectric strip waveguide are also classified into the bounded, radiation and evanescent modes (Adams, 1981; Snyder, 1983).

2.5 Fiber Bragg gratings

The modes of optical fibers and waveguides are propagated without coupling to each other in the absence of any perturbation. Coupling to the desired modes can be controlled by changing the amplitude and the phase of the perturbation in the optical fiber. The coupled mode theory (CMT) can be found in the standard text books (Huang, 1984). If the refractive index of the core varies periodically, due to the Bragg diffraction effect, scattering from different periods can be constructive for some frequencies and destructive for the other ones. Depending on the period length, the periodic structures are classified as either long period grating (LPG) or fiber Bragg grating (FBG). The period of the LPG is of the order of micrometer while in the FBG, it is of the order of nanometer. The operation of the LPG is on the basis of coupling the fundamental core mode to higher order co-propagating cladding modes. The coupling wavelength is obtained by the linear momentum conservation law or the phase matching equation $\lambda = (\beta_1 - \beta_2)\Lambda$, where β_1 and β_2 are the core and cladding mode propagation constants respectively and Λ is the period of the LPG (Kashyap, 1999).

FBG can be employed as a frequency selective reflector or a polarization selective rotator. In the reflector state, the forward modes are coupled to the backward modes. While in the polarization rotator, a mode with a definite polarization is coupled to another mode with different polarization. In the frequency selective reflector, coupling to the backward modes occurs in a narrow range of wavelengths around the wavelength for which Bragg condition is satisfied $\lambda = 2n_{eff}\Lambda$, where n_{eff} is the effective refractive index of the core. Bandwidth of FBG is typically below 1nm and depends on the amount of refractive index variation and the length of FBG. The governing equations of the FBG can be obtained from the conservation of energy and momentum (Kashyap, 1999; Chen, 2006).

Depending on the application of FBG, the period of the structure can vary in a definite way or randomly along the optical fiber core. This structure is named chirped FBG which has many applications in optical networks and sensors (Kashyap, 1999; Rao, 1997).

3. Basic optical fiber interferometer configurations

Interferometry is based on the superimposing of two or more light beams to measure the phase difference between them. Interferometer utilizes two light beams with the same frequency. Typically an incident light beam of interferometer is split into two or more parts and then recombine together to create an interference pattern. The integer number of wavelength for the optical path difference between the two paths corresponds to constructive points and odd number of half wavelengths corresponds to destructive points of the interference pattern. So in the output optical spectrum of the optical fiber interferometer (OFI), the position of minimum can be shifted to maximum position if the optical path difference varies by odd number of half wavelengths. At least two optical paths are necessary for an interfererometery experiment. These optical paths can be in one optical fiber with two or more different optical fiber modes. Each of modes defines one optical path for the interferometer such as the Sagnac interferometer where the optical paths are defined by the clockwise and counter clockwise modes. The optical paths can be defined by separate optical fibers such as Mach-Zehnder OFI. There are many interferometer configurations that have been realized with the optical fiber. To see the principle of their operation, the detail of some interferometers such as Sagnac, birefringence OFI, Mach-Zehnder, Michelson, Moiré and Fabry-Perot interferometer are presented.

3.1 Sagnac optical fiber interferometer

The configuration of a Sagnac optical fiber is illustrated by Fig. 3. The optical source is a single mode stabilized coherent semi-conductor or Erbium doped optical fiber laser. The laser output beam is assumed to be well collimated with uniform phase. The laser beam enters the lossless 3dB optical fiber coupler (OFC). At the OFC the injected light splits into two parts with equal intensity that each of them travels around single mode optical fiber coil in opposite directions. The output of Sagnac coil is guided toward a single detector.

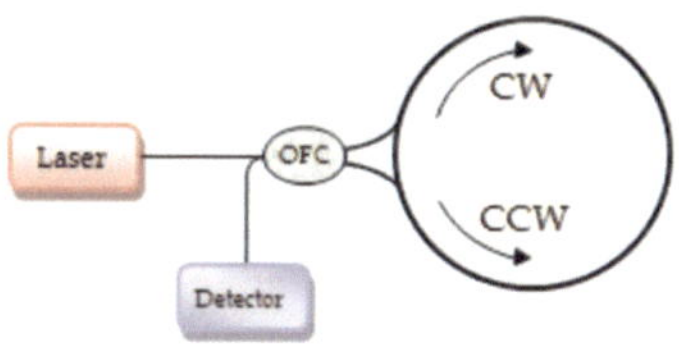

Fig. 3. A schematic diagram of Sagnac fiber interferometer.

Due to this specific configuration, fiber Sagnac interferometer has been used for rotation sensing primarily. In a non-rotating Sgnac interferometer, the clockwise (CW) and counter clockwise (CCW) modes are in phase while for a rotating Sagnac configuration due to the rotating velocity, the optical path of one of the modes is shorten and the other one is lengthen. The Sagnac effect causes the interference spectrum depends on the angular frequency of the setup (Sagnac, 1913). Analysis can be based on the Doppler frequency difference between the CW and CCW modes. The detector output frequency is the beating frequency of CW and CCW modes. When rotational axis is oriented along the optical fiber coil axis, the phase difference of CW and CCW modes is $\Delta\Phi = 8\pi NA\Omega/\lambda c$ (Burns, 1993; Vali & Shorthill, 1976), where λ is the free space optical wavelength, 'A' is the area of Sagnac coil, N is the number of the coil turn and Ω is the angular velocity. The sensitivity is the ratio of the phase difference to the angular velocity $S = 8\pi NA/\lambda c$, which is increased by increasing the coil radius, total fiber length and laser frequency. Optical fiber loss and packaging criteria limit the total fiber length and coil radius respectively.

Sagnac fiber interferometers can also be employed for sensing nonreciprocal and time-varying phenomena. So they become applicable tools for detection current, acoustic wave, strain and temperature. The optical gyroscope based on sagnac interferometer is commercially available (Bohnert et al., 2002; Lin et al., 2004; Starodumov et al., 1997; Dong & Tam, 2007; Fu et al., 2010)

3.2 Modal optical fiber interferometer

The modal interferometers are based on the difference between velocities of two different modes. Typically the first two modes of step index fiber like LP_{01} and LP_{11} or the HE_{11} and HE_{21} can be employed to design the modal interferometers. Also the two eigen polarizations of PMF are employed for modal interferometry (Bahrampour et al., 2012). The holy structure fibers have unique modal properties that are not possible with conventional optical fibers. Fig. 4 (a) and (b) show the cross section of high birefringence photonic crystal fiber (HiBi-PCF) and polarization maintaining photonic crystal fiber (PM-PCF) respectively (Villatoro, 2009).

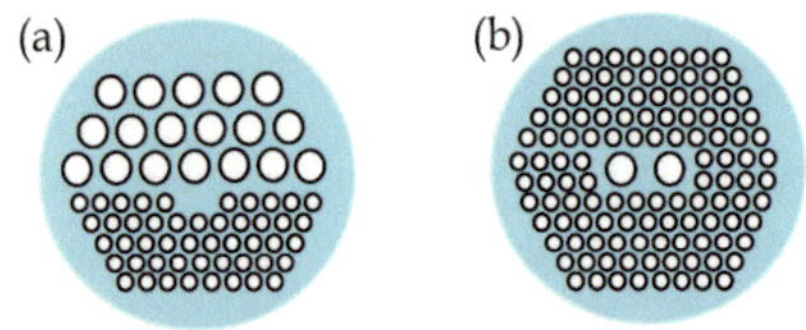

Fig. 4. Cross section of (a) HiBi-PCF (b) PM-PCF.

The PCFs have the possibility for the detection, sensing or spectroscopic analysis of gasses and liquids. In PCF a fraction of light penetrates into the voids for interaction and detection of gasses or liquids by spectroscopic methods (Villatoro et al., 2009). The holey and hollow fibers have their own advantages. Holey fiber which is filled with the desired gas or liquid, interacts with evanescent field which is only a few percent of total light power, while in hollow fiber, the fiber core is filled by gas or liquid and interacts with core light which is more than 90% of the total light power. The silica core single mode PCF bandwidth is more than one thousand nanometer which is much greater than those of an air core PCF fiber. A nano layer of rare metal coating on the surface of core and voids causes Plasmon-light interaction in PCF and extremely enhances the interferometer sensitivity (Hassani & Skorobogatiy, 2006). However the compact simple modal fiber interferometers depending on the fiber type such as Panda or birefringent PCF, can be employed in long lengths and short lengths applications (Villatoro et al., 2006).

3.3 Mach-Zehnder optical fiber interferometer

A schematic of conventional Mach-Zehnder OFI is sketched in Fig. 5.

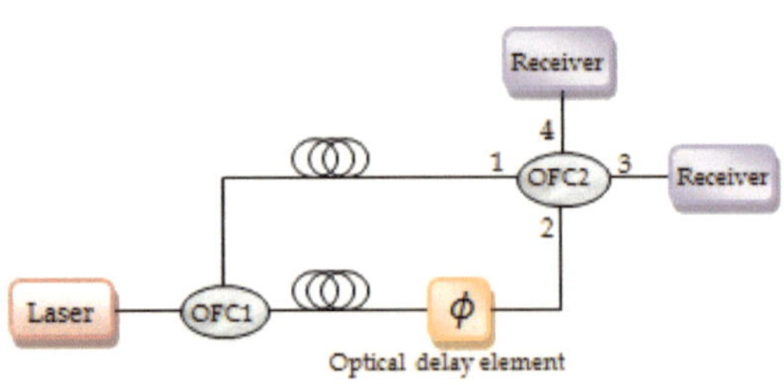

Fig. 5. A Schematic of optical fiber Mach-Zehnder interferometer.

By employing the commercial $(N \times N)$ coupler and single mode optical fibers, it is easy to construct the N-path interferometer. A schematic of N-path Mach-Zehnder interferometer is presented in Fig. 6. Each lossless linear multi port coupler is described by a $3N \times 3N$ unitary matrix. If N inputs and N outputs are linear polarized, linear coupler can be characterized by $N \times N$ matrix. As an example a symmetric 3×3 fiber coupler (tritters) which is commercially available, is described by 3×3 matrix (Weihs et al., 1996)

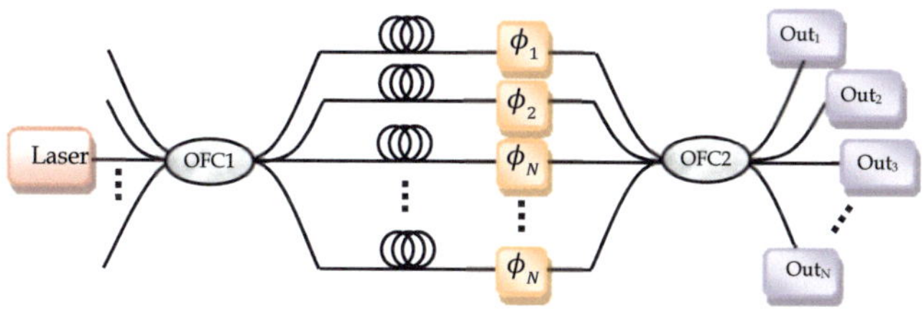

Fig. 6. A schematic configuration of N-path Mach-zehnder interferometer.

The 3-paths Mach-Zehnder interferometer is described by the product of two coupler matrices T and diagonal phase matrix $P = diag(e^{i\varphi_1}, e^{i\varphi_2}, e^{i\varphi_3})$ where φ_i(i=1,2,3) is the phase of the i-th path ($M = TPT$). This analysis is restricted to beams with identical polarization thus a scalar analysis is sufficient. It is assumed that only one of the input fields is nonzero. So the input field vector is denoted by $E_{in} = (E_{1_{in}}, 0, 0)$. The output field $E_{out} = (E_{1_{out}}, E_{2_{out}}, E_{2_{out}})$ is determined by the Mach-Zehnder transformation matrix $E_{out} = ME_{in}$. The output intensities $I_n = |E_{n_{out}}|^2$(n=1,2,3) versus the input intensity $I_o = |E_{1_{in}}|^2$ are given in the following:

$$I_n = \frac{I_0}{9}[3 + 2\cos(\varphi_{12} + \theta_n) + 2\cos(\varphi_{23} + \theta_n) + 2\cos(\varphi_{31} + \theta_n)];\ n = 1,2,3 \tag{1}$$

where $(\theta_1, \theta_2, \theta_3) = (0, -\frac{2\pi}{3}, \frac{2\pi}{3})$ and $\varphi_{ij} = \varphi_i - \varphi_j$ is the phase difference between the i-th and j-th branches. Above results are based on the loss-less fiber. For lossy fibers the phase matrix P is replaced by matrix $P' = diag(a_1, a_2\exp(i\varphi_{12}), a_3\exp(i\varphi_{13}))$ where a_n (n=1,2,3) is the transmission coefficient of the n-th optical fiber branch. The output intensities at the output of a 3×3 lossy Mach-Zehnder interferometer are:

$$\begin{aligned} I_n = \frac{I_0}{9}[&{a_1}^2 + {a_2}^2 + {a_3}^2 + 2a_1a_2\cos(\varphi_{12} + \theta_n) \\ &+ 2a_2a_3\cos(\varphi_{23} + \theta_n) + 2\,a_1a_3\cos(\varphi_{31} + \theta_n)],\quad n = 1,2,3 \end{aligned} \tag{2}$$

Similar to the interference pattern of the N-slit which is illuminated by a plane wave, there are $N - 2$ side lobes between the main peaks of interference pattern in the N-path fiber interferometer.

The sensitivity of an N-path interferometer is higher than the conventional Mach-Zehnder interferometer, because the slopes of main peaks are steeper. Mach-Zehnder interferometer can be used as a fiber sensor, because the phase difference can be changed by environmental effects such as strain. The light in the cladding is more sensitive to the surrounding changes than that in the core. The Long Period Grating (LPG) which can couple light from the core to the cladding or reverse is suitable to be employed in Mach-Zehnder fiber interferometer sensor. (Dianov et al., 1996)

3.4 Michelson optical fiber interferometer

A schematic of conventional Michelson OFI is depicted in Fig. 7. The high coherent light beam is split into two different optical paths in the upper and lower single mode optical fibers by the 2 × 2 optical fiber coupler (OFC). The light reflected back by mirrors M_1 and M_2 are recombined by the OFC to produce interference pattern at the receiver.

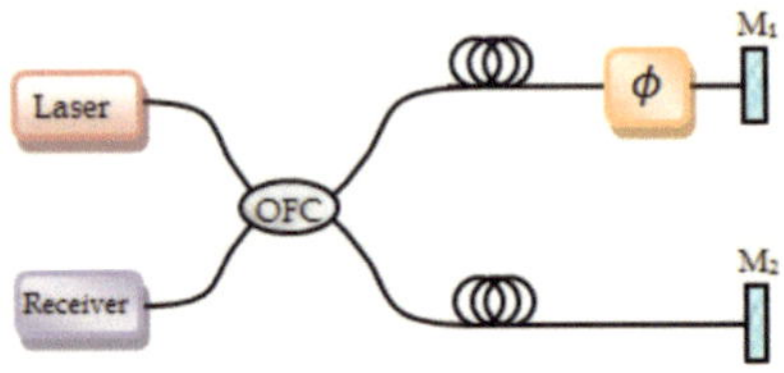

Fig. 7. A schematic configuration of Michelson OFI.

As shown in Fig. 8 by employing a $N \times N$ bidirectional coupler, the conventional Michelson OFI is generalized to the N-path Michelson OFI. Each ports of a $N \times N$ coupler can transmit incoming and outgoing waves simultaneously. Generally each linear bidirectional $N \times N$ OFC is characterized by a $6N \times 6N$ scattering matrix. In an analysis based on the identical polarization where a scalar analysis is sufficient, the scattering matrix becomes a $2N \times 2N$ matrix and denoted by Y. The incoming and outgoing electric field vectors are denoted by $E_{in} = (E_{in}^{(1)}, E_{in}^{(2)})$ and $E_{out} = (E_{out}^{(1)}, E_{out}^{(2)})$ respectively. $E_{in}^{(1)}$, $E_{out}^{(1)}$ and $E_{in}^{(2)}$, $E_{out}^{(2)}$ correspond to the $N \times 1$ vectors of the left and right ports of the $N \times N$ bidirectional coupler.

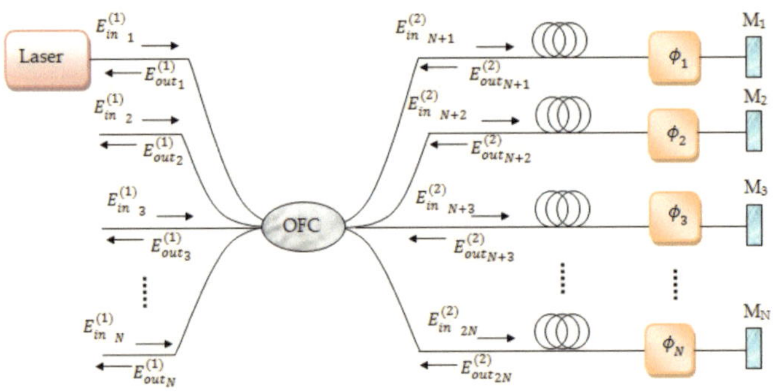

Fig. 8. A schematic configuration of N-path Michelson interferometer.

The incoming and outgoing vectors are related by $E_{out} = YE_{in}$ where Y can also be written in the block form:

$$\begin{pmatrix} E_{out}^{(1)} \\ E_{out}^{(2)} \end{pmatrix} = \begin{pmatrix} Y^{11} & Y^{12} \\ Y^{21} & Y^{22} \end{pmatrix} \begin{pmatrix} E_{in}^{(1)} \\ E_{in}^{(2)} \end{pmatrix} \quad (3)$$

where Y^{ij}(i,j=1,2) is a $N \times N$ matrix. For a lossless $N \times N$ OFC, Y is a unitary matrix. The diagonal matrix $P = diag\left(a_1^2, a_2^2 \exp(i\varphi_{12}), \ldots, a_N^2 \exp(i\varphi_{1N})\right)$ is the transfer matrix between the forward and backward waves in the optical fiber array. a_n is the transmission coefficient of the nth optical fiber and φ_{1n} is the phase difference between the phase accumulated by the field during the propagation in the nth and first optical fiber. The outgoing and incoming fields of the right ports of the $N \times N$ OFC is related by the relation $E_{in}^{(2)} = PE_{out}^{(2)}$. By combining this relation and (3), the transfer matrix of the Nth path Michelson interferometer is obtained.

$$E_{out}^{(1)} = [Y^{11} + Y^{12}Y^{21}P(1 - Y^{22}P)^{-1}]E_{in}^{(1)} \tag{4}$$

It is assumed that only one of the input fields is nonzero. The input field vector is denoted by $E_{in} = (\varepsilon_{1in}, 0,0, \dots, 0)$. The output intensities of the left ports are:

$$I_j = I_0 \left| Y_{j1}^{11} + (Y^{12}Y^{21}P(1 - Y^{22}P)^{-1})_{j1} \right|^2, \qquad j = 1, \cdots, N \tag{5}$$

As mentioned in Mach-Zehnder OFI, it is easy to show that the sensitivity of multi-paths Michelson interferometer is greater than that of conventional two paths one.

3.5 Optical fiber Moiré interferometery

Moiré interferometry is based on the fringe pattern formed by overlaying two or more gratings at different angle θ. The desired fringe pattern can also be designed by a suitable arrangement of optical fibers. The optical fiber based generator of interference grid pattern, is configured by N polarization maintained fibers. The coordinates of the center of the jth fiber in the plane $z = 0$ is denoted by $(a_j, b_j), (j = 1, \dots, N)$. The polarization angle of the jth fiber relative to the x axis is denoted by θ_j $(j = 1, \dots, N)$. The field at the point (x, y) in the $z = D$ plane is given by:

$$E = \sum \vec{E_j} e^{-i[\frac{k}{D}(xa_j + yb_j) + \varphi_j]} + c.c. \tag{6}$$

where φ_i is the phase of the jth fiber at the $z = 0$ plane. The field intensity at the point (x, y) in the observation plane is as follows:

$$I = \sum_{i=1}^{N} I_i + \sum_{i \neq j} \sqrt{I_i I_j} \cos(\theta_i - \theta_j) \cos\left\{\frac{k}{D}\left[(a_i - a_j)x + (b_i - b_j)y\right] - \varphi_{ij}\right\}, \tag{7}$$

where $I_i (i = 1, \dots, N)$ is the light intensity corresponding to the ith fiber at point (x, y), φ_{ij} is the phase difference between the ith and jth optical fibers and k is the light wave number. (Yuan et al., 2005). By suitable choosing of the parameters a_i, b_i and θ_i $(i = 1, \dots, N)$, the desired fringe configuration can be obtained. As an example consider a system of three fiber centered at $P(0,0), P(2a, 0)$ and $P(0,2a)$, where $'a'$ is the radius of the polarization maintained fiber. Fig. 9 shows the arrangement of the interference pattern generator. The interference pattern of three fibers with the same polarization direction is shown in Fig. 9 (a). The vertical and horizontal patterns correspond to the interferences of fibers 1 and 2 and fibers 1 and 3 respectively. The oblique lines families in Fig. 9 (a) are due to the interference of the fibers 2 and 3. As shown in Fig. 9 (b), by employing the vertical and horizontal polarization for the fibers 2 and 3 respectively and setting the angle 45° between the polarization of fiber 1 and x-axis, the oblique lines are eliminated. The inverse problem is to design a suitable configuration of PMF optical fibers to obtain a desired intensity distribution $I(x, y)$ or fringe pattern. By defining a suitable meter on the intensity distribution space and employing the optimization techniques such as variational method and genetic algorithm, it is possible to minimize the distance between the generated distribution and the desired distribution.

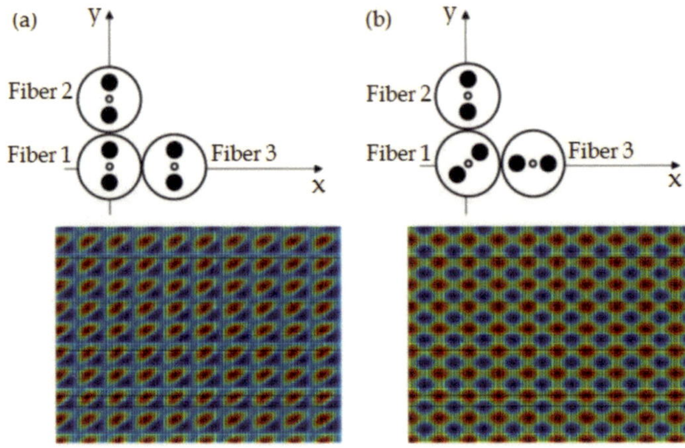

Fig. 9. Two arrangements of optical fiber Moiré interferometer and their interference patterns. a) The polarizations of the three fibers are in the same direction b) fibers 2 and 3 with vertical and horizontal polarization and fiber 3 with angle 45°.

3.6 Optical fiber Fabry-Perot interferometer

A Fabry-Perot (FP) consists of two optically parallel reflectors with reflectance $R_1(\omega)$ and $R_2(\omega)$ separated by a cavity of length L. Reflectors can be mirrors, interface of two dielectrics or fiber Bragg gratings. The cavity may be an optical fiber or any other optical medium.

Two different optical fiber Fabry-Perot interferometers are shown in Fig. 10.

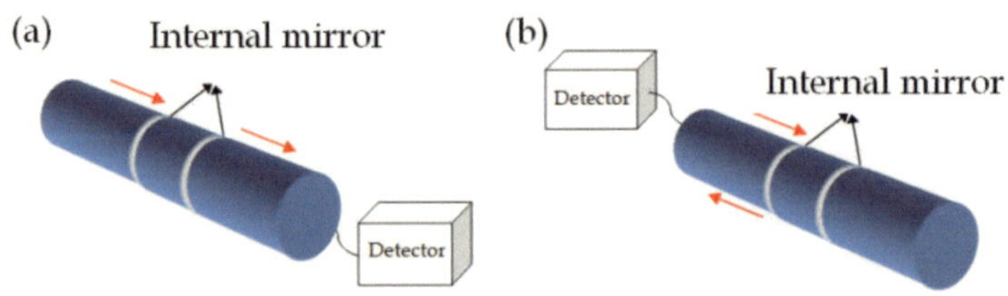

Fig. 10. (a) Fabry-perot based on the light transmission (b) Fabry-perot based on the light reflection.

One is based on the light transmission through a Fabry-Perot, while the other is based on the reflection. Due to multiple reflections, the reflected and transmitted spectrums are functions of cavity length, medium index of refraction and mirrors reflectivity. Because of energy conservation law, the transmitted spectrum is opposite to the reflected spectrum.

Optical fiber Fabry-Perots are classified as intrinsic and extrinsic types. In the intrinsic fiber FP interferometer (IFFPI), the two mirrors are separated by a single mode fiber, while in the extrinsic fiber FP interferometer (EFFPI), the two mirrors are separated by an air gap or by some solid material other than fiber. In both IFFPI and EFFPI, light from emitter to the FP and from FP to the detector are transmitted by a single mode fiber. Fig. 11 shows schematic configurations of three IFFPI. One end of the fiber shown in Fig. 11 (a) is polished as a mirror. For higher reflection the polished end is coated with switchable dielectric layers. The second mirror of IFFPI shown in Fig. 11 (a) is an internal mirror which can be produced by splicing of polished fibers or by polished coated fibers. Both mirrors of the IFFPI shown in Fig. 11 (b) are internal fiber mirrors while those are used in the IFFPI presented in Fig. 11 (c) are FBG reflectors. Depending on the application of IFFPI, one of the configurations presented in Fig. 11 can be used.

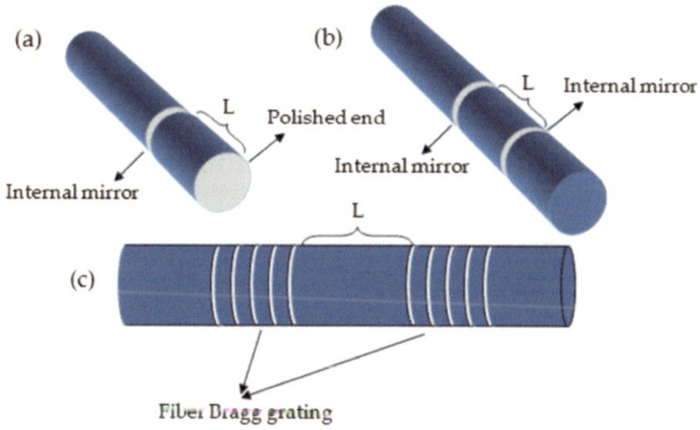

Fig. 11. Schematic configurations of three IFFPI.

Four different EFFPI configurations are shown in Fig. 12.

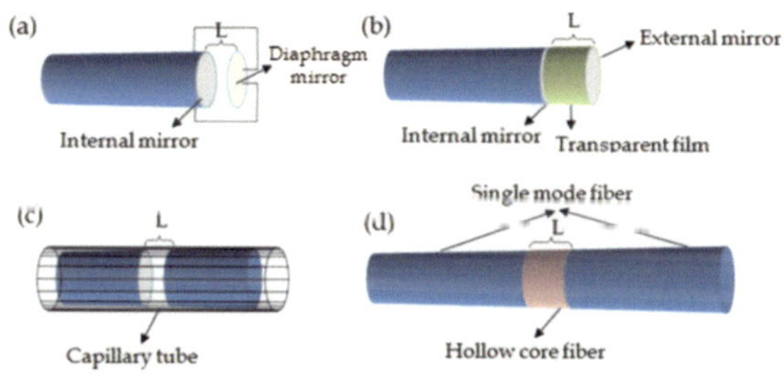

Fig. 12. Schematic configurations of four EFFPI.

In Fig. 12 (a) the air-gap cavity is bounded by the end of a polished fiber and a diaphragm mirror. The cavity length is of the order of several microns and can be increased by convex mirror diaphragm. In another configuration presented in Fig. 12 (b) a thin film of transparent solid material is coated on the end of the fiber. The air-gap cavity between two polished fiber surfaces, where the fibers are aligned in a hollow tube is another configuration of EFFPI (Fig. 12 (c)). The structure shown in Fig. 12 (d) is called the in-line fiber etalon (ILFE). The ILFE is constructed of a hollow-core fiber spliced between two single mode fibers. The diffraction loss causes to limit the practical length of EFFPI to a few hundred of microns (Shizhuo, 2008).

3.7 White light fiber interferometry

The interferometric techniques are known as the precise method for measuring physical quantities that can induce the optical path difference (OPD) in the interferometers. The coherent length of the narrow band sources such as Lasers are greater than the optical path length difference in the interferometers. Due to the periodic nature of the interferogram fringes, the interferometric measurement suffers from an integer multiple of 2π phase ambiguity. Hence interferometers driven by narrow line-width Lasers do not produce absolute data unless extra complexity is added to the interferometer. By employing the short coherent light that is illuminated from wide-band light sources, the phase ambiguity is eliminated. In wide bandwidth interferometer the fringes of the interferogram are narrowly located in the zero path length difference region (Flourney et al., 1972). So the phase difference can be determined without the phase ambiguity by measuring the fringe peak or the envelop peak of the interferogram. This type of interferometry is named as white light or low coherence interferometry. In white light interferometry (WLI) corresponding to each wavelength a separate fringe system is produced. The electric field at any point of observation is the sum of electric fields of these individual patterns. In a WLI which is adjusted such that the optical path difference is zero at the center of the field of view, the electric field of different wavelengths exhibits the maximum at the center point. The fringes of different wavelengths will no longer coincide as moving away from the center of the pattern. The fringe pattern is a sequence of colors whose saturation decreases rapidly. The central bright white light fringe can be used to adjust the WLI.

The light sources such as fluorescent lamp, SLDs, LEDs, Laser diodes near threshold, optically pumped Erbium-doped fibers and tungsten lamps, can be used in the WLI. The spectral width of SLD and LED is between 20 and 100 nm. It is expected that at the operating wavelength (1.3 μm) of these types of light sources, the coherent length is between 17 and 85 μm. Because of the wave-train damping, the Doppler effect, disturbances by neighbor atoms, noises and mode mixing effects, the practical coherent length is less than those are predicted previously.

In the WLI, one of the two arms is used as the measurement arm and the other one as the reference arm. The length of the reference arm can be controlled by different methods such as moving mirrors or Piezoelectric (PZT) devices. Generally the operation of WLI is based on the balancing the two arms of the interferometer and compensating the OPD in the measurement arm. Therefore the desired measurement can be achieved.

As the OPD between the two paths of a WLI is varied, the intensity of interference fringe drops from a maximum to a minimum value. The maximum intensity corresponds to the central white bright fringe. Measurement of the position of the central fringe in the WLI is of prime importance. Because the distance between the central fringe and its adjacent side fringes is too small and the presence of noise, the determination of the central fringe position is inaccurate, so there are some ambiguities in the central fringe identification. This problem can be solved by employing a combinational source of two or three multimode Laser diodes with different wavelengths.

White light fiber interferometers (WLFI) can be designed on different topologies of single or multi-mode fiber interferometers. Each of the single mode and multi-mode fibers has their own advantages and disadvantages. For example usually white light single mode fiber interferometer provides stable and large signal to noise ratio while in the interferometers based on multi-mode fiber, cheaper optical components are employed (Song et al., 2001; Manojlovi et al., 2010). Generally there are several WLI topologies corresponding to the standard optical fiber interferometer and their combinations (Yuan, 2002; Mercado et al., 2001;). As an example Fig. 13 shows a white light fiber optic Michelson interferometer working in the spatial domain. The LED light is coupled to the two path of Michelson interferometer through a 2 × 2 OFC without insertion loss. The reflected beams recombine on the PIN detector of the WLI. The scanning mirror is adjusted for maximum output corresponding to the position of central fringe.

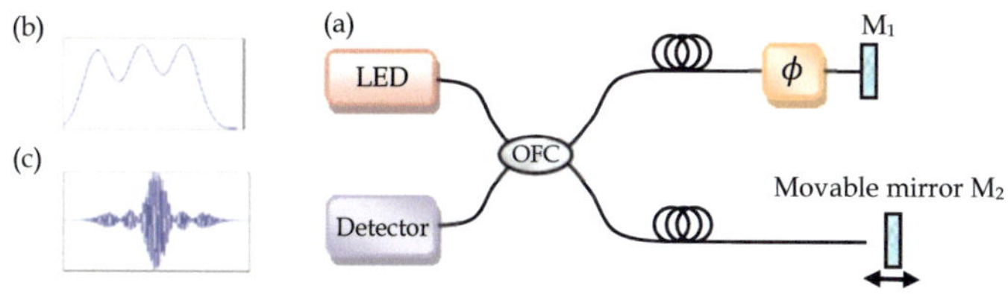

Fig. 13. (a) A schematic configuration of WLI Michelson interferometer (b) Input LED spectrum (c) Interference fringe pattern.

As shown in Fig. 13(c) for OPD less than the source coherence length, the white-light fringe pattern is produced. The position of the highest amplitude corresponds to the exactly zero optical path difference between two beams. After some mathematical manipulations for LED parameters presented in (Yuan, 1997), the normalized interference fringe pattern is calculated and result is presented in Fig. 14. The results of three peaks LED are compared with those of a normal LED to see how the multi wavelength white light source increases the precision of the central fringe position measurement relative to the single white light source.

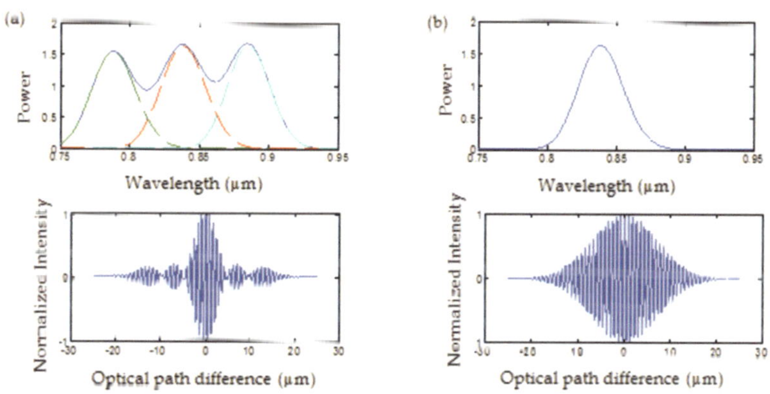

Fig. 14. The spectrum distribution of the light source (up) and their normalized interference fringe pattern (down) (a) Three peaks LED (b) Normal LED.

4. Signal recovering methods

The phase difference between two coherent light beams is detected by interferometric methods, which are most sensitive techniques for optical path difference measurement. The OPD variations have sufficiently low frequency components. So when they are converted to the light intensity on the observation plan, they can easily be detected by photo diodes, photo diode matrix or charge coupled devices (CCD). The detector output is proportional to $cos(\phi)$, where ϕ is the phase difference. In the presence of cosine and the absence of sinusoidal signal an ambiguity exists in the phase recovering. If the phase amplitude falls outside the (0 to 2π) range, in addition to sine and cosine values we must track the history of phase angle variation to know in which quadrant it precisely lie. Numerous methods have been devised for recovering the path length difference from the output signals of OFI. The methods based on the production of new frequencies are called heterodyne detection otherwise are named homodyne methods. The main three methods are briefly described in following.

4.1 Phase generated carrier (PGC) homodyne detection

In this method the interferometer Laser source is driven by combination of direct and sinusoidal current. The Laser output power and wavelength are modulated by the Laser current variation. In the presence of path length difference, the change of the wavelengths indicates itself as a change in the output phase. The current frequency can be observed in the received phase in the output. Each of the current frequency and its harmonics carries in their sidebands a copy of the created phase modulated signal. Two of these copies are chosen by band pass filters. Proper control of the amplitude of the sinusoidal current and filters configuration guaranty that the chosen copies have the same amplitude. The filters outputs are used as the inputs of an electronic mixer. One of the outputs of the mixer is proportional to sine, while the other one is proportional to the cosine of the interesting signal. Sometimes the method is also called Pseudo-Heterodyne Detection (PHD) (Jackson et al., 1982).To produce the phase shift instead of Laser frequency modulation it is possible to create phase shift with a cylindrical Piezoelectric, which is wrapped around one arm of the interferometer and is derived with a sinusoidal voltage (Hoeling et al., 2001). This case is called synthetic heterodyne method (Strauss, 1994).

4.2 Fringe-rate methods

When large phase shift is produced in an interferometer, two new methods which are called fringe-counting and fringe-rate demodulation become feasible (Barone et al., 1994; Crooker & Garrett, 1987). These methods are based on the transitions of interferometric outputs across some central value. In the fringe counting method, on a suitable period of time the transitions are counted digitally. The instantaneous frequency is determined by the ratio of the counting number to the counting time. Because in practice one must wait a short time to obtain at least one count, it is impossible to obtain an instantaneous count. However the phase can be obtained by integrating the instantaneous frequency. In the fringe-rate method, the transitions are used as inputs of a frequency to voltage converter circuit (FVC). To obtain the phase difference, the output of FVC circuit is integrated. There is no transition for weak signal. The minimum detectable signal is of the order of π radian.

4.3 Homodyne method

The operation range of the synthetic heterodyne method is limited above to π radian, while the fringe-counting and fringe-rate techniques are limited from below to π radian (Dorrer et al., 2001). A number of homodyne techniques are employed to bridge this region. All these methods are based on the use of orthogonal components without using heterodyne methods. OFIs usually have two outputs. A 2×2 OFC is employed to combine the two path beams of the interferometer and form the interference pattern. By energy conservation law it is easy to show that the two outputs are 180° out of phase from one another. When one output is dark, all energy must be presented in the other output and vice versa. So no orthogonal components can be found in the outputs. The orthogonal components are produced by the heterodyne methods. The output coupler can be modified such that the orthogonal components directly exist in the outputs. As an example a 3×3 coupler can be employed as the output coupler of the interferometer to create output with orthogonal components without employing the heterodyne detection method (Choma et al., 2003).

5. Noise sources in optical fiber interferometers

Calculation of signal to noise ratio strongly depends on the OFI topology. In principle the noise sources belong to the light source, optical fibers, detector, electronic circuits and environment (Bottacchi, 2008; Tucker & Baney, 2001). Moreover any random process in each stage of interferometry: signal generation, transmission and detection can be considered as a noise source.

The laser generation is the result of the quantum interaction of electromagnetic wave and matter. The spontaneous and stimulated emissions are quantum effects and are the noise sources in the phase and amplitude of laser output (Linde, 1986; Clark, 1999; Tsuchida, 1998). On the other hand the interaction of light with universal modes of surrounding reservoir through the mirror coupling and stimulated emission in active medium bath are also noise sources for the laser output (Scully & Zubairy, 2001). The cavity filtering and feedback can reduce the laser noise significantly (Sanders et al., 1992; Cliché et al., 2007). The phase and amplitude of laser noise cause to increase the bandwidth of the laser light. In single mode lasers by proper design of optical cavities, the bandwidth can be reduced to several kilohertz, which gives several tens of kilometers for coherent length. The mode competition and cross saturation effects are new noise sources in multimode lasers that can be employed in wide band fiber interferometry.

Rayleigh scattering, Mie scattering, core cladding interface scattering, Brillouin scattering, absorption and amplification parts in the optical fiber, are the main noise sources in OFI arms and transmission parts. Some parts of the scattered light are trapped in the guided region and travel in both direction of the fiber, contribute to the phase and amplitude noises. Other parts are scattered out of the optical fiber and affect the amplitude noise only. Except the Brillouin and Raman scatterings, all other effects are linear and do not change the light frequency. Both the Brillouin and Raman scattering have two different components Stokes and anti-Stokes frequencies. The Stokes and anti-Stokes Brillouin shifts are due to the light-acoustic phonon interaction and are about ± 25 GHz, while the Stokes and anti-Stokes Raman shift correspond to the optical phonon-photon interaction and are of the order of 13 THz (Agrawal, 2007). Beating between Stokes, anti-Stokes and direct beam can occur, but such a high beating frequencies cannot be observed at the output response of any realistic detector and are eliminated intrinsically by the low pass filter detector. The Brillouin and Raman scattering loss can be considered as a source of amplitude noise. The Rayleigh, Brillouin and Raman scattering are symmetrically distributed with respect to the forward and backward direction while those of Mie and core-cladding interference scattering are mainly in the forward direction.

Mode coupling is another source of noise in multimode fiber interferometer. In such an interferometer the mode coupling noise must be taken into account. Absorption and amplification correspond to the interaction of light with reservoirs so according to quantum Langevin equation there are some noises in the output (Scully & Zubairy, 2001). Generally Avalanch photo diode (APD), PIN diode, charge coupled device (CCD) and photo multiplier (PM) are used as the electronic detector of the OFIs. Dark current noise, shot noise, background noise, thermal noise and flicker noise are common in all of the optical detectors. The generation and recombination of electron hole are a stochastic process in semiconductor detectors and are the noise sources of such detectors. The avalanche effect, the basis of operation of APDs is a random process and causes noise generation in avalanche photo diode. The same effect on the anodes of PM can be a noise source in PM detectors. The amplifier noise which is consistent of the shot noise, Johnson noise, burst noise and flicker noise of different solid state electronic elements of the amplifier is the final intrinsic noise of OFI.

The fiber parameters can be affected by the environmental physical variations such as mechanical vibration, acoustic agitation, pressure, tension and thermal variations. In a controlled way this effects can be used to make the optical fibers as a sensor for these physical quantities, while in OFIs are noise sources. As an example the population of Stokes and anti-Stokes photons are functions of the fiber temperature and can be used to design a high precision temperature sensor for water, oil and gas leak detection systems (Harris et al., 2010; Chelliah et al., 2010). The Stokes and anti-Stokes parameters of Brillouin scattering are functions of fiber strain and fiber temperature. This effect is used to measure the strain and temperature simultaneously for structural health monitoring systems (Güemes, 2006; Bahrampour & Maasoumi, 2010). The optical fiber sensitivity to mechanical variation and acoustic waves are employed for various applications such as acoustic, vibration and ultrasonic detectors for under water sensor systems. However the output signal is affected by all the noise sources and the aim is to denoise signal by the signal processing methods. Depending on the signal, one of the denoising methods such as Fourier regularized deconvolution (ForD) and Fourier wavelet regularized deconvolution (ForWaRD) method can be employed (Bahrampour & Askari, 2006; Bahrampour et al., 2012). The wavelet deconvolution method generally use to denoise transient signals. The short time Fourier method is employed to denoise the music-like signals of a fiber intruder detector based on the birefrigent fiber interferometer (Bahrampour et al., 2012).

6. Applications of optical fiber interferometers

Optical fiber interferometers as a precise measuring interferometer or sensitive tools have many applications in all branches of science and technology (Shizhuo et al., 2008). The OFIs can be employed to design the optical components for the inline signal processing, such as band pass filters in optical communication networks. The same topologies can be easily fabricated by the light waveguides in the integrated circuits by means of photolithographic process for application in optical transmitters and receivers. Because of high sensitivity of the interferometer, the linear and nonlinear properties of optical fiber can be detected. These properties make long length and short length waveguide and fiber interferometer sensors, suitable in novel applications such as oil and gas pipeline monitoring, temperature distribution measurement in the depth of ocean and intruder sensors. Among the wide range of applications of waveguides and fiber interferometers, only a few applications in the optical communication networks and special types of fiber and waveguide interferometric sensors are mentioned in this section.

6.1 Applications in optical fiber networks

The key devices in optical DWDM communication networks are re-amplifying, re-shaping and re-timing (3R-regenerator) systems. In re-shaping and re-timing circuits the nonlinear networks such as clipper, clampers, switching and flip-flops are of prime importance. While in add-drop filters, the linear filters such as tune and notch filters have an important role. On the basis of a nonlinear Mach-Zehnder interferometer, the structure of an all optical inverter is shown in Fig. 15.

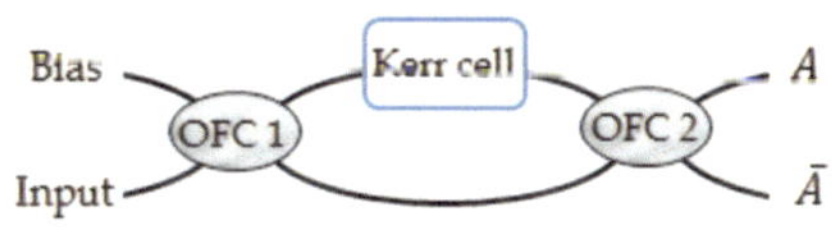

Fig. 15. A schematic of an all optical inverter. OFC is an optical 2 × 2 coupler.

An optical fiber with high nonlinear Kerr effect such as Chalcogenide glasses is employed in one of the Mach-Zehnder interferometer arms. So in the presence of a suitable light intensity at the input of optical fiber coupler 1 (OFC 1), the change of refractive index ($n = n_0 + nI$) causes a π-phase shift in the upper arm of the interferometer relative to the lower arm. It is assumed that in the absence of the input, the interferometer arms are balanced and the outputs A and $\bar{A}$ are in the constructive and destructive conditions respectively. The 0 and 1 digital states are represented by destructive and constructive output ports. In the presence of the OFC 1 input, the output A changes to 0 and the $\bar{A}$ switches to 1. This interferometer is an optical logic inverter. The structure shown in Fig. 15 is also used in quantum non-demolition experiments (Gerry & Knight, 2005). For small input intensities Fig. 15 acts as an intensity modulator circuit and $\bar{A}$ output is approximately proportional to the input intensity. By varying input intensity, the output varies from its maximum value to zero, i.e. this circuit operates as a light controlled variable attenuator.

The inverter of Fig. 16 is designed on the basis of optical waveguides to avoid the high length nonlinear optical fibers in the inverter design.

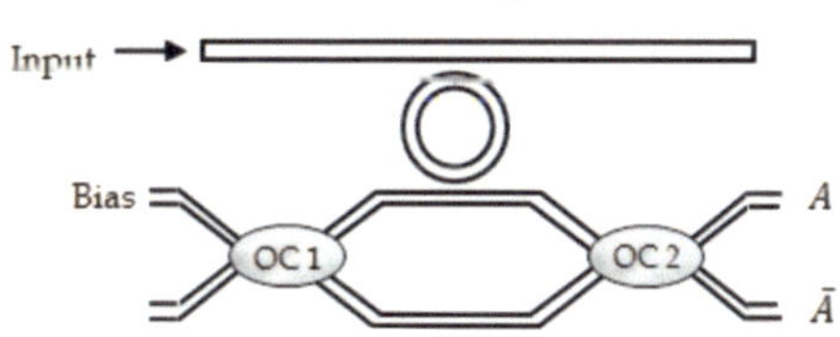

Fig. 16. A schematic of an inverter based on an optical waveguide and micro-ring resonator. OC is 2×2 optical coupler.

As shown in Fig. 16, the Kerr cell that is shown in Fig. 15 is replaced by a high dispersive nonlinear element such as micro-ring or microsphere. The phase difference between the input and output of micro-ring can be changed by the resonance frequency of the micro-ring which is controllable due to the cross-Kerr effect. The upper input to the micro-ring causes to change the micro-ring refractive index and therefore the resonance frequency is changed. So the phase difference between the input and output of the upper arm of the interferometer is changed. On the basis of optical fiber and optical waveguide interferometers in combination with ultrahigh nonlinear optical elements (UHNO), such as semiconductor optical amplifiers (SOA), different high frequency optical classical logic gates are designed and demonstrated. Also quantum interferometers such as Hong-Ou-Mandel interferometer are employed to design quantum gates (Hong et al., 1987; Olindo et al., 2006).

The bi-stability effect is the basis of the clipper and flip-flop circuits. Most bi-stability designs include both a cavity with nonlinear medium and a feedback (Bahrampour et al., 2008a, 2008b, 2008c). A novel OFI with common mode compensation is proposed by Backman (Backman, 1989). The Backman interferometer consists of a Mach-Zehnder interferometer with one nonlinear path and re-circulating delay line as shown in Fig. 17. The output intensity $|E_{out}|^2$ versus the input intensity $|E_{in}|^2$ in the steady state has the bi-stability behavior.

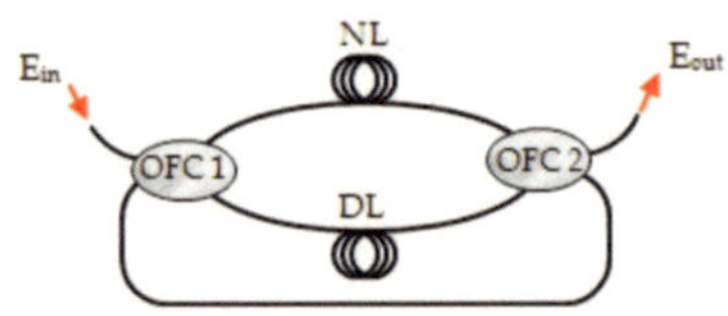

Fig. 17. A schematic of Backman interferometer. NL is nonlinear line and DL is delay line.

Flip-flops are building blocks of the sequential logic circuits such as time recovering circuits. As usual a reset-set (RS) flip-flop can be designed on the basis of regenerative feedback in the two inverter circuit. Fig.18 shows a RS flip-flop based on the two Mach-Zehnder interferometer inverters. The optical fiber couplers OFC 1 and OFC 2 are 3 × 3 couplers and OFC 3 and OFC 4 are 2 × 2 couplers. The bias light inserts to the upper and lower Mach-Zehnder interferometers (MZI 1, MZI 2) by a 2 × 2 coupler. R and S are the reset and set trigger inputs. A , $\bar{A}$ and B , $\bar{B}$ are the outputs of MZI 1 and MZI 2 respectively. Due to the energy conservation law, $\bar{A}$ and $\bar{B}$ are the logic complement of A and B outputs. In the absence of set and reset $A = B = 1$. The output complement of each inverter is connected to the control input of the other inverter. This network of Fig.18 has two stable ($A = 0, \bar{B} = 1$) and ($A = 1, \bar{B} = 0$) states. In the presence of trigger signal at S or R inputs, this system can switch between these two stable states.

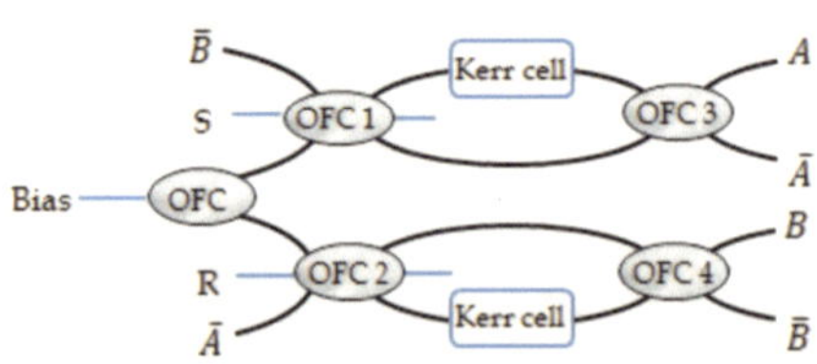

Fig. 18. A schematic of an optical Flip-Flop by combination of two Mach-Zehnder Interferometers.

In addition to the logic gates and nonlinear circuits, the fiber interferometers can be used to design linear circuits such as different types of optical filters. In many applications such as selection of a narrow spectrum from a broad band spectrum, band pass filter is of prime importance. Due to the Bragg diffraction effect, each FBG fiber can be used as a notch or band stop filter. As shown in Fig. 19, an optical coupler is employed to detect the reflected spectrum of FBG. In this filter, the input power splits into two parts by the OFC. The light reflected by a FBG is again equally split between the ports 1 and 2. Hence only 25% of the light is in the output port 2 of the band pass filter (Kashyap, 1999).

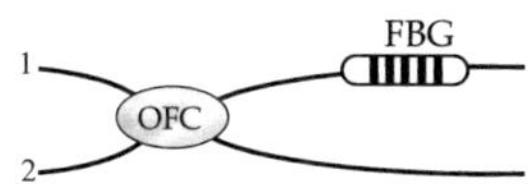

Fig. 19. A schematic of band pass filter.

To eliminate the insertion-loss of the band pass filter several interferometric methods was proposed (Kashyap, 1999). On the basis of Michelson, Mach-Zehnder and Fabry-Perot interferometers three different design of band pass filters are presented in Fig. 20(a-c) respectively(Kashyap, 1999).

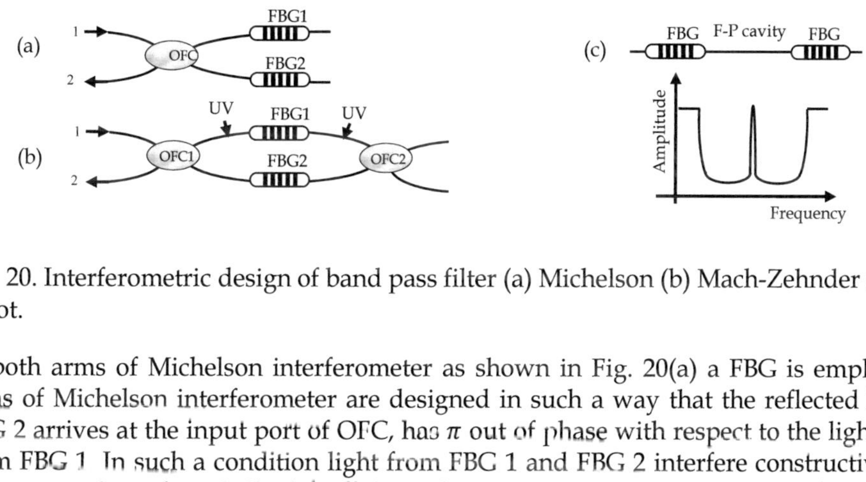

Fig. 20. Interferometric design of band pass filter (a) Michelson (b) Mach-Zehnder (c) Fabry-Perot.

In both arms of Michelson interferometer as shown in Fig. 20(a) a FBG is employed. The arms of Michelson interferometer are designed in such a way that the reflected light from FBG 2 arrives at the input port of OFC, has π out of phase with respect to the light reflected from FBG 1. In such a condition light from FBG 1 and FBG 2 interfere constructively at the output port 2, so that 100% of the light at the Bragg wavelength appears at the output port of band pass filter (Kashyap, 1999). The band pass spectrum can be designed by the profile of chirped spectrum. The dual grating Mach-Zehnder interferometer band pass filter as shown in Fig. 20 (b) is designed for the application in add-drop filters. The principle of operation is the same as that is demonstrated in Michelson band pass filter. Here "UV trimming" is used to balance the interferometer after the gratings are written. "UV trimming" relies on photo induced change in the refractive index to adjust the optical path difference. The simplest band pass filter is an inline Fabry-Perot interferometer. In distributed feedback (DFB) lasers, two FBG can be employed instead of mirrors. As shown in Fig. 20(c) a single $\lambda/4$ phase-shifted FBG has a sharp Lorentzian line shape band pass in the middle of band stop. The broader transmission band width is obtained by cascading several structure (Haus & Lai, 1992). Number of band pass peaks that they appear within the band stop increases by increasing the gap between the two grating sections.

6.2 Some applications in optical fiber sensors

The traveling wave in the dielectric medium of optical fibers and waveguides can be perturbed by their environment. This is the basic idea of the optical fiber sensors (OFS). The interaction of quantity of interest (which is called the measurand), with the optical fiber produces a modulation in the parameters of propagating light beam within the fiber. Generally there are four beam parameters for measurand modulation:

I) Intensity modulation: the intensity modulated fiber sensors are simplest and low cost fiber sensors for measuring the position, pressure and vibration in medical and industrial applications (Polygerinos et al., 2011, Jayanthkumar et al., 2006). Fig. 21 shows a distributed oil leak detection system based on intensity modulated sensor (Carrillo, 2002). The leaked oil causes to expand the polymer around the optical fiber. Due to the wrapped strain-less steel wire, the fiber bending loss increases and the oil leakage position can be measured by a commercial optical time domain reflectometer (OTDR) (Righini et al., 2009).

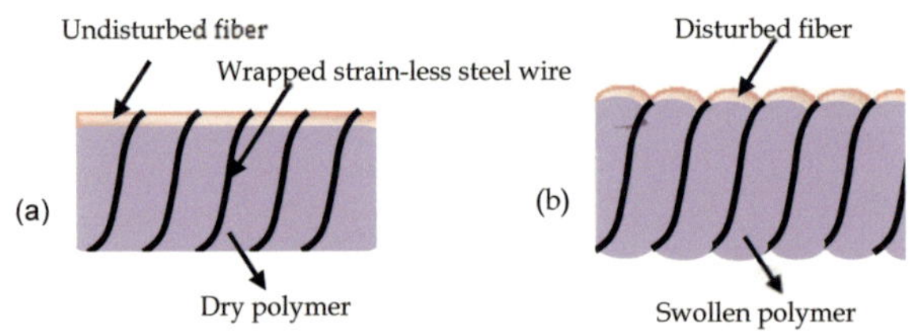

Fig. 21. Distributed Oil leak detection system.

II) Wavelength modulation: The measurands such as temperature and strain can be modulated on the resonance frequency of an inline Fabry-Perot or Bragg wavelength of an inline LPG. An optical or chemical transducers joint at the end of a fiber can be used as the wavelength modulator. Interaction of the measurand with transducer causes to change the spectral properties of transducer. The measurement of the optical spectrum of the transducer through the optical fiber makes possibility to monitor measurand status (Righini et al., 2009).

III) Polarization modulation: In birefringent optical fibers the two fundamental modes propagate with slightly different phase velocities. On the basis of high birefringent fibers several methods for current and magnetic field measurement are designed and manufactured. Fig. 22 shows the principle of operation of an intrusion sensor based on the birefringent optical fiber (Bahrampour et al., 2012).

An x-polarized ramped frequency modulated laser is injected to the birefringent fiber sensor. At the cross point of the intrusion and the fiber sensor, energy from the x-polarized mode is converted to the y-polarized mode. Due to different velocities of the x- and y-polarized modes a beating frequency is observed at the output of the detector. The intrusion position can be obtained from the output beating frequency.

The optical fiber and waveguide sensors that have been investigated and proposed for science, industrial, military, biochemical, biomedical, environment, automotive, avionic and geophysical applications are countless. One of the basic characteristics of the optical fiber sensors is their ability for long length distributed sensing. One of the most popular

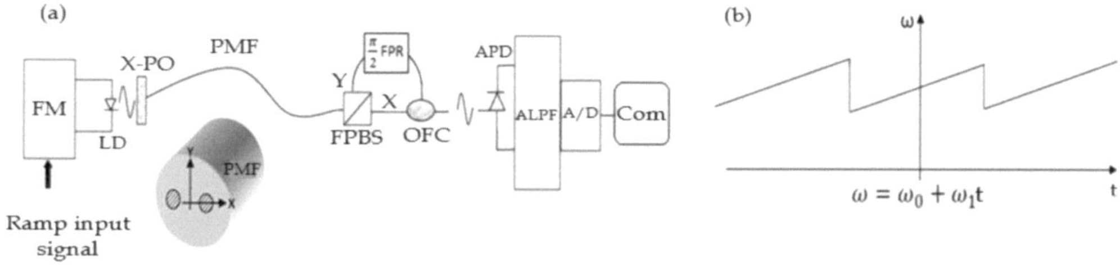

Fig. 22. A schematic of a birefringent fiber intruder detection system. FM is the frequency modulator, LD is laser diod, x-po is a x-polarizer, FPBS is a fiber polarization beam splitter, FPR is a fiber polarization rotator, APD is an avalanche photo-diode detector, ALPF is an active electronic low pass filter, A/D is an analog to digital convertor, Com. a computer system for signal processing and denoising. (b) Ramp input to the FM system (Bahrampour et al, 2012).

distributed fiber sensors is optical time domain reflectometer (OTDR) which is based on the monitoring the Rayleigh back scattering along the fiber. On the basis of Raman and Brillioulin scattering the OTDR is developed to the Raman time optical domain reflectometer (ROTDR) and Brillouin optical time domain reflectometer (BOTDR) respectively. The OTDR, ROTDR and BOTDR optical fiber sensors have applications in structural health monitoring (Glisic & Inaudi, 2008). In the absence of any intrusion, the ΦOTDR signal is saved in an electronic memory and it is compared with the ΦOTDR output continuously (Juarez et al., 2005). Due to the elasto-optic effect, in the presence of an intrusion, the fiber refractive index and hence the phase of the back scattered signal changes. This phase changes can be measured at the sensor output (Righini et al., 2009).

IV) Phase modulated sensors: Variation of the optical length of optical fiber causes a phase shift of the light beam $\Delta\phi = 2\pi(n\Delta L + L\Delta n)/\lambda_0$, where λ_0 is the free space light wavelength and n is the fiber refractive index. Phase shifts usually are measured by interferometric methods. Refractive index and fiber length can vary due to the characteristic of various measurands and therefore the cross sensitivity occurs. To avoid the cross sensitivity, special design of jacketing is necessary. The material of the jacketing is chosen such that to improve the effect of desired measurand and attenuates the others. A schematic of an optical fiber intrusion detector system is presented in Fig. 23. The light laser source through an optical circulator and a 50:50 coupler is connected to a Faraday rotating mirror (FRM). The intrusion distance (L_x) is the length between the intrusion point and FRM. The parts 3 and 4 of the coupler are connected with a fiber of length L_d to form a delay loop. The returned light through the port 1 of 50:50 coupler and circulator is transported to the detector . The intrusion point is determined after signal processing. There are four different paths in the system for transmission of light from source to detector:

Path I: $1 \to 2 - FRM - 2 \to 3 \to L_d \to 4 \to 1$

Path II: $1 \to 4 \to L_d \to 3 \to 2 \to FRM \to 2 \to 1$

Path III: $1 \to 2 \to FRM \to 2 \to 1$

Path IV: $1 \to 4 \to L_d \to 3 \to 2 \to FRM \to 2 \to 3 \to L_d \to 4 \to 1$

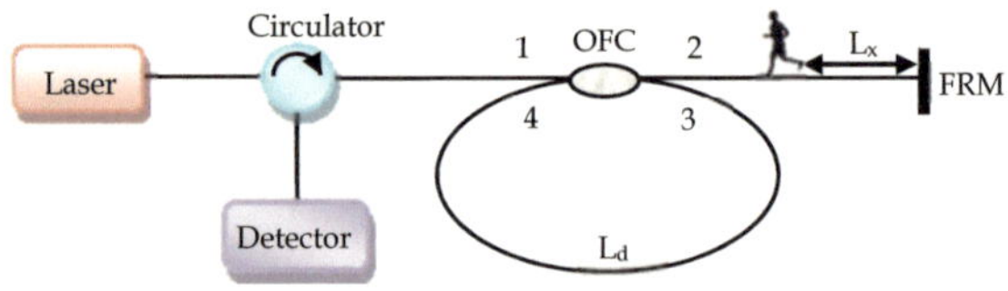

Fig. 23. A schematic of an optical fiber intrusion detector system.

All path differences except the optical path difference of the path I and II are so large, so they have no effect on the interference signal. The electric field at the photo detector is the superposition of the electric field corresponding to the four different paths. The time dependence of the strength of an intrusion at the detector and the instant t is $\Delta q = \Delta\varphi_0 sin\omega_s t$, where $\Delta\varphi_0$ and ω_s are the induced amplitude and frequency by intruder respectively. The light intensity of interference of electric field of path I and path II on the photo detector is easily obtained.

$$I_{in} = 2E_0^2 cos\left\{4\Delta\varphi_0 sin\frac{\tau_2\omega_s}{2} cos\frac{\tau_1\omega_s}{2} cos\omega_s(t - t_0)\right\}, \quad (8)$$

where $\tau_1 = 2nL_x/c$, $\tau_2 = 2nL_d/c$ and $t_0 = (\tau_1 + \tau_2)/2$. The relative amplitude of the frequency components of the detector outputs are functions of the intrusion distances and can be obtained by the Fourier transform method (Jia et al., 2008).

According to the general relativity, gravitational waves (GW) are produced when the curvature of spacetime disturbed by accelerating mass. The ripples in the curvature of spacetime propagate at the speed of light. A GW causes a tiny time dependent quadruple change of strain in the plane transverse to the wave's propagation direction. The space is stretched in one direction while is shrunk along its perpendicular direction. The strength of GW 'h' is expressed by the dimensionless strain $\delta L/L$. Due to the quadruple nature of Michelson interferometer, it is suitable device for gravitational wave detection. The interference pattern is linear measure of the strain. The amplitudes of GWs radiated from astrophysical sources at Earth are typically of the order of 10^{-21} or smaller. Detection of such weak strains needs high sensitive devices. For increasing the sensitivity of the interferometer, the lengths of the interferometer arms were increased to the order of kilometers and multi-path cell or Fabry-Perot optical cavity was used in each arm. So the light can be stored for a time comparable to the time scale of GW signal. Long base line gravitational wave detectors such as LIGO, VIRGO, GEO and TAMA are now operational. In the presence of GW, the change in the length of arms is very small. Hence many noises such as seismic noise, thermal noise and quantum noise limit the sensitivity of interferometer. Quantum noise is the fundamental and unavoidable noise in new generation of these interferometers and is due to the light-interferometer interaction. So the sensitivity of laser GW detector depends on the quantum state of light. It was shown that depending on the parameters of interferometer such as arm's lengths, frequency of laser and mass of mirrors, the optimum quantum state for the dark port is vacuum squeezed state with specific squeezing factor. By employing this optimum quantum state in the dark port, the quantum noise and optimum laser power reduce one order of magnitude relative to the conventional interferometers (Tofighi et al., 2010).

To minimize the effects of other noises, whole setup including optical elements and beam path are kept in ultra-high vacuum ($10^{-8} - 10^{-9}\ torr$) and the optical elements are suspended on top of the seismic isolation system. A highly stabilized laser and active control system for adjusting cavity length are employed in these devices. So the long baseline Laser interferometer GW detectors are high cost projects.

Fig. 24 shows another configuration for GW detection that uses optical fiber in the arms of Michelson interferometer. GW optical fiber interferometric detector is very small, cheap and simple to build and operate (Cahill, 2007; Sacharov, 2001; Cahill & Stokes, 2008).

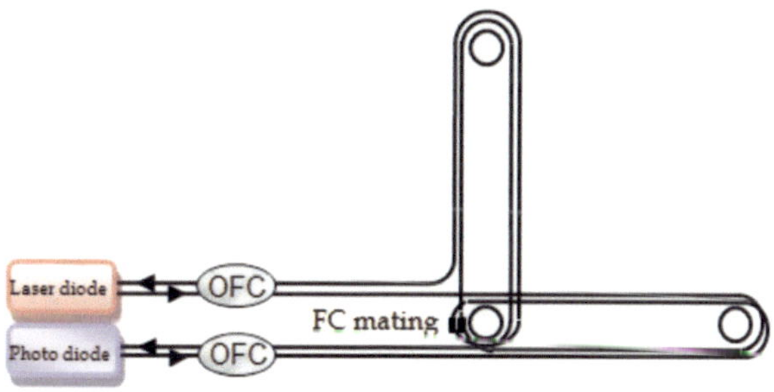

Fig. 24. A schematic of gravitational wave optical fiber interferometric detector.

7. Reference

Adams, M.J. (1981). *An Introduction to optical waveguides*. John Wiley & Sons Ltd. ISBN 0471279692

Agrawal, G. (Ed(4).). (2007). *Nonlinear fiber optics*. Academic Press, ISBN 0123695163, San Diego

Backman, A.B. (1989) *Journal of Lightwave Technology*, Vol. 7

Bahrampour, A.R. & Maasoumi, F. (2010). Resolution enhancement in long pulse OTDR for application in structural health monitoring. *Optical Fiber Technology*, Vol. 16, No. 4, pp. 240-249

Bahrampour, A.R.; Karimi, M.; Abolfazli Qamsari, M.J.; Rooholamini Nejad, H. & Keyvaninia, S. (2008). All optical set-reset flip-flop based on the passive microring resonator bistability. *Optics Communications*, Vol. 281, No. 20, pp. 5104-5113

Bahrampour, A.R.; Mohammadi Ali Mirzaee , S.; Farman ,F. & Zakeri, S.S. (2008). All optical flip flop composed of a single nonlinear passive microring coupled to two straight waveguides. *Optics Communications*, Vol. 282, No. 3, pp. 427-433

Bahrampour, A.R.; Zakeri, S.S.; Mohammadi Ali Mirzaee , S.; Ghaderi, z. & Farman ,F. (2008). All-optical set-reset flip-flop based on frequency bistability in semiconductor microring lasers. *Optics Communications*, Vol. 282, No. 12, pp. 2451-2456

Bahrampour, A. R. & Askari, A. A. (2006). Fourier-wavelet regularized deconvolution (ForWaRD) for lidar systems based on TEA CO_2 laser. Optics Communications, Vol. 257, No. 1, pp. 97-111.

Bahrampour, A. R.; Bathaee, M.; Tofighi, S.; Bahrampour, A.; Farman, F. & Vali, M. (2011).Polarization maintained optical fiber multi-intruder sensor. Submitted.

Bahrampour, A. R.; Moosavi, A.; Bahrampour, M. J. & Safaei, L. (2011). Spatial resolution enhancement in fiber Raman distributed temperature sensor by employing ForWaRD deconvolution algorithm. Optical Fiber Technology, Vol. 17, No. 2, pp. 128-134.

Barone, F.; Calloni, E.; De Rosa, R.; Di Fiore, L.; Fusco, F.; Milano, L. & Russo, G. (1994). Fringe-counting technique used to lock a suspended interferometer. *Applied Optics*, Vol. 33, No. 7, pp. 1194-119, ISSN 0003-6935

Bentley, S.J. & Boyd, R.W. (2004). Nonlinear optical lithography with ultrahigh sub-Rayleigh resolution, *OPTICS EXPRESS*, Vol. 12, No. 23, 5735-5740

Bohnert, K.; Gabus, P.; Nehring, J. & Brandle. H. (2002). Temperature and vibration insensitive fiber optic current sensor. *Journal of Lightwave Technology*, Vol. 20, No. 2, pp. 267-276

Boto, N.; Kok, P.; Abrams, D.S.; Braunstein, S.L.; Williams, C.P. & Dowling, J.P. (2000) Quantum interferometric optical lithography: exploiting entanglement to beat the diffraction limit. *Phys. Rev. Lett*, Vol. 85, No. 13, pp. 2733–2736

Bottacchi, S. (Ed(1).). (2008). *Noise and Signal Interference in Optical Fiber Transmission Systems: An Optimum Design Approach*. Wiley-Interscience, ISBN 0470060611

Brown, R.H. & Twiss, R.Q. (1956). Correlation between photons in two coherent beams of light, *Nature*, Vol. 177, No. 4497, pp. 27-29

Burns, W.K. (Ed(1).). (1993). *Optical Fiber Rotation Sensing*. Academic Press, ISBN 0121460754, San Diego

Cahill, R. T. (2007). Optical-Fiber Gravitational Wave Detector: Dynamical 3-Space Turbulence Detected. *PROGRESS IN PHYSICS*, Vol. 4, pp. 63-68, ISSN 1555-5534

Cahill, R.T. & Stokes, F. (2008). Correlated Detection of sub-mHz Gravitational Waves by Two Optical-Fiber Interferometers. *PROGRESS IN PHYSICS*, Vol. 2, pp. 103-110

Carrillo, A.; Gonzalez, E.; Rosas, A. & Marque, A.(2002). New distributed optical sensor for detection and localization of liquid leaks Part I. Experimental studies. *Sensors and Actuators A*, Vol. 99, pp. 229–235

Chelliah, P.; Murgesan, K.; Samvel, S.; Chelamchala, B.R.; Tammana, J.; Nagarajan, M. & Raj.B. (2010). Looped back fiber mode for reduction of false alarm in leak detection using distributed optical fiber sensor. *Applied Optics*, Vol. 49, No. 20, pp. 3869-3874

Chen, C.L.(2006). *Foundations for guided-wave optics*. John Wiley & Sons, ISBN 0471-75687-3

Choma, M.A.; Yang, C. & Izatt, J.A. (2003). Instantaneous quadrature low-coherence interferometry with 3×3 fiber-optic couplers. *Optics Letters*, Vol. 28, No. 22, pp. 2162-2164

Clark, T.R.; Carruthers, T. F.; Matthews, P. J. & I. N. D. III. (1999). Phase noise measurements of ultrastable 10GHz harmonically modelocked fiber laser. *Electron. Lett.*, Vol. 35, pp. 720–721

Cliche, J.F.; Painchaud, Y.; Latrasse, C.; Picard, M.J.; Alexandre, I.; Têtu, M. (2007). Ultra-Narrow Bragg Grating for Active Semiconductor Laser Linewidth Reduction through Electrical Feedback. *Proceeding in Bragg Gratings, Photosensitivity, and Poling in Glass Waveguides, OSA Technical Digest (CD) (Optical Society of America, 2007)*, paper BTuE2.

Crooker, C. M.; Garrett, S. L. (1987). Fringe rate demodulator for fiber optic interferometric sensors. Proceeding of Fiber optic and laser sensors, San Diego.

De Vos, K.; Debackere, P.; Claes, T.; Girones, J.; De Cort, W.; Schacht, E.; Baets, R. & Bienstman, P.(2009). Label-free Biosensors on Silicon-on-Insulator Optical Chips. *Proceedings of the SPIE*, Vol. 7397, pp. 739710-739710-8

Dianov, E.M.; Vasiliev, S.A.; Kurkov, A.S.; Medvedkov, O.I. & Protopopov, V.N. (1996). In-fiber Mach-Zehnder interferometer based on a pair of long-period gratings. *Proceedings of 22nd European Conference on Optical Communication - ECOC'96*, ISBN 82-423-0418-1, Oslo

Dong, X. & Tam, H.Y.(2007). Temperature-insensitive strain sensor with polarization-maintaining photonic crystal fiber based Sagnac interferometer. *APPLIED PHYSICS LETTERS*, Vol. 90, No. 15, pp. 151113

Dorrer, C.; Londero, P. & Walmsley, I.A. (2001). Homodyne detection in spectral phase interferometry for direct electric-field reconstruction. *Optics Letters*, Vol. 26, No. 19, pp. 1510-1512

Flourney, P.A.; McClure, R.W. & Wyntjes, G.(1972). White-light interferometric thickness gauge. *Appl. Opt.* Vol. 11, No. 9, pp. 1907-1915

Freitas, J.M.D.(2011). Recent developments in seismic seabed oil reservoir monitoring applications using fibre-optic sensing networks. *Measurement Science and Technology*, Vol. 22. No. 5, pp. 052001

Fu, H.Y.; Wu, C.; Tse, M.L.V.; Zhang, L.; Cheng, K.C.D.; Tam, H.Y.; Guan, B.O.; & Lu, C. (2010). High pressure sensor based on photonic crystal fiber for downhole application. Applied Optics, Vol. 49, No. 14, pp. 2639-2643

Gerry, C.C. & Knight, P.L. (2005). *Introductory quantum optics*. Cambridge University Press, ISBN 0-521-82035-9, New York

Glisic, B. & Inaudi. D.(2008). *Fibre Optic Methods for Structural Health Monitoring*. Wiley-Interscience, ISBN 0470061421

Gopel, W.; Jones, T.A.; Kleitz, M.; Lundstrom, I. Seiyama, T. (1991). *Chemical and biochemical sensors*. John Wiley and Sons, Vol. 2, ISBN 3527267697

Grattan, K.T.V. & Meggit, B. T. (Ed(1).). (1997). *optical fiber sensor technology: Devices and technology*. chapman & Hall, London

Grattan, K.T.V. & Meggit, B. T. (Ed(1).). (1999). *Optical Fiber Sensor Technology :Applications and Systems*, Kluwer Academic Publishers, ISBN 0-412-82570-8, United States

Harris, E.; Li, Yi.; Chen, L. & Bao, X. (2010). Fiber-optic Mach–Zehnder interferometer as a high-precision temperature sensor: effects of temperature fluctuations on surface biosensing. *Applied Optics*, Vol. 49, No. 29, pp. 5682-5685

Hassani, A. & Skorobogatiy, M. (2006). Design of the Microstructured Optical Fiber-based Surface Plasmon Resonance sensors with enhanced microfluidics. *OPTICS EXPRESS*, Vol. 14, No. 24, pp. 11616-11621

Haus, H.A.& Lai, Y.(1992). Theory of cascaded quarter wave shifted distributed feedback resonators. *IEEE J. Quantum Electron.* Vol. 28, No.1, pp. 205–213

Higuera, L.; Miguel, V. (2002). *Handbook of Optical Fibre Sensing Technology*. John Wiley & Sons, ISBN 0471820539, England

Hoeling, B.M.; Fernandez, A.D.; Haskell, R.C. & Petersen, D.C. (2001). Phase modulation at 125 kHz in a Michelson interferometer using an inexpensive piezoelectric stack driven at resonance. *Review of Scientific Instruments*, Vol. 72, No. 3, pp. 1630 - 1633, ISSN 0034-6748

Hong, C.K.; Ou, Z.Y. & Mandel, L. (1987). Measurement of subpicosecond time intervals between two photons by interference. *Phys. Rev. Lett.* Vol. 59, No. 18, 2044 2044–2046

Huang, H.C. (1984). *coupled mode theory as applied to microwave and optical transmission*, VNU Science Press, ISBN 90-6764-033-6, Netherlands

Jackson, D.A.; Kersey, A.D.; Corke, M. & Jones, J.D.C. (1982). Pseudo heterodyne detection scheme for optical interferometers. *Electronics Letters*, Vol. 18, No. 25, pp. 1081-1083

Jayanthkumar, A.; Gowri, N.M.; Venkateswararaju, R.; Nirmala, G.; Bellubbi, B.S. & Radhakrishna, T. (2006). Study of fiber optic sugar sensor. *PRAMANA journal of physics*, Vol. 67, No. 2, pp. 383-387

Jia, D.; Fang, N.; Wang, L. & Huang, Z. (2008). Distributed Fiber Optic In-Line Intrusion Sensor System. *Proceeding of Microwave Conference, 2008 China-Japan Joint*, pp. 608 - 611, ISBN 978-1-4244-3821-1

Juarez, J. C.; Maier, E. W.; Choi, K. N.& Taylor, H. F.(2005). Distributed fiber-optic intrusion sensor system. *Journal of Lightwave Technology*, Vol. 23, No. 6, pp. 2081 - 2087, ISSN: 0733-8724

Kashyap, R. (1999). *Fiber Bragg Gratings*. Academic Press, ISBN 0-12-400560-8, UK

Lin, W.W.; Chang, C.F.; Wu, C.W. & Chen, M.C. (2004). The Configuration Analysis of Fiber Optic Interferometer of Hydrophones. *Proceeding of OCEANS '04. MTTS/IEEE TECHNO-OCEAN '04*, Vol.2 , pp. 589 – 592, ISBN: 0-7803-8669-8

Linde, D.V. (1986). Characterization of the noise in continuously operating mode-locked lasers. *Appl. Phys. B*, vol. 39, No. 4, pp. 201–217

Manojlovi, L . (2010). A simple white-light fiber-optic interferometric sensing system for absolute position measurement. *Optics and Lasers in Engineering*, Vol. 48, No. 4, pp. 486–490

Mercado, J.T.; Khomenko, A.V. & Weidner, A.G. (2001). Precision and Sensitivity Optimization for White-Light Interferometric Fiber-Optic Sensors. *Journal of Lightwave Technology*, VOL. 19, NO. 1, pp. 70-74

Mishra, A. & Soni,A. (2011). Leakage Detection using Fibre Optics Distributed Temperature Sensing. *Proceedings of 6th Pipeline Technology Conference 2011*, Hannover, Germany

Okamoto, K. (Ed(2).). (2006). *Fundamentals of optical waveguides*. Academic Press, ISBN 0125250967, San Diego

Olindo, C.; Sagioro, M.A.; Monken, C.H. & Pádua, S. (2006). Hong-Ou-Mandel interferometer with cavities: Theory. *Physical Review A*, Vol. 73, No. 4, pp. 043806-043806.10 ISSN 1050-2947

Poli, F.; Cucinotta, A. & Selleri, S. (2007). Photonic Crystal Fibers Properties and Applications. Springer ISBN 978-1-4020-6325-1,Netherlands

Polygerinos, P.; Seneviratne, L.D. & Althoefer, K. (2011). Modeling of Light Intensity-Modulated Fiber-Optic Displacement Sensors. *IEEE TRANSACTIONS ON INSTRUMENTATION AND MEASUREMENT*, VOL. 60, NO. 4, pp. 1408-1415

Rao, Y.J. (1997). Fibre Bragg grating sensors. *Meas. Sci. Technol*, Vol. 8, pp. 355–375

Righini, G.; Tajani, A. & Cutolo, Antonello. (2009). *An Introduction to Optoelectronics Sensors*, World Scientific Publishing, ISBN 9812834125

Sacharov, V. K. (2001). Linear Relativistic Fiber-Optic Interferometer. *Laser Physics*, Vol. 11, No. 9, pp. 1014–1018

Sagnac, G. (1913). L'ether lumineux demontre par l'effet du vent relatif d'ether dans un interferometre en rotation uniforme. The demonstration of the luminiferous aether by an interferometer in uniform rotation. *C. R. Acad. Sci*, Vol. 157, pp. 708-710

Sanders, S.; Park, N.; Dawson, J.W. & Vahala, K.J. (1992). Reduction of the intensity noise from an erbium-doped fiber laser to the standard quantum limit by intracavity spectral filtering. *Appl. Phys. Lett*, Vol. 61, No. 16, pp. 1889-1891

Scully, M.O. & Zubairy. M.S. (Ed(3).). (2001). *Quantum Optics*. Cambridge University Press, ISBN 0521435951, United Kingdom

Shizhuo Yin, S.; Ruffin, P.B. & Yu, F.T.S. (Ed(2).). (2008). *Fiber Optic sensor*. CRC Press, ISBN 978-1-4200-5365-4

Snyder, A.W.; Love, J.(1983). *Optical waveguide theory* . Springer. ISBN 0412099500

Song, G.; Wang, X. & Fang, Z. (2001).White-light interferometer with high sensitivity and resolution using multi-mode fibers. *Optik*, Vol. 112, No. 6, pp. 245-249

Starodumov, A.N.; Zenteno, L.A. & De La Rosa, E.(1997). Fiber Sagnac interferometer temperature sensor. *Appl. Phys. Lett.*, Vol. 70, No. 1, pp. 19-21

Strauss, C.E.M. (1994). Synthetic-array heterodyne detection: a single-element detector acts as an array. *Optics Letters*, Vol. 19, No. 20, pp. 1609-1611

Tofighi, S.; Bahrampour, A.R. & Shojaee, F. (2010). Optimum quantum state of light for gravitational-wave interferometry. *Optics Communications*, Vol. 283, No. 6, pp. 1012-1016.

Tsuchida, H. (1998). Correlation between amplitude and phase noise in a modelocked Cr : LiSAF laser. *Opt. Lett.*, Vol. 23, pp. 1686–1688

Tucker, R.S. & Baney, D.M. (2001). Optical noise figure: theory and measurements. *Proceedings of Optical Fiber Communication Conference and Exhibit, 2001. OFC* ,Vol. 4, ISBN 1-55752-655-9, Califomia

Tucker, R.S. & Baney, D.M. (2001). Optical Noise Figure: Theory and Measurements. *Proceedings of Optical Fiber Communication Conference and Exhibit, 2001. OFC 1*, ISBN: 1-55752 655-9

Vali, V. & Shorthill, R.W. (1976). Fiber ring interferometer. *Appl. Opt*, Vol. 15, No. 5, pp. 1099-1100

Villatoro, J.; Finazzi, V.; Badenes, G. & Pruneri, V.(2009). Highly Sensitive Sensors Based on Photonic Crystal Fiber Modal Interferometers. Journal of Sensors

Villatoro, J.; Kreuzer, M.P.; Jha, R.; Minkovich, V.P.; Finazzi, V.; Badenes, G. & Pruneri, V. (2009). Photonic crystal fiber interferometer for chemical vapor detection with high sensitivity.*Optics Express*, Vol. 17,No. 3, pp. 1447-1453

Villatoro, J.; Minkovich, V.P. & Hernández, D.M. (2006). Compact Modal Interferometer Built With Tapered Microstructured Optical Fiber. *IEEE PHOTONICS TECHNOLOGY LETTERS*, VOL. 18, NO. 11, pp. 1258-1260

Wang,C.; Trivedi, S.; Kutcher,S.; Rodriguez, P.; Jin,F.; Swaminathan, V. ; Nagaraj, S.; Quoraishee, S. & Prasad. N.S. (2011). Non-Contact Human Cardiac Activity Monitoring Using a High Sensitivity Pulsed Laser Vibrometer. *Proceedings of Conference on Lasers and Electro-Optics (CLEO)* , ISBN: 978-1-4577-1223-4, Baltimore

Weihs, G.; Reck, M.; Weinfurter, H. & Zeilinger, A. (1996). All-fiber three-path Mach–Zehnder interferometer. *OPTICS LETTERS*, Vol. 21, No. 4, pp. 302-304

Yuan, L. (1997). White-light interferometric fiber-optic strain sensor from three-peak-wavelength broadband LED source. *APPLIED OPTICS*, Vol. 36, No. 25 pp. 6246-6250

Yuan, L. (2002). Multiplexed, White-Light Interferometric Fiber-Optic Sensor Matrix with a Long-Cavity, Fabry-Perot Resonator. *Applied Optics*, Vol. 41, No. 22, pp. 4460-4466

Yuan, L.; Liu, Y. & Sun, W. (2005). Fiber optic Moiré interferometric profilometry. *Proceedings of SPIE*, Vol. 5633, ISBN 9780819455888

2

The Applications of the Heterodyne Interferoemetry

Cheng-Chih Hsu
Department of Photonics Engineering, Yuan Ze University, Yuan-Tung Road, Chung-Li, Taiwan

1. Introduction

Optical interferometry is widely used in many precision measurements such as displacement[1, 2], vibration[3, 4], surface roughness[5, 6], and optical properties[7-14] of the object. For example, holographic interferomter [1-3] can be used to measure the surface topography of the rigid object. The emulsion side of the photographic plate faces the object and is illuminated by a plane wave at normal incidence. Therefore, the reflection type hologram is recorded the interference signals between the incident wave and scattered wave from the object within the emulsion layer. Then the hologram is reconstructed with laser light and the information of object surface can be obtained. The Speckle interferometry [2-4] can be used to measure the motion of the rough surface. To compare the two exposure specklegrams, then the phase difference related to the surface movement can be obtained. Abbe refractometer [7, 8] is an easy method to determine the refractive index of the material based on the total internal reflection (TIR). That means the refractive index of the testing sample will be limit by the hemisphere prism installed in the refractometer. The ellipsometer [9-12] is widely used to measure the thickness and refractive index of film or bulk materials. Typically, the optical components of ellipsometer included polarizer, compensator, sample, and analyzer. Hence, there were many different types of ellipsometer for refractive index and thickness measurement of the sample. Most popular type is rotating polarizer and analyzer ellipsometer which can be divided into rotating polarizer type and rotating analyzer type. Both of them are analysis of the ellipsometric angles (ψ, Δ) which determined directly from the adjustable angular settings of the optical components. The accuracy of the ellipsometric measurement are typically within the range 0.01° and 0.05° in (ψ, Δ) [13, 14].

Compare to previous method, the heterodyne interferometry give much more flexibility of different kinds of the measurement purposes with suitable optical configuration. In this chapter, I will review the heterodyne interferometry and focus on the applications of this kind of interferometer. First of all, I will briefly introduce the history and applications of heterodyne interferometry that will be discussed in this chapter. Before I mention the applications of the heterodyne interferometry, I would like to describe several types of heterodyne interferometry. Then I would like to describe the precision positioning with optical interferometer and focus on the heterodyne grating interferometer. After that, I will review some refractometer using heterodyne interferometer. In this section I would like to quick look some useful methods for measuring the refractive index and thickness of bulk material or thin film structure. In addition, the measurement of the optic axis and birefringence of the birefringent crystal will also be discussed in this section. The final application of the heterodyne interferometry that I would like to talk about is the concentration measurement. In this section, I will roughly classify the method into two categories. One is fiber type sensor; another is a non-fiber type sensor. And I will discuss the surface plasmon resonance (SPR) sensor in fiber-type and non-fiber type sensors. Finally, I would like to give the short conclusion, which summarized the advantages and disadvantages of the heterodyne interferometer.

2. Heterodyne interferometry

This section will introduce the development history of the heterodyne interferometry and describe the fundamental theory and basic optical configuration of the heterodyne interferometer.

2.1 History of Heterodyne light source developement

Hewlett Packard Company (HP) developed the first commercial heterodyne interferometer for precision positioning since 1966. Until now, HP systems have widely used in industry, scientific research, and education. J. A. Dahlquist, D. G. Peterson, and W. Culshaw [15] demonstrated an optical interferometer, which used Zeeman laser properties in 1966. They had the application of an axial magnetic field and resulted in the frequency difference between the right hand and left hand circular polarization states of the He-Ne laser. Because of these two polarization states are affected by equally thermal drift and mechanism vibration of the laser, the frequency difference are extremely stable. Therefore, this light source with different frequency is so called the heterodyne light source. Figure 1 showed that the first heterodyne interferometer which constructed with Zeeman laser. As you can see, the frequency shift coming from the moving mirror will be carried with ν_2. Then these two lights with different frequency will be interference at 45° and the distance-varying phase can be detected.

There are many methods can construct the optical frequency shift such as rotation or moving grating method [16, 17], accousto-optical modulator (AOM) [18, 19], electro-optical modulator (EOM) [20, 21], and modulating two slightly different wavelengths of laser diodes [22]. Suzuki and Hioki [16] proposed the idea of moving grating method for constructing the heterodyne light source in 1967. As the grating moves along y-axis with the velocity v, the frequency shift will be introduced into the ±1 order diffracted beam with $\pm\frac{v}{a}$. By suitable arrangement of the optical configuration, either one of these frequency shifted signals can be selected and to form the heterodyne light source. W. H. Stevenson [17] proposed the rotation radial grating to form the heterodyne light source in which he showed that the frequency shift were linear increased with the rotation rate of the radial grating up to 6k rpm. And the maximum frequency shift in this case was 500 kHz.

An acousto-optic modulator (AOM) uses the acousto-optic effect to diffract and shift the frequency of the light [18, 19]. The piezoelectric transducer attaches to the quartz and the

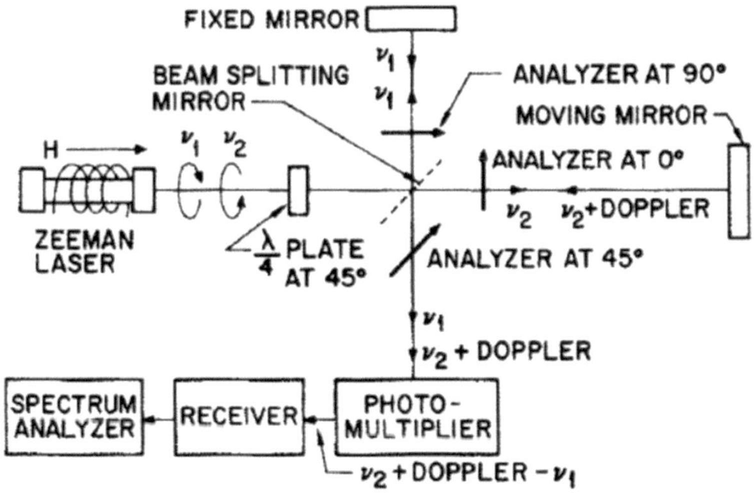

Fig. 1. The first heterodyne interfcrometer constructed by Zeeman laser [15].

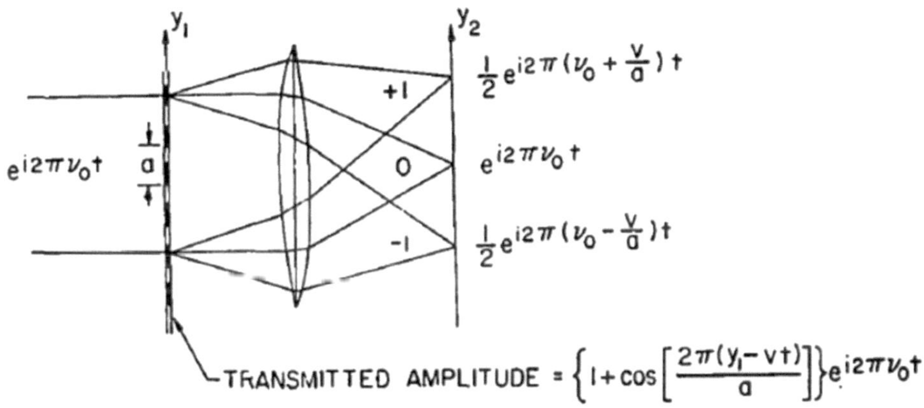

Fig. 2. The heterodyne light source constructed with moving grating [17].

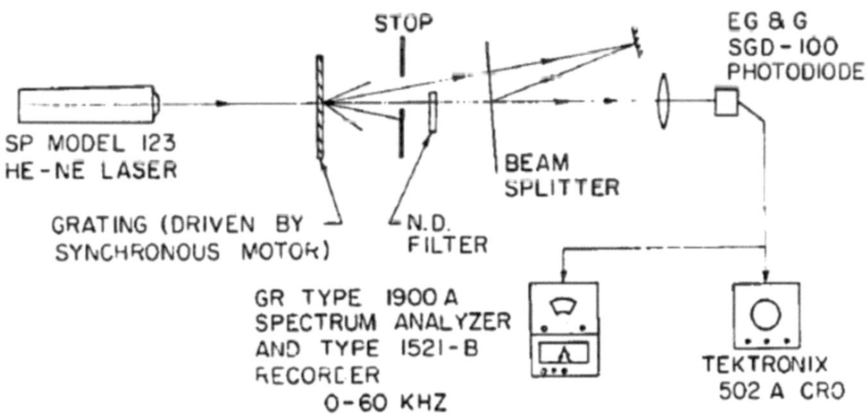

Fig. 3. The heterodyne light source constructed with rotation radial grating [17].

oscillating electric signal drives the transducer to vibrate, which creates sound wave in the quartz and changes the refractive index of the quartz as periodic index modulation. The incoming light diffracts by these moving periodic index modulation planes, which induced the Doppler-shifted by an amount equal to the frequency of the sound wave. That phenomenon is similar to the moving grating method but the fundamental concepts are momentum conservation of the phonon-photon interaction and Bragg diffraction theory. Figure 4 shows the frequency shifted by AOM that proposed in 1988 [18]. A typical frequency shifted varies from 27 MHz to 400 MHz. In the case of M. J. Ehrlich et al. [18], the frequency shifted was 29.7 MHz and the induced phase shifted over 360° by applying the voltage within 15 V.

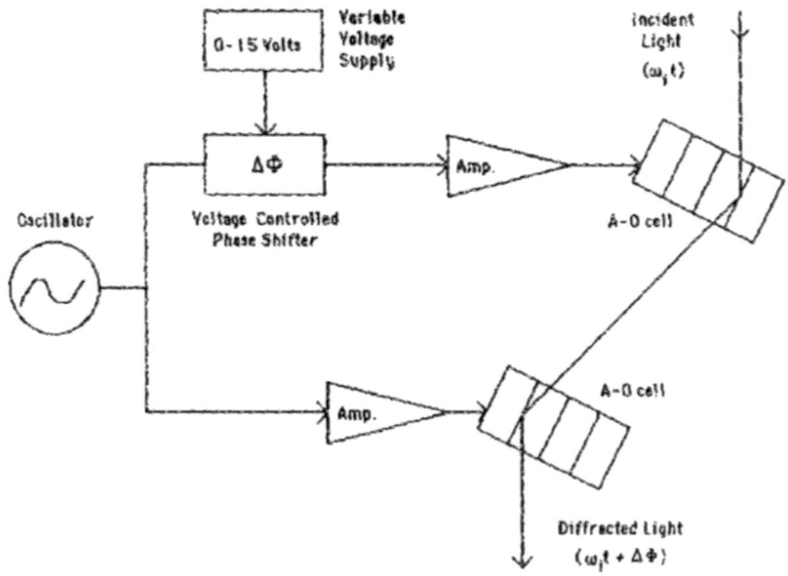

Fig. 4. The frequency shifted by AOM [18].

Electro-optic modulator is a signal-controlled optical device that based on the electro-optic effect to modulate a beam of light. The modulation may be imposed the phase, amplitude, or frequency of the modulated beam. Lithium niobate ($LiNbO_3$) is one of the electro-optic crystals that is widely used for integrated optics device because of its large-valued Pockels coefficients. The refractive index of $LiNbO_3$ is a linear function of the strength of the applied electric field, which is called Pockel effect. Figure 5 shows one of the optical configurations of the heterodyne light source constructed by EOM. The linear polarized light into the EOM, which the crystal axis is located at 45° respected to the x-axis and applied half-wave voltage $V_{\lambda/2}$ on it, the outcome light will carry the frequency shifted.

The wavelength of laser diode can be varied as the injection current and temperature of the laser diode. The wavelength increased as the injection current increased. In general, the rate of the increase is about 0.005 nm/mA at 800 nm and that will be different for different types of laser diode [22]. As the wavelength of the laser diode is changed from λ to $\lambda+\Delta\lambda$ periodically, in which the injection current is periodically changed, the frequency shift of the heterodyne signal can be obtained.

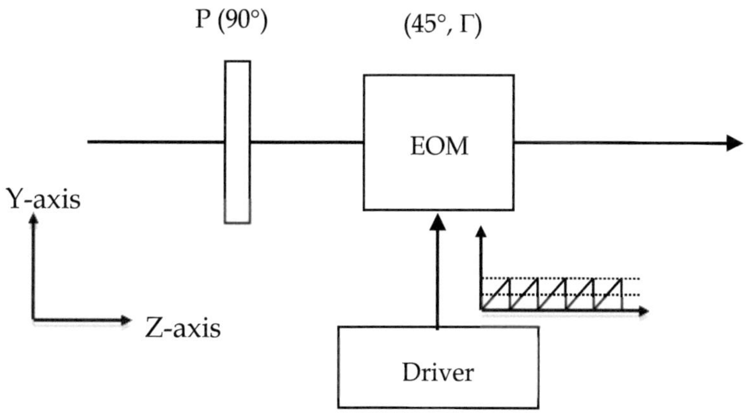

Fig. 5. The optical setup of heterodyne light source with EOM.

2.2 Type of Heterodyne interferometer

The heterodyne interferometer can be divided into two categories, one is common-path type and another is non common-path type. The common-path means that the environment influence of the polarization states of the interference signal can be ignored. Of course, one also can divide into linear polarized heterodyne and circular polarized heterodyne interferometers based on the heterodyne light source. In this section, we would like to describe that based on heterodyne light source and focus on the boundary phenomena between the heterodyne light source and testing sample.

The optical configuration of the linear polarized heterodyne light source have described in figure 5. For convenient, assume that the light propagate along z-axis and vertical direction is y-axis. If the fast axis of the EOM is located at 45° respected to the x-axis, the Jones matrix can be described [14, 23]:

$$EO(45°,\Gamma) = \begin{pmatrix} \cos 45^o & -\sin 45^o \\ \sin 45^o & \cos 45^o \end{pmatrix} \begin{pmatrix} e^{i\frac{\Gamma}{2}} & 0 \\ 0 & e^{-i\frac{\Gamma}{2}} \end{pmatrix} \begin{pmatrix} \cos 45^o & \sin 45^o \\ -\sin 45^o & \cos 45^o \end{pmatrix} = \begin{pmatrix} \cos\frac{\Gamma}{2} & i\sin\frac{\Gamma}{2} \\ i\sin\frac{\Gamma}{2} & \cos\frac{\Gamma}{2} \end{pmatrix} \quad (1)$$

where the Γ is the phase retardation of EOM and can be described $\Gamma = \frac{\pi V}{V_{\lambda/2}}$. When we applied half-wave voltage of the EOM with sawtooth electric signal, equation (1) can be approximated as

$$EO(\omega t) = \begin{pmatrix} e^{-im\pi\frac{i\omega t}{2}} & 0 \\ 0 & e^{im\pi\frac{-i\omega t}{2}} \end{pmatrix} = \begin{pmatrix} e^{i\frac{\omega t}{2}} & 0 \\ 0 & e^{-i\frac{\omega t}{2}} \end{pmatrix} \quad (2)$$

As a linear polarized light with the polarization direction at 45° pass through the EOM, then the E-field can be

$$E = EO(\omega t)\cdot E_{in} = \begin{pmatrix} e^{i\frac{\omega t}{2}} & 0 \\ 0 & e^{-i\frac{\omega t}{2}} \end{pmatrix} \cdot \frac{1}{\sqrt{2}} \begin{pmatrix} 1 \\ 1 \end{pmatrix} e^{i\omega_0 t} = \frac{1}{\sqrt{2}} \begin{pmatrix} e^{i\frac{\omega t}{2}} \\ e^{-i\frac{\omega t}{2}} \end{pmatrix} e^{i\omega_0 t} \quad (3)$$

where ω_0 and ω are optical frequency and frequency shifted between two orthogonal polarization state, respectively. Obviously, equation (3) described the linear polarized heterodyne light source.

For a circular polarized heterodyne light source, the optical configuration is showed in figure 6. As a linear polarized light pass through EOM and quarter-wave plate Q with the azimuth angle at 0°, the Jones matrix of the E-field of the outcome light can be described

$$E' = Q(0°)\cdot EO(\omega t)\cdot E_{in}$$
$$= \begin{pmatrix}1 & 0\\ 0 & i\end{pmatrix}\begin{pmatrix}\cos(\omega t/2) & i\sin(\omega t/2)\\ i\sin(\omega t/2) & \cos(\omega t/2)\end{pmatrix}\begin{pmatrix}1\\0\end{pmatrix} = \begin{pmatrix}\cos(\omega t/2)\\ -\sin(\omega t/2)\end{pmatrix}$$
$$= \frac{1}{2}\begin{pmatrix}1\\i\end{pmatrix}e^{\frac{i\omega t}{2}} + \frac{1}{2}\begin{pmatrix}1\\-i\end{pmatrix}e^{-\frac{i\omega t}{2}}. \tag{4}$$

Obviously, equation (4) describes the circular heterodyne light source that indicated the frequency shifted ω between left-hand circular polarized light and right-hand circular polarized.

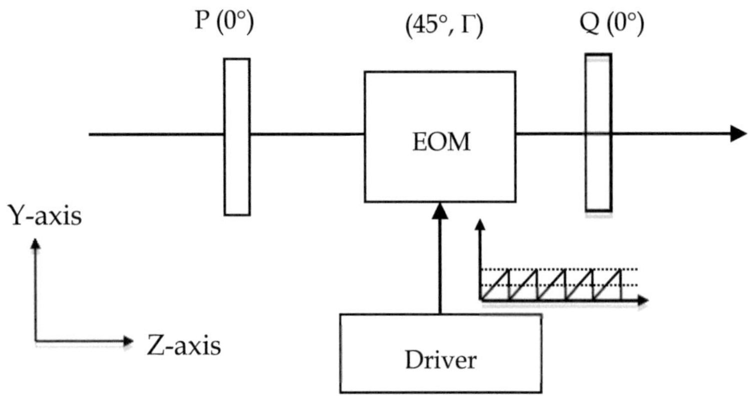

Fig. 6. The circular polarized heterodyne light source.

If the optical interferometer is constructed of the circular polarized heterodyne light source, we always call that circular heterodyne interferometer otherwise we call that heterodyne interferometer. For the specific purpose, we will arrange the tested system as transmission type, reflection type, and multi-reflection type according to the optical property of the testing sample. These types are summarized and show in figure 7. It is obvious that the polarization states (p- and s- polarization states or right-hand and left-hand circular polarization states) of the heterodyne light source are propagated at the same optical path, in which we call common-path structure. The advantage of the common-path structure is the influence of the polarization states of the heterodyne light source can be assumed and limited to the acceptable value. In general, we can ignore the error when the measurement system with common-path configuration. In figure 7, the reference signal I_r coming from the function generator can be written as

$$I_r = I'[1+\cos(\omega t)], \tag{5}$$

and direct into the lock-in amplifier. The heterodyne light source will pass through or reflect from the tested system and then pass through the analyzer AN_t with azimuth angle at α, finally detect by photodetector D_t. The tested system can be divided into three types based on the optical property of the testing sample. There are transmission, reflection, and multi-reflection types.

To consider a heterodyne light source passed through the transmission materials which induced the phase retardation φ, the E-field and intensity detected by D_t can be written as

$$E_t = AN_t(\alpha)\cdot W\cdot E_{in} = \frac{1}{2}\begin{pmatrix}1 & 1\\ 1 & 1\end{pmatrix}\begin{pmatrix}e^{\frac{i\varphi}{2}} & 0\\ 0 & e^{-\frac{i\varphi}{2}}\end{pmatrix}\begin{pmatrix}\cos\frac{\omega t}{2}\\ -\sin\frac{\omega t}{2}\end{pmatrix},$$

$$= [\cos\alpha\cos\frac{\omega t}{2}e^{i\frac{\varphi}{2}} - \sin\alpha\sin\frac{\omega t}{2}e^{-i\frac{\varphi}{2}}]\cdot\begin{pmatrix}\cos\alpha\\ \sin\alpha\end{pmatrix} \tag{6}$$

and

$$I_t = |E_t|^2 = \frac{1}{2}\left[1 + 2\sqrt{A^2+B^2}\cos(\omega t + \phi)\right] \tag{7}$$

where W is the Jones matrix of testing sample at transmission condition; A, B, and ϕ can be written as

$$A = \frac{1}{2}(\cos^2\alpha - \sin^2\alpha), \tag{8a}$$

$$B = \cos\alpha\sin\alpha\cos\varphi, \tag{8b}$$

and

$$\phi = \tan^{-1}(\frac{B}{A}) = \tan^{-1}\frac{2\cos\alpha\sin\alpha\cos\varphi}{(\cos^2\alpha - \sin^2\alpha)}, \tag{8c}$$

It is obvious that the phase retardation φ will be carried by the testing signal I_t. To compare I_r and I_t with lock-in amplifier, the phase difference ϕ coming from the testing sample can be obtained. Substitute the phase difference into equation (8c), the phase retardation of the sample can be determined.

Of course, if the testing sample is not transparence, the reflection type or multi-reflection type can be applied to measure the optical property of the testing sample. To consider a circular heterodyne light source is reflected by the testing sample, passed through the analyzer with the azimuth angle α, and finally detected by photodetector. According to Jones calculation, the E-field and intensity can be expressed as

$$E_t = AN(\alpha)\cdot S\cdot E'$$

$$= \begin{pmatrix}\cos^2\alpha & \sin\alpha\cos\alpha\\ \sin\alpha\cos\alpha & \sin^2\alpha\end{pmatrix}\begin{pmatrix}r_p & 0\\ 0 & r_s\end{pmatrix}\begin{pmatrix}\cos\frac{\omega t}{2}\\ -\sin\frac{\omega t}{2}\end{pmatrix}$$

$$= \left(r_p\cos\alpha\cos\frac{\omega t}{2} - r_s\sin\alpha\sin\frac{\omega t}{2}\right)\begin{pmatrix}\cos\alpha\\ \sin\alpha\end{pmatrix} \tag{9}$$

and

$$I_t = |E_t|^2 = I_0[1 + \cos(\omega t + \phi)] \tag{10}$$

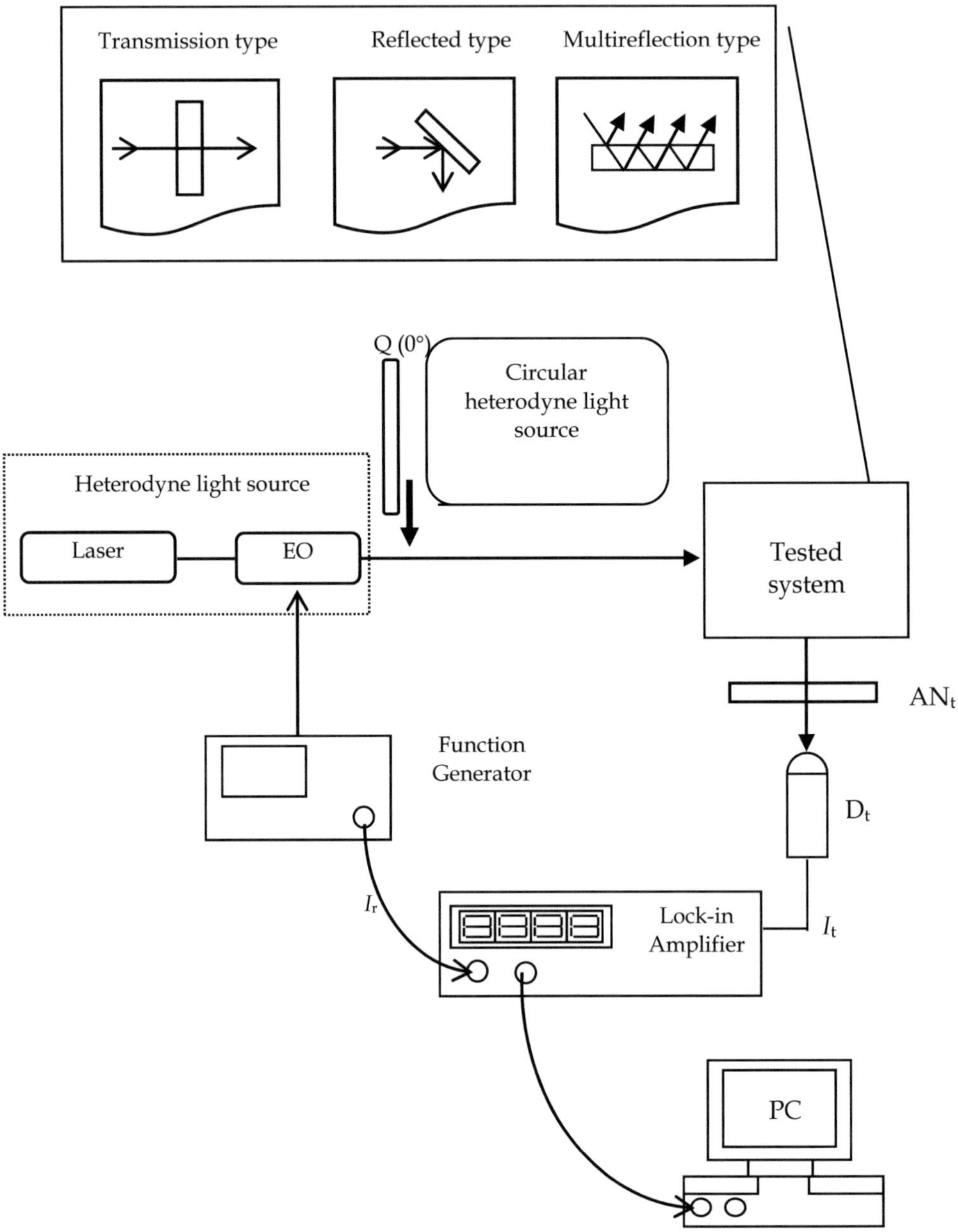

Fig. 7. Types of the heterodyne interferometer with common-path structure.

Where **S** is the Jones matrix of testing sample at reflection condition, r_p and r_s are the reflection coefficients, I_0 and ϕ are the average intensity and phase difference coming from the sample between p- and s- polarizations, which can be written as

$$I_0 = \frac{(r_p{}^2 \cos^2 \alpha + r_s{}^2 \sin^2 \alpha)}{2} \tag{11a}$$

and

$$\phi = \tan^{-1}\left(\frac{2\sin\alpha\cos\alpha \cdot r_p r_s}{r_p{}^2\cos^2\alpha + r_s{}^2\sin^2\alpha}\right) \tag{11b}$$

The reflection coefficients in the reflection matrix of the sample can be expressed by Fresnel equation that can be divided into single reflection and multi-reflection depended on the testing structure. Hence, the r_p and r_s can be written as [14, 23]

(1) single reflection

$$r_p = \frac{n_2\cos\theta - n_1\cos\theta_t}{n_2\cos\theta + n_1\cos\theta_t}, \tag{12a}$$

$$r_s = \frac{n_1\cos\theta - n_2\cos\theta_t}{n_1\cos\theta + n_2\cos\theta_t}, \tag{12b}$$

(2) multi-reflection

$$r_p(\beta) = \frac{r_{1p} + r_{2p}\exp(i\beta)}{1 + r_{1p}r_{2p}\exp(i\beta)}, \tag{13a}$$

$$r_s(\beta) = \frac{r_{1s} + r_{2s}\exp(i\beta)}{1 + r_{1s}r_{2s}\exp(i\beta)}, \tag{13b}$$

and

$$\beta = \frac{2\pi n_2 d\cos(\theta_t)}{\lambda}. \tag{13c}$$

Where n_1 and n_2 are the refractive indices of air and testing sample, θ and θ_t are the incident angle and refracted angle, β is the phase difference coming from the optical path difference in the testing sample, λ is the wavelength of the heterodyne light source. It is obvious that the optical properties of the testing sample can be obtained by substitute phase difference into the equations (10) ~ (13).

On the other hand, the typical optical configuration of the non-common path is shown in figure 8. It is clear that p- and s- polarizations will be propagated at two different paths when they passed through the polarization beam splitter (PBS). In practice, the environment disturbance will not be neglected in non-common path configuration because of these two orthogonal polarization states will have different influence at different path. Therefore, the non-common path optical interferometry using for precision measurement should be seriously taken consideration of stability of the environment disturbance. Figure 8 shows the optical configuration of the displacement measurement. The p- and s- polarizations will reflect by mirrors M_1 and M_2, then pass through the analyzer with azimuth angle at 45°. Therefore, the E-field and intensity of the interference signal between two arms can be written as

$$E_t = \begin{pmatrix} e^{\frac{i\omega t}{2} - ik(2d_p)} \\ e^{\frac{i\omega t}{2} - ik(2d_s)} \end{pmatrix} e^{i\omega_0 t}, \tag{14}$$

and

$$I_t = \frac{1}{2}\left\{1 + \cos\left[\omega t - \frac{4\pi}{\lambda}(d_p - d_s)\right]\right\}. \tag{15}$$

If the mirror M_1 is moved with time, the phase difference $\frac{4\pi d_p}{\lambda}$ will be changed and the displacement variation can be measured by comparing the testing signal and reference signal with lock-in amplifier.

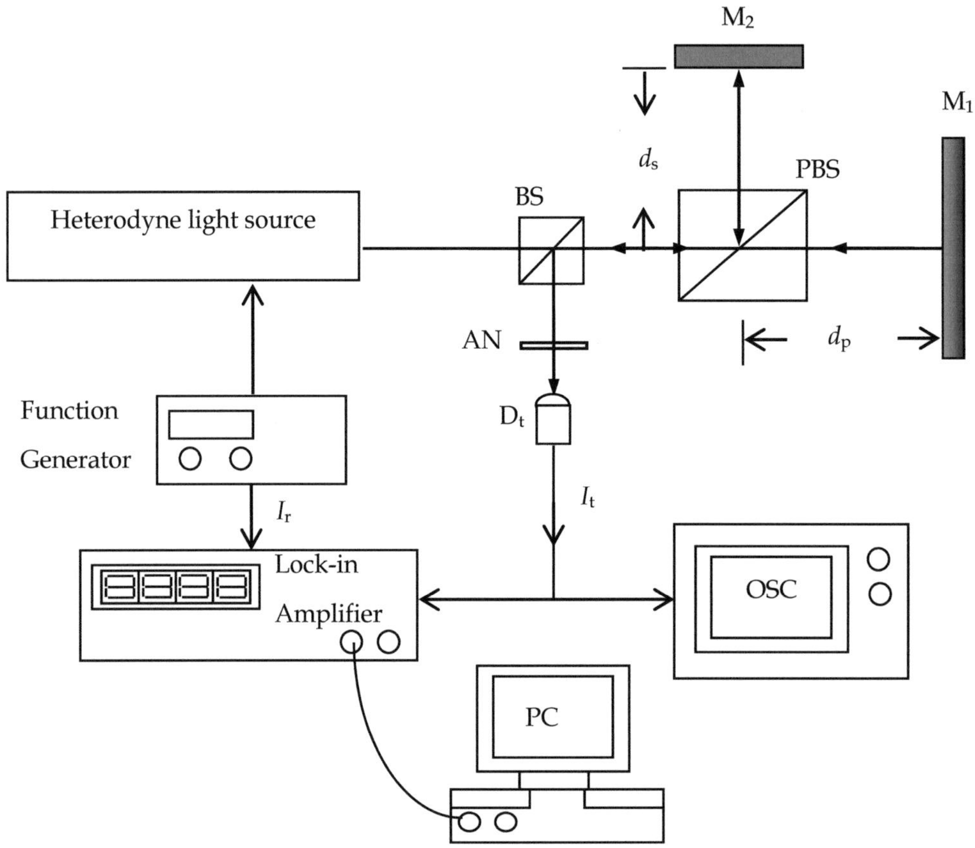

Fig. 8. The optical configuration of the non-common path displacement measurement.

3. Accurate positioning with heterodyne interferometer

Nano-scale positioning devices have become a significant requirement in scientific instruments used for nanotechnology applications. These devices can be applied to nano-handling, nanomanipulation, and nanofabrication. In addition, they are an essential part of the scanning probe microscopy (SPM) and widely used in many research fields. The precision positioning devices consist of three principle parts, which are the rolling component, the driving system and the position sensor. Piezoelectric actuator is the most popular method for driving system and commercial products have been on the market for a few decades. Therefore, the piezoelectric actuator and the position sensor will play the role of the positioning and the feedback control of the rolling element. To achieve the high resolution positioning, the sensing methods of position sensor become more important and have attracted great attention over the past two decades. In this section, we will introduce a few of typical precision positioning methods [24-30] which used heterodyne interferometry.

C. C. Hsu [29] proposed the grating heterodyne interferometry (GHI) to measure the in-plane displacement. The schematic diagram of this method is shown in figure 10. The diffracted grating has mounted on the motorized stage and four diffracted lights will diffract and propagate in the x-z and y-z planes which are for measuring the displacement in x- and y- directions respectively. Based on the Jones calculation, the E-field of the ±1 diffracted beams can be expressed

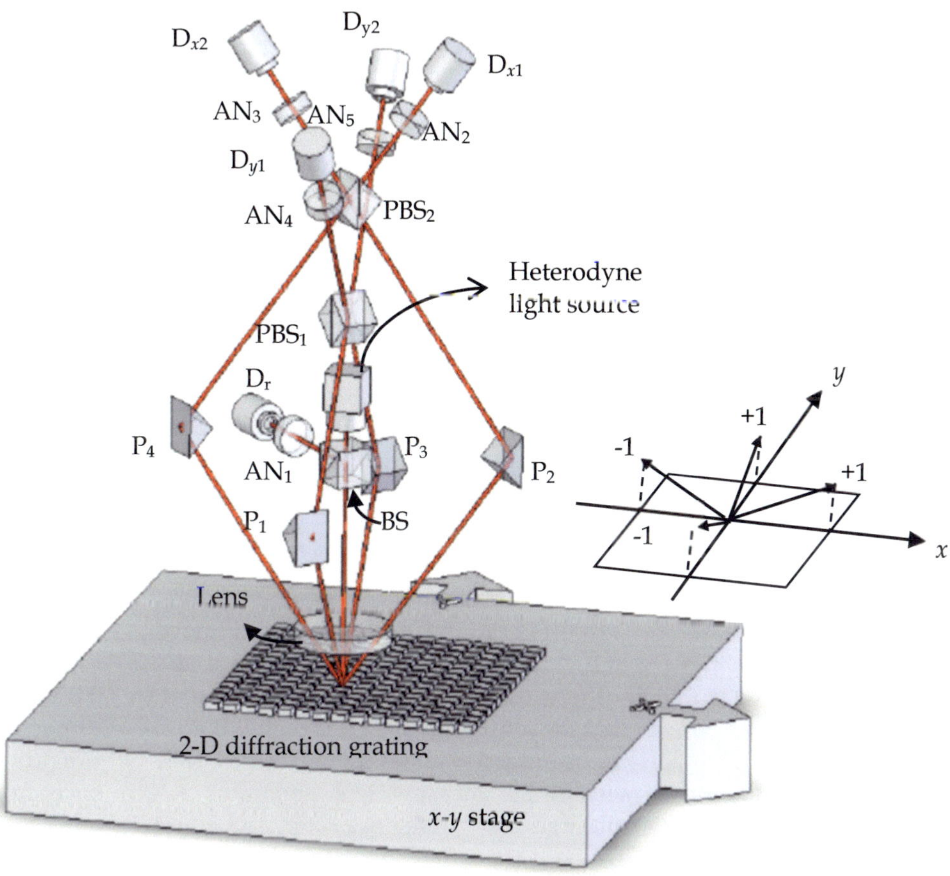

Fig. 10. 2-D displacement measurement system with GHI [29].

$$E_{x\pm1} \propto \exp\left(ikl_{x\pm1} \pm i\frac{2\pi}{g_x}d_x\right) \cdot E_H = \exp\left(ikl_{x\pm1} \pm i\frac{2\pi}{g_x}d_x\right)\begin{pmatrix} e^{i\omega t/2} \\ e^{-i\omega t/2} \end{pmatrix}, \text{(x-direction)} \quad (16a)$$

and

$$E_{y\pm1} \propto \exp\left(ikl_{y\pm1} \pm i\frac{2\pi}{g_y}d_y\right) \cdot E_H = \exp\left(ikl_{y\pm1} \pm i\frac{2\pi}{g_y}d_y\right)\begin{pmatrix} e^{i\omega t/2} \\ e^{-i\omega t/2} \end{pmatrix}, \text{(y-direction)} \quad (16b)$$

where g_x and g_y are the grating pitch in x- and y- directions, d_x and d_y are the displacement along the x- and y- directions respectively. To consider x-direction displacement measurement, the ±1st order diffracted lights will be collected by a lens L and propagate into two paths: (1) prism P_2 → polarization beam splitter PBS_2 → analyzer AN_2 (45°) → detector

D_{x1}, (2) prism P_4 → polarization beam splitter PBS_2 → analyzer AN_3 (45°) → detector D_{x2}. It is similar to the y-direction displacement measurement, the ±1st order diffracted lights will be propagated into (3) prism P_1 → polarization beam splitter PBS_2 → analyzer AN_4 (45°) → detector D_{y1}, (4) prism P_3 → polarization beam splitter PBS_2 → analyzer AN_5 (45°) → detector D_{y2}. After Jones calculation, they can be written as

$$I_{x1} \propto \left|E_{+1s,x} + E_{+1p,x}\right|^2 = \frac{1}{2}\left[1 + \cos\left(\omega t + k\left(l_{-1,x} - l_{+1,x}\right) - \frac{4\pi}{g_x} d_x\right)\right], \tag{17a}$$

$$I_{x2} \propto \left|E_{-1s,x} + E_{+1p,x}\right|^2 = \frac{1}{2}\left[1 + \cos\left(\omega t + k\left(l_{+1,x} - l_{-1,x}\right) - \frac{4\pi}{g_x} d_x\right)\right], \tag{17b}$$

$$I_{y1} \propto \left|E_{+1s,y} + E_{-1p,y}\right|^2 = \frac{1}{2}\left[1 + \cos\left(\omega t + k\left(l_{-1,y} - l_{+1,y}\right) - \frac{4\pi}{g_y} d_y\right)\right], \tag{18a}$$

and

$$I_{y2} \propto \left|E_{-1s,y} + E_{+1p,y}\right|^2 = \frac{1}{2}\left[1 + \cos\left(\omega t + k\left(l_{+1,y} - l_{-1,y}\right) - \frac{4\pi}{g_y} d_y\right)\right]. \tag{18b}$$

To compare the equations (17) and (18), the phase difference coming from the movement in x- and y- directions can be obtained and expressed as

$$\phi_i = \frac{8\pi}{g_i} d_i + 2kl_i \quad (i = x, y), \tag{19}$$

where l_x and l_y are the path difference between grating and PBS_1 and PBS_2 respectively. In practice, the second term in equation (19) can be assumed the initial phase. Therefore, the displacement can be obtained as phase difference is measured and grating pitch is given.

Figure 11 shows that the 2-D displacement measurement with 2-D grating. The movement of the stage is toward to 45° respected to the x-direction and moved 180 nm. The displacement projection in the x- and y-direction are about 120 nm and 140 nm respectively. It is obvious that there are small difference between the results measured by GHI and HP 5529A. Hsu's results can observe that the sensitivity of GHI is higher than HP 5529A and the smallest displacement variation can be judged is about 6 pm. Besides, GHI can provide the 2-D displacement monitoring with single measurement apparatus which have many advantages such as easy alignment, high cost/preference ratio, and easy integrated to the motorized system.

Recently, J. Y. Lee [30] proposed a novel method to measure the 2-D displacement which have quasi-common optical path (QCOP) configuration. The optical structure is shown in figure 12. Based on the clever arrangement, the expanded heterodyne beam is divided into 4 parts *A*, *B*, *C* and *D*. According to the Jones calculation, the amplitudes of these 4 parts are given by

$$E_B = E_C = J(180°) \cdot E_H = \begin{pmatrix} e^{-i\omega t/2} \\ e^{i\omega t/2} \end{pmatrix}, E_A = E_D = J(0°) \cdot E_H = \begin{pmatrix} e^{i\omega t/2} \\ e^{-i\omega t/2} \end{pmatrix}. \tag{20}$$

The expanded heterodyne beam will reflect by a mirror and focus by a lens with suitable focal length, in which can make the zero order (m=0) beam overlap with the ±1 order diffracted beams. The beam distribution is shown in detail in the inset. When the grating moves along the x direction, the interference phase changes can be observed from the

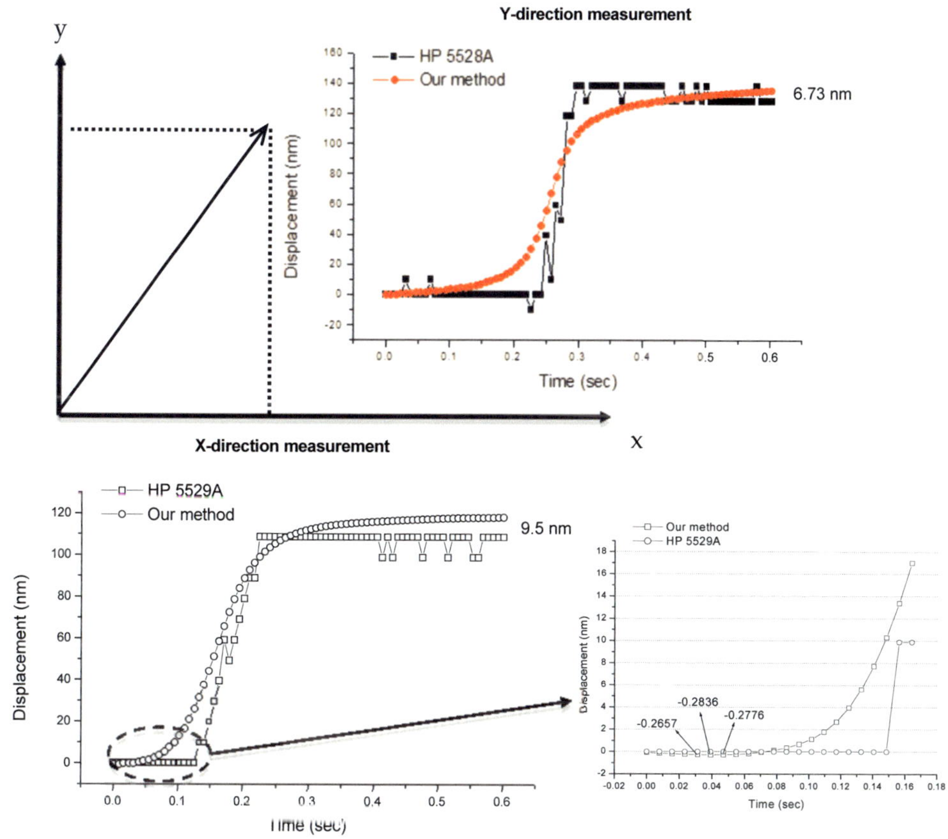

Fig. 11. Displacement measurement of 2-D movement of motorized stage with GHI and HP 5529A.

overlapping area O_1 to O_4; when the grating moves along the y direction, the interference phase changes can be observed from the overlapping area O_5 to O_8. One can use an iris before the focus lens to control the overlapping area. The overlapping areas (O_1 and O_5) are chosen to pass through two polarizers P_1 and P_2 with transmittance axes at 0°. The interference of the light is detected using two detectors D_1 and D_2. The interference signal I_1 and I_2 measured by the detectors D_1 and D_2 can be written as

$$I_1 = 1 + \cos(\omega t - \phi_{x1}), \tag{21a}$$

$$I_2 = 1 + \cos(\omega t - \phi_{y1}). \tag{21b}$$

A polarizer P_3 for which the transmittance axis is at 45° and a detector D_3 are used to measure the intensity of the non-overlapping areas which can be a reference signal I_3 (measured by D_3) and written as

$$I_3 = 1 + \cos(\omega t), \tag{22}$$

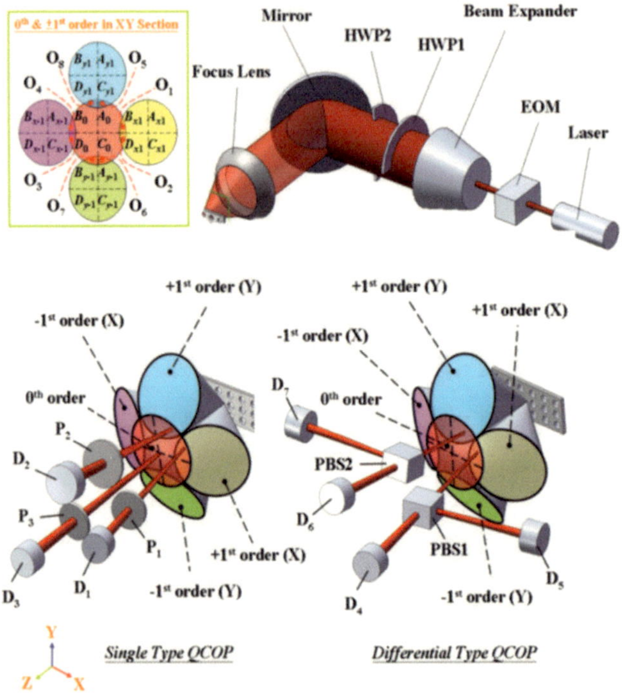

Fig. 12. The schematic of the single type and differential type QCOP heterodyne grating interferometer.

These three signals I_1, I_2, and I_3 are sent into the lock-in amplifier and the phase differences $\Phi_x = \phi_{x1}$ (between I_1 and I_3) and $\Phi_y = \phi_{y1}$ (between I_2 and I_3) are given by

$$\phi_i = \frac{2\pi}{g_i} d_i \quad (i = x, y), \tag{23}$$

where d_i is the displacement in x- and y- directions; g_i is the grating pitch of 2-D grating in x- and y- directions. It is obvious that the 2-D displacement can be obtained as the phase difference and grating pitch of the 2-D grating are given. In the differential type QCOP method, two polarization beam splitters (PBSs) are used to separate the two overlapping beams into four parts. Therefore, the interference signals detected by D_4, D_5, D_6, and D_7 can be written as

$$I_4 = 1 + \cos(\omega t - \phi_{x1}) \text{ and } I_5 = 1 + \cos(\omega t + \phi_{x1}); \quad \text{(for x- direction)} \tag{24a}$$

$$I_6 = 1 + \cos(\omega t - \phi_{y1}) \text{ and } I_7 = 1 + \cos(\omega t + \phi_{y1}). \quad \text{(for y- direction)} \tag{24b}$$

These two pairs signal are sent into the multi-channel lock-in amplifier, the phase differences $\Phi_x = \phi_{x1} - (-\phi_{x1})$ (between I_4 and I_5) and $\Phi_y = \phi_{y1} - (-\phi_{y1})$ (between I_6 and I_7) are 4 times of ϕ_i.

Figure 13 shows a top view of the experimental results in the XY section and the XY stepper moves with a displacement of 1 mm. It is clear that the slight difference between the results measured by the laser encoder and QCOP method. The difference is coming from a tiny angle between the moving direction and the grating, which can be alignment by mounting 2D grating on the rotation stage. In their case, the larger difference was about 12 μm in the y-direction and the smallest difference was about 29 nm in the x- direction for a displacement of 1 mm. Based on the error analysis, if the phase resolution (0.001°) of the lock-in amplifier is considered, the corresponding displacement resolutions of the differential and single type interferometers are estimated to be 9 pm and 4.5 pm for a grating pitch of 3.2 μm, respectively. If only high frequency noise is considered, the measurement resolution of the differential and single type QCOP interferometers can be estimated to be 1.41 nm and 2.52 nm.

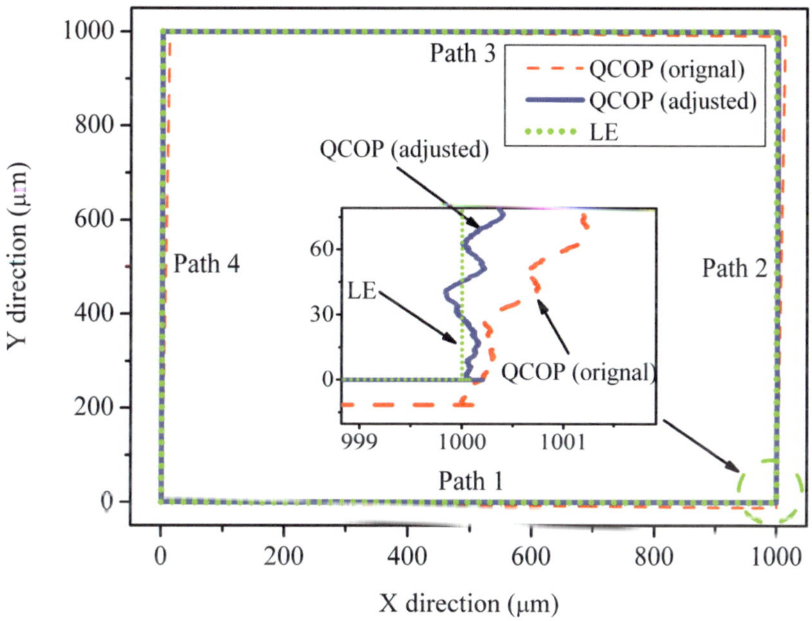

Fig. 13. Displacement measurement of the quadrangular motion with the QCOP method [30].

4. Optical constants measurement with heterodyne interferometer

Optical constants of the materials such as refractive index, birefringence, optical activity, and thickness are significant parameters in material science. There are many methods [9-12, 31-40] can determine those factors, most popular method is ellipsometer [9-12]. Recently, these factors can be obtained by heterodyne interferometer. In this section, we will review some novel methods [31-40] for optical constants measurement with heterodyne interferometer.

C. C. Hsu [32] proposed a novel method for determine the refractive index of the bulk materials with normal incident circular heterodyne interferometer (NICHI) and the schematic diagram is shown in figure 14. The circular heterodyne light source was incident into the modified Twyman-Green interferometer, in which the testing signal reflected from the sample can be interfered and carried by the circular heterodyne light beams. Based on Jones calculation, the interference signal measured by D can be written as

$$I_t = |E_1 + E_2|^2 = I_0[1 + \gamma\cos(\omega t + \phi)], \quad (25)$$

where E_1 and E_2 are the E-field coming from the optical path 1 and path 2 respectively. I_0, γ, and ϕ are the mean intensity, the visibility and the phase of the interference signal, respectively. In additions, they can derive from the Jones calculation and written as

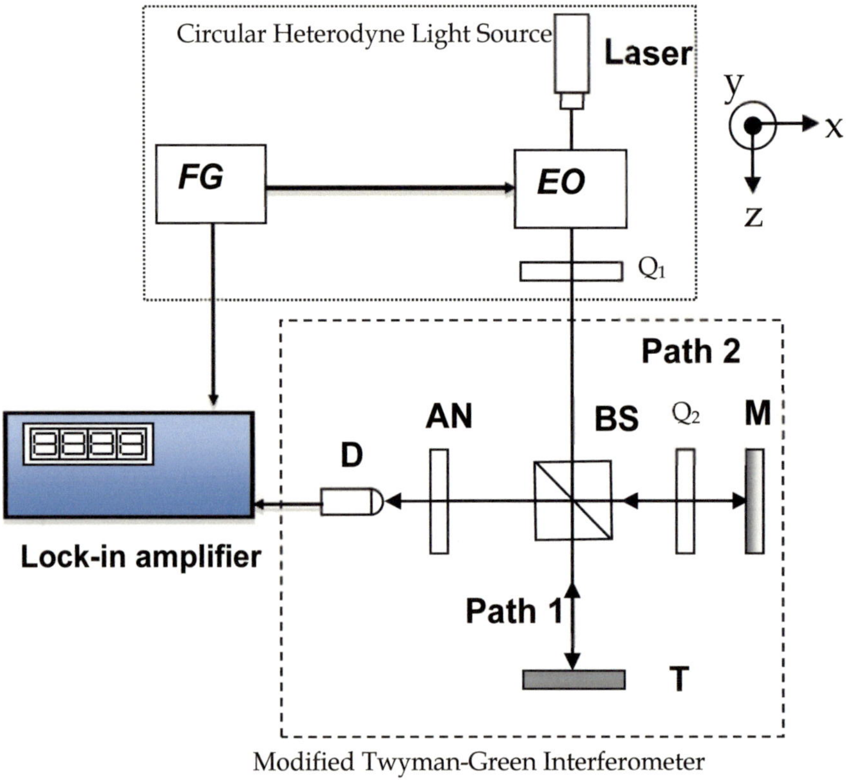

Fig. 14. The optical configuration of normal incident circular heterodyne interferometer [32].

$$I_0 = \left[\frac{1}{2}\left(\frac{n-1}{n+1}\right)^2 + 2{r_m}^2 + 2\left(\frac{n-1}{n+1}\right)r_m \sin 2\alpha \sin(\phi_{d1} - \phi_{d2})\right], \quad (26a)$$

$$\gamma = \frac{\sqrt{A^2+B^2}}{\left[\frac{1}{2}\left(\frac{n-1}{n+1}\right)^2 + 2{r_m}^2 + 2\left(\frac{n-1}{n+1}\right)r_m \sin 2\alpha \sin(\phi_{d1} - \phi_{d2})\right]}, \quad (26b)$$

$$\phi = \tan^{-1}\left(\frac{B}{A}\right), \quad (26c)$$

where the symbols A and B can be written as

$$\mathrm{A} = \frac{1}{2}\left[\left(\frac{n-1}{n+1}\right)^2 - 4{r_m}^2\right]\cdot(\cos^2\alpha - \sin^2\alpha), \quad (27a)$$

$$\mathrm{B} = \frac{1}{2}\left[\left(\frac{n-1}{n+1}\right)^2 - 4{r_m}^2\right]\cdot\sin 2\alpha - 2\frac{n-1}{n+1}r_m, \quad (27b)$$

where r_m is the normal reflection coefficients of the test medium. If the phase can be measured and the reflectivity of mirror is given, the refractive index of the testing sample will be obtained. Furthermore, it is clear that the resolution of refractive index is strongly related to the azimuth angle of analyzer and the reflectivity of the mirror. To derive equation (26) to n, the resolution of refractive index can be written as

$$\Delta n = \frac{1}{\frac{d\phi}{dn}}|\Delta\phi| = [\frac{ac-b}{cd}]|\Delta\phi|, \tag{28}$$

where a, b, c, and d are

$$a = \frac{2[-\frac{4r_m}{(1+n)^2}+\frac{2(n-1)}{(1+n)^3}\sin 2\alpha]}{[\left(\frac{n-1}{1+n}\right)^2-4{r_m}^2][\cos^2\alpha-\sin^2\alpha]}, \tag{29a}$$

$$b = 2\frac{4(n-1)}{(1+n)^3}\left\{\frac{-2(n-1)r_m}{1+n}+\frac{1}{2}[\left(\frac{n-1}{n+1}\right)^2-4{r_m}^2]\cdot\sin 2\alpha\right\}, \tag{29b}$$

$$c = [\left(\frac{n-1}{n+1}\right)^2-4{r_m}^2]^2(\cos^2\alpha-\sin^2\alpha), \tag{29c}$$

$$d = 1+\frac{4\left\{\frac{-2(n-1)r_m}{n+1}+\frac{1}{2}[\left(\frac{n-1}{n+1}\right)^2-4{r_m}^2]\sin 2\alpha\right\}^2}{[\left(\frac{n-1}{n+1}\right)^2-4{r_m}^2]^2[\cos^2\alpha-\sin^2\alpha]^2}. \tag{29d}$$

The simulation results were shown in figure 15 and resolution of the refractive index can be reached 10^{-5} as the suitable experimental conditions were chosen.

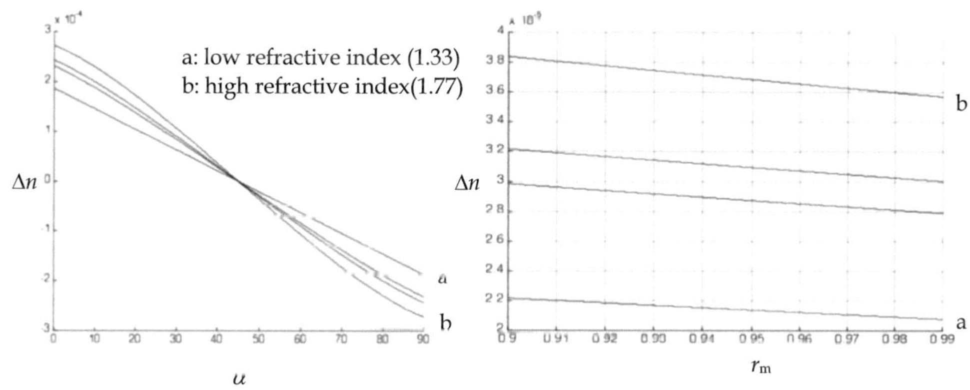

Fig. 15. The relationship between the azimuth angle, reflectivity of the mirror, and resolution of the refractive index [32].

In 2010, Y. L. Chen and D. C. Su [38] developed a full-field refractive index measurement of gradient-index lens with normal incident circular heterodyne interferometer (NICHI). They used high speed CMOS camera to record 2D interference signal and the optical configuration was shown in figure 16.

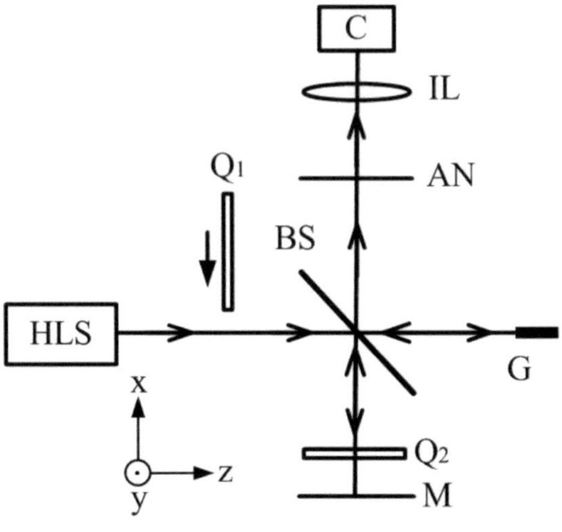

Fig. 16. Optical configuration of full-field normal incident circular heterodyne interferometer [38].

In this interferometer, one is for reference beam (BS→Q$_2$→M→Q$_2$→BS→AN→IL→C) and one is for testing beam ((BS→G→BS→AN→IL→C). Here, G means GRIN lens. They were interfere with each other after passing through AN. Before insert the Q$_1$, the interference signal can be written as

$$I_1 = I_0 + \gamma_1 \cos(2\pi f t + \phi_1)$$

$$= \frac{1}{2}\left\{r^2 + {r_m}^2 - 2 r r_m \cos\left[2\pi f t + \frac{\pi}{2} - (\phi_{d1} - \phi_{d2} + \phi_r)\right]\right\} \quad (30)$$

where I_0 , γ_1 , and ϕ_1 are the mean intensity, visibility, and phase of the interference signal, respectively. Then insert the Q$_1$, the interference signal can be written as

$$I_2 = [r r_m \sin(\phi_{d1} - \phi_{d2} + \phi_r)]\cos(2\pi f t) + \left[\frac{1}{2}({r_m}^2 - r^2)\right]\cos(2\pi f t) + C, \quad (31a)$$

and

$$\phi_2 = \cot^{-1}\left[\frac{2 r r_m \sin(\phi_{d1} - \phi_{d2} + \phi_r)}{({r_m}^2 - r^2)}\right]. \quad (31b)$$

It is obvious that the refractive index of the GRIN lens G can be the function of ϕ_1 and ϕ_2 which expressed as

$$n = \frac{\cot\phi_2 - r_m \cos\phi_1 + r_m\sqrt{\cos^2\phi_1 + \cot^2\phi_2}}{\cot\phi_2 + r_m \cos\phi_1 - r_m\sqrt{\cos^2\phi_1 + \cot^2\phi_2}}. \quad (32)$$

Therefore, for a specified r_m, ϕ_1 and ϕ_2 are given by the measurement, the refractive index of GRIN lens can be obtained.

For full-filed heterodyne phase detection can be realized with three-parameter sine wave fitting method that proposed by IEEE standards 1241-2000. The fitting equation has the form of

$$I(t) = \sqrt{{A_0}^2 + {B_0}^2}\cos(2\pi f t + \varphi) + C_0, \tag{33a}$$

and

$$\varphi = \tan^{-1}\left(\frac{-B_0}{A_0}\right). \tag{33b}$$

where A_0 , B_0 , and C_0 are real numbers and they can be derived with the least-square method. And finally the phase of the all pixels on the CCD camera can be obtained. Based on their method, they demonstrated the two dimensional refractive index distribution of the GRIN lens and showed in figure 17.

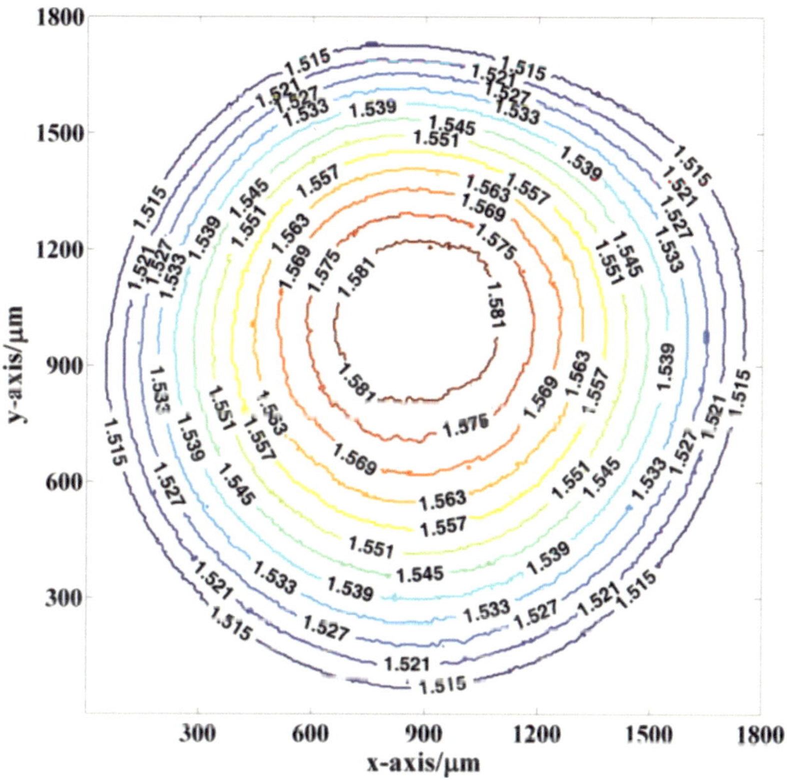

Fig. 17. The refractive index contour of GRIN lens measured by full-field NICHI [38].

For the measurement of the optical constants of the thin film, K. H. Chen and C. C. Hsu [33] proposed a circular heterodyne refractometer. The optical configuration was shown in figure 18. The circular heterodyne light source was incident onto the sample at θ_0 and the light will be partially transmitted and reflected at the interface between the thin film and substrate. If the transmission axis of *AN* is located at α with respect to the x-axis, then the E-field of the light arriving at *D* is given

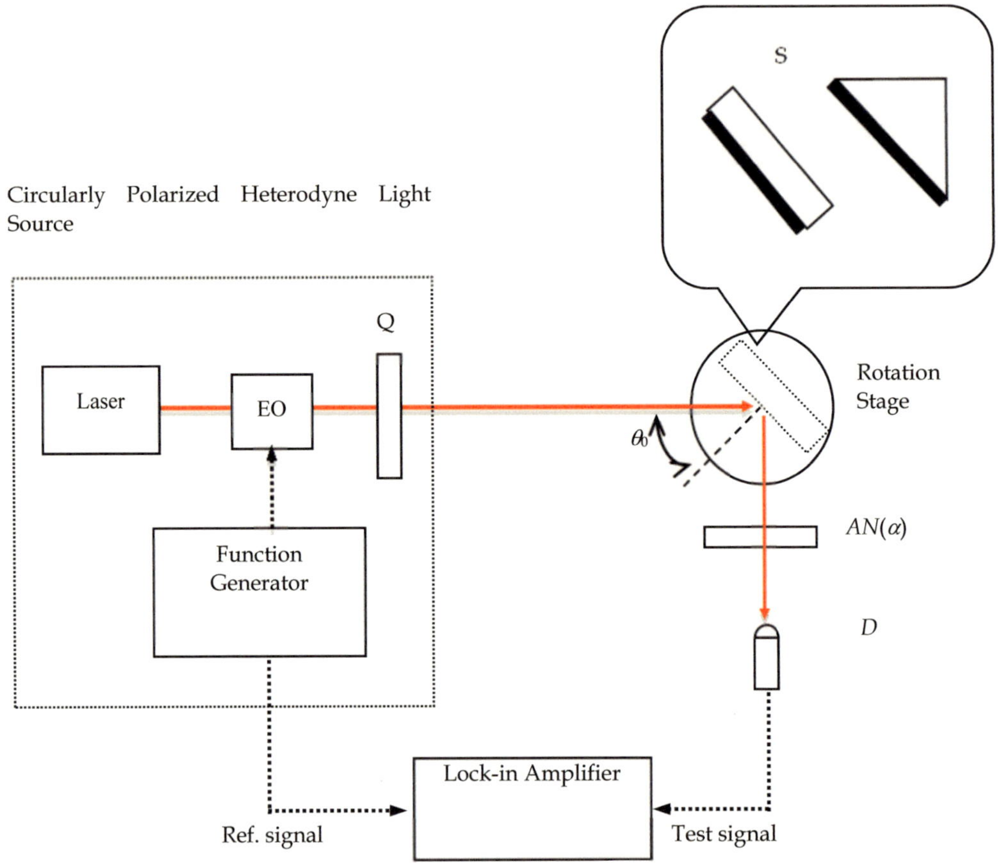

Fig. 18. The circular heterodyne refractor.

$$E_t = \begin{pmatrix} \cos^2 \alpha & \sin\alpha\cos\alpha \\ \sin\alpha\cos\alpha & \sin^2\alpha \end{pmatrix} \begin{pmatrix} r_p & 0 \\ 0 & r_s \end{pmatrix} \begin{pmatrix} \cos\frac{\omega t}{2} \\ -\sin\frac{\omega t}{2} \end{pmatrix} = \begin{pmatrix} r_p \cos^2\alpha \cos\frac{\omega t}{2} - r_s \sin\alpha\cos\alpha\sin\frac{\omega t}{2} \\ r_p \sin\alpha\cos\alpha\cos\frac{\omega t}{2} - r_s \sin^2\alpha \sin\frac{\omega t}{2} \end{pmatrix}. \tag{34}$$

Therefore, the testing signal detected by the detector D can be written as

$$I_t = |E_t|^2 = I_0[1 + \frac{\sqrt{A^2+B^2}}{I_0}\cos(\omega t + \phi)], \tag{35}$$

where I_0 and γ are the bias intensity and the visibility of the signal, and ϕ is the phase difference between the p- and s- polarizations coming from the reflection of the sample. They can be expressed as

$$I_0 = \frac{1}{2}(|r_p|^2 \cos^2\alpha + |r_s|^2 \sin^2\alpha), \tag{36a}$$

$$A = \frac{1}{2}(|r_p|^2 \cos^2\alpha - |r_s|^2 \sin^2\alpha), \tag{36b}$$

$$B = \frac{1}{2}(r_p r_s^* + r_s r_p^*) \sin\alpha\cos\alpha, \tag{36c}$$

and

$$\phi = \tan^{-1}(\frac{B}{A}) = \tan^{-1}[\frac{(r_p r_s^{*} + r_s r_p^{*}) \sin\alpha \cos\alpha}{(|r_p|^2 \cos^2\alpha - |r_s|^2 \sin^2\alpha)}]. \tag{36d}$$

The symbols r_p and r_s are the Fresnel equation (as equation 13); r_p^* and r_s^* are the conjugates of r_p and r_s, respectively. It is obvious that the phase difference coming from the samples are function of the incident angle and the transmission axis of the analyzer *AN*. In practice, one can adjust three transmission axis of the analyzer at fixed incident angle or change three different incident angles with fixed transmission axis of the analyzer to get the corresponding phase difference ϕ. Therefore, substitute the phase difference into equation (36d) the optical constants of the sample can be obtained. According to Chen's results, they can successfully measure the thin metal film deposited on the glass substrate with lower measurement errors, which are 10^{-3} for the complex refractive index and 10^{-1} nm for the thickness.

Birefrigent crystals (BC) have been used to fabricate polarization optical components for a long time. To enhance their qualities and performances, it is necessary to determine the optical axis (OA) and measure the extraordinary index n_e and the ordinary index n_o accurately. There are many methods proposed to measure these parameters of the birefrigent crystal. Huang *et al.* [39] measured (n_e, n_o) of the wedge-shaped birefrigent crystal with transmission-type method. Therefore, the accuracy of thickness, flatness and parallelism of the two opposite sides of the birefrigent crystals are strongly required. D. C. Su and C. C. Hsu [37] proposed a novel method for measuring the extraordinary index n_e, the ordinary index n_o, and the azimuth angle of the birefrigent crystal with single apparatus which described in figure 19. Based on the circular heterodyne interferometer (CHI) and replaced the sample by the birefrigent crystal, the Jones vector of the E-field detected by *D* can be written as

$$E_t = AN(\beta) \cdot \begin{bmatrix} r_{pp} & r_{ps} \\ r_{sp} & r_{ss} \end{bmatrix} \cdot E_i$$

$$= \left(\left(r_{pp} \cos\beta + r_{sp} \sin\beta \right) \cos\frac{\omega t}{2} - \left(r_{ps} \cos\beta + r_{ss} \sin\beta \right) \sin\frac{\omega t}{2} \right) \begin{pmatrix} \cos\beta \\ \sin\beta \end{pmatrix} \tag{37}$$

where S is the Jones matrix for BC, r_{pp} and r_{ss} are the direct-reflection coefficients, and r_{ps} and r_{sp} are the cross-reflection coefficients [14], respectively. Based on Fresnel equations, r_{pp}, r_{ss}, r_{ps} and r_{sp} are the function of the n_e, n_o, and azimuth angle α of the birefrigent crystal. Therefore, the intensity of the testing signal can be expressed

$$I_t = |E_t|^2 = \frac{\left(r_{pp} \cos\beta + r_{sp} \sin\beta \right)^2 + \left(r_{ps} \cos\beta + r_{ss} \sin\beta \right)^2}{2} [1 + \cos(\omega t + \phi)], \tag{38a}$$

and

$$\phi = \tan^{-1}\left(\frac{2\left(r_{pp} \cos\beta + r_{sp} \sin\beta \right)\left(r_{ps} \cos\beta + r_{ss} \sin\beta \right)}{\left(r_{pp} \cos\beta + r_{sp} \sin\beta \right)^2 + \left(r_{ps} \cos\beta + r_{ss} \sin\beta \right)^2} \right). \tag{38b}$$

Theoretically, it is difficult to obtain the n_e, n_o, and azimuth angle α of the birefrigent crystal by substituting the phase difference, which is arbitrary choose the measurement conditions

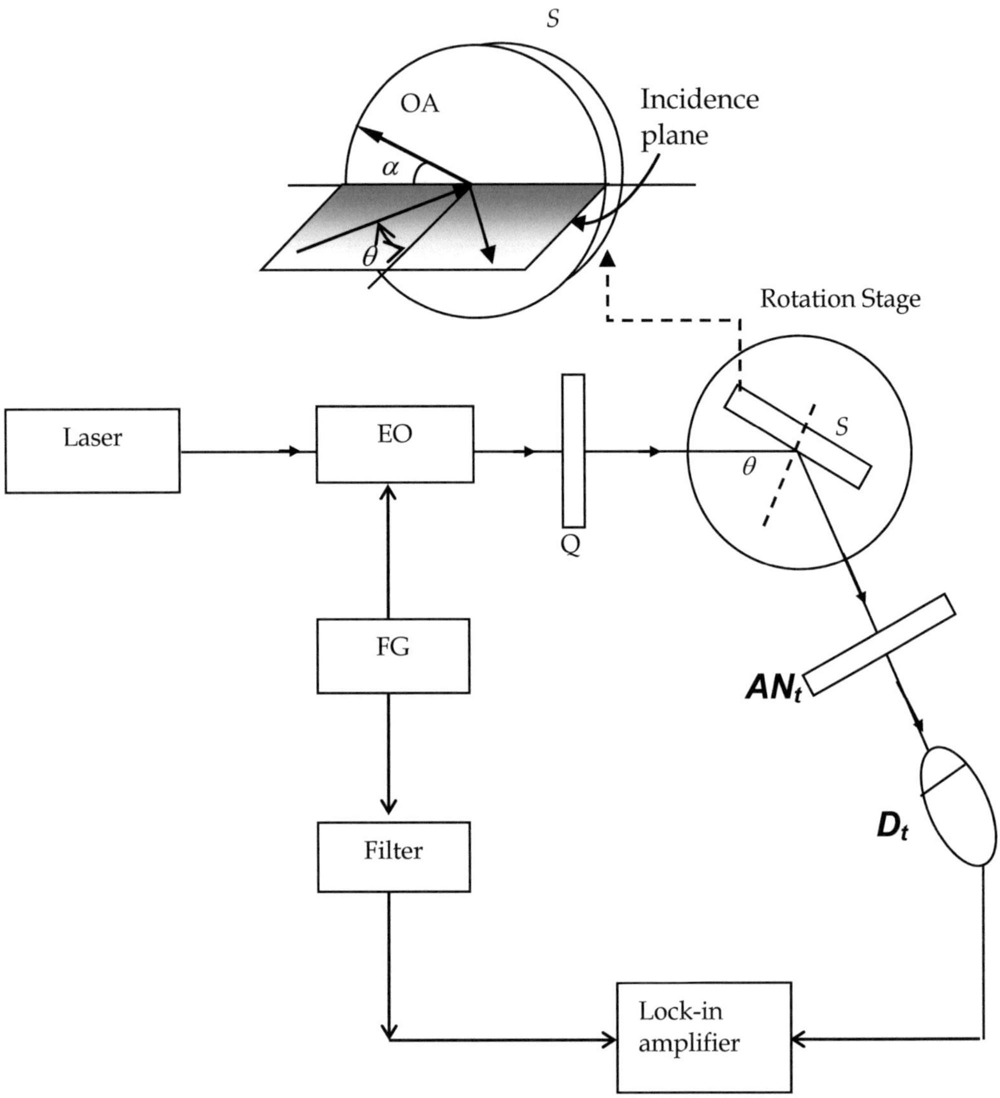

Fig. 19. Optical configuration of the determination of the optical properties of birefrigent crystal with CHI.

of the incident angle and azimuth angle of the analyzer, into equation (38b). Therefore, Su developed a sequence for determining these parameters. First, let azimuth angle β of analyzer equal to 0° and phase difference can be written as

$$\phi = \tan^{-1}\left(\frac{2r_{pp}r_{ps}}{r_{pp}{}^2 - r_{ps}{}^2}\right). \tag{39}$$

As azimuth angle α of the birefrigent crystal at 0° or 90°, r_{ps} and r_{sp} will be equal to 0, and phase difference ϕ is equal to 0. But in this period, one cannot determine the azimuth angle α of brifrigent crystal exactly at 0° or 90°.

Second, fixed the azimuth angle of the birefrigent crystal and rotated azimuth angle β of analyzer to nonzero position. The phase difference ϕ can be expressed as

$$\phi = \tan^{-1}\left(\frac{\sin 2\beta \cdot r_{pp} r_{ps}}{r_{pp}{}^2 \cos^2\beta - r_{ss}{}^2 \sin^2\beta}\right). \tag{40}$$

At this period, the r_{pp} and r_{ss} will be under one of the conditions (i) α=90°or (ii) α=0°. Hence, we solved the n_e and n_o under conditions (i) and (ii) with two different incident angles.

Third, determine the correct solution with two justifications. (1) Rationality of the solution: In general, both n_e and n_o are within the range 1 and 5. If any estimated data of n_e and n_o is not within this range, it is obvious that the estimated data may be incorrect. (2) Comparison between n_e and n_o: Either a positive or negative crystal is tested, all two pairs of solutions of either group should meet with only either $n_e > n_o$ or $n_e < n_o$. If not, then that group is incorrect.

Based on Su's procedure, they have successfully determined the n_e, n_o, and azimuth angle α of the birefrigent crystal, which were positive crystal (quartz) and negative crystal (calcite), with lower error of the refractive index (~ 10^{-3}) and azimuth angle (~0.1°) of BC.

J. F. Lin *et al* [40] proposed a transmission type circular heterodyne interferometer to determine the rotation angle of chiral medium (glucose solution). The optical setup was shown in figure 20 and the E-field of the testing signal derived from Jones calculation is given

$$\mathrm{E} = \mathrm{A}(0)\cdot \mathrm{S}(\theta)\cdot Q_1(45)\cdot EO(90)\cdot P(45)\cdot E_{in}$$

$$= \begin{bmatrix}1 & 0\\ 0 & 0\end{bmatrix}\begin{bmatrix}\cos\theta & \sin\theta\\ -\sin\theta & \cos\theta\end{bmatrix}\begin{pmatrix}\frac{1-i}{2} & \frac{1+i}{2}\\ \frac{1+i}{2} & \frac{1-i}{2}\end{pmatrix}\begin{bmatrix}e^{-i\omega t/2} & 0\\ 0 & e^{-\omega t/2}\end{bmatrix}\begin{bmatrix}\frac{1}{2} & \frac{1}{2}\\ \frac{1}{2} & \frac{1}{2}\end{bmatrix}\begin{bmatrix}0\\ E_0 e^{iw_0 t}\end{bmatrix}, \tag{41}$$

where θ is the optical rotation angle of the chiral medium. And the intensity of the testing signal detected by the photodetector can be derived and written as

$$I_t = I_{dc}[1 - \sin(\omega t - 2\theta)], \tag{42}$$

Compare with the reference signal by lock-in amplifier, the phase difference between the reference and testing signals can be obtained. Theoretically, the optical rotation angle is strongly related to the concentration, temperature, and propagation length of the chiral medium (glucose solution). And that can be expressed as

$$C = \frac{100\theta}{L[\theta]}, \tag{43}$$

where the glucose concentration C (g/dl) in a liquid solution, θ (degree) is optical rotation angle, L (decimeter) is the propagation length in chiral medium.

Figure 21 showed that the optical rotation angle of the glucose solution which the concentration was varied from 0 to 1.2 g/dl. Their results showed the good linearity and high sensitivity which can achieve 0.273° /g/dl.

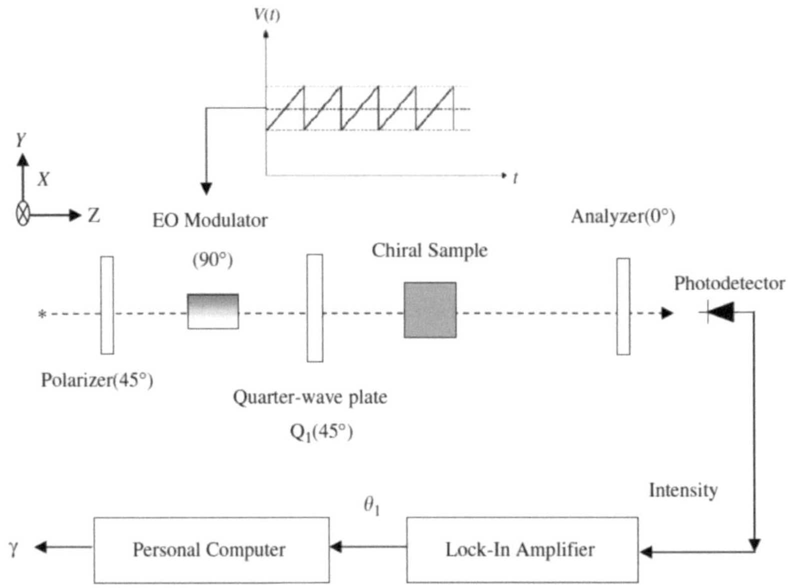

Fig. 20. Schematic diagram of circular heterodyne interferometer for measuring the optical rotation angle in a chiral medium [40].

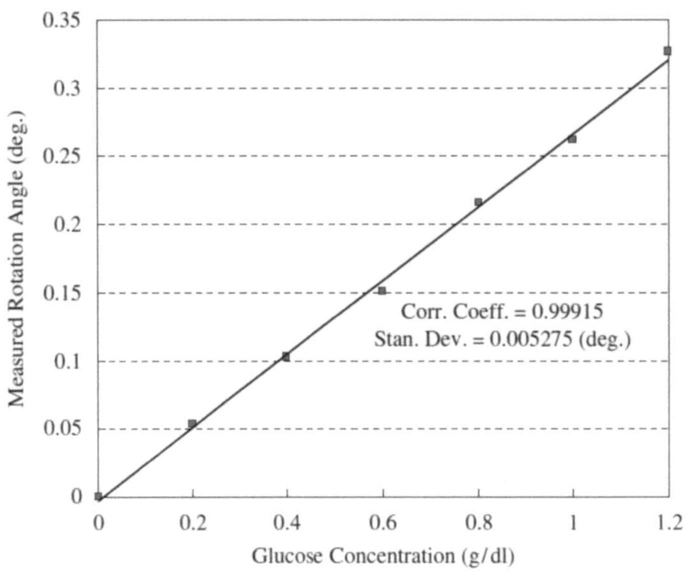

Fig. 21. Optical rotation angle of glucose solution with concentration variation [40].

5. Concentration measurement with heterodyne interferometer

The concentration of solution is an important factor in food, chemical and biochemical industrial, especially in health care and disease prevention. For example, the blood glucose concentration is related to the diabetes. To control the blood glucose concentration within the normal level is critical issue to the diabetic daily care. Therefore, many researchers developed novel methods for measuring the solution concentration. Because of the advantages of the optical method such as high sensitivity, high resolution, non-contact, and quick response, optical measurement method become more popular in past few decades. And these methods can roughly divide into fiber type sensor [41, 44-45, 49] and non-fiber type sensor [42-43, 46-48, 50]. In this section, we will review some recent development in both type sensors for measuring concentration of the specific chemical compound [41-50].

M. H. Chiu [41] developed a D-shape fiber sensor with SPR property and integrated with heterodyne interferometer which could detect variation in the alcohol concentration of 2%. The optical setup was shown in figure 22. The heterodyne light source was guided into the sensor and suffered the attenuate total reflection (ATR) at the sensing region. Because of the refractive index of the sample will be varied as the concentration changed. And induce the phase difference between the p- and s- polarizations. To measure the phase difference can be obtained the concentration variation of the sample.

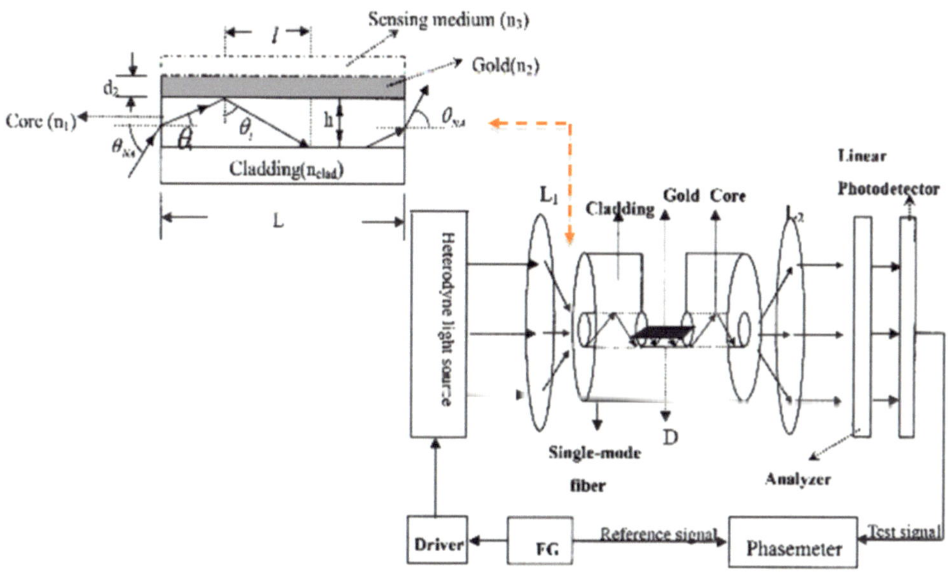

Fig. 22. The scheme of the D-shape fiber sensor [41].

Figure 22 shows that the testing signal detected by photodetector and sent into the phasemeter. Therefore, the interference signal can be written as

$$I(t) = I_0\left\{1 + V\cos\left[\omega t + \left(\frac{L(\phi_p-\phi_s)}{2h\tan\theta_i}\right)\right]\right\}, \tag{44}$$

where I_0 and V are the average intensity and visibility; L and h are the sensing length and core diameter; $(\phi_p$-$\phi_s)$ is the phase difference between p- and s- polarizations; θ_i is the incident angle at the interface between fiber core and metal film. Based on the Fresnel equation, one can derive the $(\phi_p$-$\phi_s)$ from the amplitude reflection coefficient under ATR condition and it is obvious that $(\phi_p$-$\phi_s)$ is the function of the refractive index of the sensing medium. Figure 23 shows that the results measured by the D-shape fiber sensor for different concentration of the alcohol.

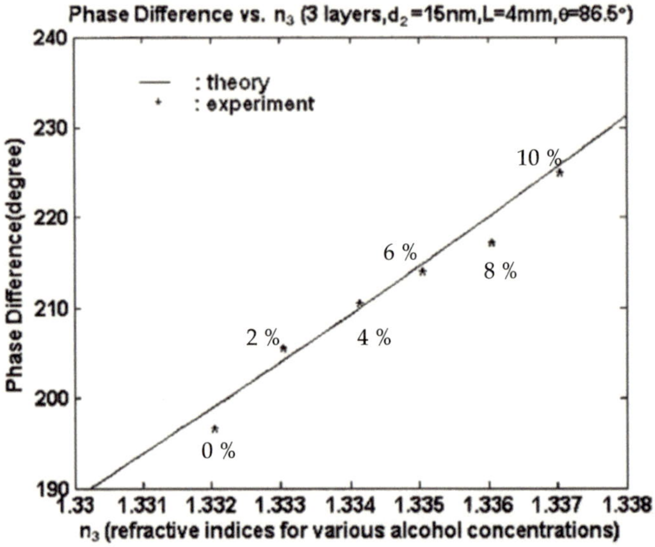

Fig. 23. The experiment result of different concentration of alcohol measured by D-shape fiber sensor [41].

In Chiu's results, they can observe the concentration variation 2 %, in which the corresponding refractive index variation is about 0.0009. Based on error analysis, their method can be reached 2×10^{-6} refractive index unit.

Recently, T. Q. Lin [44] and C. C. Hsu [45] developed a fiber sensor which immobilized glucose oxidase (GOx) on the fiber core for measuring the glucose concentration in serum and phosphate buffer solution (PBS). Their measurement method integrated the fiber sensor and heterodyne interferometer which showed in figure 24.

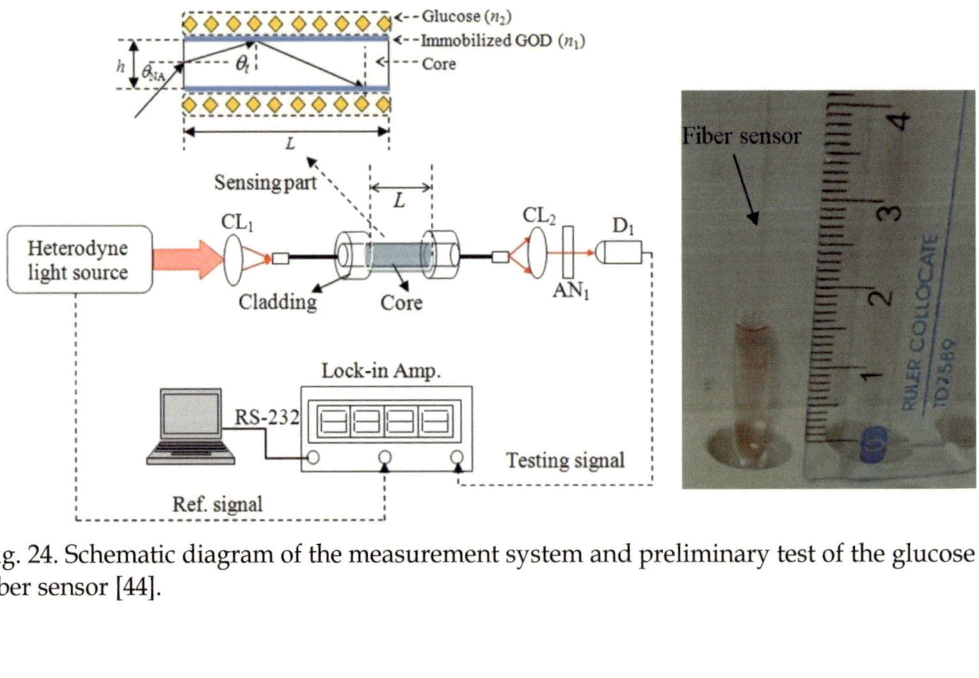

Fig. 24. Schematic diagram of the measurement system and preliminary test of the glucose fiber sensor [44].

As the heterodyne light source enters the sensing part, the light beam undergoes total internal reflection (TIR) and the phase difference between the p- and s- polarization states can be written as

$$\phi_t = m\phi_{TIR} = \frac{L}{h\tan\theta_t}\cdot\tan^{-1}\left(\frac{\sqrt{\sin^2\theta_t-(\frac{n_2}{n_1})^2}}{\tan\theta_t\cdot\sin\theta_t}\right), \tag{45}$$

where n_1 and n_2 are the refractive indices of the immobilized GOx and the testing solution. θ_t and m are the incident angle and the number of TIRs that occur at the interface between the GOx and the testing solution. After dripping the testing sample onto the sensor, the phase will vary as the glucose reacts with the GOx to be converted into gluconic acid and hydrogen peroxide. The chemical reaction can be formulated as follows:

$$\text{Glucose} + O_2 \xrightarrow{\text{GOx}} \text{gluconic acid} + H_2O_2. \tag{46}$$

It means that the refractive index (n_2) will change and consequently the phase will change. Besides, the refractive index n_2 is a function of the concentration of the testing sample. Different concentration of the solution has different refractive index. Therefore, one can determine the concentration variation by measuring the phase variation. In their methods, the phase difference can be carried in the heterodyne interference signal and written as

$$I_t = I_0[1+\cos(\omega t)+\phi_t].$$

To deserve to be mentioned, they found that the pH property between the testing sample and sensor is critical issue for rapid measurement. Figure 25 shows that the response time and response efficiency of the fiber sensor. It is clear that the response time for measuring glucose solution was shorter than those for serum measurement. And the response efficiency for measuring glucose solution was faster than those for serum measurement at different GOx concentration.

Based on their results, this fiber sensor has good linearity of the calibration curve for glucose solution and serum sample. And they showed the best resolutions were 0.1 and 0.136 mg/dl for glucose solution and serum based sample, respectively.

One of the non-fiber type sensors is SPR (surface Plasmon resonance) sensor which has been applied in field such as pharmaceutical development and life sciences. And SPR provides ultra high sensitivity for detecting tiny refractive index (RI) changes or other quantities which can be converted into an equivalent RI. The heterodyne interferometer detects the SPR phase by using a Zeeman laser or optical modulator, such as an acousto-optic modulator or electro-optic modulator and has been reported in the literature. Heterodyne phase detection techniques offer the high measurement performance high sensitivity and high resolution in real-time. J. Y. Lee [43] proposed wavelength-modulation circular heterodyne interferometer (WMCHI) with SPR sensor for measuring the different concentration of alcohol. The diagram of the WMCHI is shown in figure 26.

The SPR sensor had the Kretschmann configuration consists of a BK7 prism coated with a 50 nm gold film and integrated with micro-fluid channel which used to inject the testing

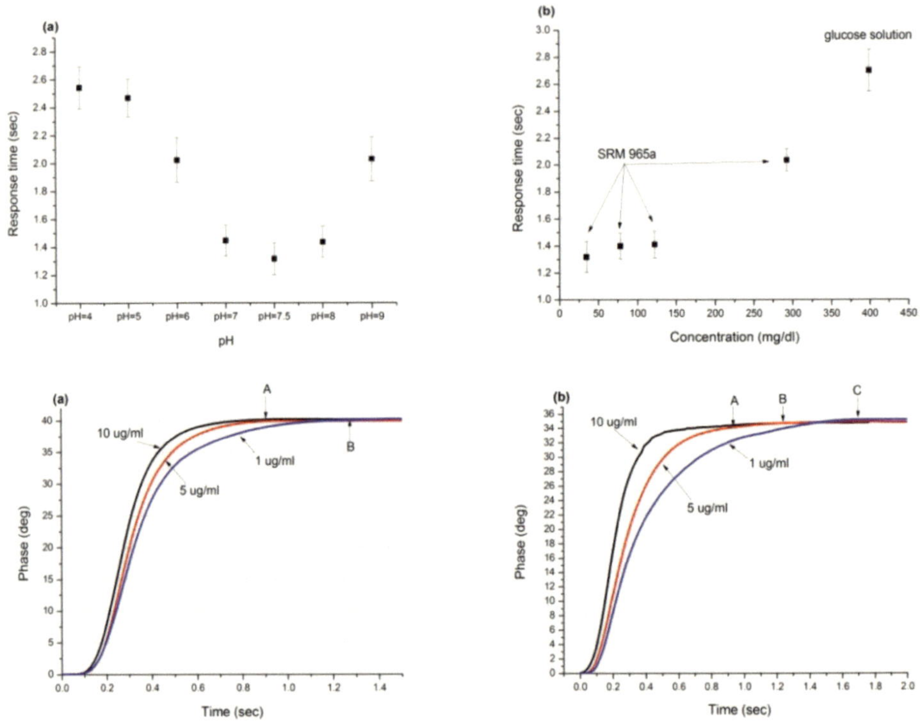

Fig. 25. The response time and response efficiency of the fiber sensor for measuring both glucose solution and human serum [44].

sample. The E-field detected by D_1 and D_2 can be written as

$$E_1 = P(45°) \cdot J_{SPR} \cdot J_Q(45°) \cdot E_h$$

$$= \frac{1}{2}\begin{bmatrix} 1 & 1 \\ 1 & 1 \end{bmatrix} \begin{bmatrix} |r_p|e^{i\phi p} & 0 \\ 0 & |r_s|e^{i\phi s} \end{bmatrix} \frac{1}{\sqrt{2}} \begin{bmatrix} 1 & i \\ i & 1 \end{bmatrix} \begin{bmatrix} e^{i(\phi_0-\omega t)/2} \\ e^{-i(\phi_0-\omega t)/2} \end{bmatrix}$$

$$= \frac{1}{2\sqrt{2}}\left[\left(|r_p|e^{i\phi p} + i|r_s|e^{i\phi s}\right)e^{\frac{i(\phi_0-\omega t)}{2}} + \left(i|r_p|e^{i\phi p} + |r_s|e^{i\phi s}\right)e^{-\frac{i(\phi_0-\omega t)}{2}}\right]\begin{bmatrix} 1 \\ 1 \end{bmatrix}, \quad (47a)$$

and

$$E_2 = P(-45°) \cdot J_{SPR} \cdot J_Q(45°) \cdot E_h$$

$$= \frac{1}{2\sqrt{2}}\left[\left(|r_p|e^{i\phi p} + i|r_s|e^{i\phi s}\right)e^{\frac{i(\phi_0-\omega t)}{2}} + \left(i|r_p|e^{i\phi p} + |r_s|e^{i\phi s}\right)e^{-\frac{i(\phi_0-\omega t)}{2}}\right]\begin{bmatrix} 1 \\ -1 \end{bmatrix}, \quad (47b)$$

where J_{SPR} is the Jones matrix of SPR sensor. They became two testing signals and sent into lock-in amplifier which can obtained the phase difference between them. The phase difference Φ of these two signals is obtained as

$$\Phi = \left(\Phi_0 + \tan^{-1}\frac{B}{A}\right) - \left(\Phi_0 - \tan^{-1}\frac{B}{A}\right) = 2\tan^{-1}\left(\frac{|r_p|^2 - |r_s|^2}{2|r_p||r_s|\cos\phi}\right). \quad (48)$$

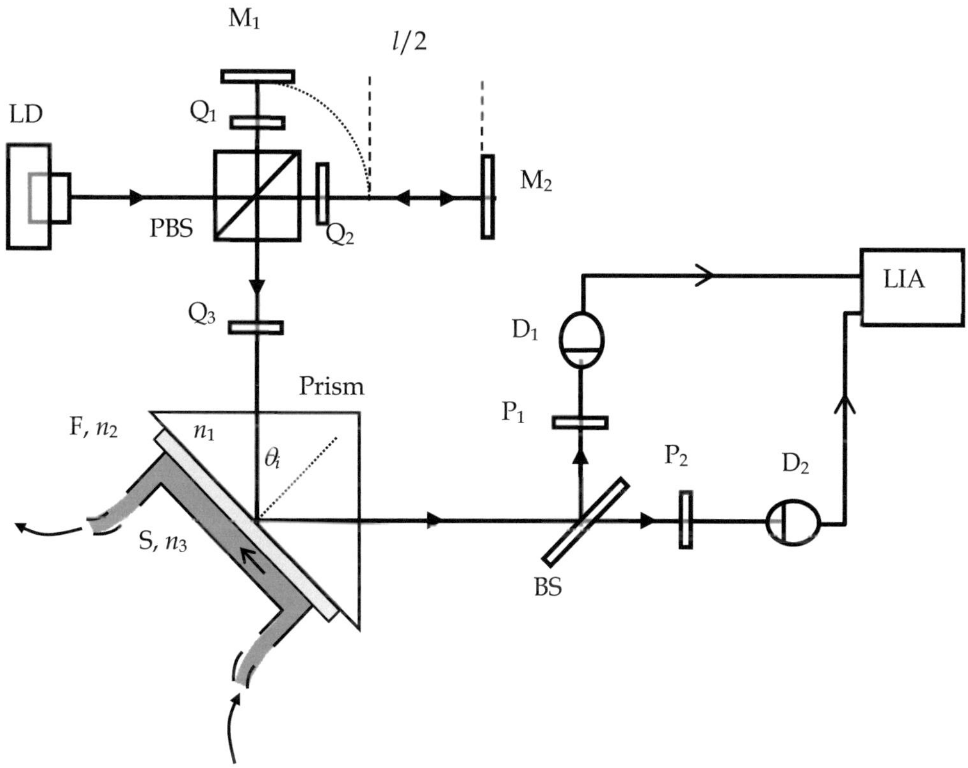

Fig. 26 Schematic diagram of WMCHI for different concentration measurement [43].

Based on equation (44), it is clear that the resonant angles for ϕ and Φ are different. Obviously, Φ is the function of r_p, r_s and ϕ which varies with the refractive index of sample n_3 and the incident angle. The relationship between ϕ and Φ and the incident angle was shown in figure 27 It is obvious that the maximum of Φ is larger than that of ϕ. On the other hand, Φ can be larger than 10,000 in the incident angle interval between 66.25° and 66.75°. This means that Lee's method has a high angle toleration and larger dynamic range.

C. Chou [42] proposed a novel pair surface plasma wave biosensor which provided a larger dynamic measurement range for effective refractive index. In their system, it can avoid excess noise coming from laser intensity fluctuation and environment disturbance. It is important to retain the amplitude stability in this method for high detection sensitivity. Figure 28 showed the amplitude sensitivity PSPR method. In this figure, PBS separated the pair of p-polarization waves and the pair of s-polarization waves, which can be optical heterodyne interference signal at the photodetectors D_p and D_s. Then these two signals can be expressed as

$$I_{p1+p2}(\Delta\omega t) = A_{p1}A_{p2}\cos(\Delta\omega t + \phi_p), \tag{49a}$$

$$I_{s1+s2}(\Delta\omega t) = A_{s1}A_{s2}\cos(\Delta\omega t + \phi_s), \tag{49b}$$

where A_{P1} and A_{P2} are the attenuated amplitudes of the reflected P_1 and P_2 waves respectively; A_{S1} and A_{S2} are the attenuated amplitudes of the reflected S_1 and S_2 waves

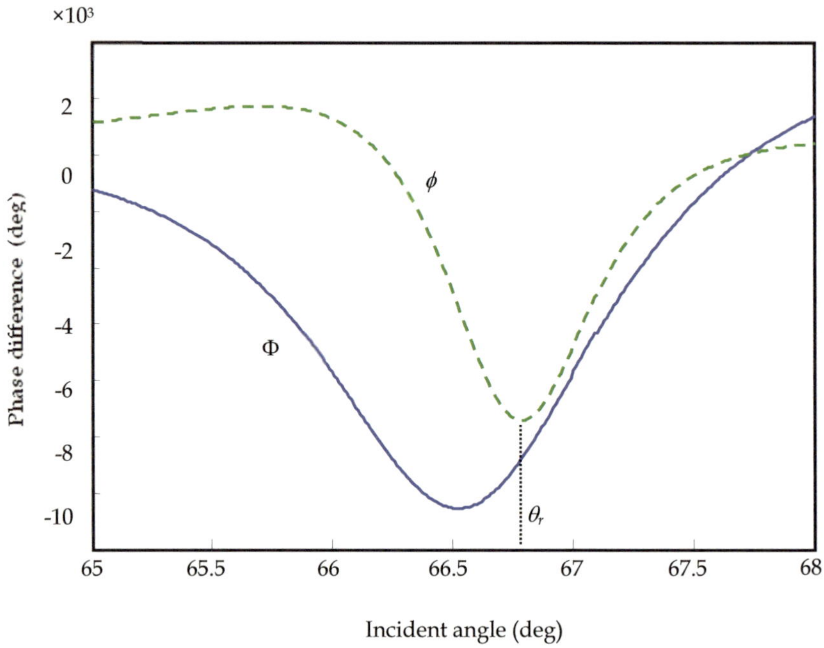

Fig. 27. The relationship between ϕ and Φ and the incident angle [43].

respectively. ϕ_P and ϕ_S are the phase differences of the reflected P and P and the reflected S and S waves respectively. In equation (49), ϕ_P and ϕ_S are equal to 0 and these two interference signals will remain at maximum intensity under the SPR proceeded.

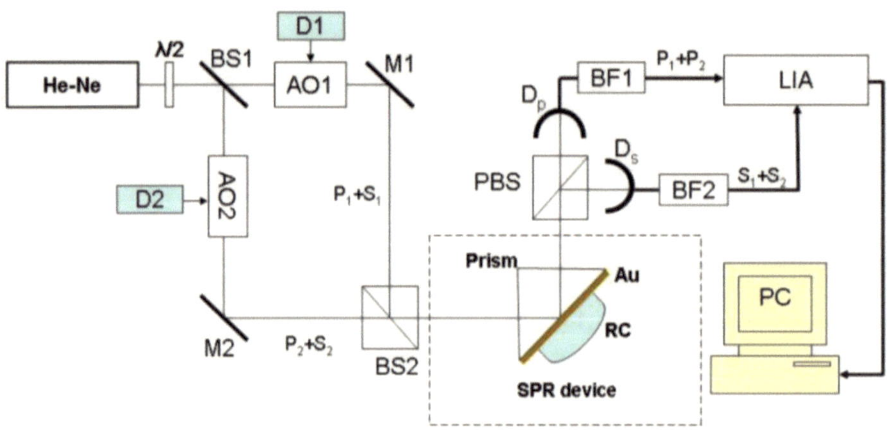

Fig. 28. The schematic of the amplitude sensitivity PSPR [42].

Based on this method, Chou demonstrated three different testing samples with concentration variation which were sucrose, glycerin-water solution, and rabbit anti-mouse IgG. In figure 29, the best of these sample are 8×10^{-8}, 7.6×10^{-7}, and 2×10^{-9}, respectively.

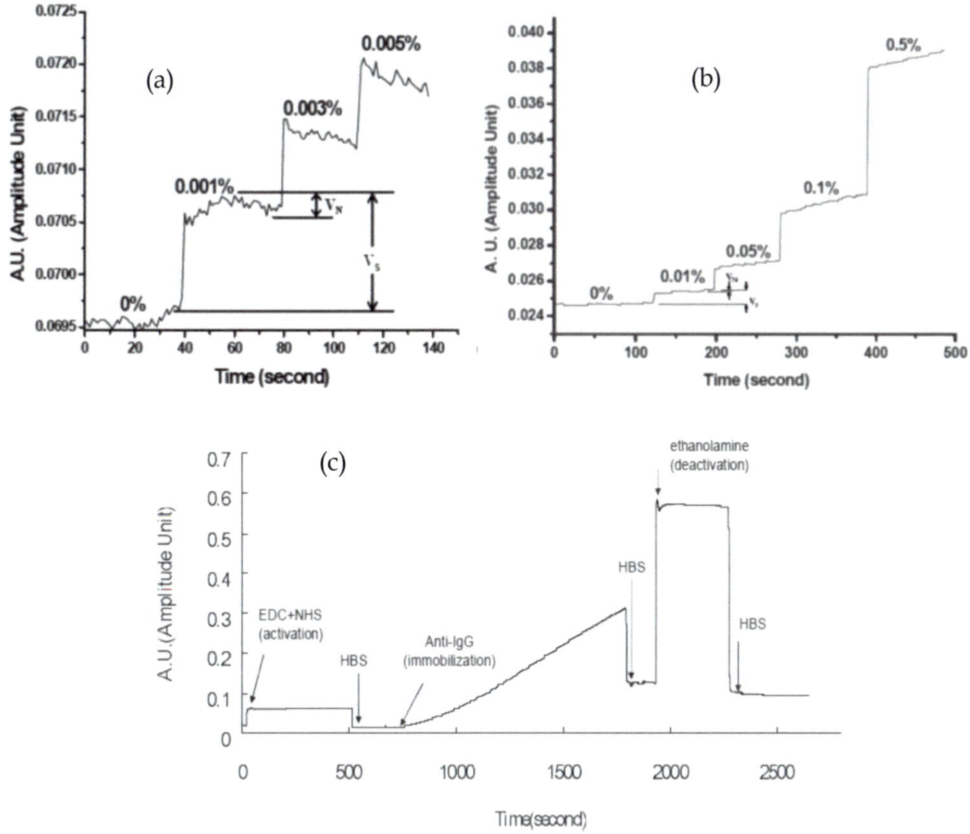

Fig. 29. The measurement results of the PSPR method [42].

6. Conclusion

In this chapter, we reviewed some recent development or state of the art techniques. It shows that the heterodyne interferometry is a mature technique and can be applied to many different aspects. For example, the diffraction grating heterodyne interferometry (DGHI) provided nanometer resolution for precision positioning which can be integrated with motorized stage. Full-field circular heterodyne interferometry (FFCHI) can be used to investigate the two-dimensional optical properties, such as refractive index and birefrigence of testing sample. For this point of view, the heterodyne interferometry can be a refractometer with high accuracy. To integrate with optical sensor, the heterodyne interferometry can be used to diagnose the concentration of the body fluid such as blood glucose and glycerin or the protein interaction between body-antibody. Therefore, the heterodyne interferometry is a powerful, flexible, integrable, and reliable technique for precision metrology.

7. References

[1] L. C. De Backer, "In-plane displacement measurement by speckle interferometry," Non-destructive Testing 8, pp. 177-180 (1975).

[2] B. Pan, K. Qian, H. Xie, and Anand Asundi, "Two-dimensional digital image correlation for in-plane displacement and strain measurement: a review," Measurement Science and Technology 20, pp. 062001 (2009).

[3] S. Nakadate, "Vibration measurement using phase shifting time average holographic interferometry," Applied Optics 25, pp. 4155-4161 (1986).

[4] G. Pedrini, W. Osten, and M. E. Gusev, "High-speed digital holographic interferometry for vibration measurement," Applied Optics 45, pp. 3456-3462 (2006).

[5] W. B. Ribbens, "Surface roughness measurement by two wavelength holographic interferometry," Applied Optics 13, pp. 1085-1088 (1974).

[6] C. J. Tay, M. Thakur, C. Quan, "Grating projection system for surface contour measurement," Applied Optics 44, pp. 1393-1400 (2005).

[7] Haller, H. A. Huggins, and M. J. Freiser, "On the measurement of indices of refraction of nematic liquid," Molecular Crystals and Liquid Crystal 16, pp. 53-59 (1972).

[8] W. Lukosz and P. Pliska, "Determination of thickness and refractive index of SiO2 films on silicon wafers using an Abbe refractometer," Optics Communications 85, pp. 381-384 (1991).

[9] J. A. D. Feijter, J. Benjamins, and F. A. Veer, "Ellipsometry as a tool to study the adsorption behavior of synthetic and biopolymers at the air-water interface," Biopolymers 17, pp. 1759-1772 (1978).

[10] F. L. McCrackin, E. Passaglia, R. R. Stromberg, and H. L. Steinberg, "Measurement of the thickness and refractive index of very thin films and the optical properties of surfaces by ellipsometry," Journal of Research of the National Bureau of Standards A. Physics and Chemistry 67A, pp. 363-377 (1963).

[11] B. T. Liu, S. J. Tang, Y. Y. Yu, and S. H. Lin, "High refractive index polymer/inorganic hybrid films containing high TiO2 contents," Colloids and Surfaces A: Physicochemical and Engineering Aspects 377, pp. 138-143 (2011).

[12] C. M. Jan, Y. H. Lee, K. C. Wu, and C. K. Lee, "Integrating fault tolerance algorithm and circularly polarized ellipsometer for point of care applications," Optics Express 19, pp. 5431-5441 (2011).

[13] H. G. Tompkins and E. A. Irene, *Handbook of Ellipsometry*, Willian Andrew, New York, 2010

[14] R. M. A. Azzam and N. M. Bashara, *Ellipsometry and Polarized Light*, Elsevier Science Publisher, Netherlands, 1992

[15] J. A. Dahlquist, D. G. Peterson, and W. Culshaw, "Zeeman laser interferometer," Applied Physics Letters 9, pp. 181-183 (1966).

[16] T. Suzuki and R. Hioki, "Translation of light frequency by a moving grating," Journal of the Optical Society of America 57, pp. 1551 (1967).

[17] W. H. Stevenson, "Optical frequency shifting by means of a rotating diffraction grating," Applied Optics 9, pp. 649-652 (1970).

[18] M. J. Ehrlich, L. C. Phillips, and J. W. Wagner, "Voltage-controlled acousto-optic phase shifter," Review of Scientific instrument 59, pp. 2390-2392 (1988).

[19] T. Q. Banh, Y. Ohkubo, Y. Murai, and M. Aketagawa, "Active suppression of air refractive index fluctuation using a Fabry-Perot cavity and a piezoelectric volume actuator," Applied Optics 50, pp. 53-60 (2011).

[20] D. C. Su, M. H. Chiu, and C. D. Chen, "A heterodyne interferometer using an electro-optic modulator for measuring small displacement," Journal of Optics 27, pp. 19-23 (1996).
[21] W. K. Kuo, J. Y. Kuo, and C. Y. Huang, "Electro-optic heterodyne interferometer," Applied Optics 46, pp. 3144-3149 (2007).
[22] H. Kikuta, K. Iwata, and R. Nagata, "Distance measurement by the wavelength shift of laser diode light," Applied Optics 25, pp. 2976-2980 (1986).
[23] M. Born and E. Wolf, *Principles of Optics*, Cambridge University Press, UK, 1999.
[24] C. K. Lee, C. C. Wu, S. J. Chen, L. B. Yu, Y. C. Chang, Y. F. Wang, J. Y. Chen, and W. J. Wu, "Design and construction of linear laser encoders that possess high tolerance of mechanical runout," Applied Optics 43, pp. 5754-5762 (2004).
[25] C. H. Liu, W. Y. Jywe, and C. K. Chen, "Development of a diffraction type optical triangulation sensor," Applied optics 43, 5607-5613 (2004).
[26] C. F. Kao, C. C. Chang, and M. H. Lu, "Double-diffraction planar encoder by conjugate optics," Optical Engineering 44, pp. 023603-1 (2005).
[27] J. Y. Lee, H. Y. Chen, C. C. Hsu and C. C. Wu, "Optical heterodyne grating interferometry for displacement measurement with subnanometric resolution," Sensors and Actuators A: Physical 137 pp. 185-191 (2007).
[28] C. C. Hsu, C. C. Wu, J. Y. Lee, H. Y. Chen, and H. F. Weng, "Reflection type heterodyne grating interferometry for in-plane displacement measurement," Optics Commun. 281 pp. 2583-2589 (2008).
[29] C. C. Wu, C. C. Hsu, J. Y. Lee, and C. L. Dai, "Optical heterodyne laser encoder with Sub-nanometer resolution," Measurement Science and Technology 19 pp. 045305-045313 (2008).
[30] H. L. Hsieh, J. C. Chen, G. Lerondel, and J. Y. Lee, "Two-dimensional displacement measurement by quasi-common-optical path heterodyne grating interferometer," Optics Express 19, pp. 9770-9782 (2011).
[31] Y. L. Lo and P. F. Hsu, "Birefringence measurements by an electro-optic modulator using a new heterodyne scheme," Optical engineering 41, pp. 2764-2767 (2002).
[32] C. C. Hsu, K. H. Chen, and D. C. Su, "Normal incidence refractometer", Optics Communications 218, pp. 205-211 (2003).
[33] K. H. Chen, C. C. Hsu, and D. C. Su, "A method for measuring the complex refractive index and thickness of a thin metal film," Applied Physics B 77, pp.839-842 (2003).
[34] C. C. Hsu, J. Y. Lee, and D. C. Su, "Thickness and optical constants measurement of thin film growth with circular heterodyne interferometry", Thin Solid Film 491, pp. 91-95 (2005).
[35] C. Y. Hong, J. J. Chieh, S. Y. Yang, H. C. Yang, and H. E. Horng, "Simultaneous identification of the low field induced tiny variation of complex refractive index for anisotropic and opaque magnetic fluid thin film by a stable heterodyne Mach-Zehnder interferometer," Applied Optics 48, pp. 5604-5611 (2009).
[36] J. Y. Lin, "Determination of the refractive index and the chiral parameter of a chiral solution based on chiral reflection equations and heterodyne interferometry," Applied Optics 47, pp. 3828-3834 (2008).
[37] C. C. Hsu and D. C. Su, "Method for determining the optic axis and (ne, no) of a birefringent crystal", Applied Optics 41, pp. 3936-3940, (2002).

[38] Y. L. Chen, H. C. Hsieh, W. T. Wu, W. Y. Chang, and D. C. Su, "Alternative method for measuring the full field refractive index of a gradient index lens with normal incidence heterodyne interferometry," Applied Optics 49, pp. 6888-6892 (2010).

[39] Y. C. Huang, C. Chou, and M. Chang, "Direct measurement of refractive indices of a linear birefringent retardation plate," Optics Communications 133, pp. 11-16 (1997).

[40] J. F. Lin, C. C. Chang, C. D. Syu, Y. L. Lo, and S. Y. Lee, "A new electro-optic modulated circular heterodyne interferometer for measuring the rotation angle in a chiral medium," Optics and Laser in Engineering 47, pp. 39-44 (2009).

[41] M. H. Chiu, S. F. Wang, and R. S. Chang, "D-type fiber biosensor based on surface Plasmon resonance technology and heterodyne interferometry," Optics Letters 30, pp. 233-235 (2005).

[42] C. Chou, H. T. Wu, Y. C. Huang, Y. L. Chen, and W. C. Kuo, "Characteristics of a paired surface plasma waves biosensor," Optics Express 14, pp. 4307-4315 (2006).

[43] J. Y. Lee and S. K. Tsai, "Measurement of refractive index variation of liquids by surface plasmon resonance and wavelength modulated heterodyne interferometry," Optics Communications 284, pp. 925-929 (2011).

[44] T. Q. Lin, Y. L. Lu, and C. C. Hsu, "Fabrication of glucose fiber sensor based on immobilized GOD technique for rapid measurement," Optics Express 18 pp. 27560-27566 (2010).

[45] C. C. Hsu, Y. C. Chen, J.Y. Lee, and C. C. Wu, "Reusable glucose f iber sensor for measuring glucose concentration in serum," Chinese Optics Letters, 9 pp. 100608-1~100608-3 (2011).

[46] J. Y. Lin, J. H. Chen, K. H. Chen, and D. C. Su, "A new type of liquid refractometer," Physics Status Solidi C 5, pp. 1020-1022 (2008).

[47] P. Nath, H. K. Singh, P. Datta, K. C. Sarma, "All fiber optic sensor for measurement of liquid refractive index," Sensors and Actuators A: Physical 148, pp. 16-18 (2008).

[48] Y. L. Yeh, "Real time measurement of glucose concentration and average refractive index using a laser interferometer," Optics and Laser in Engineering 46, pp. 666-670 (2008).

[49] K. S. Kim, Y. Mizuno, M. Nakano, S. Onoda, and K. Nakamura, "Refractive index sensor for liquids and solids using dielectric multilayer films deposited on optical fiber end surface," Photonics Technology Letters 23, pp. 1472-1474 (2011).

[50] H. H. Hamzah, N. A. Yusof, A. B. Salleh, and F. A. Baker, "An optical test strip for the detection of Benzonic acid in food," Sensors 11, pp. 7302-7313 (2011).

3

One-Shot Phase-Shifting Interferometry with Phase-Gratings and Modulation of Polarization Using N≥4 Interferograms

Gustavo Rodríguez Zurita, Noel-Ivan Toto-Arellano and Cruz Meneses-Fabián
Benemérita Universidad Autónoma de Puebla, México

1. Introduction

Phase-shifting interferometry requires (PSI) of several interferograms of the same optical field with similar characteristics but shifted by certain phase values to retrieve the optical phase. This task has been usually performed by stages with great success and requires of a series of sequential shots [Meneses et al., 2006a]. However, time-varying phase distributions are excluded from this schema. Several efforts for single-shot phase-shifting interferometry have been tested successfully [Novak et al.,2005 ;Rodriguez et al., 2008a], but some of them require of non-standard components and they need to be modified in some important respects in order to get more than four interferograms. Two-windows grating interferometry, on the other hand, has been proved to be an attractive technique because of its mechanical stability as a common-path interferometer [Arrizón and De La Llave 2004]. Moreover, gratings can be used as convenient phase modulators because they introduce phase shifts through lateral displacements. In this regard, phase gratings offer more multiplexing capabilities than absorption gratings (more useful diffraction orders because higher diffraction efficiencies can be achieved). Furthermore, with two phase gratings with their vector gratings at 90° (grids) there appear even more useful diffraction orders[Toto et al., 2008]. Modulation of polarization can be independently applied to each diffraction order to introduce a desired phase-shift in each interference pattern instead of using lateral translations. These properties combine to enable phase-shifting interferometric systems that require of only a single-shot, thus enabling phase inspection of moving subjects. Also, more than four interferograms can be acquired that way. A simple interferogram processing enables the use of interference fringes with different fringe modulations and intensities. In this chapter, the basic properties of two-windows phase grating interferometry (TWPGI) and modulation of polarization is reviewed on the basis of the far-field diffraction properties of phase gratings and grids. Phase shifts in the diffraction orders can be used as an advantage because they simplify the needed polarization filter distributions. It is finally remarked, that these interferometers are compatible with interference fringes exhibiting spatial frequencies of relative low values and, therefore, no great loss of resolution is related with several interferograms when simultaneously using the same image field of the camera. To extract optical phase distributions which evolve in time, the capture of the *n* shifted interferograms with one shot is desirable. Some approaches to perform this task have been already demonstrated [Wyant, 2004; Rodriguez et al., 2009], although only for n = 4 to our knowledge. Among these systems, the one using two windows in the object plane of a 4f system with a phase-grating in the Fourier plane and modulation of polarization (TWPGI) is a very simple possibility [Rodriguez 2008a]. In this communication, the capability of TWPGI to capture more than four interferograms in one shot is demonstrated with the introduction of a phase grid in place of the grating. To test TWPGI for more than four interferograms, the case of n=(N+1) interferograms has been chosen. This method reduces errors in phase calculations when noisy interferograms are involved [Malacara D, 1998]. Experimental results for n = 5, 7, 9 interferograms are shown.

2. Experimental setup

A The Fig. 1 shows the arrangement of an ideal one-shot phase-shifting grating interferometer incorporating modulation of polarization. A combination of a quarter-wave plate Q and a linear polarizing filter P generates linearly polarized light at an appropriate azimuth angle (45°) entering the interferometer. Two quarter-wave plates (Q_L and Q_R) with their orthogonal fast axes are placed in front of the two windows of the common-path interferometer so as to generate left and right circularly polarized light as the corresponding beam leaves each window, see Fig. 1(a). A phase grating is placed at the system´s Fourier plane as the pupil. In the image plane, Fig. 1(b) superimposition of diffraction orders result, causing replicated images to interfere.

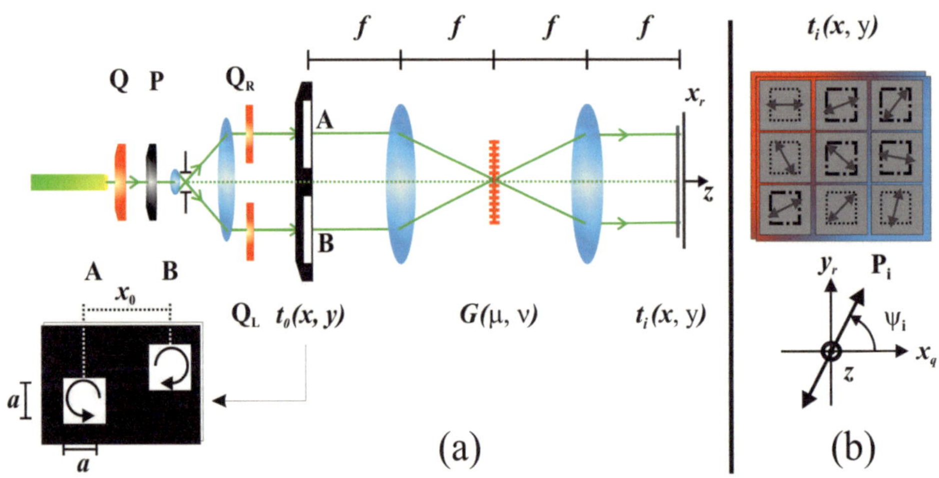

Fig. 1. (a) One-shot phase-shifting grating interferometer with modulation of polarization. A, B: windows. PBS: Polarizing Beam splitter; M_i: Mirrors; G(μ,ν): Phase grid. (b) ψ_i: polarizing angles (with i = 1,2,3,4 to obtain phase-shifts ξ_i =0°, 90°, 180°, 270° respectively). Translation of coordinates around the order position: $x_q = x - qF_0$ and $y_r = y - rF_0$. α'= 1.519 *rad.*

The phase shifting ξ_i , i= 1…4, results after placing a linear polarizer to each one of the interference patterns generated on each diffracting orders in the exit plane (P_1, P_2, P_3, P_4).

Each polarizing filter transmission axis is adjusted at different angle ψ_i, so as to obtain the desired phase shift ξ_i for each pair of orders. For a 90° phase-shift ξ_i between interfering fields, the polarization angles ψ_i in each diffraction order must be 0°, 45°, 90° and 135° for the case of ideal quarter-wave retardation (α= 90°). In the next sections, some particularities arising from the optical components available for our set-up are discussed. Among these, the calculation of ψ_i for the case of a non exact quarter-wave retardation is considered through an example.

3. Interference patterns with polarizing filters and retarding plates

3.1 Phase grids

Object and image planes are described by (x,y) coordinates. A periodic phase-only transmittance $G(\mu,\nu)$ is placed in the frequency plane (u,v). Then $\mu=u/\lambda f$ and $\nu=\upsilon/\lambda f$ are the frequency coordinates scaled to the wavelength λ and the focal length f. In the plane (u,v), the period of G is denoted by d (the same in both axis directions) and thus, its spatial frequency by $\sigma= 1/d$. Two neighboring diffraction orders have a distance of $X_0=\lambda f/d$ in the image plane. Then, $\sigma\cdot u= X_0\cdot\mu$. Taking the rulings of one grating along the μ direction and the rulings of the second grating along the ν direction, the resulting centered phase grid can be written as

$$G(\mu,\varsigma)=e^{i2\pi A_g \sin[2\pi\cdot X_x\mu]}\,e^{i2\pi A_g \sin\left[2\pi\cdot X_y\nu\right]}=\sum_{q=-\infty}^{\infty} J_q\left(2\pi A_g\right)e^{i2\pi\cdot qX_0\mu}\sum_{r=-\infty}^{\infty} J_r\left(2\pi A_g\right)e^{i2\pi\cdot rX_0\nu} \quad (1)$$

where the frequencies along each axes directions are taken of the same value The Fourier transform of the phase grid becomes

$$\tilde{G}(x,y)=\sum_{q=-\infty}^{q=\infty}\sum_{r=-\infty}^{r=\infty} J_q\left(2\pi A_g\right)J_r\left(2\pi A_g\right)\delta\left(x-qX_0,y-rX_0\right) \quad (2)$$

which consists of point-like diffraction orders distributed in the image plane on the nodes of a lattice with a period given by X_0.

3.2 Two-window phase-grating interferometry: fringe modulation

Phase grating interferometry is based on a phase grating placed as the pupil of a 4f Fourier optical system [Rodriguez et al., 2008a;Thomas and Wyant ,1976]. The use of two windows at the object plane in conjunction with phase grating interferometry allows interference between the optical fields associated to each window with higher diffraction efficiency [Arrizon and Sanchez, 2004; Ramijan, 1978]. Such a system performs as a common path interferometer (Fig. 1). When using birefringent plates which do not perform exactly as quarter-wave plates for the wavelength employed, the polarization angles of the linear polarizing filters to obtain 90° phase-shifts must change[Rodriguez et al., 2008a].

To calculate the phase shifts induced in a more general polarization states by linear polarizers, consider two fields whose Jones vectors are described respectively by

$$\vec{J}_L(x,y)=\frac{1}{\sqrt{2}}\begin{pmatrix}1\\ e^{i\alpha'}\end{pmatrix}\quad \vec{J}_R(x,y)=\frac{1}{\sqrt{2}}\begin{pmatrix}1\\ e^{-i\alpha'}\end{pmatrix}, \tag{3}$$

These vectors represent the polarization states of two beams emerging from a retarding plate with phase retardation $\pm\alpha'$. Each beam enters the plate with linear polarization at ±45° with respect to the plate fast axis. Due to their orientations, the electric fields of the beams rotate in opposite directions, thereby the indices *L* and *R*. A convenient window pair for a grating interferometer implies an amplitude transmittance given by

$$\vec{t}_0(x,y)=\vec{J}_L\cdot w(x-\frac{x_0}{2},y-\frac{y_0}{2})+\vec{J}_R w'(x+\frac{x_0}{2},y+\frac{y_0}{2}), \tag{4}$$

and x_0 and y_0 give the mutual separations between the centers of each window along the coordinate axis. One rectangular aperture can be written as w(x,y)= rect[x/a]· rect[y/b] whereas the second one, as w´(x,y)= w(x,y) exp{ iϕ(x,y)}, a relative phase between the windows being described with the function . a represents the side length of each window. The image $\vec{t}_i(x,y)$ formed by the system consists basically of replications of each window at distances X_0, that is, the convolution of $\vec{t}_0(x,y)$ with the point spread function of the system, defined by the inverse Fourier transform of

$$\tilde{\Gamma}_2(\mu,\nu)=\frac{1}{2}\begin{pmatrix}1 & 0\\ 0 & 1\end{pmatrix}G_2(\mu,\nu). \tag{5}$$

This results into the following

$$\begin{aligned}\vec{t}_f(x,y)&=\vec{t}_2(x,y)*\Im^{-1}\{\tilde{\Gamma}_2(\mu,\nu)\}\\ &=\frac{1}{2}\begin{pmatrix}1 & 0\\ 0 & 1\end{pmatrix}\cdot\vec{t}_2(x,y)*\Im^{-1}\{\tilde{G}_2(\mu,\nu)\}.\end{aligned} \tag{6}$$

Assuming $y_0=x_0$, by invoking the condition of matching first-neighboring orders, $X_0=x_0$, $q'=q+1$ and $r'=r+1$, and the image is then basically described by

$$\begin{aligned}&\vec{t}_0(x,y)*\tilde{G}(x,y)=\\ &\quad\vec{J}_L\sum_{q,r}^{\infty}J_qJ_r\cdot w(x-(q+\tfrac{1}{2})x_0,y-(r+\tfrac{1}{2})x_0)\\ &\quad+\vec{J}_R\sum_{q',r'}^{\infty}J_{q'}J_{r'}\cdot w'(x-(q'+\tfrac{1}{2})x_0,y-(r'+\tfrac{1}{2})x_0)=\\ &\sum_{q=-\infty}^{\infty}\sum_{r=-\infty}^{\infty}\left\{\vec{J}_LJ_qJ_r+\vec{J}_RJ_{q+1}J_{r+1}\cdot\exp\left[i\phi\left(x-(q+\tfrac{1}{2})x_0,y-(r+\tfrac{1}{2})x_0\right)\right]\right\}\end{aligned} \tag{7}$$

where some inessential constants are dropped. By selecting the diffraction term of order *qr*, after placing a linear polarizing filter with transmission axis at the angle ψ, $\vec{J}_{\psi}^{L}$, its irradiance results proportional to

$$\left\| \vec{J}_L{}' J_q J_r + \vec{J}_R{}' J_{q+1} J_{r+1} \cdot \exp[i\varphi(x',y')] \right\|^2 = \\ A(\psi,\alpha') \cdot \left[\left(J_q J_r\right)^2 + \left(J_{q+1} J_{r+1}\right)^2 + 2 J_q J_r J_{q+1} J_{r+1} \cdot \cos[\xi(\psi,\alpha') - \varphi(x',y')] \right], \tag{8}$$

with

$$\vec{J}_\psi^L = \begin{pmatrix} \cos\psi & -\sin\psi \\ \sin\psi & \cos\psi \end{pmatrix}, \ \vec{J}_L{}' = J_\psi^L \vec{J}_L, \ \vec{J}_R{}' = J_\psi^L \vec{J}_R, \tag{9}$$

and also [4]

$$A(\psi,\alpha') = 1 + \sin(2\psi)\cdot\cos(\alpha'), \ \xi(\psi,\alpha') = ArcTan\left[\frac{\sin(\alpha')}{\dfrac{\cot(2\psi)}{1+\tan(\psi)\cdot\cos(\alpha')} + \cos(\alpha')} \right]. \tag{10}$$

Plots of $\xi(\psi,\alpha')$ and $A(\psi,\alpha')$ are shown in Fig. 2 for several values of α'.

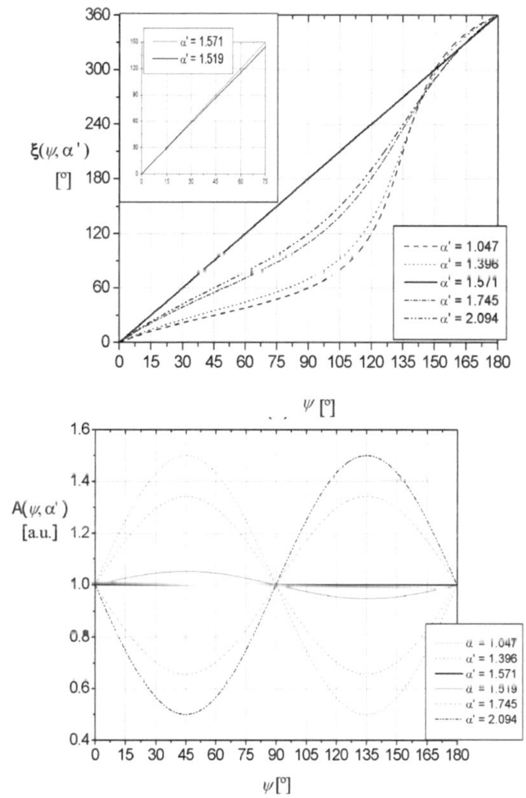

Fig. 2. (a) Phase shift $\xi(\psi,\alpha')$ as a function of ψ for several values of α'. Insert: α' for ideal retardation and experimental retardation. (b) Amplitude $A(\psi,\alpha')$ as a function of ψ for several values of α'.

Thus, an interference pattern between fields associated to each window must appear within each replicated window. It is shifted by an amount ξ induced by polarization. For the case of exact quarter-wave retardation, $A(\psi,\pi/2)=1$ and $\xi(\psi,\pi/2)=2\psi$. Otherwise, these quantities must be evaluated with Eqs. (8). The fringe modulation m_{qr} of each pattern would be of the form

$$m_{qr} = \frac{2J_q J_{q-1} J_r J_{r-1}}{J_q^2 J_r^2 + J_{q-1}^2 J_{r-1}^2} \tag{11}$$

The Fourier spectrum of the grid in our tests behaves as sketched in Fig. 3, where two equal phase gratings are shown with their respective +4th diffraction order assumed negative [Fig. 3(a)]. Thus, the -4th diffraction order results also negative. A phase grid is formed with the gratings at 90° and the resulting Fourier spectrum forms a rectangular reticule [Fig. 3(b)]. Due to the π phase difference between orders, there are orders pointing out toward the reader (circles) or away (crosses). Because the window are displaced, two Fourier spectra become shifted from the origin diagonally and in opposite directions [Fig. 3(c)]. Similar rows and columns are encircled within the dotted lines. Under our matching condition, the order qr superimposes with the order $(q\text{-}1)(r\text{-}1)$. Thus, some orders are in phase (dots with dots or crosses with crosses, but only one symbol depicted) and others out of phase (dot with cross). Then, only one symbol means positive contrast, while both symbols mean contrast reversal [Fig. 3(c)].

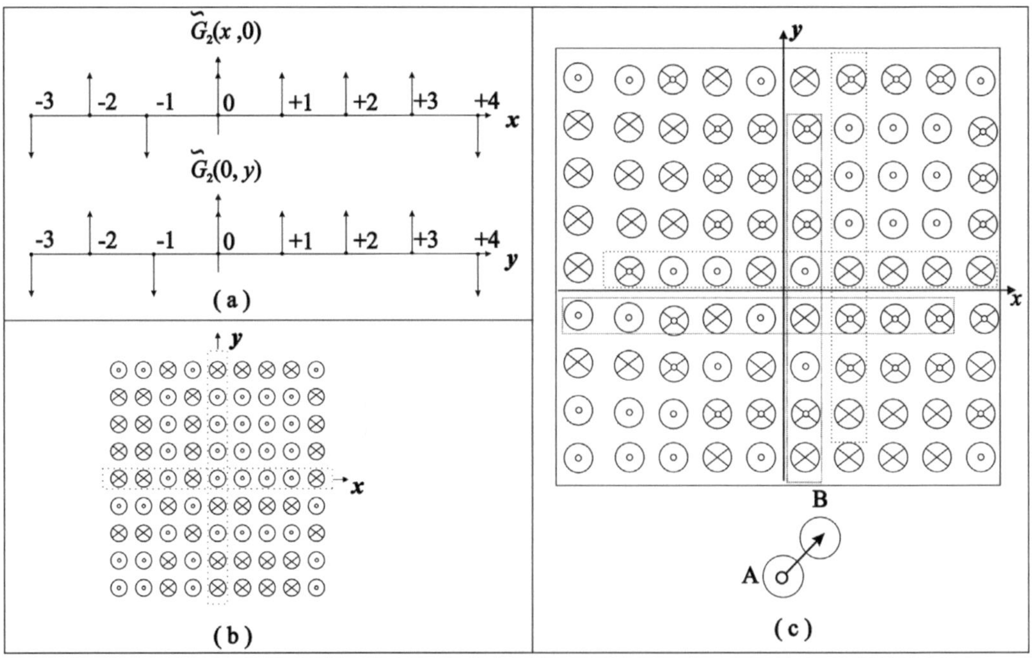

Fig. 3. a) One dimensional spectra of identical phase gratings to be crossed in order to construct a grid. b) Corresponding image plane for the phase grid. c) Two shifted Fourier spectra are superimposed according to the windows displacement A-B.

4. Experimental testing of the phase-shifts in phase grids

For the case of the diffractions orders belonging to a phase-grid constructed with two crossed gratings of equal frequency, the corresponding interference patterns are shown in Fig.4. Each grating gives patterns as in Fig. 4(a) when placed alone in the system of Fig. 1 with no plate retarders neither linear polarizing filters. The whole figure is a composite image because patterns of higher order have lower intensities. The fringe modulation signs are in agreement with the conclusions derived from Fig. 2. The relative phase values of the 16 patterns within the square (drawn with dotted lines in the patterns of Fig. 4) employing the method from Kreis, 1986 can be seen in Table 1.

Any grating displacement on its plane only introduces a constant phase term in Eqs. (6) and (8) which, in turn, only shifts each interference pattern by the same amount independent of the diffraction order [Arrizon and Sanchez 2004; Meneses et al., 2006b]. Modulation of polarization employed to attain the needed shifts in each interferogram is described in the next sections taking only four of 16.

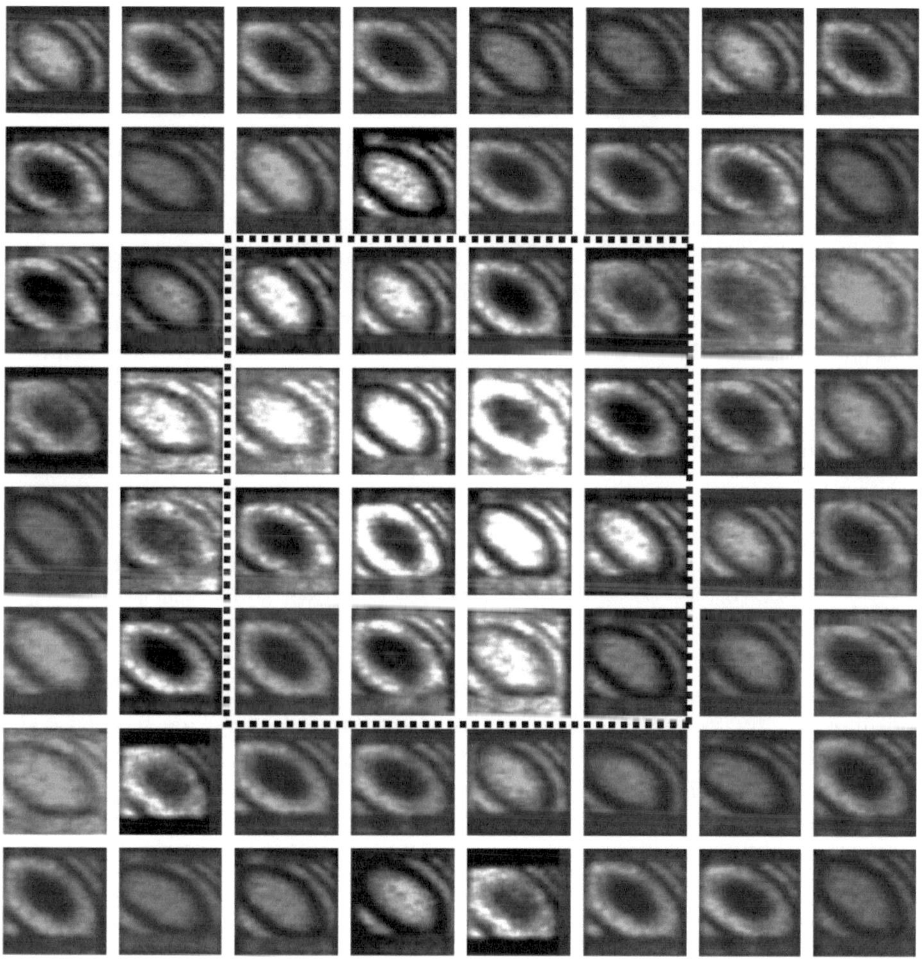

Fig. 4. Experimental patters for a phase-grid.

Shifts (rad)			
0.016	0.040	3.129	3.173
0.009	0.016	3.150	3.187
3.144	3.173	0.000	0.010
3.137	3.122	0.017	0.037

Table 1. Phase shifts of the 16 patterns within the dotted square of Fig. 3, as measured by the method from Kreis 2986.

5. Phase-grid interference patterns with modulation of polarization

Incorporating modulation of polarization, a TWPGI can be used for dynamic interferometry. This system is able to obtain four interferograms 90° phase-apart with only one shot. Phase evolving in time can then be calculated and displayed on the basis of phase-shifting techniques with four interferograms. The system performs as previous proposals to attain four interferograms with a single shot [Barrientos et al., 1999; Novak et al., 2005]. In the following sections, a variant of a TWPGI able to capture $N \geq 4$ interferograms in one shot is described. It consists of the set-up shown in Fig. 1. The system uses a grid as a beam splitter in a way that resembles the well-known double-frequency shearing interferometer as proposed by Wyant,2004, but our proposal differs from it not only because of its modulation of polarization, the use of a single frequency and the use of two windows, but also in the phase steps our system introduces. Besides, ours is not a shearing interferometer of any type.

The Fig. 1 shows the arrangement of a one-shot phase-shifting grid interferometer including modulation of polarization with retarders for the windows and linear polarizers on the image plane. The system generates several diffraction orders of similar irradiances in the average but not equal fringe modulations, as expected. Each interferogram image was scaled to the same values of grey levels (from 0 to 255). Previous reports show that a simplification for the polarizing filters array can be attained when using the phase shifts of π [10] to obtain values of ξ of 0, $\pi/2$, π and $3\pi/2$, due to the π-shifts, only two linear polarizing filters have to be placed (instead of four filters, without the π-shifts). The transmission axes of the filter pairs P_1, P_3 and P_2, P_4 can be the same for each as long as they cover two patterns 180° phase apart (Fig. 4). The needed values of ψ have to be of $\psi_1=0°$ and $\psi_2=45°$ with ideal quarter-wave retarders. But considering the retarders at disposal, it can be shown with Eq. (10) that ψ can be of $\psi_1=0°$ and $\psi_2=46.577°$. They are sketched in Fig. 4. The square enclosing the 16 windows replicas in the same figure is to be compared with the similar square of Fig. 5 (dotted lines).

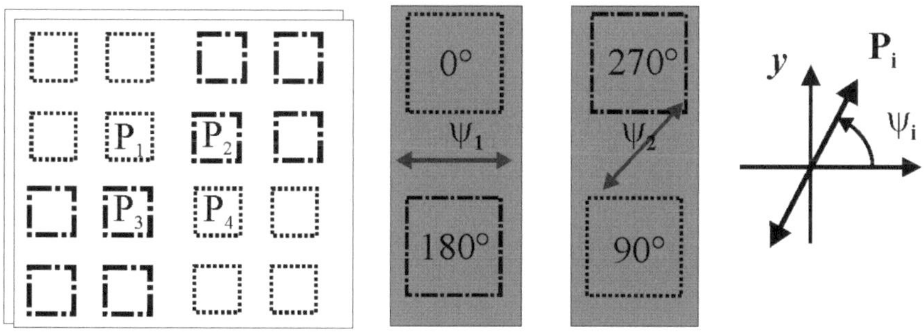

Fig. 5. Polarizing filters array for 90° phase stepping.

5.1 Case of four interferograms

For phase-shifting interferometry with four patterns, four irradiances can be used, each one taken at a different ψ angle. The relative phase can be calculated as [Schwieder, 1983]

$$\tan\phi = \frac{\left\|\vec{J}_1\right\|^2 - \left\|\vec{J}_3\right\|^2}{\left\|\vec{J}_2\right\|^2 - \left\|\vec{J}_4\right\|^2} \tag{12}$$

where $\left\|\vec{J}_1\right\|^2$, $\left\|\vec{J}_2\right\|^2$, $\left\|\vec{J}_3\right\|^2$ and $\left\|\vec{J}_4\right\|^2$ are the intensity measurements with the values of ψ such that $\xi(\psi_1)=0, \xi(\psi_2)=\pi/2, \xi(\psi_3)=\pi, \xi(\psi_4)=3\pi/2$ respectively. Note that $\xi(\psi,\pi/2)=2\psi$ and $A(\psi,\pi/2)=1$, so a good choice for the retarders is quarter-wave retarders, as is well known. Dependence of ϕ on the coordinates of the centered point has been simplified to x,y. The same fringe modulation m_q results as in Eq. (8). Therefore, the discussion about fringe modulation given in previous sections is retained when introducing the modulation of polarization. Such polarization modulation can be made also for grids, resulting in similar conclusions.

5.2 Case of five, seven, and nine interferograms

To demonstrate the use of the several interferograms obtained to extract phase under the conditions as described above, we choose the symmetrical N+1 phase steps algorithms for data processing in the cases N= 4, 6, 8. The phase for N shifts is given by [Malacara, 1998]:

$$\tan\phi(x,y) = \frac{\sum_{i}^{N+1} I_i \sin\left(2\pi\frac{i-1}{N}\right)}{\sum_{i}^{N+1} I_i \cos\left(2\pi\frac{i-1}{N}\right)} \tag{13}$$

where N+1 is the number of interferograms. The Fig. 6 shows the polarizing filters employed. For the case of five interferograms, only three linear polarizing filters have to be placed. The transmission axes of the filter pairs P_n can be the same for each as long as they cover two patterns with 180° phase shift in between.

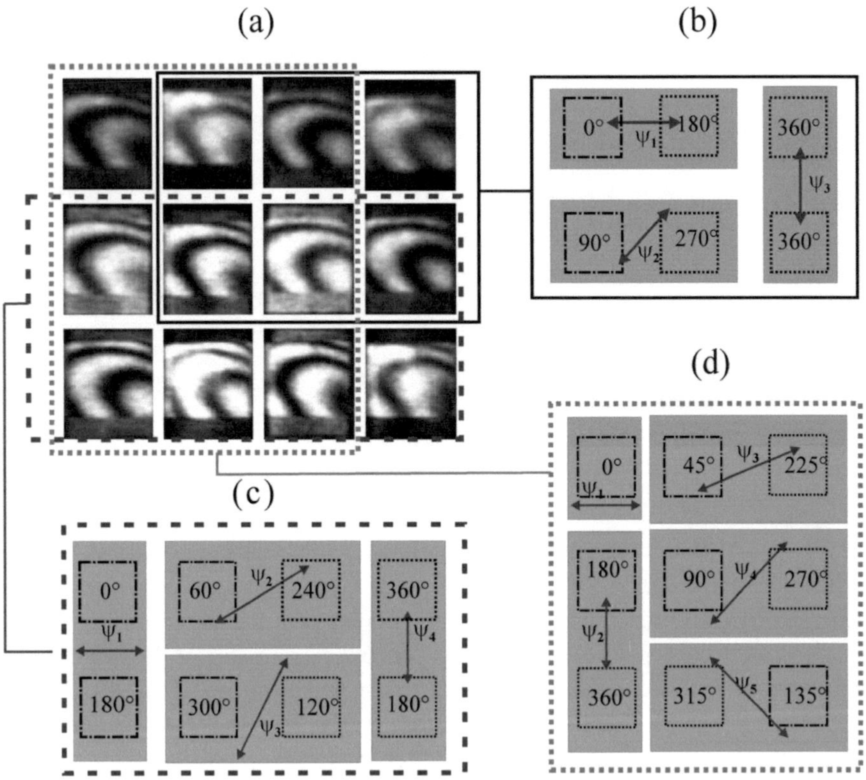

Fig. 6. Polarizing filters array for five, seven and nine interferograms. a) 12 interference patterns detected with a polarizing filter at ψ=35°covering all of them. Some π-shifts can be recognized when reversal contrasts are present. b) N=4, symmetrical five. c) N= 6, symmetrical seven. d) N= 8, symmetrical nine.

6. Experimental results

Two objects for testing are a phase disk and a phase step. When each object was placed separately in one of the windows using the TWPGI with the polarizers array, the interferograms of Fig. 7 were obtained. For each object, the four interferograms are shown together with the calculated unwrapped phase. However, more than four interferograms could be used, whether for N-steps phase-shifting interferometry [Shwieder et al., 1983] or for averaging images with the same shift. Examples, some typical raster lines for each unwrapped phase are shown in Fig. 8(in arbitrary phase units).

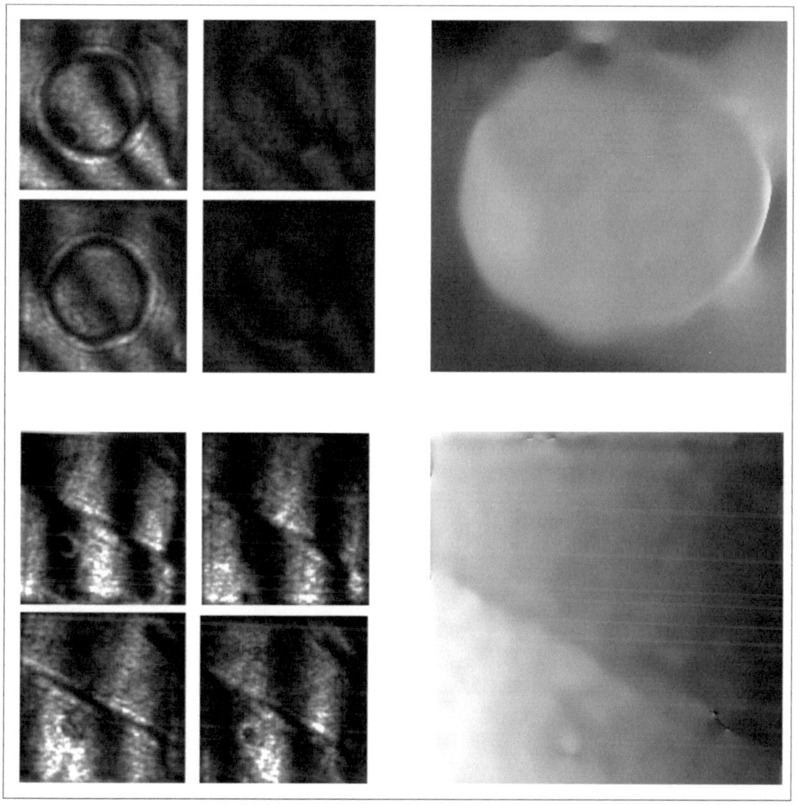

Fig. 7. Upper row: phase dot. Four 90° phase-shifted interferograms and unwrapped phase. Lower row: phase step. Four 90° phase-shifted interferograms and unwrapped phase.

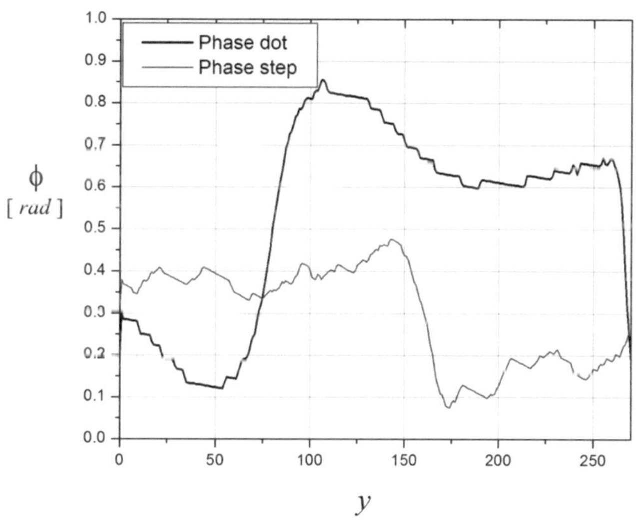

Fig. 8. Unwrapped calculated phases along typical raster lines of each object of Fig. 5. Scale factor : 0.405 rad.

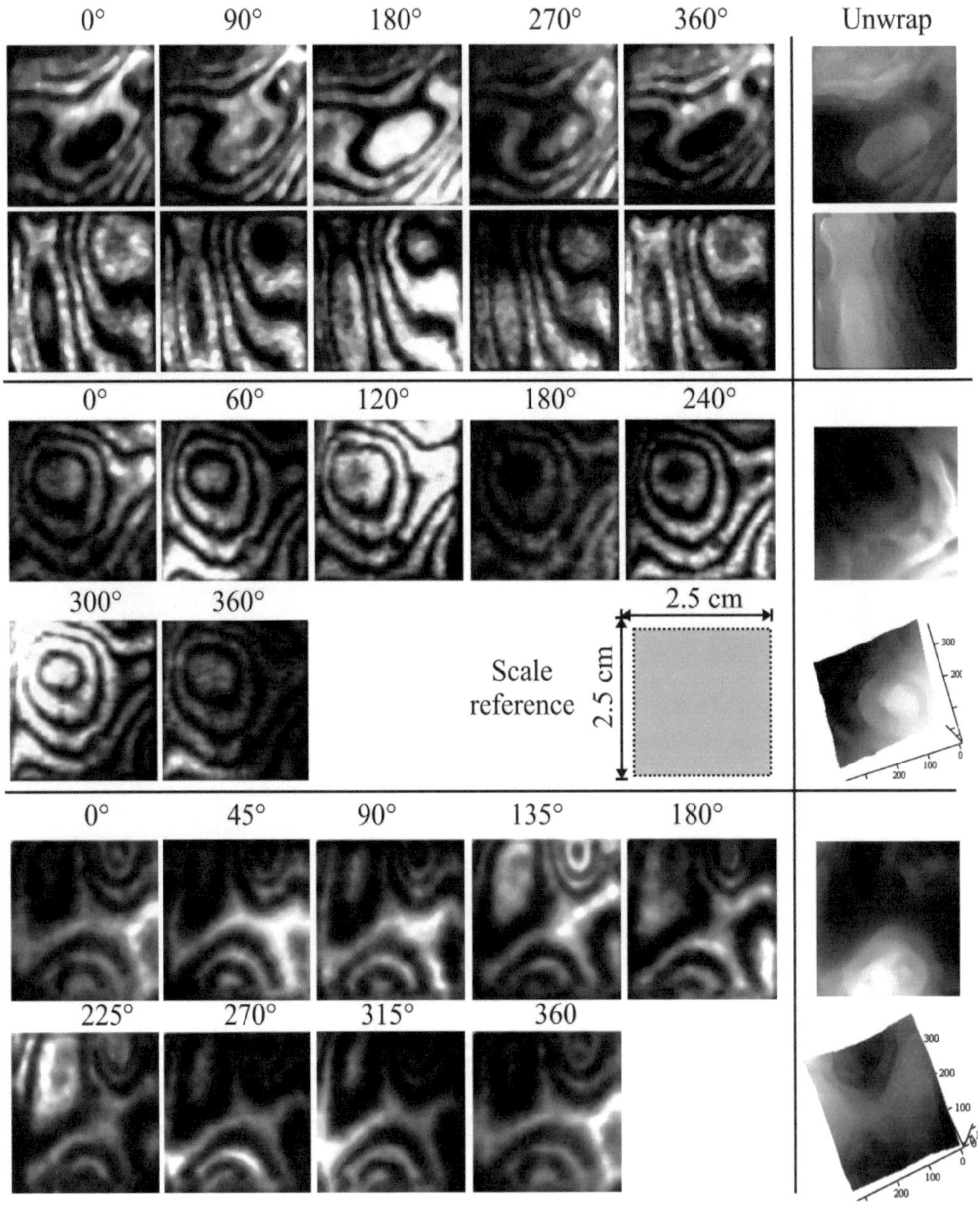

Fig. 9. Flow of oil drops on glass. Phase-shifted interferograms and unwrapped phases. Upper two rows: two examples of five 90° phase-shifts. Center rows: seven 60° phase-shifts. Lower rows: nine 45° phase-shifts. Reference square for scale dimensions.

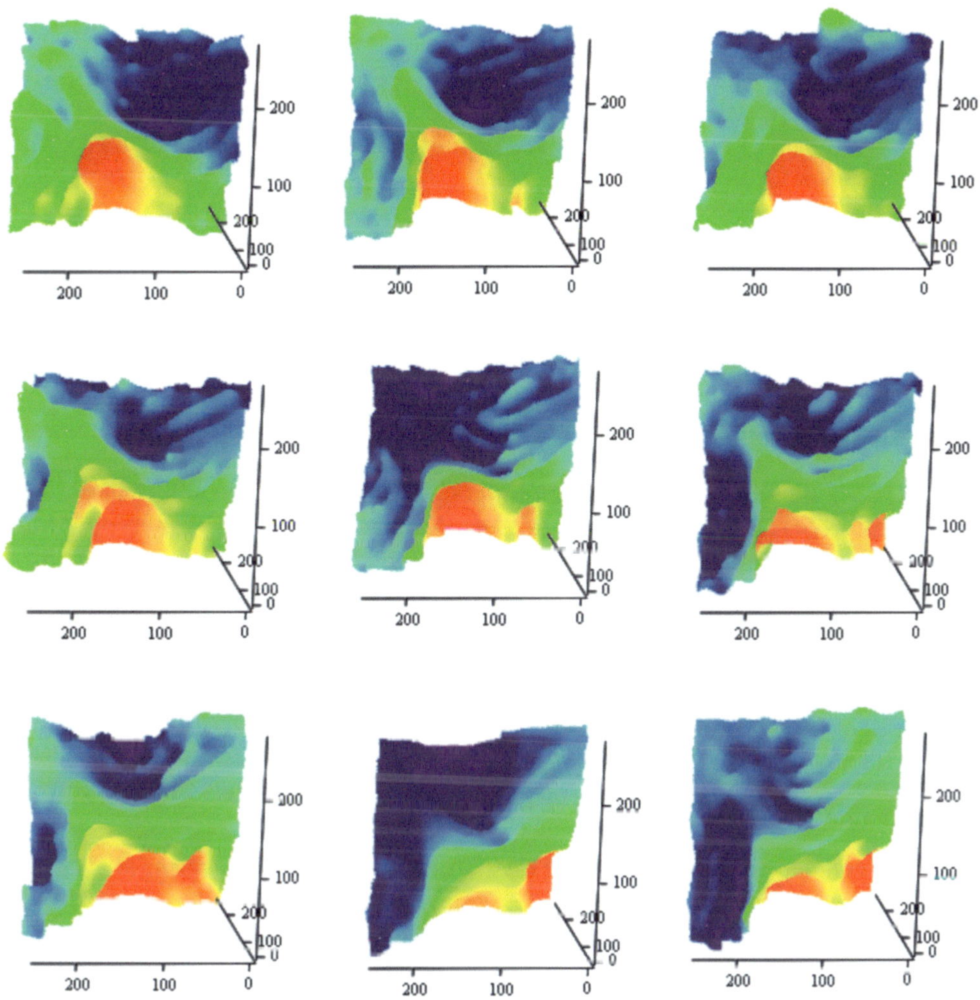

Fig. 10. Typical frames from an unwrapped phase from interferograms of oil flowing.

Considering the retarders at disposal, according to Eq. (10) it can be shown that ψ_n can be of $\psi_1 = 0°$, $\psi_2 = 46.577°$ and $\psi_3 = 92.989°$, each step ξ being of 90°. For the case of symmetrical seven, each step ξ is of 60°, so it can be shown that ψ_n are $\psi_1 = 0°$, $\psi_2 = 30.800°$, $\psi_3 = 62.330°$ and $\psi_4 = 92.989°$. For symmetrical nine, $\psi_1 = 0°$, $\psi_2 = 92.989°$, $\psi_3 = 22.975°$, $\psi_4 = 46.577°$ and $\psi_5 = 157.903°$. In this case, each step ξ has to be 45°. The corresponding results are shown in Fig. 9, where the object was an oil drop running down over a microscope slide. Each interferogram was subject to rescaling and normalization and then, a filtering process prior to phase calculation through Eq. (13).

6.1 Moving distributions

Immersion oil was applied to a glass microscope slide and allowed to flow under the effect of gravity by tilting the slide slightly. The slide was put in front of one of the object windows of the system of Fig. 1. Fig. 10 shows the resulting unwrapped phase evolution of oil flow (Case N=4).

7. Final remarks

The experimental set-up for a polarizing two-window phase-grating common-path interferometer has been described. This system is able to obtain four interferograms 90° phase-apart in only one shot. Therefore, it is suitable to carry out phase extraction using phase shifting techniques. Phase evolving in time can then be calculated and displayed. The system is considerably simpler than previous proposals to attain four interferograms with only one shot. In its present form, it is, however, best suited to relative small objects which do not introduce polarization changes. Because it works with interferograms placed relatively far from the optical axes, experimental results suggest that some method has to be introduced to compensate mainly for distortion, among other off-axis aberrations. This compensation could be optical (as a better design of the optical imaging system) or digital (fringe distortion compensation by inverse transformation). In the experiments, the use of is described retarding plates which are not quarter-wave plates. Although they can perform well enough in principle, it seems better to use quarter-wave plates because no additional variations of the interferogram amplitude arise. Also in this case, a simpler polarization filter array can be used taking advantage of the diffraction properties of a phase grating. Some special phase gratings design could optimize the interferometric system described.

This system is able to obtain n = (N+1) interferograms with only one shot ($n \leq 16$). Tests with $2\pi / N$ phase-shifts were presented, but other approaches using different phase-shifts could be attained using linear polarizers with their transmission axes at the proper angle before detection. The phase shifts of π due to the grid spectra allows the use of a number of polarizing filters which is less than the number of interferograms, simplifying the filter array. Other configurations for the window positions which are different as the one reported in this communication can also be possible. The accuracy in measurements is the one typical of phase-shifting. Some trade-offs appear while placing several images over the same detector field, but for low frequencies interferograms (with respect to the inverse of the pixel spacing) the influence of these factors seems to be rather small if noticeable. The interferometer could be used for objects with no changes of polarization.

8. Acknowledgements

Partial support from Benemérita Universidad Autónoma de Puebla (BUAP), project: 154984 (CONACYT-BUAP) is also acknowledged. Author N.-I. Toto-Arellano expresses sincere appreciation to Luisa, Miguel and Damian for the support provided, and to CONACYT for grant 102137/43055.

9. References

Arrizón V. and Sánchez-De-La-Llave D., (2004). Common-Path Interferometry with One-Dimensional Periodic Filters, *Opt. Lett.* Vol. 29, pp. 141-143.

Barrientos-García B., Moore A. J., Pérez-López C., Wang L., and Tschudi T.,(1999) Spatial Phase-Stepped Interferometry using a Holographic Optical Element, *Opt. Eng.* Vol. 38, pp. 2069-2074

Malacara D., Servin M., and Malacara Z.; Marcel Dekker (1998). *Interferogram Analysis for Optical Testing.*

Meneses-Fabian C., Rodriguez-Zurita G., and Arrizon V. (2006a). Common-Path Phase-Shifting Interferometer with Binary Grating, *Opt. Commun.* Vol. 264, pp. 13-17.

Meneses-Fabian C., Rodriguez-Zurita G., and Arrizon V., (2006b). Optical Tomography of Transparent Objects with Phase-Shifting Interferometry and Stepping-Wise Shifted Ronchi Ruling, *J. Opt. Soc. Am. A*, Vol. 23, pp. 298-305.

Novak M., Millerd J., Brock N., North-Morris M., Hayes J. and Wyant J.C., (2005). Analysis of a micropolarizer array-based simultaneous phase-shifting interferometer, *Appl. Opt.*, Vol. 44, pp. 6861-6868.

Kreis T., (1986). Digital Holographic Interference Phase Measurement Using the Fourier-Transform Method, *J. Opt. Soc. Am. A*, Vol. 3, pp. 847-855.

Rodriguez-Zurita G., Meneses-Fabian C., Toto-Arellano N., Vázquez-Castillo J. F. and Robledo-Sánchez C.,(2008). One-Shot Phase-Shifting Phase-Grating Interferometry with Modulation of Polarization: case of four interferograms, *Opt. Express*, Vol. 16, 7806-7817.

Rodriguez-Zurita G., Toto-Arellano N. I., Meneses-Fabian C. and Vázquez-Castillo J. F., (2008). One-shot phase-shifting interferometry: five, seven, and nine interferograms, *Opt Letters*, Vol. 33, pp. 2788-2790.

Rodríguez-Zurita G., Toto-Arellano N. I., Meneses Fabian C. and Vázquez-Castillo J. F., (2009).Adjustable lateral-shear single-shot phase-shifting interferometry for moving phase distributions, *Meas. Sci. Technol.* Vol. 20, pp. 115902.

Schwieder J., Burow R., Elssner K.-E., Grzanna J., Spolaczyk R. and Merkel K., (1983). Digital Wave-Front Measuring Interferometry: some systematic error sources, *Appl. Opt.*, Vol. 22, pp. 3421-3432.

Thomas D. A., and Wyant J. C.,(1976). High Efficiency Grating Lateral Shear Interferometer, *Opt. Eng.*, Vol. 15, pp. 477.

Toto-Arellano N. I., Rodríguez-Zurita G., Meneses-Fabian C., Vazquez-Castillo J. F., (2008). Phase shifts in the Fourier spectra of phase gratings and phase grids: an application for one shot phase-shifting interferometry, Opt. Express, Vol. 16, pp. 19330-19341

Wyant J.C. (2004). Vibration insensitive interferometric optical testing, in Frontiers in Optics, OSA Technical Digest, OTuB2.

4

Phosphor-Based White Light Emitting Diode (LED) for Vertical Scanning Interferometry (VSI)

Wee Keat Chong[1], Xiang Li[1] and Yeng Chai Soh[2]
[1]*Singapore Institute of Manufacturing Technology, A*STAR*
[2]*Nanyang Technological University*
Singapore

1. Introduction

Vertical scanning interferometry (VSI) is an established optical method for surface profile measurement by analyzing a series of interference patterns of low coherence light with known optical path difference among them. As white light is commonly used as low coherence light source, vertical scanning interferometry is also known as white light interferometry (WLI).

Vertical scanning interferometry is most commonly used as surface profilometer which is considered as an enabling and supporting technology to other fields such as surface finishing, machining and material science. It is a non-contact three-dimensional surface measurement technique that provides accuracy up to nanometer level and measurement range up to a few hundred micrometers.

Fig. 1 graphically illustrates the schematic diagram of vertical scanning interferometry in Michelson interferometer configuration; the light beam from the light source is split into two: one to reference surface and one to measurement surface, then these light beams reflect and interfere with each other. Interference pattern occurs when the optical path difference (OPD) between these two light beams is small, within the coherence length of the light source. The interference pattern is known as interferogram (as shown in Fig. 2), it is recorded by area-based photo-sensitive sensor such as CCD camera. Correlogram is the function of intensity response of each pixel against optical path difference, and it is further processed for height profile measurement. Fig. 3 graphically illustrates correlogram and coherence peak function of vertical scanning interferometry.

Apart from the light source, the hardware of vertical scanning interferometry has not changed much in the past one decade. As phosphor-based white light emitting diode (LED) promises greater power, longer lifetime, low heat dissipation and compactness, it is replacing the conventional light source in vertical scanning interferometry.

The conventional light source for white light has a very broad and smooth spectrum, for example Fig. 4 shows the spectrum of Quartz Tungsten Halogen Lamps (model no 6315 from NewPort). However due to the spectral response of photo detector, the effective spectrum of conventional white light is considered as single Gaussian function in visible light spectrum. On the other hand, phosphor-based white LED consists of single color LED

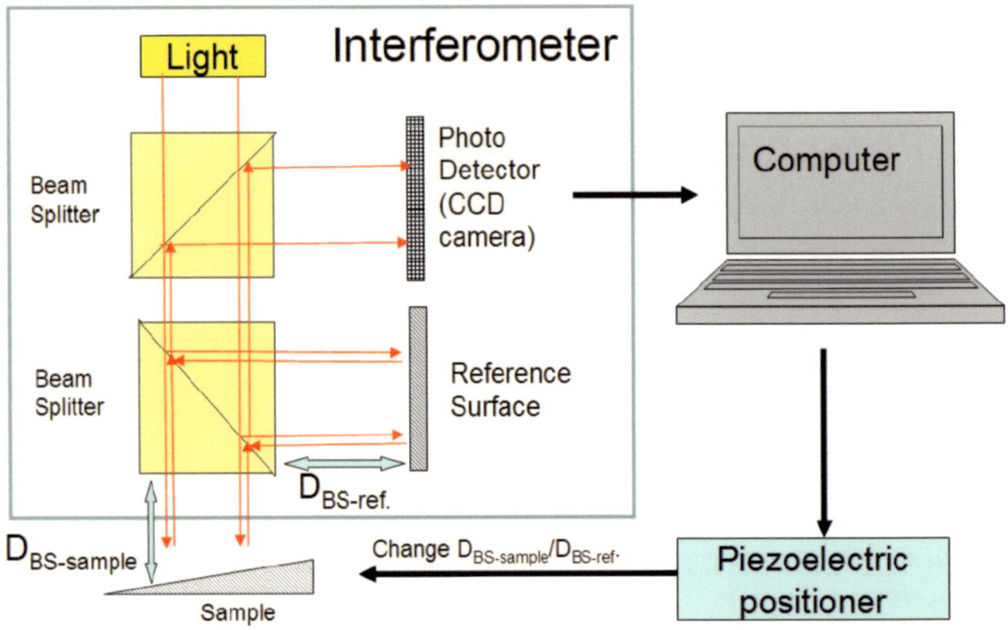

Fig. 1. Schematic diagram of vertical scanning interferometry in Michelson interferometer configuration.

Fig. 2. Example of interferogram of vertical scanning interferometry: Fringe on a tilted flat surface captured using a CCD camera.

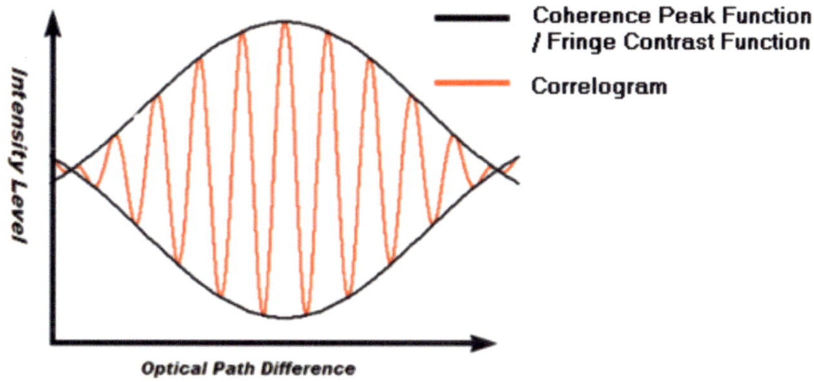

Fig. 3. Example of correlogram and corresponding fringe contrast function (also known as coherence peak function) in vertical scanning interferometry.

(normally blue) and phosphor of different color (normally yellow) to produce white light, so there are two peaks in its spectrum. Fig. 5 compares the effective intensity spectrum of conventional white light and phosphor-based white LED, the major difference between these two light sources is the number of peaks in spectral domain.

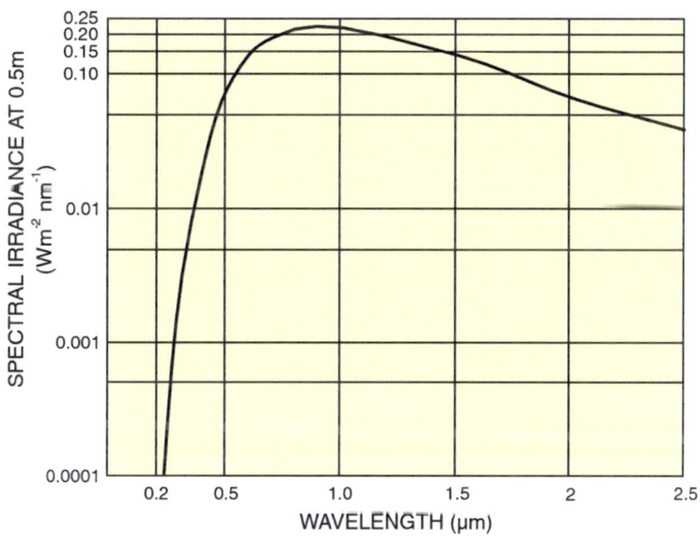

Fig. 4. Spectral irradiance at 0.5 m from the 6315 1000W QTH Lamp3 (by NewPort).

As most prior works (Guo, Zhao, & Chen, 2007; Gurov, Ermolaeva, & Zakharov, 2004; Mingzhou, Chenggen, Cho Jui, Ivan, & Shihua, 2005; Pavli?ek & Soubusta, 2004) assume the use of conventional white light, the effects of phosphor-based white LED on vertical scanning interferometer is the focus on this chapter. Other than that, this chapter also covers a computationally efficient signal modelling method for vertical scanning interferometry and method to improve performance of vertical scanning interferometry with phosphor-based white LED.

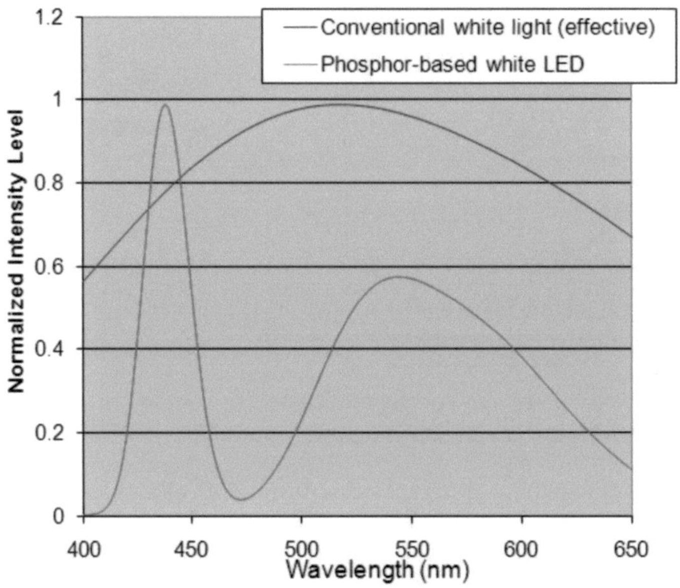

Fig. 5. Comparing effective intensity spectrum of conventional white light and phosphor-based white LED (LXHL-LW6C by LumiLEDs).

2. Theory

Besides optical path difference, the correlogram is affected by the following factors: (1) spectrum of light source (2) numerical aperture of objective (3) optical transfer function of imaging system (4) reflectivity of reference and sample surface (5) phase change at reference and sample surface. The intensity response can be formulated as following:

$$I_{\text{interference}}(z) = C_1 \int_{bandwidth} \int_0^{\theta_0} \{k^2 \times \cos[2k(z - z_0)\cos\theta + \phi] \times \sin\theta \cos\theta d\theta\} F(k) dk \qquad (1)$$

where

C_1 is a constant,
z is a independent variable which corresponds to height change by piezoelectric positioner,
z_0 is the height which corresponds to surface profile,
k is angular wave number ($k=2\pi/\lambda$),
$\sin\theta_0$ is numerical aperture (NA) of objective lens,
$\emptyset$ is the phase offset and $F(k)$ is the intensity spectrum of light source.

Detail on derivation and modeling of the intensity response can be found in literature by Chim and Kino (1990), Kino and Chim (1990), de Groot and Lega (2004) and Sheppard and Larkin (1995).

Equation (1) is a generalized model that can simulate the effects of changing spectrum of light source and numerical aperture of objective lens. There is strong correlation between model parameters and physical setting of the system, for example, the term *F(k)* in Equation (1) is equal to the intensity spectrum which can be measured by optical spectrum analyzer. The drawback of this generalized model is intensive computational load: as the intensity spectrum of light source is either difficult to be expressed in symbolic form or modelled as Gaussian, Equation (1) has to be solved by numerical integration which is a computationally intensive and time consuming process. Therefore there is a simplified version of the generalized model and it can be expressed as follows:

$$I(z) = I_{dc}(z) + I_{amplitude}(z)\exp\left(\frac{-(z-z_0)^2}{\sigma^2}\right)\cos\left(\frac{2\pi(z-z_0)}{\lambda_m}+\phi\right) \quad (2)$$

where

z is defocus position (related to optical path difference),
z_0 is related to the profile of sample surface,
I_{dc} is constant value and not related to interference,
$I_{amplitude}$ is amplitude of interference signal,
λ_m is equivalent wavelength of light,
σ is related to coherent length of light and
$\emptyset$ is phase offset.

This simplified model is based on assumptions that the numerical aperture of objective is small and the intensity spectrum of light, *F(k)* in Equation (1), is single Gaussian function. The advantage of such simplification is computational efficiency, and the drawback is poor correlation between the model parameters and theoretical parameters, for example, the parameter σ in Equation (2) does not have physical meaning (such as the spectrum of light source). As such, the parameters of the simplified model have to be determined empirically, and it is impossible to simulate the correlogram based on specified spectrum and/or numerical aperture value.

To reduce computational load, de Groot and Lega (2004) proposed simplification in frequency domain: Equation (1) is first transformed into frequency domain, followed by simplification, applying numerical integration and lastly inverse Fourier transform back to its original domain. This approach is 200 times faster than direct numerical integration of the generalized model, but it is still a time consuming process due to (1) the use of Fourier and inverse Fourier transform and (2) it still requires numerical integration process.

2.1 Computationally efficient signal modeling

Before looking into the computationally efficient signal modelling, let's look into the cause of intensive computational load of using Equation (1) - numerical integration. Numerical integration is a process calculating the approximated value of a definite integral which can be expressed as following:

$$\int_a^b f(x)dx \approx \frac{(b-a)}{n}\left(\frac{f(a)+f(b)}{2} - \sum_{k=1}^{n-1} f\left(a+k\frac{b-a}{n}\right)\right) \quad (3)$$

where

n is the number of interval

The accuracy of numerical integration process is proportional to the parameter n in Equation (3). However, as n goes up, the computational load increases.

There are two cases which numerical integration is required: (1) the integrand may be known for certain region only and/or (2) the anti-derivative of the integrand does not exist. For signal simulation of vertical scanning interferometry, the intensity spectrum, $F(k)$ in Equation (1), it is either sampled by spectrum analyzer or modelled as Gaussian (which is not an explicit integral). As such, the generalized model, i.e. Equation (1), has to be solved by numerical integration which is computationally intensive.

Chong et al. (2010a) proposed to remove the numerical integration process by representing a single Gaussian function as a sum of two piecewise cosine functions, which can be expressed as following:

$$a\exp\left(\frac{-(x-x_m)^2}{\sigma^2}\right)=\begin{cases} a_1\cos\left(\dfrac{2\pi(x-x_m)}{b_1}\right)+a_2\cos\left(\dfrac{2\pi(x-x_m)}{b_2}\right) & \text{for } (x_m - c)\le x\le(x_m+c) \\ 0 & \text{else}\end{cases} \tag{4}$$

The transformation from single Gaussian (with parameters of a, x_m and σ) to a sum of two piecewise cosine functions (with parameters of a_1,a_2,b_1,b_2,x_m and c) is modeled as a linear transformation and solved by trust region approach (for generating data) and linear regression, followed by minimizing error of fitting with respect to C_{range}. As a result, the unknown in Equation (4) can now be expressed as following:

$$\begin{cases} a_1 = 0.7888a \\ a_2 = 0.2049a \\ x_m = x_m \\ b_1 = 7.6777\sigma - 0.0078 \\ b_2 = -2.4769\sigma + 0.0024 \\ c = 0.852\min(|b_1|,|b_2|) \end{cases} \tag{5}$$

Next, Equation (1) is derived to elementary form as follows:

$$I_{\text{interference}}(z) = C\int_{bandwidth} k^2F(k)\int_0^{\theta_0}\cos[2k(z-z_0)\cos\theta+\phi]\sin\theta\cos\theta\, d\theta\, dk$$

$$= C\int_{bandwidth} k^2F(k)\left[-\frac{(2kz\cos\theta-2kz_0\cos\theta)\sin(2kz\cos\theta - 2kz_0\cos\theta+\phi)+\cos(2kz\cos\theta-2kz_0\cos\theta+\phi)}{4k^2z^2-8k^2z_0z+4k^2z_0^2}\right]_0^{\theta_0} dk$$

$$= C\int_{k_{ll}}^{k_{ul}} \frac{a\cos\left(2\pi\dfrac{k-k_m}{b}\right)}{8z_0z-4z^2-4z_0^2}[(2kz\cos\theta_0-2kz_0\cos\theta_0)\sin(2kz\cos\theta_0 - 2kz_0\cos\theta_0+\phi)+\cos(2kz\cos\theta_0-2kz_0\cos\theta_0+\phi) - (2kz-2kz_0)\sin(2kz-2kz_0+\phi)-\cos(2kz-2kz_0+\phi)]dk$$

$$
\begin{aligned}
&= C\int_{k_{ll}}^{k_{ul}} \frac{a\cos\left(2\pi\frac{k-k_m}{b}\right)}{8z_0z-4z^2-4z_0^2}[2k\cos\theta_0(z-z_0)\sin(2k\cos\theta_0(z-z_0)+\phi)+ \\
&\qquad \cos(2k\cos\theta_0(z-z_0)+\phi)-\cos\phi]dk \qquad (6)\\
&= C\left[\left(g(z,z_0,\theta_0,k_{ul},b)-g(z,z_0,0,k_{ul},b)\right)-\left(g(z,z_0,\theta_0,k_{ll},b)-g(z,z_0,0,k_{ll},b)\right)\right]
\end{aligned}
$$

Let $g(z,z_0,\theta,k,b)$

$$
\begin{aligned}
&= \frac{D}{4}\left(bU\left(\left(2(\pi-bU)k\cos\left(\frac{(b\phi+2k_m\pi-2\pi k+2bUk)}{b}\right)+b\sin\left(\frac{b\phi+2k_m\pi-2\pi k+2bUk}{b}\right)\right)\Big/(\pi-bU)^2+\right.\right.\\
&\left(-2(\pi+bU)k\cos\left(\frac{b\phi-2k_m\pi+2\pi k+2bUk}{b}\right)+b\sin\left(\frac{b\phi-2k_m\pi+2\pi k+2bUk}{b}\right)\right)\Big/(\pi+bU)^2\Bigg)+\\
&\left.\left(b\left(\sin\left(\frac{b\phi+2k_m\pi-2\pi k+2bUk}{b}\right)\Big/(-\pi+bU)+\sin\left(\frac{b\phi-2k_m\pi+2\pi k+2bUk}{b}\right)\Big/(\pi+bU)\right)\right)\right)
\end{aligned}
$$

where

$D=1/\left(8zz_0-4z^2-4z_0^2\right)$

$U=(z-z_0)\cos\theta$

Chong et al.'s model (2010a) reduces the computational time by 256800 times compared to conventional direct numerical integration on Equation (1), and it is 2784 times faster than de Groot and Lega (2004) 's approach.

2.2 Phosphor-based white LED

Phosphor-based white LED consists of single color LED (normally blue) and phosphor of different color (normally yellow) to produce white light, so there are two peaks in its spectrum and the intensity spectrum can be expressed as follows:

$$f(k)=BYratio\times e^{-\left(\frac{k-k_{blue}}{\sigma_{blue}}\right)^2}+e^{-\left(\frac{k-k_{yellow}}{\sigma_{yellow}}\right)^2} \qquad (7)$$

where

k is angular wave number ($=2\pi/\lambda$)

k_{blue} and k_{yellow} indicate the peak angular wave number of blue and yellow light

σ_{blue} and σ_{yellow} indicate the spread of blue and yellow light in spectrum domain

In general, the angular wave number of blue and yellow light are 14.37 rad/nm (438nm) and 11.25 rad/nm (558nm) respectively, the spread of blue and yellow light in spectrum domain are 0.4941 rad/nm and 1.439 rad/nm; these values vary slightly across manufacturers/model.

Compared to conventional white light (with reference to Fig. 5), the intensity spectrum of phosphor-based white LED is significantly different.

3. Effects of phosphor-based white LED on vertical scanning interferometry

As mentioned earlier, the intensity spectrum of phosphor-based white LED is significantly different from conventional lighting which most prior arts adopted. In this section, the effects of phosphor-based white LED on vertical scanning interferometry in correlogram and reconstructed surface profile level are investigated by simulation, followed by experimental verification.

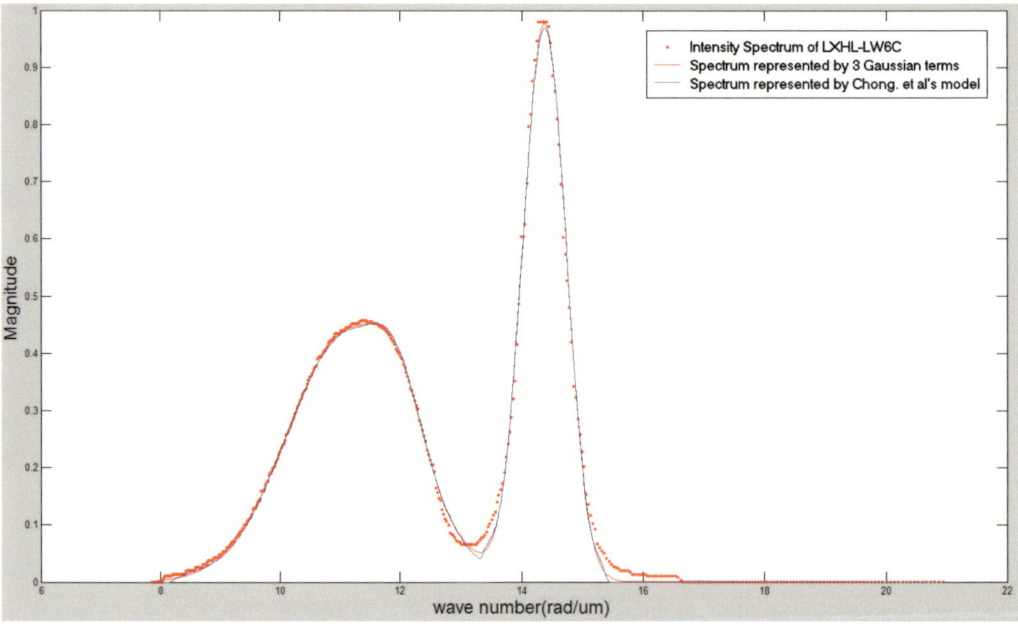

Fig. 6. Intensity spectrum of phosphor-based LED, LXHL-LW6C by LumiLEDs and comparison between its presentations by three Gaussian functions and Chong et al.'s method.

To study the effects of phosphor-based white LED, a commercial off the shelf phosphor-based white LED, LXHL-LW6C by LumiLEDs is selected, and the numerical aperture of objective is assumed to 0.4 which is a typical value for 20x objectives. This configuration is adopted for both simulation and experimental verification.

To simulate the correlogram of LXHL-LW6C by LumiLED, the intensity spectrum (as shown in Fig. 6) is first fitted to three Gaussian functions by non-linear least square fitting method; next each Gaussian terms is replaced with a sum of two piecewise cosine functions according Equation (4) and Equation (5); by Equation (6), contribution of each cosine term is calculated; lastly, the resultant intensity response is the sum of contribution by six cosine terms.

Fig. 6 shows that Chong et al.'s method represents the intensity spectrum of LXHL-LW6C by LumiLED well compared to representation by three Gaussian functions and the original intensity spectrum.

3.1 Effects on correlogram

Based on the intensity spectrum of LxHL-LW6C and numerical aperture of 0.4, the corresponding correlogram is simulated using the computational efficient signal modelling by Chong et al and shown in Fig. 7 (b). The distinctive feature highlighted in Fig. 7 (b) is the result of having two peaks in the spectrum of phosphor-based white LED.

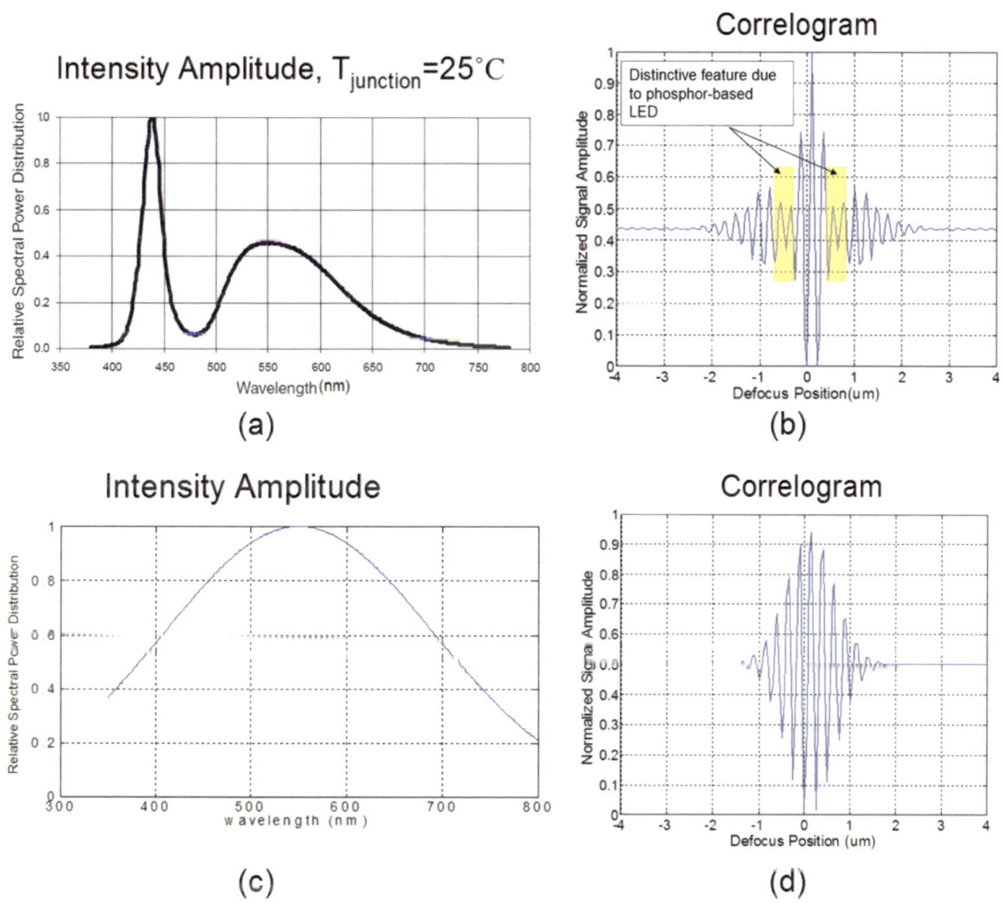

Fig. 7. Comparing spectrum and correlogram of phosphor-based white LED and conventional white light: (a) spectrum of phosphor-based LED, LXHL-LW6C; (b) simulated correlogram based on spectrum of (a); (c) effective spectrum of conventional white light; (d) simulated correlogram based on spectrum of (c).

As shown in Fig. 7, the correlogram of vertical scanning interferometry using phosphor-based white LED is significantly different from that using light source of Gaussian spectrum (as shown in Fig. 7(c) and (d)).

Next, an experimental verification is conducted to verify the simulation result. The experiment configuration is as follows: a 40x Nikon mirau-based interferometric objective with numerical aperture of 0.4 and phosphor-based white LED from LumiLEDs LXHL-LW6C was used.

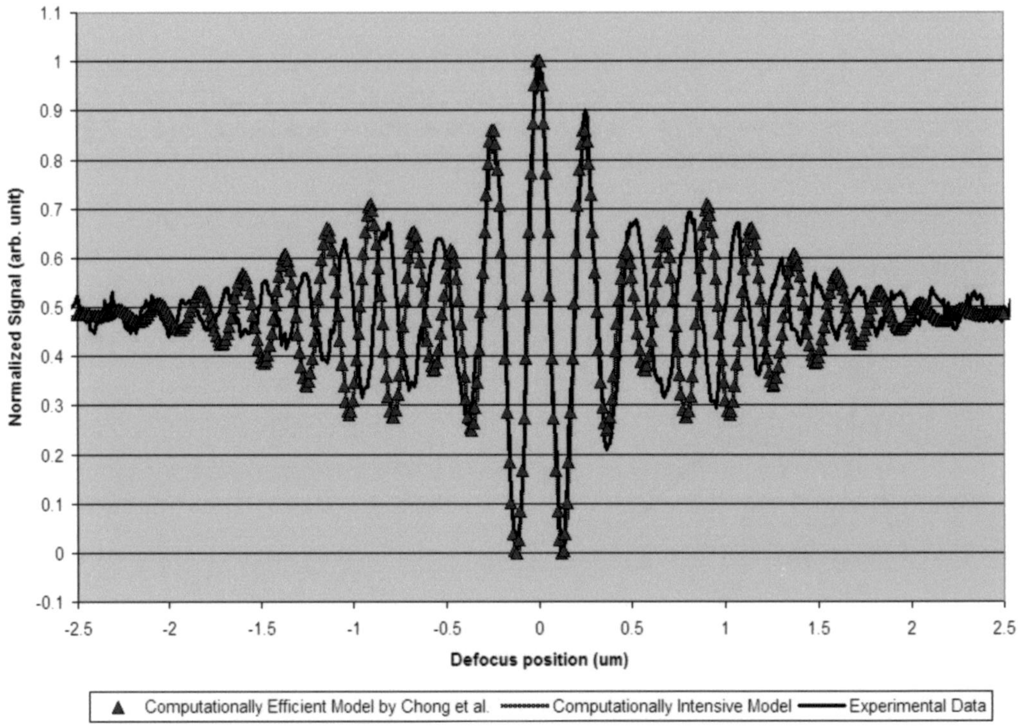

Fig. 8. Graphical comparison between computationally efficient signal modeling by Chong et al., computationally intensive model and experimental data.

Fig. 8 graphically compares the experimental data with simulation data, it shows that the distinctive feature due to the use of phosphor-based white LED (highlighted in Fig. 7(b)) is consistent with the experimental data.

By experiment and simulation, it is shown that the intensity response (also known as correlogram) of vertical scanning interferometry using phosphor-based white LED significantly different from light source of Gaussian spectrum. The fringe contrast function (envelope function of correlogram) is not longer a single Gaussian function (Chong, Li, & Wijesoma, 2010b).

3.2 Effects of reconstructed height profile

As correlogram alone does not provide information on height, a reconstruction algorithm is required to transform correlogram into height information. There are two categories for surface height profile reconstruction algorithm: (1) fringe contrast based approach and (2) phase-based approach. Fringe contrast based approach finds the maximum of fringe contrast function which corresponds to the height profile; while phase-based approach transforms intensity response into frequency domain, followed by phase signal analysis.

This section investigates the effects of phosphor-based LED on three reconstruction algorithms proposed by Gaussian fitting method (Mingzhou et al., 2005), Centroid approach (Ai & Novak, n.d.), and Frequency domain analysis (FDA) (P. J. de Groot & Deck, 1994) respectively. Among these three algorithms, only FDA is phase-based approach; the other two are fringe contrast based approach. Among these two methods, the major difference is that Gaussian fitting method by Mingzhou et al assumes the fringe contrast function is a Gaussian signal while the centroid approach does not.

Mingzhou et al.'s method recovers height by first finding the envelope of the correlogram, followed by Gaussian fitting. Ai and Novak proposed that the centroid of the correlogram corresponds to the maximum of fringe. As phase-based approach, Groot and Deck's method breaks down the white light into multiple single wavelength components and applies phase signal analysis similar to phase shifting interferometry.

Simulation is used for investigating the effects of phosphor based white LED on reconstructed height profile. For comparison purpose, two sets of data are simulated: one is based on light with Gaussian spectrum; another one is based on phosphor-based white LED, LXHL-LW6C by LumiLEDs. A line profile of 1um step height is selected, the line profile consists of 256 surface points and each surface point has a corresponding intensity response. The sampling interval of the intensity response is 50nm, and each intensity response is corrupted by Gaussian white noise (zero mean, variance of 0.05). Next we reconstructed these two sets of data using these three construction algorithms mentioned earlier, and the reconstructed profiles are shown in Fig. 9 and Fig. 10. The repeatability (in term of standard deviation) and accuracy of reconstructed profiles are further analyzed.

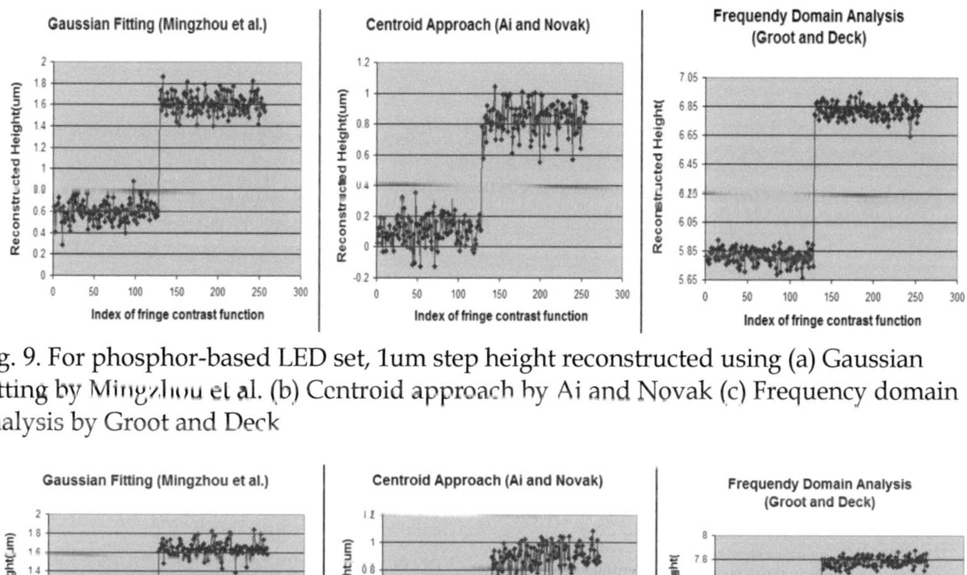

Fig. 9. For phosphor-based LED set, 1um step height reconstructed using (a) Gaussian Fitting by Mingzhou et al. (b) Centroid approach by Ai and Novak (c) Frequency domain analysis by Groot and Deck

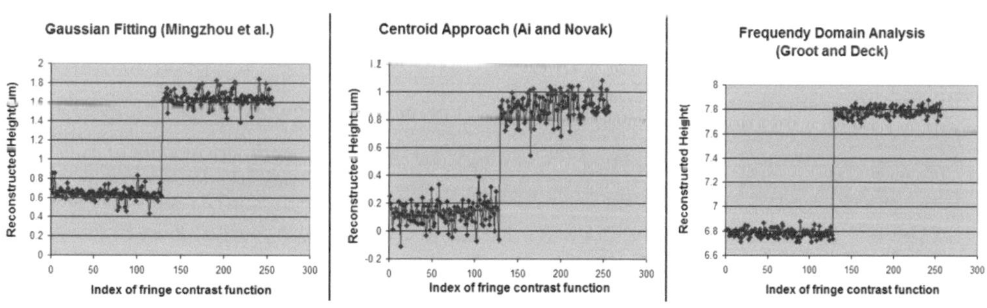

Fig. 10. For light source of Gaussian spectrum, 1um step height reconstructed using (a) Gaussian Fitting by Mingzhou et al. (b) Centroid approach by Ai and Novak (c) Frequency domain analysis by Groot and Deck

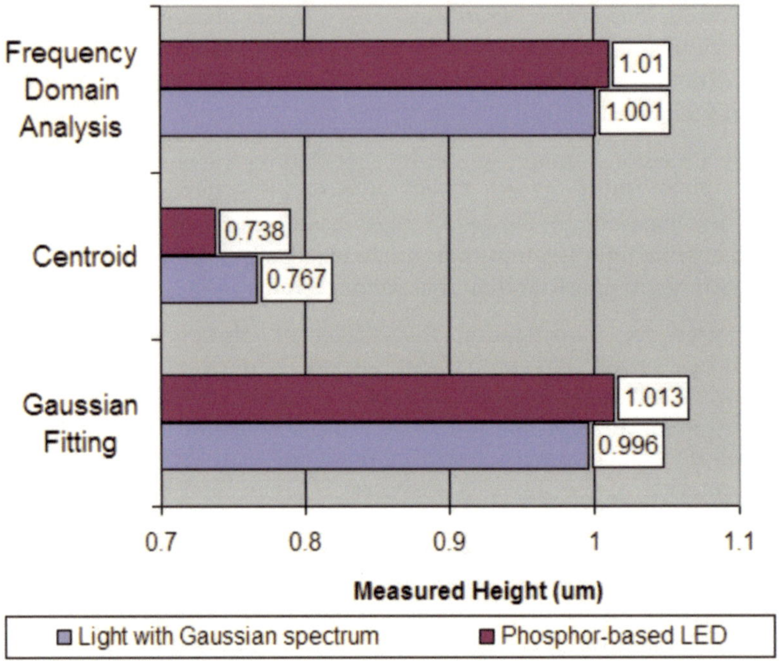

Fig. 11. Comparing measurement accuracy of different algorithms between different light sources, ideal value is 1um.

Fig. 11 shows the comparison of accuracy among different algorithms using different lighting. As the accuracy of individual algorithm is highly dependent on the complexity of the algorithm, it is beyond the scope of this investigation. However it is noticeable that the accuracy of all three algorithms decreases as the light is switched to phosphor-based white LED.

Fig. 12 shows the comparison of measurement repeatability among different algorithms using different lighting, and it is clear that the use phosphor-based LED decreases the repeatability of measurement (standard deviation is inversely proportional to repeatability) at different scale: Gaussian fitting method by Mingzhou et al. suffers most, followed by frequency domain analysis by Groot and Deck, lastly the centroid approach by Ai and Novak. This observation can be explained as follows:

- For the centroid approach by Ai and Novak: This method does not make assumption on the fringe contrast function, so the change in fringe contrast function has relatively little effect on its reconstruction.
- For Gaussian fitting by Mingzhou et al.: This method assumes that the fringe contrast function is a Gaussian function, so it suffers the worst repeatability. As the assumption on the fringe contrast function is not valid, the fitting process is unable to produce good result.
- For the frequency domain analysis by Groot and Deck: Although this method processes the correlogram in the frequency domain only, the change in fringe contrast function does affect the amount and the quality of information selected for frequency domain analysis. This effect leads to improvement discussed in next session.

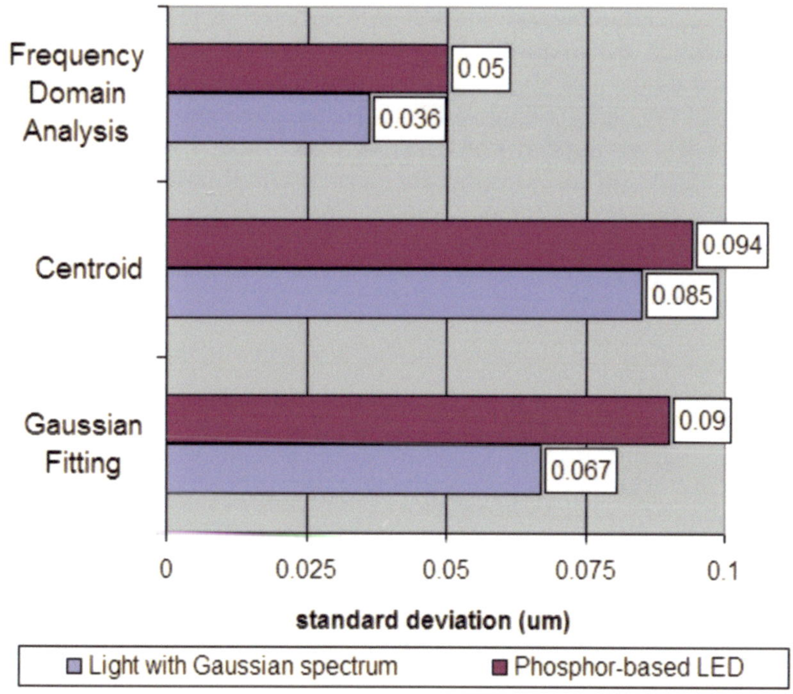

Fig. 12. Comparing the measurement repeatability of different algorithms between different light sources: measurement repeatability is inversely proportional to the standard deviation of perfectly flat surface, ideal value is zero.

4. Modification and improvement for phosphor-based white LED on VSI

As identified earlier, phosphor-based white LED degrades the performance of two reconstruction algorithms (Gaussian fitting method and Frequency domain analysis approach) as it breaks the assumption adopted by these reconstruction algorithms. These undesired effects of phosphor-based white LED can be avoided by either redesigning reconstruction algorithm that does not assume the distribution of intensity spectrum or setting constraint on the input to existing reconstruction algorithm such that the assumption is valid. Redesigning a reconstruction algorithm is beyond the scope of this chapter, so a constraint is applied to make assumption required by Gaussian fitting and frequency domain analysis approach valid.

For Gaussian fitting method, the two valleys (distinctive features highlighted in Fig. 7 (b)) make the fringe contrast function has 3 peaks and can not be modelled as single Gaussian function. However as shown in Fig. 13, the envelope of correlogram between two valleys (distinctive features highlighted in Fig. 7 (b)) can be reasonably modelled as single Gaussian, setting a constraint selecting only these data for single Gaussian fitting would fulfil the assumption of the original Gaussian fitting approach. So, instead of fitting the whole correlogram to a Gaussian, Gaussian fitting approach is modified such that it fit subset of correlogram to a Gaussian.

For frequency domain analysis approach, the required assumption is that there should be only one peak in the spatial frequency ranges from 20rad/um to 30 rad/um which correspond to wavelength of 628nm and 419nm. However for phosphor-based white LED, there are two peaks in the frequency ranges of interest (as shown in Fig. 14(a)), so we applied a constraint on the input for frequency domain analysis such that it would not have two peaks in spatial frequency domain analysis. As the distinctive features are the result of phosphor-based LED, we applied only data between these two features for frequency domain analysis. Fig. 14 (b) confirms that the assumption of frequency domain analysis is met by reducing the amount of data for frequency domain analysis, there is only one distinct peak in spatial frequency domain.

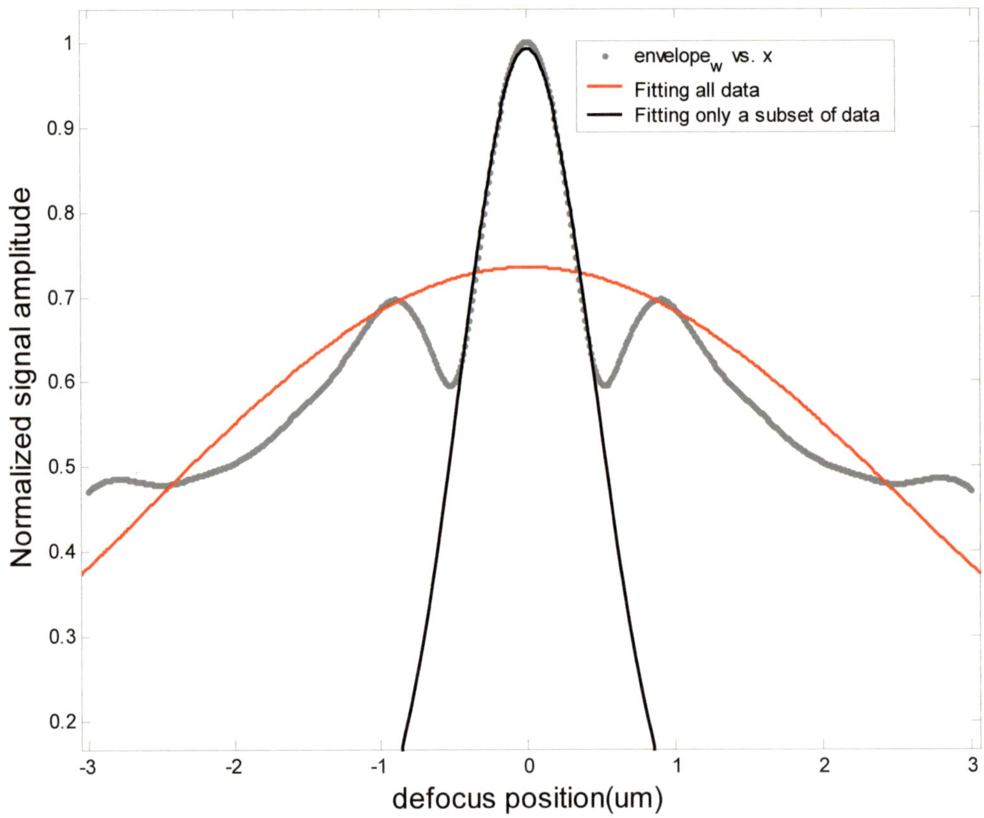

Fig. 13. Fitting the fringe contrast function with single Gaussian: Selecting a subset of data (such as -0.5um<=defocus position<= 0.5um) would lead to a good fit compared to selecting all data.

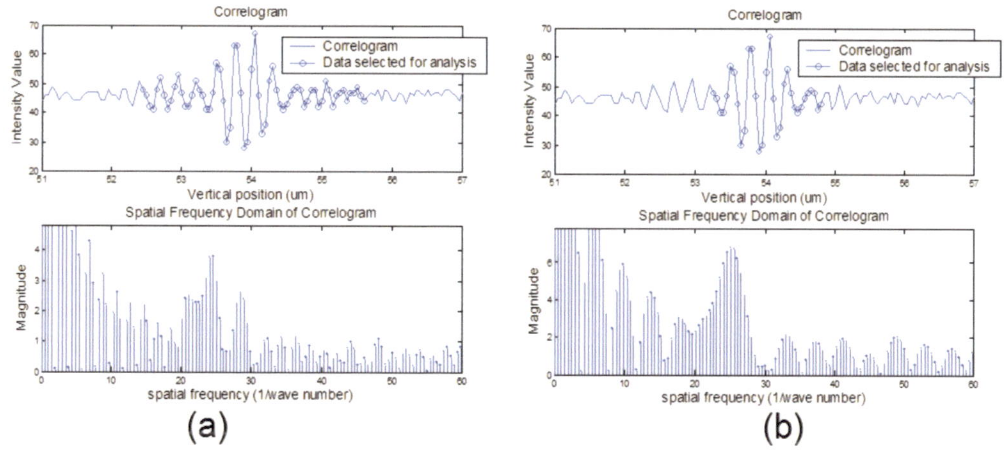

Fig. 14. Illustrating the effects of selecting (a) 64 data (b) 32 data around the maximum of correlogram in spatial frequency domain. According to FDA approach, there should be only 1 peak (in the region around 20 to 30 rad/um), this assumption is met in (b) which 32 data around peak are selected.

Fig. 15 shows that the proposed modification on Gaussian fitting approach and frequency domain analysis works, the constraint on the input to reconstruction algorithms improves the measurement repeatability.

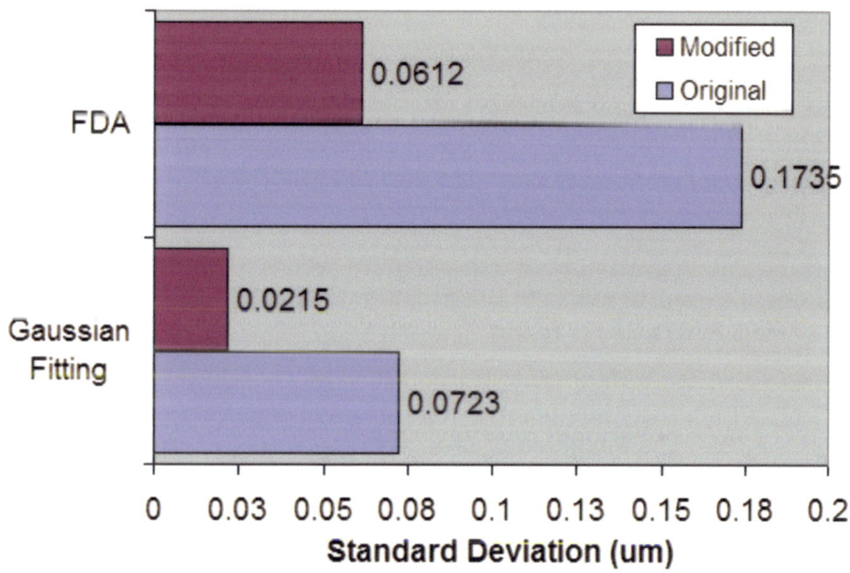

Fig. 15. Simulation verification on the proposed modification: Comparing standard deviation of measuring perfectly flat surface reproduced by proposed modification and original.

As an experimental verification, the configuration is identical to the earlier simulation but measured a 10um±80nm standard step height. 256 correlograms are collected at sampling interval of 50nm, the repeatability of measuring optically flat surface was used to measure the performance of modified Gaussian fitting, modified frequency domain analysis, original Gaussian fitting and original frequency domain analysis.

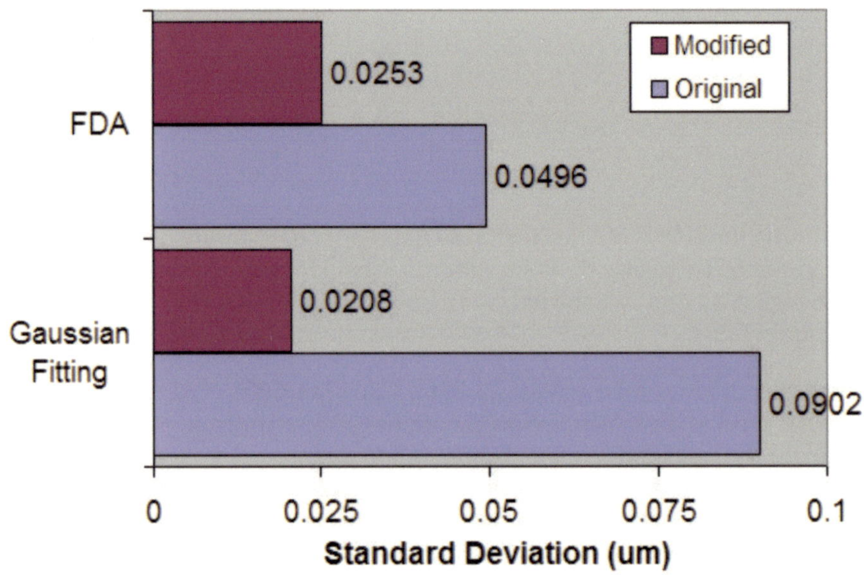

Fig. 16. Experimental verification on the proposed modification: Comparing standard deviation of measuring perfectly flat surface reproduced by proposed modification and original.

Fig. 16 shows that the repeatability of both modified reconstruction algorithm are better than original one, the result agrees with simulation result, but it is controversial to common norm that more data leads to better result.

The result shows that for surface reconstruction of vertical scanning interferometry, it is more important to fulfil the assumption of reconstruction algorithm rather than fitting as many data as possible. The modification of inputting a subset of correlogram (which is around the maximum) improves the performance of reconstruction algorithm for phosphor-based LED, and it is applicable to reconstruction algorithm of both fringe contrast based and phase-based approach.

5. Conclusion

In this chapter, it is shown that the use of phosphor-based LED on vertical scanning interferometry affect the repeatability and accuracy of vertical scanning interferometry, especially repeatability. The effect on the correlogram is inevitable, as the fringe contrast function can not longer be modelled as single Gaussian function. The effect on the reconstructed height profile varies depending on the assumption adopted by individual reconstruction algorithm. However the undesired effects of phosphor-based white LED can be removed by applying a constraint on the input to existing reconstruction algorithm.

6. Acknowledgment

I would like to thanks my supervisors, Prof Soh Yeng Chai, Dr Li Xiang Leon and Dr Sardha WIJESOMA, whose help, stimulating suggestion and encouragement helped me in my research.

I would also like to thanks my colleagues from Singapore Institute of Manufacturing Technology (SIMTech), especially Dr Zhang Ying for their support, guidance and valuable advice

7. References

Ai, C., & Novak, E. K. (n.d.). Centroid approach for estimating modulation peak in broad-bandwidth interferometry. United States Patent. Retrieved from http://www.freepatentsonline.com/5633715.html

Chim, S. S. C., & Kino, G. S. (1990). Correlation microscope. *Optics Letters, 15*(10), 579-581. doi:10.1364/OL.15.000579

Chong, W. K., Li, X., & Wijesoma, S. (2010a). Computationally efficient signal modeling for vertical scanning interferometry. *Applied Optics, 49*(26), 4990-4994. doi:10.1364/AO.49.004990

Chong, W. K., Li, X., & Wijesoma, S. (2010b). Effects of phosphor-based LEDs on vertical scanning interferometry. *Optics Letters, 35*(17), 2946-2948. doi:10.1364/OL.35.002946

de Groot, P., & Colonna de Lega, X. (2004). Signal Modeling for Low-Coherence Height-Scanning Interference Microscopy. *Applied Optics, 43*(25), 4821-4830. doi:10.1364/AO.43.004821

Groot, P. J. de, & Deck, L. L. (1994). Surface profiling by frequency-domain analysis of white light interferograms. In C. Gorecki & R. W. T. Preater (Eds.), *Optical Measurements and Sensors for the Process Industries* (Vol. 2248, pp. 101-104). SPIE. Retrieved from http://dx.doi.org/10.1117/12.194308

Guo, H., Zhao, Z., & Chen, M. (2007). Efficient iterative algorithm for phase-shifting interferometry. *Optics and Lasers in Engineering, 45*(2), 281-292. doi:10.1016/j.optlaseng.2005.11.002

Gurov, I., Ermolaeva, E., & Zakharov, A. (2004). Analysis of low-coherence interference fringes by the Kalman filtering method. *Journal of the Optical Society of America A, 21*(2), 242-251. doi:10.1364/JOSAA.21.000242

Kino, Gordon S., & Chim, S. S. C. (1990). Mirau correlation microscope. *Applied Optics, 29*(26), 3775-3783. doi:10.1364/AO.29.003775

Mingzhou, L., Chenggen, Q., Cho Jui, T., Ivan, R., & Shihua, W. (2005). Measurement of transparent coating thickness by the use of white light interferometry. *Third*

International Conference on Experimental Mechanics and Third Conference of the Asian Committee on Experimental Mechanics (Vol. 5852, pp. 401-406). SPIE.

Pavliček, P., & Soubusta, J. (2004). Measurement of the Influence of Dispersion on White-Light Interferometry. *Applied Optics, 43*(4), 766-770. doi:10.1364/AO.43.000766

Sheppard, C. J. R., & Larkin, K. G. (1995). Effect of numerical aperture on interference fringe spacing. *Applied Optics, 34*(22), 4731-4734. doi:10.1364/AO.34.004731

5

Similariton-Based Spectral Interferometry for Signal Analysis on Femtosecond Time Scale

Levon Mouradian, Aram Zeytunyan and Garegin Yesayan
Ultrafast Optics Laboratory, Faculty of Physics,
Yerevan State University,
Armenia

1. Introduction

The *signal analysis problem on the femtosecond time scale* employs the powerful arsenal of contemporary optics, involving the methods of nonlinear and adaptive optics, Fourier optics and holography, spectral interferometry, etc. The nonlinear-optical techniques of FROG and its modifications [1], the most popular and commercialized, provide accurate and complete determination of the temporal amplitude and phase by recording high-resolution spectrograms, which are further decoded by means of iterative phase-retrieval procedures. The approach of *spectral interferometry (SI)* [2] and its developments to the methods of SPIDER [3], SPIRIT [4], and SORBETS [5] have the advantage of non-iterative phase retrieval. The recently developed and effectively applied MIIPS technique [6] operates with spectral phase measurement through its adaptive compensation up to transform-limited pulse shaping by using feedback from the SHG process. All these methods are based on the spectral phase determination, spectrum measurement, and reconstruction of a temporal pulse. The pulse direct measurement is possible by the transfer of temporal information to the space or frequency domain, or to the time domain with a lager - measurable scale. The time-to-space mapping through Fourier holography, implemented in real time by using a special multiple-quantum-well photorefractive device [7] or second-harmonic-generation crystal [8], provides a temporal resolution given by the duration of reference pulse. In an alternative approach, the 103× up-conversion time microscope demonstrates a 300 fs resolution [9]. In the time-to-frequency conversion approach, the methods of optical frequency inter-modulation by sum-frequency generation [10,11] and electro-optical modulation [12–13] are limited to the picosecond domain. The technique of optical chirped-pulse gating [14] demonstrates sub-picosecond resolution. The method of pulse spectro-temporal imaging through temporal lensing is more promising, having as a principal limit of resolution the ~1 fs nonlinear response time of silica [15–20]. Its recent modifications, implemented in the silicon chip [21] and similariton-induced parabolic temporal lens [22], provide accurate, high-resolution direct measurement of a pulse in the spectrometer as in the femtosecond optical oscilloscope.

Many modern scientific and technological problems, such as revealing the character and peculiarities of nonlinear-dispersive similariton [23], studies of the nonlinear-dispersive regime of spectral compression and shaping of transform-limited rectangular pulses [24], characterization of prism-lens dispersive delay line [25], etc, however, along with the amplitude information, demand also the phase information, possible through additional interferometric measurements, motivating the urgency of SI methods for the complete characterization of femtosecond signal. The *classic method of SI* is based on the interference of the signal and reference beams spectrally dispersed in a spectrometer, with the spectral fringe pattern caused by the difference of spectral phases [2]. The known spectral phase of the reference permits to retrieve the spectral phase of signal, and, together with the spectrum measurement, to recover the complex temporal amplitude of the signal through Fourier transformation. The setup of classic SI is rather simple, and the measurement is accurate as any interferometric one, but its application range is restricted by the bandwidth of the reference. The SI characterization of a signal that has undergone a nonlinear interaction with medium requires a special broadband reference to fully cover the broadened signal spectrum. To avoid this restriction, the self-referencing methods of spectral shearing interferometry are developed [3-5]. This improvement promotes the SI-methods to the class of the most popular and commercialized methods of accurate measurements on the femtosecond time scale, making them compatible with the FROG techniques [1], at the expense of a complicated optical arrangement. Our developed method of similariton-based SI, along with its self-referencing performance, keeps the simplicity of the principle and configuration of the classic SI [2].

Similaritons, pulses with the distinctive property of self-similar propagation, recently attract the attention of researchers, due to fundamental interest and prospective applications in ultrafast optics and photonics, particularly for high-power pulse amplification, optical telecommunications, ultrafast all-optical signal processing, etc [26,27]. The self-similar propagation of the high-power pulse with parabolic temporal, spectral, and phase profiles was predicted theoretically in the 90's [28]. In practice, the generation of such parabolic similaritons is possible in active fibers, such as rare-earth-doped fiber amplifiers [29-31] and Raman fiber amplifiers [32], as well as in the laser resonator [33]. The generation of parabolic similariton has been also proposed in a tapered fiber with decreasing normal dispersion, using either passive dispersion-decreasing fiber [34] or a hybrid configuration with Raman amplification [35]. Another type of similariton is generated in a conventional uniform and passive (without gain) fiber under the combined impacts of Kerr-nonlinearity and dispersion [23]. In contrast to the parabolic similariton with parabolic amplitude and phase profiles, this nonlinear-dispersive similariton has only parabolic phase but maintains its temporal (and spectral) shape during the propagation, as well. Our SI studies of this type of pulses [36,37] show the linearity of their chirp (parabolic phase), with a slope given only by the fiber dispersion. This property leads to the spectrotemporal similarity and self-spectrotemporal imaging of nonlinear-dispersive similariton, with accuracy given both by spectral broadening and pulse stretching. The relation of such a similariton with the flat-top "rectangular" pulse, shaped in the near field of nonlinear-dispersive self-interaction, leads to its important peculiarity: the bandwidth of nonlinear-dispersive similariton is given by the input pulse power, with slightly varying coefficients given by the input pulse shape [23, 38,39]. Both the parabolic similariton of active fiber and nonlinear-dispersive similariton of passive fiber are of interest for applications in ultrafast optics, especially for pulse compression [40,41] and shaping [42], similariton-referencing temporal lensing and spectrotemporal imaging [23], and SI [43]. The applications to ultrafast optics, however, demand the generation and study of broadband similariton. Particularly, the resolution of the femtosecond oscilloscope, based on the similariton-induced parabolic lens, is given by the bandwidth of similariton [22], and the application range for similariton-based SI [43] is as large as broadband the similariton-reference is. The pulse compression ratio is also as high as the spectral broadening factor is [40,41]. Experimentally, our SI study permitted the complete characterization of the nonlinear-dispersive similariton of up to 5 THz bandwidths [36,37]. For broadband similaritons (of up to 50-THz bandwidths), we applied the chirp measurement technique through spectral compression and frequency tuning in the sum-frequency generation process [22,44,45].

In our method of *similariton-based SI*, the part of signal is injected into a fiber to generate the nonlinear-dispersive similariton-reference. The residual part of the signal, passing an optical time delay, is coupled with the similariton in a spectrometer. The spectral fringe pattern, on the background of the signal and similariton spectra, completely covers the signal spectrum, and the whole phase information becomes available for any signal. The known spectral phase of the similariton-reference allows to retrieve the signal spectral phase, and by measuring also the signal spectrum, to reconstruct the complex temporal amplitude of the signal through Fourier transformation. Thus, the method of similariton-based SI joins the advantages of both the classic SI [2] and spectral shearing interferometry [3-5], combining the simplicity of the principle and configuration with the self-referencing performance [46, 47]. Experimentally examining the similariton-based SI, we carried out the comparative measurements with a prototype of the femtosecond oscilloscope (FO) based on the pulse spectrotemporal imaging in the similariton-induced temporal lens. The resolution of such a similariton-based FO is given by the transfer function of the similariton's spectrum [22], and FO with a similariton-reference of the bandwidth of 50 THz bandwidths provides the direct measurement of temporal pulse in a spectrometer of 7 fs temporal resolution. In the comparative experiments, we demonstrate and study the methods of similariton-based SI and spectrotemporal imaging as two applications of similariton. The reference-based methods become self-referencing by the use of similariton. The results of the measurements carried out by similariton-based SI and FO are in quantitative accordance. The similariton-based spectrotemporal imaging has the advantage of direct pulse measurement leading to the development of a femtosecond optical oscilloscope, but it does not give phase information without additional interferometric measurement. The similariton-based SI provides the complete (amplitude and phase) characterization of femtosecond signal.

Concluding the chapter introduction, our studies on the generation of nonlinear-dispersive similaritons of up to 50 THz bandwidths, and their experimental characterization by means of the methods of SI and chirp measurement through the technique of frequency tuning in spectral compression process are presented below. Our studies state that only fiber dispersion determines the phase of broadband nonlinear-dispersive similariton. Afterwards, our demonstration and study of the similariton-based SI for femtosecond signal complete characterization and the comparative experiments with FO based on the similariton-induced temporal lens are presented, carried out together with theoretical check and autocorrelation measurements, evidencing the quantitative accordance and high precision of both the similariton-referencing methods of SI and temporal lensing.

Outlining this chapter, after a brief review of modern methods of signal analysis on the femtosecond time scale, we present our studies of the nature and peculiarities of similariton generated in passive fiber carried out by classic SI, then – research of broadband similariton in view of problems of femtosecond signal analysis-synthesis, and finally - comparative studies on the development of novel method of self-referencing similariton-based SI.

2. Spectral interferometric study of nonlinear-dispersive similariton

In this section, our spectral interferometric studies for the complete characterization of the similariton generated in passive fiber due to the combined impacts of nonlinearity and dispersion are described. Studies of the generation of nonlinear-dispersive similariton of passive fiber, its distinctive properties, especially its origin, nature and relation with the spectron and rectangular pulses, and the temporal, spectral and phase features in view of potential applications are presented. The nonlinear-spectronic character of such a similariton, with the key specificity of linear chirping, leads to important applications to the signal analysis - synthesis problems in ultrafast optics.

The outline of this section is the following: first, a rough analytical discussion and numerical studies are given to reveal the features of nonlinear-dispersive similariton, afterwards, experiments for the spectral interferometric characterization of nonlinear-dispersive similariton are presented to demonstrate the peculiarities of similariton predicted by the theory, then the technique for the measurement of similariton chirp by the use of spectrometer and autocorrelator is described, and finally, the study of the bandwidth (and duration) rule of nonlinear-dispersive similariton is presented.

2.1 Similariton self-shaped (generated) in passive fiber

In a *rough analysis* of nonlinear-dispersive similariton, first we consider the pulse propagation through a pure dispersive medium. In the far field of dispersion a *spectron* pulse is shaped [23], which repeats its spectral profile, in the temporal analogy of the Fraunhofer zone diffraction, and therefore propagates self-similarly. Mathematically, the solution of the dispersion equation for temporal amplitude $A(f,t) = FT^{-1}[\tilde{A}(0,\omega)\exp(-i\beta_2 f\omega^2/2)]$ at the propagation distance $f >> L_D$ obtains the form

$$A(f,t) \propto \tilde{A}(0,\omega)\exp(i\beta_2 f\omega^2/2)\Big|_{\omega=Ct} = \tilde{A}(0,\omega)\,|_{\omega=Ct}\exp(iCt^2/2)\,. \qquad (1)$$

In Eq. (1), $\tilde{A}(\omega) \equiv FT[A(t)]$ is the complex spectral amplitude, FT - the operator of Fourier transformation, $C \equiv d\omega/dt \approx -[\phi''(\omega_0)]^{-1} = (\beta_2 f)^{-1}$ - chirp slope; $L_D \equiv (\beta_2\Delta\omega_0^2)^{-1}$ - dispersive length, $\phi''(\omega_0)$ - second derivative of the dispersion-induced spectral phase $\phi(\omega)$ at the central frequency ω_0, β_2 - coefficient of second order dispersion, and $\Delta\omega_0$ - input spectral bandwidth. The condition of temporal Fraunhofer zone, meaning enough large pulse stretching $s \equiv \Delta t/\Delta t_o \approx \Delta\omega_o^2/C >> 1$, gives the $1/s \approx C/\Delta\omega_o^2$ precision of the spectron's spectrotemporal similarity $|A(f,t)| \propto |\tilde{A}(0,\omega)|_{\omega=Ct} = |\tilde{A}(f,\omega)|_{\omega=Ct}$. For a 100 fs pulse propagating in a standard single-mode fiber L_D is of ~10 cm, and at the output of 1-m fiber we will have pulse stretching of $s \approx f/L_D$ ~ 10, and spectrotemporal similarity of spectron of the $1/s$ ~10% precision.

For a nonlinear-dispersively propagating pulse, the nonlinear self-interaction broadens the spectrum and increases the impact of dispersion, leading to a higher-precision spectron-similariton shaping. Quantitatively, for 100-fs pulse radiation in a single mode fiber with average power $p \sim 100$ mW at a 76 MHz repetition rate, the nonlinear interaction length $L_{NL} \equiv (\beta_0 n_2 I_0)^{-1}$ is much shorter than the dispersive one: $L_{NL} \sim 1$ cm << $L_D \sim 10$ cm. This allows us to roughly split the impacts of nonlinear self-interaction and dispersive deformation of the pulse, assuming that first we have pure nonlinear self-phase modulation of the pulse and spectral broadening, and afterwards pure dispersive stretching and dispersive-spectronic propagation. For the phase of output pulse, we have $\varphi_D(f,t) = Ct^2/2$, and an additional phase term φ_{NL} of $FT[A(f \sim L_{NL}, t)]|_{\omega=\gamma t}$, come from the pulse initial propagation step of nonlinear self-interaction. Assuming it parabolic at the central energy-carrying part of the pulse at the propagation distances of $\sim L_{NL}$, we have the phase $\varphi_{NL}(f,t) = C_{NL}^{-1}\omega^2/2|_{\omega=\gamma t} = (C^2/C_{NL})t^2/2$, with $C_{NL} \equiv \varphi_{NL}''$. For the overall output phase $\varphi_\Sigma = \varphi_D + \varphi_{NL}$, we have $\varphi_\Sigma(z,t) = Ct^2(1+C/C_{NL})/2$. Considering the nonlinear spectral broadening and dispersive pulse stretching factors ($b \equiv \Delta\omega/\Delta\omega_0$ and $s \equiv \Delta t/\Delta t_0$), we have for the nonlinear, dispersive, and overall chirp slopes at the output, respectively: $C_{NL} = \Delta\omega/\Delta t_0 = \Delta\omega_0^2 b$, $C = \Delta\omega_0/\Delta t = \Delta\omega_0^2/s$, and $C_\Sigma = C(1+C/C_{NL}) = C[1+(sb)^{-1}]$. Since $C/C_{NL} = (sb)^{-1}$, for spectral broadening $b \sim 10$ and pulse stretching $s \sim 10$ ($\Delta t_0 \sim 100$ fs, $p \sim 100$ mW average power at a 76 MHz repetition rate, $f \sim 1$ m of fiber), we will have $C_\Sigma = C(1+C/C_{NL}) \approx C$, with the accuracy of $C/C_{NL} \sim 1\%$.

Thus, for the femtosecond pulse nonlinear-dispersive self-interaction at $f \sim 1$ m of fiber, we have spectron of $\sim 1/sb = C/\Delta\omega^2 \sim 1\%$ precision. Considering the key peculiarity of the nonlinear-dispersive spectron-similariton, that practically the fiber dispersion determines the chirp slope, we can describe it following way:

$$A(f,t) \propto |\tilde{A}(f,\omega)|_{\omega=Ct} \exp(iCt^2/2). \qquad (2)$$

Another interesting issue is the relation of nonlinear-dispersive similariton with the rectangular pulses, shaped due to the pulse nonlinear-dispersive self-interaction at the fiber lengths $f \sim 2\sqrt{L_D L_{NL}}$. For such nonlinear-dispersive rectangular pulses, the temporal stretching and spectral broadening are up to $\Delta t \approx 2\Delta t_0$ and $\Delta\omega \approx 2\Delta\omega_0(L_D/f)$, respectively, since the pulse optimal compression ratio is $\Delta t_0/\Delta t_c \approx \sqrt{L_D/L_{NL}}/2 = L_D/f$, and $\Delta t_0/\Delta t_c \approx 2\Delta\omega/\Delta\omega_0$. Thus, in this case the chirp slope obtains the value $C = \Delta\omega/\Delta t \approx (2\Delta\omega_0 L_D/f)/(2\Delta t_0) = (\beta_2 f)^{-1}$. Therefore, during the pulse nonlinear-dispersive self-interaction in fiber, the chirp slope becomes equal to the one of pulse dispersive propagation $C \approx (\beta_2 f)^{-1}$, starting from the fiber lengths $f \sim 2\sqrt{L_D L_{NL}}$, and nonlinear-dispersive rectangular pulses can be considered as an earlier step of the shaping of nonlinear-dispersive similariton.

Summarizing our rough analysis, we can expect the spectronic nature of nonlinear-dispersive similariton of passive fiber, its spectrotemporal similarity and imaging with the accuracy $\sim 1/sb = C/\Delta\omega^2$, and with the scaling coefficient of the chirp slope $C \approx (\beta_2 f)^{-1}$, given by dispersion only.

To check the terms and conclusions of the rough analytical discussion above, a quantitative analysis of the process is carried out through *numerical modeling* based on the complete wave pattern. The mathematical description of the pulse nonlinear-dispersive self-interaction in fiber is based on the standard nonlinear Schrödinger equation (NLSE) with the terms of Kerr nonlinearity and second-order dispersion, adequate to the pulse durations of ≥ 50 fs [23]. The split-step Fourier method is applied to solve the NLSE. In simulations, the pulse propagation distance is expressed in dispersive lengths L_D ($L_D \sim 10$ cm for 100-fs input pulses); the power of radiation in fiber is given by the nonlinear parameter $R \equiv L_D / L_{NL} = (\beta_2 \Delta\omega_0^2)^{-1} \beta_0 n_2 I_0 \sim I_0$ ($p = 100$ mW average power of a 100-fs pulse radiation at a 76 MHz repetition rate in a standard single-mode fiber corresponds to $R = 6$; $p \leq$ 100uW is adequate to the pulse pure dispersive propagation of $R = 0$). The dimensionless running time t and centralized frequency Ω are normalized to the input pulse duration Δt_0 and bandwidth $\Delta\omega = 1 / \Delta t_0$, respectively.

Fig. 1 and 2 show the dynamics of similariton shaping: pulse (top row), chirp (middle row) and spectrum (bottom row) are shown during the pulse propagation in fiber. Fig. 1 illustrates the first step of nonlinear-dispersive self-interaction when, typically, rectangular pulses are shaped, and Fig. 2 shows the step of similariton shaping. The spectral broadening and decreasing of the pulse peak power lead to the "activation" of dispersion; the pulse obtains a linear chirp (parabolic phase), and the self-spectrotemporal similarity of nonlinear-dispersive similariton takes place (Fig. 2).

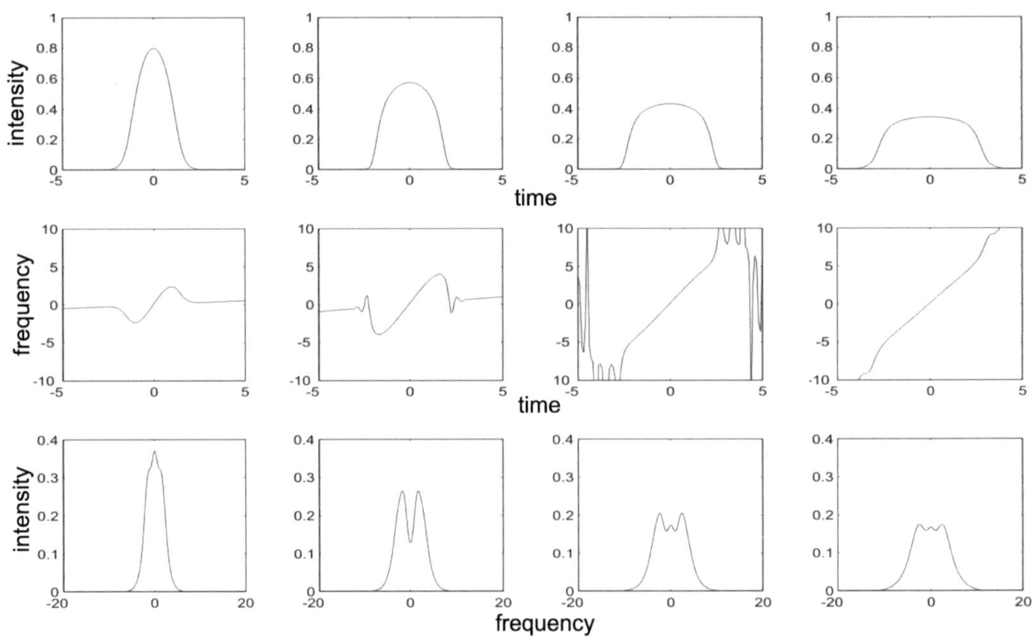

Fig. 1. Shaping of rectangular pulses ($R = 30$). From left to right: pulse evolution in fiber for $f / L_D = 0.1$; 0.2; 0.3; and 0.4. From top to bottom: pulse, chirp, and spectrum.

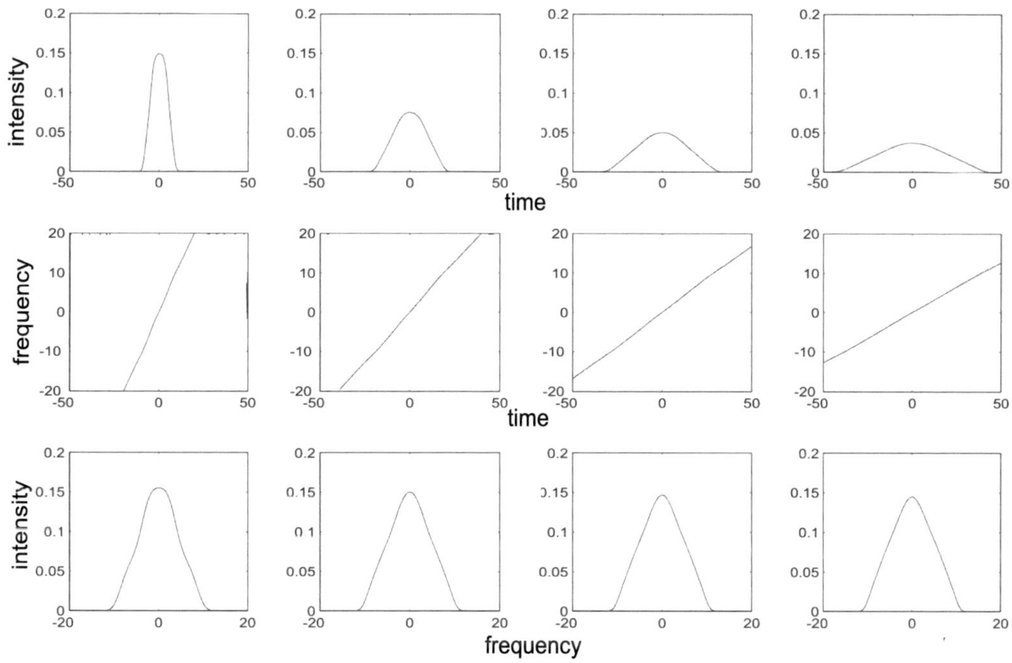

Fig. 2. Shaping and evolution of nonlinear-dispersive similariton in the fiber ($R = 30$). From left to right: $f / L_D = 1; 2; 3; 4$. From top to bottom: pulse, chirp, and spectrum.

Our simulations show that even in case of pulses with complex initial forms the output pulse has nearly parabolic form at its central energy-carrying part. The irregularities of the temporal and spectral profiles are forced out to the edges during the nonlinear-dispersive self-interaction, and pulse and spectrum become more and more parabolic. However, the pulse dispersive stretching decreases its peak intensity, finally minimizing the impact of nonlinear self-phase modulation. The spectrum does not change any more and the further alteration of the pulse shape has a dispersive character only.

It is important that our simulations for the nonlinear-dispersive similariton at a given fiber length and different power values confirm also the prediction of the analytical discussion: the output chirp slope is practically independent of the input pulse intensity, and depends only on the fiber length. The increase of input pulse power leads to the spectral broadening and temporal stretching of pulse, keeping the chirp coefficient unchanged: the chirp slope is the same in all cases, even in case of pulse pure dispersive propagation. It allows extracting the full information on the nonlinear-dispersive similariton having the spectrum and fiber length. This statement is checked for sufficiently powerful pulses, when the character of pulse self-interaction is nonlinear-dispersive (but not pure dispersive), and the result is the same: the chirp slope is independent of the power.

We studied also the chirp of nonlinear-dispersive similariton versus the chirp of input pulse: the chirp slope of similariton is practically constant, when the pulse intensity is high enough. In case of the dispersive propagation, the induced chirp simply imposes on the initial chirp according to Eq. (1). In case of nonlinear-dispersive propagation, the chirp “forgets” about the initial chirp according to analytic discussion above: it becomes independent of the input chirp according to Eq. (2).

2.2 Spectral interferometric characterization of nonlinear-dispersive similariton

We carried out experimental studies to check-confirm the predictions of our rough discussion and numerical analysis above. We applied the classic method of SI [2] to completely characterize and study the generation process and peculiarities of nonlinear-dispersive similariton of a passive fiber.

Fig. 3 schematically illustrates our experiment. Using a Mach-Zender interferometer, we split the input radiation of a standard Coherent Verdi V10-Mira 900F femtosecond laser system into two parts.

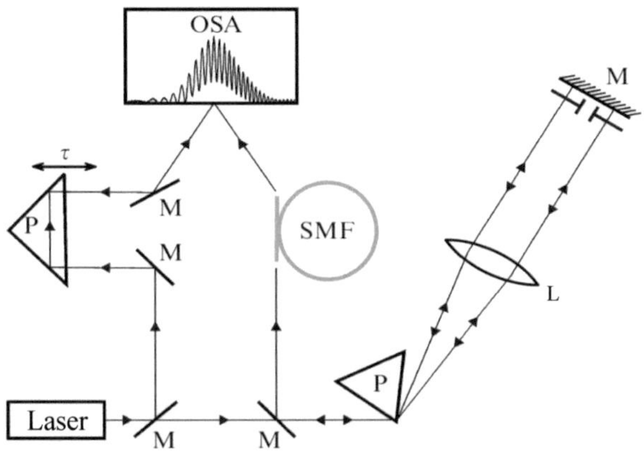

Fig. 3. Schematic of the experiment. Laser - Coherent Verdi V10-Mira 900F femtosecond laser system, M - mirrors, P - prisms, L - lens, OSA - optical spectral analyzer, and SMF - single-mode optical fiber.

The low-power pulse serves as a reference. For the high-power pulse, first we filter its spectrum of the bandwidth $\Delta\lambda = 11$ nm down to the value $\Delta\lambda = 2$ nm. We use standard polarization-preserving fibers Newport F-SPF @820 nm and ThorLabs HP @ 780 nm of different lengths - 1 m, 9 m, and 36 m. The spectra of the pulses at the output of fiber are broadened, however, the spectrum of the reference pulse covers them completely. This allows measuring the spectral phase of similariton within the whole range of its spectrum. The spectral interferometric fringe pattern is recorded by an optical spectrum analyzer (OSA Ando 6315), and the spectral phase is retrieved. Having the spectrum and retrieved spectral phase, the temporal profile of the similariton is reconstructed by Fourier transformation.

The performance of the experiment is given schematically in Fig. 4 by means of the spectrograms of relevant steps. Fig. 4(a) shows the spectrum of the laser pulse, (b) is the spectrum of spectrally filtered and shaped pulse, (c) is the spectrum of nonlinear-dispersive similariton and (d) is the SI fringe pattern. Fig. 4(e) shows the measured spectral phases of the similaritons generated from different input pulses. The spectral phases are parabolic ($\phi = -\alpha\omega^2 / 2$) and their coefficients α have nearly the same values in all cases of dispersive and nonlinear-dispersive propagations: $\alpha = 0.32$ ps^2 for the pure dispersive propagation of single-peak pulse, and 0.33 ps^2, 0.328 ps^2, 0.35 ps^2 for the nonlinear-dispersive propagations of single-, double- and distant double-peak pulses, respectively. The parabolic phase (linear chirp) leads to the self-spectrotemporal imaging of similariton. The accuracy of imaging increases with the decreasing of the chirp slope, which is approximately equal to $C \approx \alpha^{-1}$. Fig. 4 and 5 illustrate the typical behavior of nonlinear-dispersive similariton in case of f = 9 m.

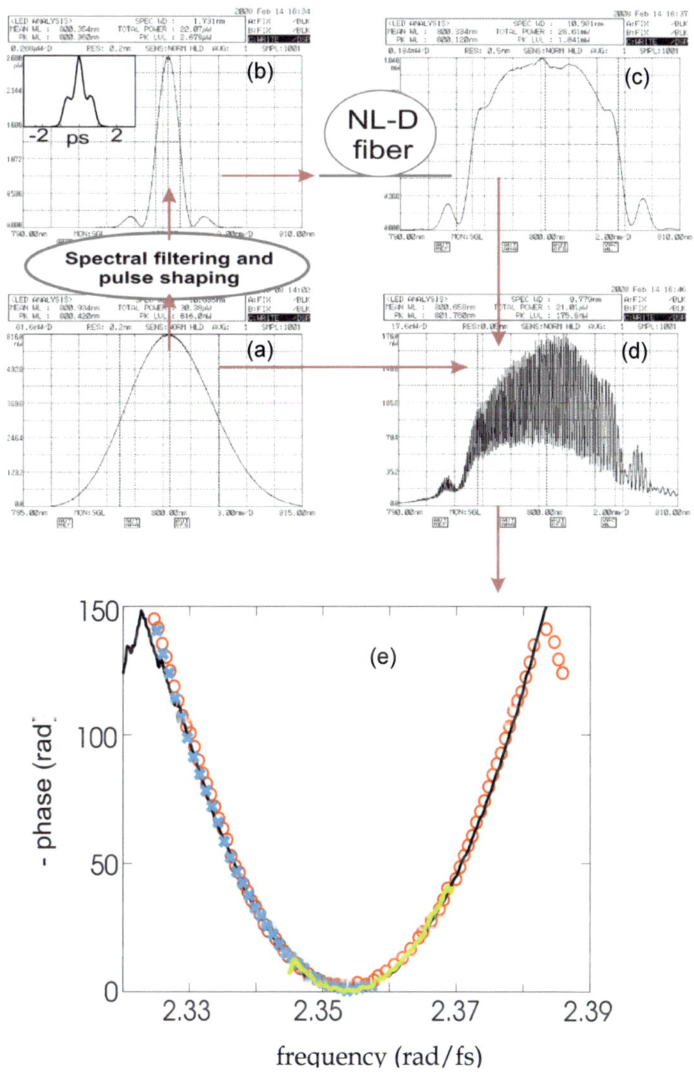

Fig. 4. (a-d) Schematic of the experiment given by the spectrograms of the relevant steps; (e) spectral phases of nonlinear-dispersive similaritons generated from input single- (thin black line), double- (red ∘) and distant double-peak (blue ×) pulses in comparison with the one for pure dispersively propagated single-peak pulse (thick yellow line).

Having the spectral phase and spectral profile, the temporal profile of nonlinear-dispersive similariton is retrieved. Fig. 5 shows the spectral and temporal profiles of the similaritons with the spectral phases of Fig. 4(e). The black curves are the spectra and the gray-dotted curves are the pulses. They coincide with each other, that is, takes place the self-spectrotemporal imaging of nonlinear-dispersive similariton. A good spectrotemporal similarity is seen in case of input single-peak pulse (a). For the input double-peak pulse of (b), the matching between the spectral and temporal profiles of similaritons is qualitative only. To obtain a quantitative agreement, one must use a longer fiber, increasing the α coefficient: the spectral and temporal profiles of similaritons practically coincide for the thick orange line of (b), showing the temporal profile of similariton for the increased α coefficient.

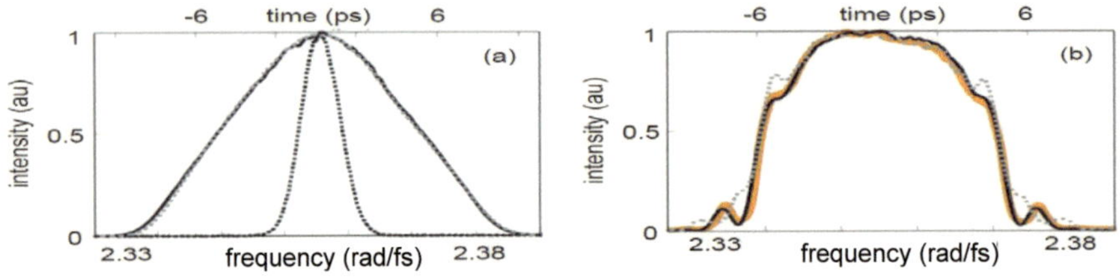

Fig. 5. Self-imaging of nonlinear-dispersive similaritons generated from input (a) single- , and (b) double-peak pulses. Black solid curves show the spectra of nonlinear-dispersively propagated pulses (the black dotted one of (a) stands for pure dispersively propagated pulse), gray dotted lines show the retrieved temporal profiles. The thick orange curve of (b) shows the temporal profile of similariton for 4α increased coefficient of spectral phase.

To show the relation between the nonlinear-dispersive similariton and the rectangular pulse of fiber, spectral interferometric measurements are carried out using a short fiber (f = 1 m). Fig. 6 illustrates the shaping of a rectangular pulse in a nonlinear-dispersive fiber.

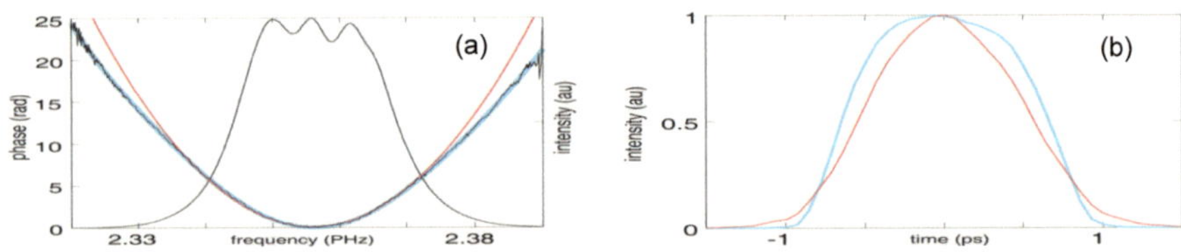

Fig. 6. (a) Spectrum of nonlinear-dispersive rectangular pulse with the relevant spectral phase (black solid line) and fitted parabola (pink). The blue line is a high-order polynomial fit. (b) Temporal profile of the rectangular pulse (blue) in comparison with the pulse retrieved by the fitted parabolic spectral phase (pink).

Here the black curves are the spectrum and spectral phase measured, the blue curve is a high-order polynomial fit and the pink is a fitted parabola. The measured spectral phase has a parabolic shape only at the central energy-carrying part of spectrum. Deviation from the parabola at the wings leads to the shaping of a rectangular pulse shown in Fig. 6(b) with the blue curve. Even in this case the chirp slope at the central energy-carrying part of the pulse/spectrum is also determined only by the fiber length (α = 0.0465 ps^2). The pink curve of Fig. 6(b) is the retrieved pulse by the fitted parabolic spectral phase.

The complete and precise SI study confirmed the principal description of nonlinear-dispersive similariton by Eq. (2), leading to its self-spectrotemporal similarity and imaging by the scaling coefficient of the chirp slope $C = (\beta_2 z)^{-1}$. This allows carrying out the similariton chirp studies by a simpler way of the spectrum and autocorrelation track measurements, and afterwards calculation of the spectrum autocorrelation. The comparison of the measured and calculated autocorrelations extracts the chirp slope. A good accordance between the chirp slope values measured by this and SI methods occurs. This simple method permits checking easily the results on similariton chirp numerical study. The experimental results, in agreement with the theory, show that the chirp slope of nonlinear-dispersive similariton is practically independent of the input pulse phase modulation in a wide range of $\alpha_0\Delta\omega_0^2$ from -3 to +3.

Taking into account the relation between nonlinear-dispersive similariton and rectangular pulse discussed, it seems reasonable to expect that the *bandwidth of similariton* is equal to the one for rectangular pulse. To determine the bandwidth (and afterwards the duration) of nonlinear-dispersive similariton, the relation for the pulse optimal compression can be used [23]. This gives the following relation for the spectral broadening of nonlinear-dispersive similariton: $b \equiv \Delta\omega / \Delta\omega_0 \approx \sqrt{R} \equiv \sqrt{L_D / L_{NL}} = k\sqrt{W / \Delta t_{in}} / \Delta\omega_0 = k\sqrt{P} / \Delta\omega_0$, where P is the input pulse power, Δt_{in} - input pulse duration, $W = p / v$ - pulse energy, p - average power of pulse radiation with a repetition rate v, and $k \equiv \sqrt{n_2\beta_0(\beta_2 S)^{-1}}$ is a coefficient given by the fiber parameters (n_2 - coefficient of the Kerr nonlinearity, $\beta_0 = 2\pi / \lambda_0$ - wave number, β_2 - group-velocity dispersion coefficient, S - fiber mode area). This thesis (Fig. 7) is checked numerically (a) and experimentally (b). The confirmation of the truthfulness of the brief discussion above gives the following rule for the $\Delta\omega$ bandwidth and Δt duration of nonlinear-dispersive similariton:

$$\Delta\omega = k\sqrt{P}, \qquad \Delta t = \Delta\omega / C = k\beta_2 f\sqrt{P}. \tag{3}$$

For comparison, the spectral bandwidth of the similariton generated in a fiber amplifier is $\Delta\omega(z) = [(g\beta_0 n_2 PW) / (2\beta_2^2 S)]^{1/3} \exp(gz/3)$, where W and P are the input pulse energy and power, and g is the gain coefficient [31].

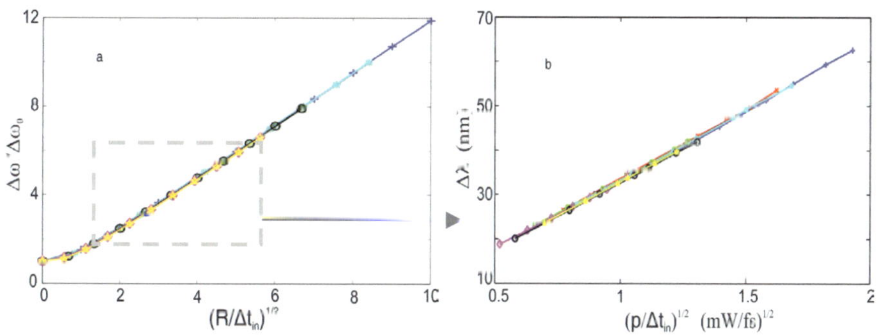

Fig. 7. (a) Simulation: spectral broadening versus R; (b) Experiment: similariton bandwidth $\Delta\lambda$ versus $(p / \Delta t_{in})^{1/2}$. Blue lines correspond to the transform-limited pulse of 100 fs duration, red and cyan - to 140 fs pulse, green and black - to 225 fs pulse, magenta and yellow - to 320 fs pulse. Red, green and magenta are related to pulses stretched in a medium with normal dispersion, and cyan, black, and yellow - to pulses stretched in a medium with anomalous dispersion.

This revealed property of nonlinear-dispersive similariton of Eq. (3) can be used for the measurement of femtosecond pulse duration, alternatively to the autocorrelation technique.

Concluding, our spectral interferometric studies demonstrate the following properties of nonlinear-dispersive similariton generated in a passive fiber:

- linear chirp, with a slope given only by the fiber dispersion and independent of the amplitude, chirp and power of the input pulse;
- relation with the rectangular pulse of nonlinear-dispersive fiber;
- property of spectrotemporal similarity / self-spectrotemporal imaging, with accuracy determined by spectral broadening and pulse stretching together;
- only initial pulse power determines the spectral bandwidth of similariton.

3. Broadband similariton for femtosecond signal analysis and synthesis

The applications to ultrafast optics demand the generation and study of *broadband similariton* [43,44]. Particularly, the pulse compression ratio is as high as the spectral broadening factor is [41], the resolution of the femtosecond oscilloscope, based on the similariton-induced parabolic lens is given by the bandwidth of similariton [22], and for the similariton-based SI, the application range is as large as broadband the similariton-reference is [43,46].

In this section, our studies on the generation and characterization of a broadband, 50-THz bandwidth nonlinear-dispersive similariton are presented with the objective to reveal its distinctive properties in view of its applications to the signal synthesis and analysis problems on the femtosecond time scale. According to the spectral-interferometric characterization of nonlinear-dispersive similariton of the bandwidths of 5 THz ($\Delta\lambda$ ~10 nm at $\lambda \sim 800$ nm), it is described by Eq. (2), or for its slowly varying amplitude $A(f,t)$ and phase $\varphi_f(f,t)$ we have:

$$A(f,t) \propto \tilde{A}^*(f,\omega)\big|_{\omega=Ct},\ \varphi_f(t) = -\phi_f(\omega)\big|_{\omega=Ct} = \beta_2 f\omega^2/2\big|_{\omega=Ct} = Ct^2/2. \tag{2'}$$

For comparison, the spectron pulse has the same dispersion-induced phase [Eq. (1)]. The applications of similariton demand to check and generalize this key peculiarity for broadband pulses.

First, the *numerical modeling* is carried out to have the complete physical pattern and reveal the distinctive peculiarities of the generation and propagation of broadband similariton. The mathematical description, based on the generalized nonlinear Schrödinger equation, considers the high-order terms of third-order dispersion (TOD), shock wave (self-steepening) and delayed nonlinear response (related to the Raman gain), together with the principal terms of self-phase modulation and second order dispersion (SOD) [44,47]. The split-step Fourier method is used in the procedures of numerical solution of the equation in simulations for 100-fs pulses of a standard laser with up to 500 mW of average power at a 76 MHz repetition rate (70 kW of peak power) in a few meters of standard single-mode fiber (losses are negligible). Simulations in these conditions show that the high-order nonlinear factors of shock wave and delayed nonlinear response do not impact on the process under study; however, the impact of TOD is expressed in the generated broadband similaritons of

50 THz bandwidth. It is conditioned by the physical pattern of the process: the nonlinear self-interaction of powerful pulses leads to large spectral broadening (with the factor of b ~10 just at the first ~1 cm of propagation in fiber, in our case), substantially increasing the impact of dispersion (with the factors of b^2 ~100 for SOD, and b^3 ~1000 for TOD), resulting in the pulse stretching and peak intensity decreasing (with the factor of b^2 ~100), and thus essentially decreasing and blocking out the high-order nonlinear effects. The impact of high-order nonlinear effects can be significant in case of high-power pulse propagation in low-dispersion fibers, e.g., for photonic crystal fibers with all-normal flattened dispersion, where a longer and more efficient nonlinear self-interaction results in the generation of octave spanning supercontinuum [48].

The numerical analysis shows that the fiber TOD is expressed additively in the $\phi(f,\omega)$ and $\varphi(f,t)$ parabolic spectral and temporal phase profiles due to its small impact as compared to the SOD, and its value is the same for the nonlinear-dispersive similariton and the spectron pulse of pure dispersive propagation:

$$\Delta\varphi(f,t) \approx \Delta\phi(f,\omega)|_{\omega=Ct} = -\frac{\beta_3 f}{6}\omega^3\bigg|_{\omega=Ct} \approx -\frac{\beta_3}{6\beta_2^3 f^2}t^3, \qquad (4)$$

where β_3 is the TOD coefficient. The precision of Eq. (4) is of ~4%, according to the simulations for the SOD and TOD coefficients of fused silica (β_2 = 36.11 fs²/mm and β_3 = 27.44 fs³/mm). Thus, the chirp measurement of broadband similariton becomes urgent, since it gives the TOD of fiber and permits to generalize the description of Eq. (2′). The SI study permitted the complete characterization of nonlinear-dispersive similariton of up to 5 THz bandwidths [23]. For broadband similaritons (of up to 50-THz bandwidths), the chirp measurement technique through spectral compression and frequency tuning in the sum-frequency generation process is applied [44,45].

In the *experiment*, a broadband nonlinear-dispersive similariton of 50-THz FWHM-bandwidth is generated in a piece of passive fiber and the chirp measurement is carried out through frequency tuning in the SFG-spectral compression process [44-47]. Fig. 8 shows the schematic of the experimental setup. The Coherent Verdi V10 + Mira 900F femtosecond laser system is used, with the following parameters of radiation: 100 fs pulse duration, 76 MHz repetition rate, 1.6 W average power, 800 nm central wavelength. Beam-splitter (BS) splits the laser radiation into high- and low-power parts (80%+20%). The high-power pulse (100 fs FWHM-duration and pulse energy of up to 7 nJ, corresponding to 70 kW peak power) is injected into a standard single-mode fiber (1.65 m Newport F-SPF PP@820 nm) by means of a 10× microscope objective and a broadband nonlinear dispersive similariton is generated.

Fig. 9(a) shows the spectrum of 107-nm (50 THz) FWHM-bandwidth similariton, recorded by the optical spectrum analyzer Ando 6315 (OSA). This spectral profile represents the spectrotemporal image of the generated similariton in the approximation of its linear chirp. The asymmetry in this spectrotemporal profile evidences the impact of TOD. Although the initial pulse asymmetry can also cause the spectral asymmetry in the near field of dispersion, typically for the picosecond-scale experiments, the impact of the possible asymmetry of the initial laser pulse on the spectrotemporal shape of similariton becomes insignificant, according to our simulations.

For the characterization of broadband similariton, its intensity profile can be measured by means of the cross-correlation technique, using the SFG-interaction of similariton with the laser pulse (as a reference). A spectral detection of the SFG-signal will give also information on the chirp of similariton, modifying the cross-correlation technique to the cross-correlation frequency-resolved optical gating (XFROG) [49,50]. Our additional modification is the use of a dispersively chirped reference pulse, which provides a spectrally compressed SFG-signal in a wider spectral range, and thus, more efficient measurement [22,44]. Experimentally, the low-power pulse is directed into the dispersive delay line with anomalous dispersion (D-line; conventional prism compressor consisting of a 3.5-m separated SF11 prism pair with a reverse mirror) and stretch it 22 times, resulting in the pulse autocorrelation duration of 3.1 ps. Then, using a lens, we direct the similariton and the dispersively stretched pulse to the nonlinear β-barium borate crystal (BBO, type 1 - ooe, 800 nm operating wavelength) to have SFG-spectral compression. The SFG-interaction of up- and down-chirped pulses results in the chirp cancellation and spectral compression, and a temporal delay between these two pulses leads to the frequency shift of the SFG-signal, according to the concept of the temporal lens [22]. In the experiment, the temporal delay is provided by shifting the reverse mirror of the D-line and recording the relevant SFG-compressed spectra by OSA. Measurements with D-line and without it are carried out, replacing the D-line with a simple temporal delay (TD), to compare the techniques of SFG-spectral compression and XFROG. Fig. 9(b) shows the relevant 3D frequency tuning patterns for the chirp measurement.

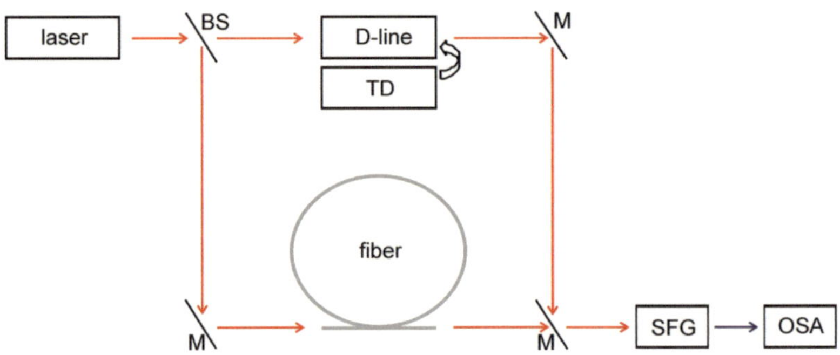

Fig. 8. Schematic of experimental setup: laser - Coherent Verdi V10 + Mira 900F femtosecond laser system, BS - beam splitter, M - mirrors, D-line - dispersive delay line (conventional prism compressor consisting of a SF11 prism pair with a reverse mirror), TD - temporal delay, fiber - Newport F-SPF PP@820 nm, SFG - BBO crystal for SFG, OSA - optical spectrum analyzer.

The technique of SFG-spectral compression, as compared to XFROG, is more efficient, providing sharper spectral signal in a wider spectral (and temporal) range. The temporal delay between the SFG-interacting pulses in the range of ± 16 ps results in a ± 20 nm wavelength shift for the 22 times SFG-spectrally compressed signal (down to 0.12 nm at 400 nm central wavelength), corresponding to the chirp measurement of similariton in the span range of 160 nm (75 THz) at 800 nm central wavelength. This 3D pattern completely characterizes the generated broadband similariton. Its projections represent the temporal and spectral profiles of the intensity $I(t)$ and $I(\lambda)$, and the curve $\lambda(t)$ connected with the chirp $\omega(t) = \dot{\varphi}_f(t)$, as well as with the derivative of spectral phase $\dot{\phi}_f(\omega)$, according to the spectrotemporal similarity of broadband similariton described by Eqs. (2′) and (4). In general, through the Fourier transformation of complex temporal amplitude, given by the measured temporal pulse and chirp, the spectral complex amplitude and afterwards the spectral phase could be retrieved. For our case of the broadband similariton, the Eq. (2′) permits to have the spectral phase information by a simple scaling $\omega = Ct$, and we have $\dot{\phi}_f(\omega) \approx -\dot{\varphi}_f(t) / C$ for the derivative of spectral phase.

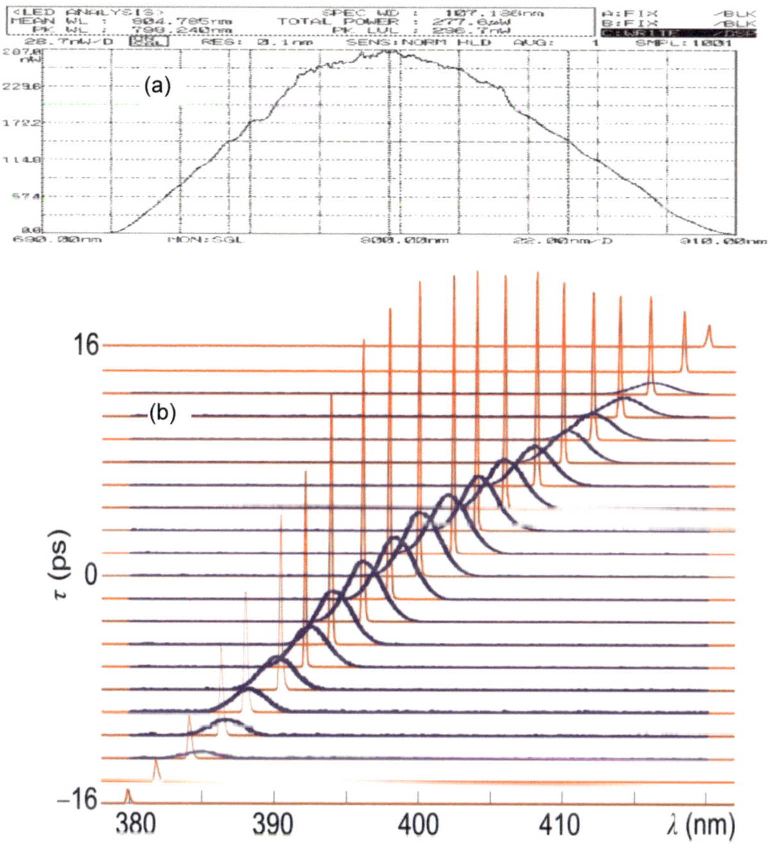

Fig. 9. (a) Measured spectrum of broadband similariton, and (b) 3D frequency tuning patterns with spectral compression (sharp peaks) and without it (thick lines) for the chirp measurement: the 40 nm (75 THz) frequency tuning at 400 nm is adequate to the 160 nm spectral range of similariton at 800 nm for the 32 ps range of temporal delay between SFG-interacting pulses.

Fig. 10 shows the $\dot{\phi}_f(\omega)$ curve obtained this way. It is described by the polynomial $\dot{\phi}_f(\omega) = -15.06 \times 10^3 \text{ fs}^3 \times \omega^2 - 81.13 \times 10^3 \text{ fs}^2 \times \omega$. The quadratic component of $\dot{\phi}_f(\omega)$ is shown separately (inset). The circles in Fig. 10 are the measured experimental points; the dashed and solid curves are for the linear and parabolic fits, respectively.

Concluding, we generated nonlinear-dispersive similariton of 50 THz bandwidth, and carried out its complete characterization through the chirp measurement, using the technique of frequency tuning in the SFG-spectral compression. Our studies state that only fiber dispersion determines the phase (chirp) of broadband nonlinear-dispersive similariton. The fiber TOD results in the same additional phase for broadband nonlinear-dispersive similariton and spectron. The ~1% accuracy of the linear fit for the chirp of the 50-THz bandwidth similariton gives the range of applications for aberration-free similariton-based spectrotemporal imaging [22] and spectral interferometry [45-47]. The described approach to the generation and characterization of broadband similariton can be helpful also for its applications in pulse compression [40,41] and CARS microscopy [51]. For these applications, the low value of TOD can impact significantly and it should be considered more carefully.

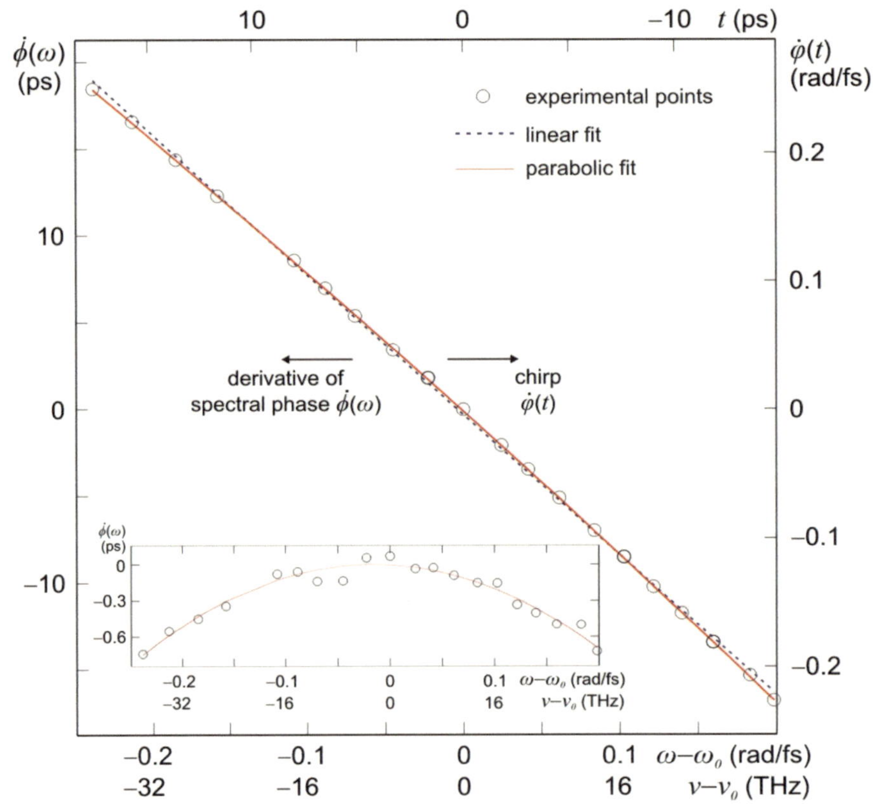

Fig. 10. Broadband similariton's chirp $\dot{\varphi}_f(t)$ and derivative of spectral phase $\dot{\phi}_f(\omega)$ with its quadratic component, separately (inset). The ~1% difference from the linear fit in the range of ~50 THz gives the range of applications for the similariton-based methods of signal characterization (Section 4).

4. Similariton based self-referencing spectral interferometry for femtosecond pulse characterization

Our studies on broadband similariton serve as a basis for development of a novel method of similariton-based of SI for the femtosecond pulse complete characterization. Below we present its demonstration in comparison with the method of pulse spectrotemporal imaging in the similariton-induced temporal lens. Application of similariton to the reference-based methods upgrades them up to the self-referencing ones, substantially improving their performance due to the enlarged application range along with the simplicity of the principle and configuration.

Our experimental study of nonlinear-dispersive similariton, described in sec. 2.2, as a remarkable example for the demonstration-application of SI-approach, clearly shows that the classic SI provides accurate measurement with a rather simple setup, but its application range is restricted by the bandwidth of the reference. As mentioned above, the SI characterization of a signal that has undergone a nonlinear interaction with medium requires a special broadband reference to fully cover the broadened signal spectrum. To avoid this restriction, the self-referencing methods of spectral shearing interferometry, such as SPIDER [3] and SPIRIT [4], are developed. This improvement promotes the SI to the class of the most popular and commercialized methods of accurate measurements on the femtosecond time scale, making it compatible with the FROG [1] and MIIPS [6] techniques, at the expense of a more complicated optical arrangement. Our proposed method of similariton-based SI, along with its self-referencing performance, keeps the simplicity of the principle and configuration of the classic SI.

Describing the principle of *similariton-based SI*, for its implementation the setup of Fig. 8 is used, removing the nonlinear BBO crystal and D-line. Splitting the signal beam, its part is injected into a fiber to generate the nonlinear-dispersive similariton-reference, with the complex spectral amplitude $\tilde{A}(f,\omega) = |\tilde{A}(f,\omega)|\exp[i\phi(f,\omega)]$. The residual part of the signal of a complex spectral amplitude $\tilde{A}(0,\omega) = |\tilde{A}(0,\omega)|\exp[i\phi(0,\omega)]$, is coupled with the similariton in a spectrometer with an appropriate time delay. The spectral fringe pattern $S_{SI}(\omega) = 2|\tilde{A}(0,\omega)||\tilde{A}(f,\omega)|\cos[\phi(0,\omega)-\phi(f,\omega)]$, on the background of the signal and similariton spectra, completely covers the signal spectrum $S(0,\omega) = |\tilde{A}(0,\omega)|^2$, and the whole phase information becomes available, for any signal. The known spectral phase of the similariton-reference allows to retrieve the signal spectral phase $\phi(0,\omega)$, and by measuring also the signal spectrum, to reconstruct the complex temporal amplitude $A(0,t)$ of the signal through Fourier transformation. Thus, the method of similariton-based SI joins the advantages of both the classic SI [2] and spectral shearing interferometry [3-5], combining the simplicity of the principle and configuration with the self-referencing performance. Examining the similariton-based SI, we compare its measurements with the ones carried out by a prototype of the femtosecond oscilloscope (FO) based on the pulse spectrotemporal imaging in the similariton-induced temporal lens in the SFG process. Our comparative study, involving also theoretical and autocorrelation check, along with the demonstration and study of the similariton-based SI, serves also for the inspection of the prototype of similariton-based FO, the measurements of which previously were compared with the autocorrelation only [22].

The method of SFG-*spectrotemporal imaging* for direct femtosecond scale measurements is based on the conversion of temporal information to the spectral domain in a similariton-induced parabolic temporal lens [22,46,47]. The setup of the similariton-based SI is modified to FO by replacing the temporal delay (TD) with a dispersive delay line (D-line) and placing a nonlinear crystal for SFG at the system output (returning to the initial configuration of Fig. 8). In the spectral domain, the dispersive delay works as a parabolic phase modulator, and the signal $\tilde{A}(0,\omega)$ passed through is described as $\tilde{A}(d,\omega)=\tilde{A}(0,\omega)\exp(i\ddot{\phi}_d\omega^2/2)$, with the given coefficient $\ddot{\phi}_d \approx -C_d^{-1}$. In the fiber arm, we have a nonlinear-dispersive similariton with the known parameters as in the case of similariton-based SI. In both arms of the setup, we have practically linearly chirped pulses, and the temporal and spectral complex amplitudes repeat each other in the temporal Fraunhofer zone, i.e. spectron pulses are formed [47]: $A(d,t)\propto \tilde{A}(d,\omega)$, and $A(f,t)\propto \tilde{A}(f,\omega)$ with $\omega = C_{d,f}t$. Under the conditions of the opposite and same value chirps $C_f=-C_d\equiv C$, and constant similariton spectrum throughout the signal spectrum, the output temporal SFG-signal repeats the input spectral amplitude: $A_{SFG}(t)\propto A(d,t)\times A(f,t)\propto \tilde{A}(0,\omega)$. Accordingly, the output spectral and input temporal amplitudes repeat each other $\tilde{A}_{SFG}(\omega)\propto A(0,t)$, and the output SFG-spectrum displays directly the input temporal pulse: $S_{SFG}(\omega)=|\tilde{A}_{SFG}(\omega)|^2\propto |A(0,t)|^2=I(0,t)$, with the scale $\omega = Ct$. The resolution of such a similariton-based FO is given by the transfer function of the similariton's spectrum [22,47], and FO with a similariton-reference of the bandwidth of a few tens of nanometers provides the direct measurement of temporal pulse in a spectrometer, exceeding the resolution of the achievement of silicon-chip-based ultrafast optical oscilloscope [21] by an order of magnitude.

In the *experiment*, different amplitude- and phase-modulated pulses at the setup input are shaped and the signal radiation is split by a beam-splitter (80% + 20%). The low-power part is directed to the TD or D-line (SF 11 prism pair with the reverse mirror) for similariton-based SI and spectrotemporal imaging, respectively. In the second path, the high-power pulse (with average power of up to 500 mW) is injected into a standard single-mode fiber (1.65 m Newport F-SPF PP@820 nm) by a microscope objective (10$\times$) to generate broadband nonlinear-dispersive similaritons. For the SI-measurements, these two pulses are coupled directly into the OSA and the SI fringe pattern and signal spectrum are registered. To retrieve the spectral phase, the Fourier-transform algorithm of the fringe-pattern analysis are used [44-47]. For the FO-measurements, a BBO crystal at the input of OSA is placed, and the SFG-spectrotemporal image is registered directly. The similariton-based SI and FO measurements are carried out together with the autocorrelation check by a standard APE PulseCheck autocorrelator.

First, the similariton-based SI for the laser pulses stretched and chirped in SF11 glasses of different thickness is tested, comparing the results with the autocorrelation measurements (Fig. 11). For the dispersion-induced parabolic spectral phases of the stretched pulses, the coefficients of the dispersion-induced parabolic spectral phases [Fig. 11(a)] are the following: $\alpha \equiv \phi''(\omega)=1.94\times10^{-3}$, 4.94×10^{-3}, 6.34×10^{-3}, and 10.78×10^{-3} ps^2 for the 0, 2, 3, and 5-cm glasses, respectively. The SI-reconstructed pulses, correspondingly, have durations of 108, 197, 252, and 365 fs, in a good accordance with the autocorrelation durations of 156, 298, 369 and 539 fs of the measurements shown in Fig. 11(b).

Afterwards, measurements for multi-peak pulses are carried out together with the autocorrelation check. The SI calibrating measurement of the α coefficient for similariton, using the known laser pulse as a reference, gives the value $\alpha = 2.1\times10^4$ fs^2, in accordance with the

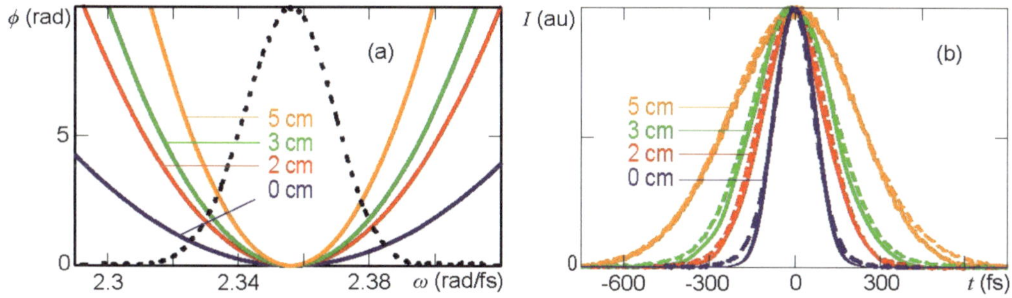

Fig. 11. Similariton-based SI for pulses dispersively stretched and chirped in SF11 glasses of different thickness: (a) retrieved spectral phases with the measured spectrum (dotted line), and (b) autocorrelation functions of SI-reconstructed temporal pulses (solid) in comparison with autocorrelation traces measured (dashed).

expression $\alpha = \beta_2 f$ with the values f = 49 cm and β_2 = 40 fs^2/mm (D_λ = –103 ps/nm/km at the wavelength 850 nm, according to the data provided by the fiber manufacturer). Then different multi-peak signal pulses are shaped inserting thin glass plates in parts of the beam. The beam parts passed through the plates obtain time delay with respect to the free-propagated part. The movement of the plates along the vertical axis adjusts the power proportion among the peaks. The thicknesses of the plates give the time delay between the peaks; e.g. a 0.12 mm thick glass plate gives a 200 fs delay, if assuming the refractive index of the plate equal to 1.5. Using double- and triple- peak signal pulses, we carry out SI-measurements and compare the results with measured autocorrelation tracks.

The application range and limitations of similariton-based SI are conditioned by the parameters of input radiation and fiber, which are necessary to generate the similariton-reference with parabolic phase. Fig. 12 and 13 illustrate the experiment for the typical regime preceding the similariton shaping: results for double- and triple-peak signal pulses are shown. As Fig. 12 illustrates, having the spectrum (b, black) and SI-retrieved spectral phase (b, blue solid line) of pulse, its temporal profile (c, blue solid line) is reconstructed through Fourier transformation. To check the precision of our measurements through similariton-based SI, we calculate the autocorrelation of reconstructed pulse (d, blue solid line) and compare it with the intensity autocorrelation measured at the input of the system (d, black). The spectral shape of nonlinear-dispersive similariton (a) ensures the fulfillment of necessary conditions for the parabolicity of the spectral phase of similariton (according to [23]). The structure of the similariton spectrum (a), strange at first glance, is typical for short lengths of nonlinear-dispersive interaction, and is observed also for parabolic similaritons generated in fiber amplifiers [52]. The blue solid and red dashed curves in Fig. 12 correspond to the pulse reconstruction with the spectral phase coefficients $\alpha = 2.1 \times 10^4$ fs^2 and $\alpha = 1.995 \times 10^4$ fs^2 (5% difference), respectively. Fig. 13 shows the analogue procedures for a pulse with more complex sub-structure.

Finally, we compare the measurements of the similariton-based SI and FO, together with a theoretical check. The double-peak signal pulses are shaped with the spectral domain amplitude- and phase-modulation given by the peaks' temporal distance T and their proportion μ. The temporal amplitude $A(t) = A_0(t) + \mu A_0(t+T)$ corresponds to the complex

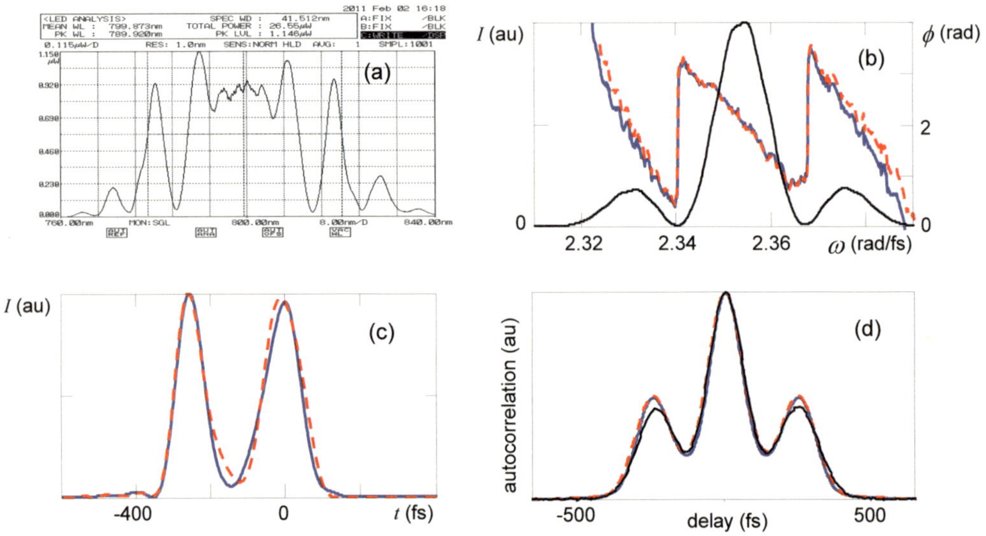

Fig. 12. Reconstruction of double-peak pulse (shaped by means of a 130- μ m thick glass) through similariton-based SI in comparison with autocorrelation measurement: (a) spectrum of nonlinear-dispersive similariton; (b) retrieved spectral phase and measured spectrum; (c) reconstructed pulse temporal profile; and (d) autocorrelation tracks. Blue solid and red dashed curves are for $\alpha = 2.1 \times 10^4$ fs^2 and $\alpha = 1.995 \times 10^4$ fs^2 (5% difference), respectively, and the black one in (d) is the measured autocorrelation track.

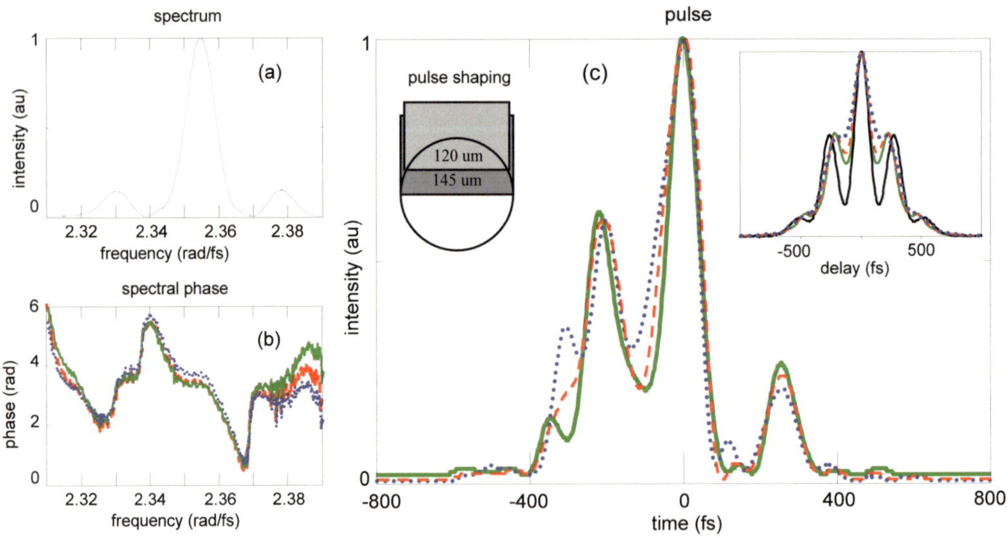

Fig. 13. Reconstruction of three-peak pulse (shaped by means of 145 and 120-um thick glasses) through similariton-based SI in comparison with autocorrelation: (a) measured spectrum; (b) retrieved spectral phases; (c) reconstructed pulse temporal profiles and autocorrelation tracks (inset). Blue dotted, red dashed and green solid curves are for $\alpha = 2.1 \times 10^4$ fs^2, 1.995×10^4 fs^2, and 1.89×10^4 fs^2, respectively, and the black one of the inset is the measured autocorrelation.

spectral amplitude $\tilde{A}(\omega) = \tilde{A}_0(\omega)\rho(\omega)\exp[i\phi(\omega)]$, with the $\rho(\omega) = \sqrt{1+\mu^2+2\mu\cos(\omega T)}$ amplitude- and $\phi(\omega) = \arctan[(\sin\omega T)/(\mu^{-1}+\cos\omega T)]$ phase-modulation. To shape such double-peak pulses, the laser beam is expanded and a thin glass plate in its part is placed, as described above. The thickness of the plate gives the time delay between the peaks; e.g. a 0.12 mm thick glass plate gives a 200 fs delay, according to autocorrelation check. The similariton-based SI and FO are comparatively experimented using the double-peak signal pulse: the SI-reconstructed pulses are compared with spectrotemporal images of the signal.

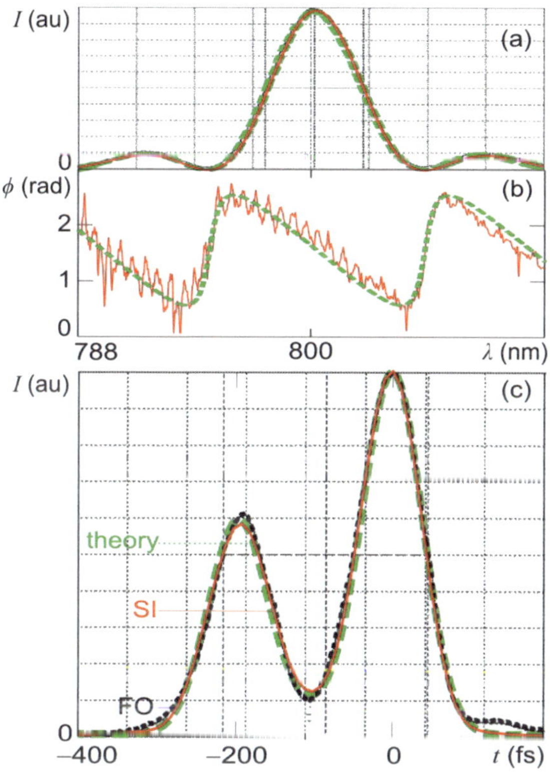

Fig. 14. Comparison of similariton-based SI and FO for a double-peak signal pulse: (a)measured spectrum, (b) retrieved spectral phase, and (c) pulse. Dashed, solid and dotted curves are for the theory, similariton-based SI and FO, respectively.

Fig. 14 illustrates this experiment by the results for a double-peak signal pulse: the quantitative accordance of the measured spectrum (a) and retrieved spectral phase (b) with the theoretical curves (dashed) leads to an accurate pulse reconstruction through similariton-based SI (c, solid). An accurate spectrotemporal imaging (c, dotted) is ensured by the similariton of the bandwidths of ≥ 40 nm. The differences between these independent SI- and FO-measurements and theoretical curve are hardly seen, evidencing both the accuracy of the mentioned measurements and the potential of similariton-based methods.

Obviously, the demonstrated methods of femtosecond signal characterization can be implemented also by the use of "standard" parabolic similaritons generated in active or dispersion decreasing fibers. In a recent progress in the generation of parabolic broadband similaritons, bandwidths of up to 11 THz (40 nm at 1050 nm central wavelength) are achieved [52]. However, the use of the nonlinear-dispersive similariton generated in a piece of standard passive fiber currently is more beneficial, providing larger bandwidths and thus larger application ranges with technically simpler experimental arrangement.

Thus, the methods of similariton-based SI and spectrotemporal imaging are experimentally demonstrated as two applications of similariton. The reference-based methods become self-referencing by the use of similariton. The described comparative study, involving also theoretical check and autocorrelation measurements, ensures the quantitative accordance and high precision of both the similariton-referencing methods. While the method of similariton-based spectrotemporal imaging has the advantage of direct pulse measurement, and thus leads to the development of a femtosecond optical oscilloscope, it does not give phase information. The method of similariton-based SI provides the complete (amplitude and phase) characterization of femtosecond signal. The method of similariton-based SI provides the complete (amplitude and phase) characterization of femtosecond signal.

5. Conclusion

Our spectral interferometric studies demonstrate the spectronic nature and distinctive properties of nonlinear-dispersive similariton, of up to 5-THz bandwidth, generated in a passive fiber. The key property of nonlinear-dispersive similariton of having a parabolic phase (linear chirp), given by the fiber dispersion only, leads to its spectrotemporal similarity and thus to its self-spectrotemporal imaging, with the accuracy given by spectral broadening and pulse stretching together.

Generating similaritons of 50-THz bandwidth, we carry out their complete characterization through the chirp measurement, using the technique of frequency tuning in the process of spectral compression by sum-frequency generation. The studies permit to generalize the description of nonlinear-dispersive similaritons, verifying that only fiber dispersion determines the phase (chirp) of such broadband similaritons. The third order dispersion of fiber results in the same additional phase for broadband nonlinear-dispersive similariton and spectron. The ~1% accuracy of the linear fit for the chirp of the 50-THz bandwidth similariton gives the range of applications for aberration-free similariton-based spectrotemporal imaging and spectral interferometry. The described approach to the generation and characterization of broadband similariton can be helpful also for its applications in pulse compression and CARS microscopy.

We develop and implement a similariton based self-referencing method of spectral interferometry for the complete characterization of femtosecond signal. The method is based on the similariton generation from the part of signal and its use as a reference for the interference with the signal in the spectrometer. Therefore, the method of similariton-based spectral interferometry combines the advantage of the simple principle and configuration with the self-referencing performance. We experiment the similariton-based method of spectral interferometry in comparison with the measurements carried out with the prototype of femtosecond oscilloscope based on the spectrotemporal imaging in a similariton-induced temporal lens. Our comparative study, carried out together with theoretical check and autocorrelation measurements, evidences the quantitative accordance and high precision of both the similariton-referencing methods of spectral interferometry and spectrotemporal imaging for accurate femtosecond-scale temporal measurements. The similariton-based spectrotemporal imaging has the advantage of direct pulse measurement leading to the development of a femtosecond optical oscilloscope, but it does not give the phase information without additional interferometric measurement. The novel method of similariton-based spectral interferometry, with a rather simple setup and self-referencing performance, provides the complete (amplitude and phase), high-resolution characterization of femtosecond signal.

6. Acknowledgments

The work was carried out within the framework of the 978027 project of Science for Peace Programme of North Atlantic Treaty Organization, collaborative programme project IE007 of Centre National de la Recherche Scientifique (CNRS), France - State Committee of Science (SCS), Armenia, and ANSEF grant # PS-opt-2903. A. Zeytunyan also acknowledges SCS, National Foundation of Science and Advanced Technology (NFSAT), and Civilian Research and Development Foundation (CDRF) for financial support in the framework of the Early Career Support Program (grant numbers A-16 and ECSP-09-50).

7. References

[1] D.J.Kane, R.Trebino "Single-shot measurement of the intensity and phase of an arbitrary ultrashort pulse by using frequency-resolved optical gating" Opt.Lett.18, 823–825 (1993).

S. Akturk, M. Kimmel, P. O'Shea, R. Trebino "Measuring spatial chirp in ultrashort pulses using singleshot frequency-resolved optical gating" Opt. Express 11, 68–78 (2003).

[2] J. Piasecki, B. Colombeau, M. Vampouille, C. Froehly, J.A. Arnaud, "Nouvelle méthode de mesure de la réponse impulsionnelle des fibres optiques," Appl. Opt. 19, 3749 (1980).

F. Reynaud, F. Salin, and A. Barthélémy, "Measurement of phase shifts introduced by nonlinear optical phenomena on subpicosecond pulses," Opt. Lett. 14, 275–277 (1989).

[3] C. Iaconis and I. A. Walmsley, "Spectral phase interferometry for direct electric-field reconstruction of ultrashort optical pulses," Opt. Lett. 23, 792–794 (1998).

[4] V.Messager, F.Louradour, C.Froehly, A.Barthélémy "Coherent measurement of short laser pulses based on spectral interferometry resolved in time" Opt.Lett.28, 743–745 (2003).

M. Lelek, F. Louradour, A. Barthélémy, C. Froehly, T. Mansourian, L. Mouradian, J.-P. Chambaret, G. Chériaux, B. Mercier, "Two-dimensional spectral shearing interferometry resolved in time for ultrashort optical pulse characterization," J. Opt. Soc. Am. B 25, A17–A24 (2008).

[5] P.Kockaert, M.Haelterman, Ph.Emplit, C.Froehly, "Complete characterization of (ultra)-short optical pulses using fast linear detectors," IEEE J. Sel.Top.Quantum Electron. 10, 206–212 (2004).

[6] V.V. Lozovoy, I. Pastirk, M. Dantus, "Multiphoton intrapulse interference. IV. Ultrashort laser pulse spectral phase characterization and compensation," Opt. Lett. 29, 775 (2004).

B.Xu, J.M.Gunn, M.Dela Cruz, V.V.Lozovoy, M.Dantus "Quantitative investigation of the multiphoton intrapulse interference phase scan method for simultaneous phase measurement and compensation of femtosecond laser pulses" J.Opt.Soc.Am. B 23, 750-759 (2004).

[7] M. C. Nuss, M. Li, T. H. Chiu, A. M. Weiner, A. Partovi, "Time-to-space mapping of femtosecond pulses," Opt. Lett. 19, 664–666 (1994).

[8] P.C.Sun, Y.T.Mazurenko, Y.Fainman, "Femtosecond pulse imaging: ultrafast optical oscilloscope," J. Opt. Soc. Am. A 14, 1159–1170 (1997).

[9] C.V.Bennett, B.H.Kolner, "Upconversion time microscope demonstrating 103_ magnification of femtosecond waveforms," Opt. Lett. 24, 783–785 (1999).

[10] M.Vampouille, A.Barthélémy, B.Colombeau, C.Froehly "Observation et applications des modulations de fréquence dans les fibers unimodales" J. Opt. (Paris) 15, 385–390 (1984).

[11] M. Vampouille, J. Marty, C. Froehly, "Optical frequency intermodulation between two picosecond laser pulses," IEEE J. Quantum Electron. 22, 192–194 (1986).

[12] M. T. Kauffman, W. C. Banyai, A. A. Godil, D. M. Bloom, "Time-to-frequency converter for measuring picosecond optical pulses," Appl. Phys. Lett. 64, 270–272 (1994).

[13] J. Azana, N. K. Berger, B. Levit, B. Fischer, "Time-tofrequency conversion of optical waveforms using a single time lens system," Phys. Scr. T118, 115–117 (2005); IEEE Photon. Technol. Lett. 16, 882–884 (2004); Opt. Commun. 217, 205–209 (2003).

[14] E.Arons, E.N.Leith, A.Tien, R.Wagner, "Highresolution optical chirped pulse gating," Appl. Opt. 36, 2603–2608 (1997).

[15] L.Kh.Mouradian, F.Louradour, V.Messager, A.Barthélémy, C.Froehly "Spectro-temporal imaging of femtosecond events" IEEE J. Quantum Electron. 36, 795–801 (2000).

[16] L.Kh.Mouradian, A.V.Zohrabian, C.Froehly, F.Louradour, A.Barthélémy "Spectral imaging of pulses temporal profile," in *Conference on Lasers and Electro-Optics (CLEO / Europe)*, OSA Tech. Digest Series, paper CMA5 (1998).

[17] L.Kh.Mouradian, F.Louradour, C.Froehly, A.Barthélémy "Self- and cross-phase modulation of chirped pulses: spectral imaging of femtosecond pulses" in *Nonlinear Guided Waves and Their Applications*, OSA Tech. Digest Series, v.5, paper NFC4 (1998).

[18] L.Kh.Mouradian, A.V.Zohrabyan, V.J.Ninoyan, A.A.Kutuzian, C.Froehly, F.Louradour, A.Barthélémy, "Characterization of optical signals in fiber-optic Fourier converter," Proc. SPIE 3418, 78–85 (1998).

[19] A.V.Zohrabyan, A.A.Kutuzian, V.Zh.Ninoyan, L.Kh.Mouradian "Spectral compression of picosecond pulses by means of cross phase modulation" AIP Conf. Proc. 406, 395–401 (1997).

[20] N. L. Markaryan L. Kh. Muradyan, "Determination of the temporal profiles of ultrashort pulses by a fibre-optic compression technique," Quantum Electron. 25, 668–670 (1995).
[21] M.A.Foster, R.Salem, D.F.Geraghty, A.C.Turner-Foster, M.Lipson, A.L.Gaeta, "Silicon-chip based ultrafast optical oscilloscope," Nature 456, 81–84 (2008).
[22] T. Mansuryan, A. Zeytunyan, M. Kalashyan, G. Yesayan, L. Mouradian, F. Louradour, A.Barthélémy, "Parabolic temporal lensing and spectrotemporal imaging: a femtosecond optical oscilloscope," J. Opt. Soc. Am. B 25, A101–A110 (2008).
[23] A.Zeytunyan,G.Yesayan,L.Mouradian, P.Kockaert,P.Emplit, F.Louradour, A.Barthélémy "Nonlinear-dispersive similariton of passive fiber," J. Europ. Opt. Soc. Rap. Public. 4, 09009 (2009).
[24] M.A.Kalashyan, K.A.Palandzhyan, G.L.Esayan, L.Kh.Muradyan "Generation of transform-limited rectangular pulses in a spectral compressor" Quntum Electron. 40 (10), 868–872. 2010,
[25] M.A.Kalashyan, K.H.Palanjyan, T.J.Khachikyan, T.G.Mansuryan, G.L.Yesayan, L.Kh.Mouradian "Prism -Lens Dispersive Delay Line" Tech.Phys.Lett. 35 (3), 211-214 (2009).
[26] J.M.Dudley, C.Finot, D.J.Richardson, G. Millot, "Self-similarity and scaling phenomena in nonlinear ultrafast optics," Nature Physics 3, 597–603 (2007).
[27] C.Finot, J.M.Dudley, B.Kibler, D.J.Richardson, G.Millot, "Optical parabolic pulse generation and applications," IEEE J. Quantum Electron. 45, 1482–1489 (2009).
[28] D.Anderson, M.Desaix, M.Karlson, M.Lisak, M.L.Quiroga-Teixeiro, "Wave-breaking-free pulses in nonlinear optical fibers," J. Opt. Soc. Am. B 10, 1185–1190 (1993).
[29] M.E. Fermann, V.I. Kruglov, B.C. Thomsen, J.M. Dudley, J.D. Harvey, "Self-similar propagation and amplification of parabolic pulses in optical fibers," Phys. Rev. Lett. 84, 6010–6013 (2000).
[30] V.I. Kruglov, A.C. Peacock, J.M. Dudley, J.D. Harvey, "Self similar propagation of high-power parabolic pulses in optical fiber amplifiers," Opt. Lett. 25, 1753–1755 (2000).
[31] V.I.Kruglov, A.C.Peacock, J.D.Harvey, J.M.Dudley "Self-similar propagation of parabolic pulses in normal-dispersion fiber amplifiers," J. Opt. Soc. Am. B 19, 461–469 (2002).
[32] C. Finot, G. Millot, C. Billet, J.M. Dudley, "Experimental generation of parabolic pulses via Raman amplification in optical fiber," Opt. Express 11, 1547–1552 (2003).
[33] F.Ö. Ilday, J.R. Buckley, W.G. Clark, F.W. Wise, "Self-similar evolution of parabolic pulses in a laser," Phys. Rev. Lett. 92, 213902 (2004).
[34] T. Hirooka, M. Nakazawa, "Parabolic pulse generation by use of a dispersion-decreasing fiber with normal group-velocity dispersion," Opt. Lett. 29, 498–500 (2004).
[35] C. Finot, B. Barviau, G. Millot, A. Guryanov, A. Sysoliatin, S. Wabnitz, "Parabolic pulse generation with active or passive dispersion decreasing optical fibers," Opt. Express 15, 15824–15835 (2007).
[36] A.S.Zeytunyan, K.A.Palandjyan, G.L.Yesayan, L.Kh.Mouradian, "Nonlinear dispersive similariton: spectral interferometric study" Quantum Electron. 40, 327–328 (2010).

[37] A.S.Zeytunyan, K.A.Palanjyan, G.L.Yesayan, L.Kh.Mouradian, "Spectral-interferometric study of nonlinear-spectronic similariton:" J. Contemp. Phys. 45, 64–69 (2010).
[38] C.Finot, B.Kibler, L.Provost, S.Wabnitz, "Beneficial impact of wave-breaking for coherent continuum formation in normally dispersive nonlinear fibers," J. Opt. Soc. Am. B 25, 1938–1948 (2008).
[39] A.S.Zeytunyan, H.R.Madatyan, G.L.Yesayan, L.Kh.Mouradian "Diagnostics of femto-second laser pulses based on generation of nonlinear-dispersive similariton" J. Contemp. Phys. 45, 169–171 (2010).
[40] V.I. Kruglov, D. Méchin, J.D. Harvey, "High compression of similariton pulses under the influence of higher-order effects," J. Opt. Soc. Am. B 24, 833–838 (2007).
[41] K.Palanjyan, A.Muradyan, A.Zeytunyan, G.Yesayan, L.Mouradian, "Pulse compression down to 17 femtoseconds by generating broadband similariton," Proc. SPIE 7998, 79980N (2010).
[42] C. Finot, G. Millot "Synthesis of optical pulses by use of similaritons" Opt. Express 12, 5104–5109 (2004).
[43] A.Zeytunyan, G.Yesayan, L.Mouradian, F.Louradour, A.Barthélémy, "Applications of similariton in ultrafast optics: spectral interferometry and spectrotemporal imaging," in *Frontiers in Optics*, OSA Tech. Digest, paper FWI5 (2009).
[44] A.Zeytunyan, A. Muradyan, G. Yesayan, L. Mouradian, "Broadband similariton" Laser Physics 20, 1729–1732 (2010).
[45] A.Zeytunyan, A.Muradyan, G.Yesayan, L.Mouradian, F.Louradour, A.Barthélémy "Measuring of Broadband Similariton Chirp" Nonlinear Photonics OSA Tech. Digest, paper NME46 (2010).
[46] A.Zeytunyan, A.Muradyan, G.Yesayan, L.Mouradian, F.Louradour, A.Barthelemy, Similariton for Femtosecond Optics Proc. ECOC 2010 (19-23 Sep. 2010), Torino, Italy, paper Mo.2.E.5 (2010).
[47] A.Zeytunyan, A.Muradyan, G.Yesayan, L.Mouradian, F.Louradour, A.Barthélémy "Generation of broadband similaritons for complete characterization of femtosecond pulses" Opt. Commun. v. 284, pp. 3742–3747 (2011).
[48] A.M. Heidt, "Pulse preserving flat-top supercontinuum generation in all-normal dispersion photonic crystal fibers," J. Opt. Soc. Am. B 27, 550-559 (2010).
[49] D.T. Reid, P. Loza-Alvarez, C.T.A. Brown, T. Beddard, W. Sibbett, "Amplitude and phase measurement of mid-infrared femtosecond pulses by using cross-correlation frequency-resolved optical gating," Opt. Lett. 25, 1478-1480 (2000).
[50] J. Dudley, X. Gu, L. Xu, M. Kimmel, E. Zeek, P. O'Shea, R. Trebino, S. Coen, R. Windeler, "Cross-correlation frequency resolved optical gating analysis of broadband continuum generation in photonic crystal fiber: simulations and experiments," Opt. Express 10, 1215-1221 (2002).
[51] A.F. Pegoraro, A. Ridsdale, D.J. Moffatt, Y. Jia, J.P. Pezacki, A. Stolow, "Optimally chirped multimodal CARS microscopy based on a single Ti:sapphire oscillator," Opt. Express 17, 2984-2996 (2009).
[52] W.H. Renninger, A. Chong, F.W. Wise, "Self-similar pulse evolution in an all-normal-dispersion laser," Phys. Rev. A 82, 021805(R) (2010).

6

Spectral Low Coherence Interferometry: A Complete Analysis of the Detection System and the Signal Processing

Eneas N. Morel and Jorge R. Torga
Universidad Tecnológica Nacional, Facultad Regional Delta
Campana, Buenos Aires
Argentina

1. Introduction

Low coherence interferometry (LCI) is an optical technique in which light is used as an instrument to obtain high resolution optical images in a great diversity of materials. It is possible to find in the literature, a variety of configurations with different names, based on the same principle: low coherence reflectometry (OLCR), optical coherence tomography (OCT), white light interferometry (WLI), are only some examples of this. The main idea of this technique is to measure the echo time delay of backscattered light in the sample through the characterization of the interference intensity obtained when the light coming from the sample and the light reflected in a reference surface overlap. When a low coherence source is used, the interference signal is temporally and spatially localized, so it is possible to use this property to obtain distance values or parameters related to the time of flight of the light, reflected in different sections of a sample.

LCI has been proposed and studied since the beginning of the optical science. However, for the last 20 years, there has been a dramatically increase in its applications, mainly due to the development of optical coherence tomography and the evolution of new light sources. The first works of in- surface and optical material characterization was reported in 1960. In 80´s some applications in fibre optics characterization (Takada et al, 1987), topography surfaces and internal structures in transparent media (Youngquist et al, 1987) were proposed with a LCI set-up. It is generally accepted that first biological application was reported by (Fercher, 1988). After these first works, optical coherent tomography (OCT) became a powerful technique for medical diagnosis, (Huang, 1991), in which images of the human retina and coronary artery were obtained. Since then, OCT has evolved, and nowadays it is a well-established technique for ophthalmic diagnosis and other biological tissues (Brezinski, 2006).

New developments in light sources, fibre optic elements, detectors and processing techniques allow a dramatic increase in resolution and speed of image acquisition. These advances transform this technique in a powerful three-dimensional visualization method which has a wide diversity of applications. Over the last ten years its growth has been explosive; proof of this is the increasing number of publications, patents and companies involved in this subject (www.octnews.org). There is no doubt that this evolution has been the result of the numerous successful OCT applications in the medical field, especially in ophthalmology (Drexler et al, 2008). Over the last years its application in material characterization and non-destructive techniques is another field that is growing fast (Goode, 2009). Micro structures and MEMS characterization, surface topography, structural parameters in semitransparent materials, analysis and visualization of structural vibrations, are only a few examples (Bruce et al, 1991), (Wiesauer et al, 2005), (Wiesner et al, 2010).

Nowadays, there are several companies that offer commercial systems based on this tecnique. Polytec®, Thorlabs®, Carl Zeiss® are only some examples. Figure 1 show an image obtained with the Stratus from Carl Zeiss®. This equipment is capable of exploring the eye at a speed of several thousand lines per second.

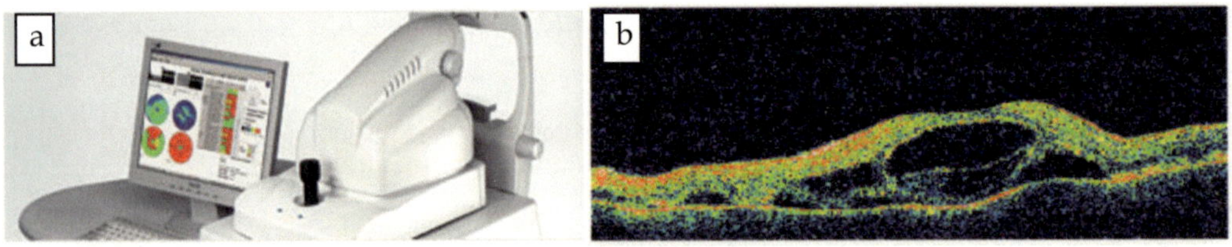

Fig. 1. a) Stratus 3000 Carl Zeiss®. b) OCT image of retina, taken with Stratus 3000 (Michael P Kelly, Duke Eye Imaging, Duke University Eye Center, Durham, NC). Photos used with permission of Carl Zeiss®.

Another example is shown in figure 2; TSM-1200 TopMap® μ.Lab from Polytec® that can be used to acquire high-resolution topographical maps of functional surfaces and microstructures.

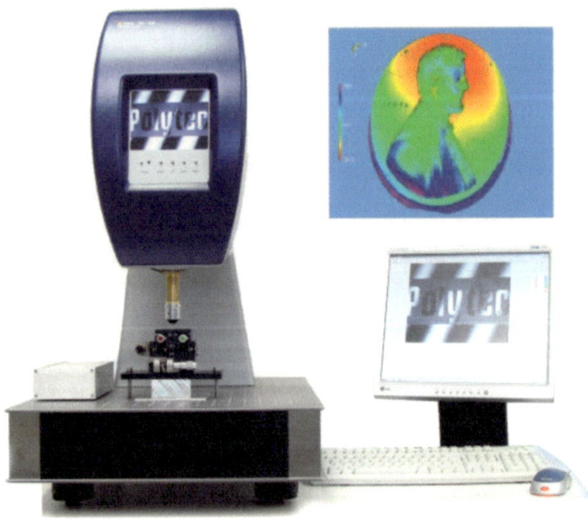

Fig. 2. TMS-1200 TopMap μ.Lab. Photos used with permission of Polytec®.

High-speed 3D OCT imaging can provide comprehensive data that combines the advantages of optical coherence tomography and microscopy in a single system. Shown below are some 3D image data sets of two samples; in figures 4 it is shown how this technique can obtain the surface topography of a screw.

Fig. 3. 3D optical profiling of an M2 metal screw. Photos used with permission of Thorlabs®.

In figure 4 it is shown a tomography of a finger skin.

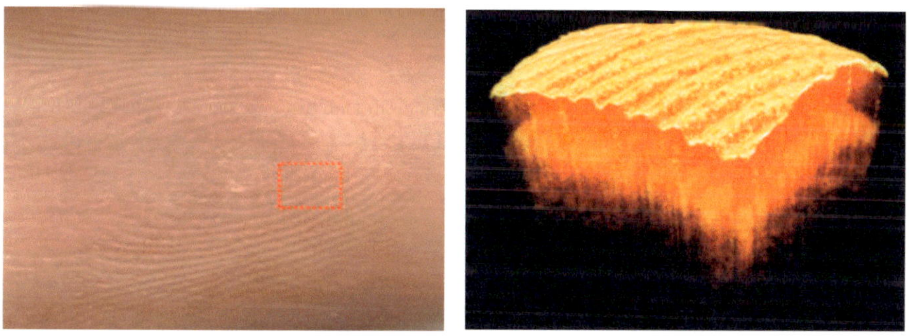

Fig. 4. Human finger pad. Photos used with permission of Thorlabs®.

1.1 Basis of low coherence Interferometry

Interference can be considered as the wave pattern that is obtained after the superposition of two or more waves. In low coherence interferometry, superposition is obtained with a broadband light source, and the wave pattern in this situation is considerably different from that obtained with a standard monochromatic source. (Born, M. & Wolf, E., 1999).

In interferometry techniques the interference signal is usually obtained from the superposition between light backscattered from a reference arm and a sample arm as it is shown in figure 5 below.

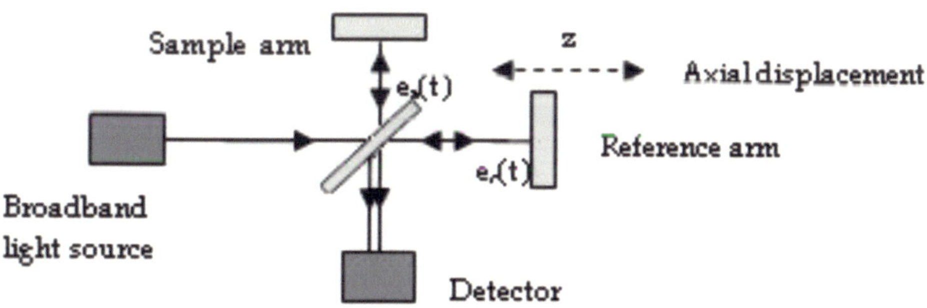

Fig. 5. Michelson interferometer set-up.

The total intensity at the detector at a time t, is obtained as the superposition of $e_r(t) = Re(Er(t))$ and $e_s(t) = Re(E_s(t+\tau))$, the electric field from the reference and from the sample respectively. $Re(E(t))$, indicates the real part of the complex expression of the electric field, $E(t)$. τ is the difference in time of flight of light in each arm and it is a consequence of the different optical path length. The symbol < > indicates temporal average, K is a constant factor.

$$I = K\left\langle \left| E_r + E_s \right|^2 \right\rangle = I_r + I_s + 2K\langle \mathrm{Re}(E_r(t+\tau)E_s(t)) \rangle \quad (1)$$

The reflectivity in the interface at the end of the reference and the sample arm are considered by the coefficient *R*, which is assumed constant. If *Io* is the intensity at the output of the light source, the intensity of each arm after the reflections is: $I_r = R_r I_o$ and $I_s = R_s I_o$. Then, equation 1 can be written:

$$I = R_r I_o + R_s I_o + 2\sqrt{R_s R_r}\,\mathrm{Re}(\Gamma(\tau)) \quad (2)$$

In the last term in equation 2, usually named the interference term, the complex coherence function $\Gamma(\tau) = K\langle E_o(t)\, E_o(t+\tau)\rangle$ is defined (Goodman, 1984); that is the correlation of the electric field. This function is closely related to the concept of coherence time -*T* - a measure of the delay between two beams which is necessary to blur the interference term. A useful way to define this concept is (Goodman, 1984):

$$T = \frac{1}{\left|\Gamma(0)\right|^2} \int_{-\infty}^{\infty} \left|\Gamma(\tau)\right|^2 d\tau \quad (3)$$

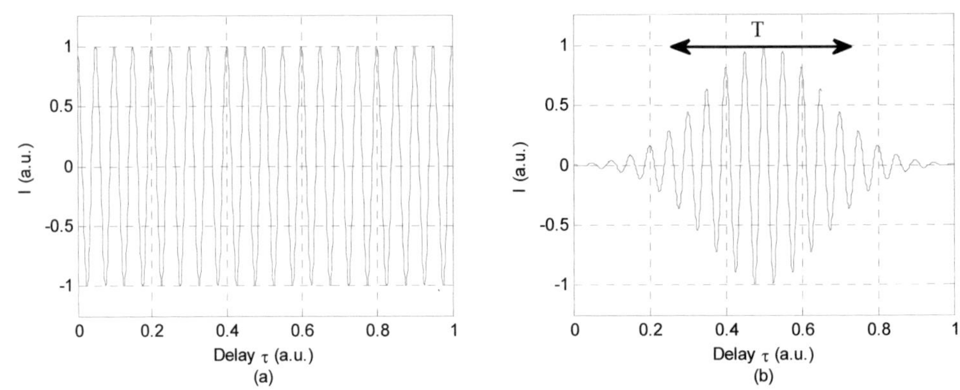

Fig. 6. Total intensity (I), as a function of time delay(τ) for a coherent source a) and for a low coherence source b).

It is possible to illustrate this concept -close to the basic idea of low coherence interferometry- considering the total intensity *I* as a function of τ, the time delay. To do this, we perform an axial scan of the reference sample. In the simplified interferometer scheme of figure 5 this can be done with a movement of the reference surface with a constant velocity (axis z figure 5).

When the light source is monochromatic, the total intensity shows the same variation (figure 6-a). Using a low coherence source, interference is only observed when the path length is within the coherence length of the light source (figure 6-b).

1.2 Experimental configurations

We will show a generic system of low coherence interferometry with a brief description of the experimental set-up and the function of each component. The general configuration can be separated in four well differentiated sections: the broadband light source, the interferometer, the sample with its holder and the detection system.

Figure 7 shows the basic set-up used in most of the low coherence applications. The initial system is the broadband light source [1], that is a critical factor since an appropriate selection in its characteristics is crucial in the final quality of the images obtained (Drexler et al, 2008). Nowadays there is an important offer of broadband sources with different center wavelengths, power, and beam quality (stability, noise, single transverse mode, etc.).

Diode systems are probably the most utilized light sources, as the superluminiscent LED`s (SLD). They have the advantage to be easy to hand and operate; also, their spatial dimension is reduced, their price is usually lower than other systems, and they are appropriated for no-laboratory applications. Amplified spontaneous emission (ASE) sources are an interesting alternative as they offer a broad bandwidth in the infrared region. Finally there is a group based on ultra-short laser pulses and supercontinuum light sources which is an attractive option as it offers high power, broad bandwidth and good quality in spatial mode. These systems are expensive comparatively.

The output beam from the source is directed to the interferometer. There is a great variety of configurations proposed in the literature, but in most of these works a Michelson type is used (Schmitt et al, 1999). A typical configuration is shown in figure 7. The output beam is split in the beamsplitter [4]. One of the beams goes to the reference surface [2]-usually a mirror- and the other beam goes to the sample [3]. After the reflection on each surface, both beams are sent to the detector [5]. The superposition generates the interference signal.

With this signal it is obtained a measurement of the OPD between both arms. In most cases the light beam is focused on the sample so that the measurement process is performed point to point on the desired region. For opaque samples, the light is reflected only on the surface, so a topography measurement is obtained. For transparent or semi-transparent media, light is reflected from sub surface structures in the sample, so a tomography measurement is performed.

Usually the sample-holder or the light beam can be displaced in a 2-D lateral movement (axis x and y in the figure 7). This way the sample can be inspected in the region of interest.

Although there are many configurations proposed based on the general configuration described before, it is possible to make a division into three main groups, each one with its particular characteristics. They are commonly known as:

1. Time domain low coherence interferometry (TDLCI)
2. Spectral domain low coherence interferometry (SDLCI)
3. Sweep source domain low coherence interferometry (SSDLCI)

In TDLCI mode the total interference is obtained with a one element detector([5] in figure 7), usually a photodiode, and the total intensity is registered while the optical path of the reference arm changes from a maximum to a minimum value that is predefined for each set-up (displacement in the z direction - figure 7). While the reference arm is moving the sample is maintained in the same position (Drexler et al, 2008).

In SDLCI mode the detection [5] is performed with a spectrometer. In this scheme the interference signal is obtained with the superposition of the spectral intensity of both arms. (Drexler et al, 2008).

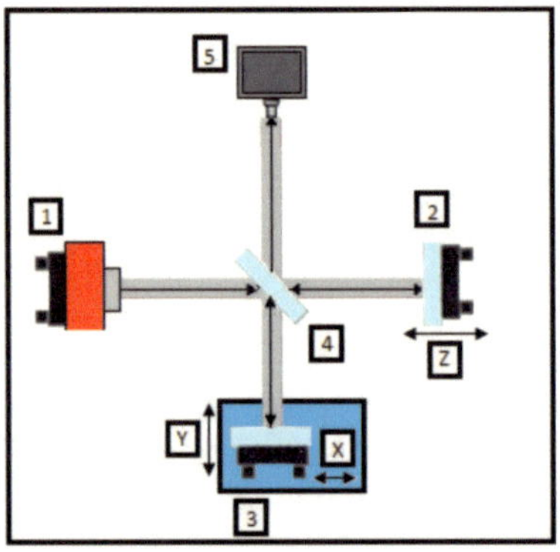

Fig. 7. A typical low coherence interferometry experimental set-up.

Based on the same idea, in the SSDLCI configuration, a tuneable laser and a one element photodetector are used. The light source is swept in wavelength as a function of time, so the spectral components are not encoded by spatial separation, they are encoded in time. After a complete swept we obtain the spectrum one by one wavelength (Choma et al 2003), and consequently the interference signal.

2. Spectral domain low coherence interferometry

2.1 Introduction

Spectral domain low coherence interferometry (SDLCI) -also known as Fourier domain low coherence interferometry (FDLCI)- is, as mentioned before, one of the configurations commonly used in low coherence interferometry. The first result using this technique (Fercher et al, 1995) was for ophthalmological measurements and some years later (Hasler et al, 1998) in dermatological applications. After these initial works, it began to be shown as a competitive technique with the time domain method (TD-OCT), a well established method for optical coherence tomography (OCT) applications at that time. Since then, SDLCI has shown several advantages which have favoured its development and the high level of acceptance that it has nowadays.

These advantages can be summarized in the lack of need for a fast mechanical scanning mechanism (Drexler et al, 2008) that brings the simplicity of no moving parts in the reference arm of the interferometer, and the superior sensitivity of the detection (Leitgeb et al, 2003). The SDLI typical configuration is shown in figure 8. The light source and the interferometer scheme follow the same characteristics described before, but the detection system is a distinct point in this technique. The main idea is that the optical path difference between the sample and the reference is obtained from the analysis of the superposition of the spectrum of the light source reflected in each arm of the interferometer. This spectrum is commonly measured with a spectrometer. Most of the detection systems employ linear arrays as a sensor element. Some works with 2-D CCD systems and individual processing of each pixel, has been already presented (Vakhtin et al, 2003).

Figure 8 below illustrates the basic experimental set-up assuming a simple situation in which there are only two reflections, one at the end of each of the arms of the interferometers. The interference signal is measured with the spectrometer sensor where the total intensity is obtained as a function of the wavelength (λ).

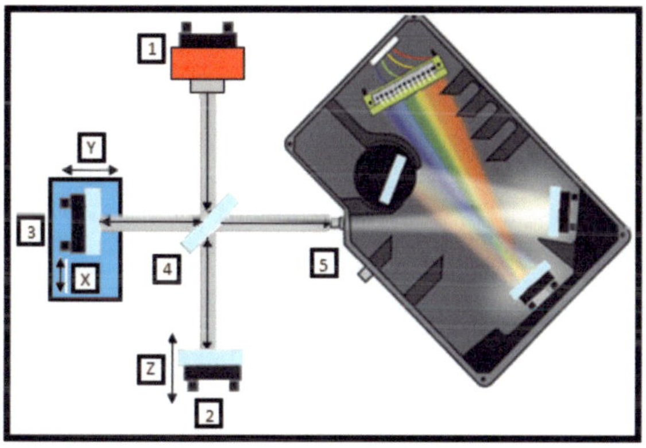

Fig. 8. A typical spectral doamin low coherence interferometry experimental set-up.

2.2 The interference signal and the detection system

It is assumed that $I_o(k)$ encodes the power spectral dependence of the light source ([1] in figure 8), were $k=2\pi/\lambda$, is the wavenumber. The interference fringes surges from the superposition of the spectrum of the light coming from the two reflections. We call $I_{r1}(k)$ to the intensity coming from the reference arm and $I_{s1}(k)$ to the intensity coming from the sample arm . The expression for the total intensity is then:

$$I(k) = I_{r1}(k) + I_{s1}(k) + 2\sqrt{I_{r1}(k)I_{s1}(k)}\cos(k\Delta x_{r1s1}) \quad (4)$$

The first two terms are known as DC intensities in the literature (Drexler et al, 2008). The last term in equation 4 is the interference component, and it includes the OPD dependence that, in this simple situation, is given by the value $\Delta X_{r1s1} = x_{r1}-x_{s1}$, where x_{r1} and x_{s1} are the total path length in each arm, both measured from the beamsplitter.

To simplify the expression, we use $\beta^2 = R_{s1}/R_{r1}$, the reflectivity coefficients ratio, and it is assumed that these coefficients have no dependence on the wavenumber, that is: $I_{r1}(k) = R_{r1}I_o(k)$ and $I_{s1}(k) = R_{1s}I_o(k)$, as mentioned in equation 5, so:

$$I(k) = R_{r1}I_o(k)\left(1 + \beta^2 + 2\beta\cos(k\Delta x_{r1s1})\right) \quad (5)$$

The figure 9 shows an example of an image obtained in a 2D-CCD sensor at the end of the spectrometer (see set-up picture). The image corresponds to the total intensity described before in equation 4. In this particular example, the source is a superluminescent diode (λo = 840 nm, $\Delta\lambda$ = 20nm) and $R_{1s} = R_{1r} \approx 0.08$.

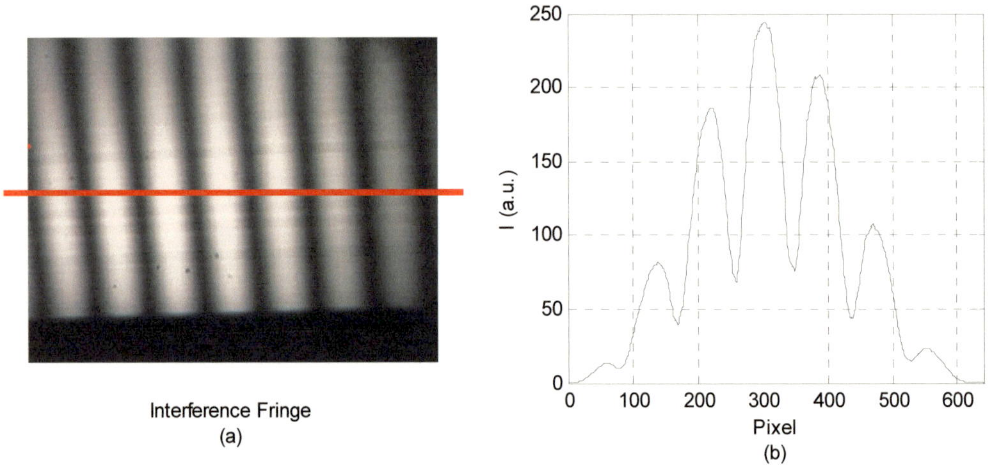

Fig. 9. a) Interferences fringes obtained with a 2-D spectrometer sensor b) Intensity as a function of k obtained from a row of pixel from the image a).

Figure 9-b represents total intensity, obtained from a line of pixels of the image shown in figure 9-a. The modulation in the curve is produced by the interference term over the Gaussian spectrum of the light source.

In a general situation in which it is assumed that there are several reflections in the reference arm and in the sample arm, the total intensity expression can be written as:

$$I(k) = \sum_{i=1}^{N} I_i(k) + \sum_{i,j=1}^{N} 2\sqrt{I_i(k)I_j(k)}\cos\left(\Delta x_{ij}\right) \tag{6}$$

N represents the total number of reflectors in the sample and in the reference. The sub index i or j identifies the region where the reflection is produced in the reference (r_1, r_2, r_3, …) or in the sample arm (s_1, s_2, s_3, …). To illustrate this point, we show a typical application in which the reference arm is a mirror (1r) and the sample is a slab in air. The slab has two interfaces (1s and 2s), an inner group index *n*, and a thickness *d*.

$$\begin{aligned} I(k) = I_{r1}(k) + I_{s1}(k) + I_{s2}(k) \\ + 2\sqrt{I_{r1}(k)I_{s1}(k)}\cos\left(\Delta x_{r1s1}\right) \\ + 2\sqrt{I_{r1}(k)I_{s2}(k)}\cos\left(\Delta x_{r1s2}\right) \\ + 2\sqrt{I_{s1}(k)I_{s2}(k)}\cos\left(\Delta x_{s1s2}\right) \end{aligned} \tag{7}$$

The first three terms are the DC terms mentioned before. The second and the third term correspond to the cross correlation between the reference and each of the sample surfaces; the forth term is the correlation component that surges from the reflections on both interfaces of the sample. (see figure 10).

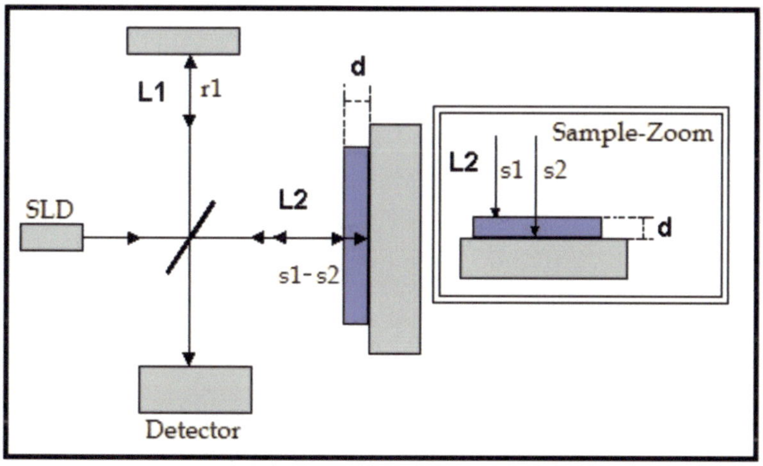

Fig. 10. Experimental set-up for thickness measurement of a slab.

So in this situation the OPD values in equation 7 are: $\Delta x_{r1s1} = x_{r1} - x_{s1}$, $\Delta x_{r1s2} = x_{r1} - (x_{s1} + 2.\ n.d)$ and $\Delta x_{s1s2} = 2.n.d$. Where x_{s1}, x_{s2}, x_{r1} and 2.nd, are the total optical path distances corresponding to each interface indicated in figure 10.

The usual way to obtain the OPD values from the total intensity data is to apply the Fourier transform. We call *FI(x)* to the Fourier Transform of *I(k)*, where x is the conjugate variable of the wavenumber *k*.

As an example it is shown the expression for *FI(x)* in the situation mentioned before (equation 4), where the interference signal is obtained from only two reflections (r_1 and s_1). So:

$$FI(x) = F(I_0(k)) \otimes F\left(R_{r1} + R_{s1} + 2\sqrt{R_{r1}R_{s1}}\cos(\Delta x_{r1s1}k)\right) \tag{8}$$

We denote the convolution as ($\otimes$).

According to the Wiener-Kinchin theorem (Goodman, 1984), the first term in the convolution is the coherence function $\Gamma(x)$ and the second term yields two delta functions located at $\pm\Delta x_{r1s1}$ (Papoulis, 1962).

Therefore, the last expression can be written as:

$$FI(x) = \alpha.\Gamma(x) \otimes \left[\delta(x)(R_{r1} + R_{s1}) + 2\sqrt{R_{s1}R_{r1}}\left(\delta(x - \Delta x_{r1s1}) + \delta(x + \Delta x_{r1s1})\right)\right] \tag{9}$$

Being α a constant factor.

Figure 11-a and 11-b illustrate the interference signals in two situations that correspond to different deep modulation (*β = 1, β = 0.5*). Figures 11-c and 11-d show the absolute value of the corresponding Fourier transform for each signal. This graph shows three "interference peaks". One is centred in the origin of the *x* axis and the "lateral peaks", denoted as P and Q in the figures, are centred in $\pm\Delta x_{r1s1}$.

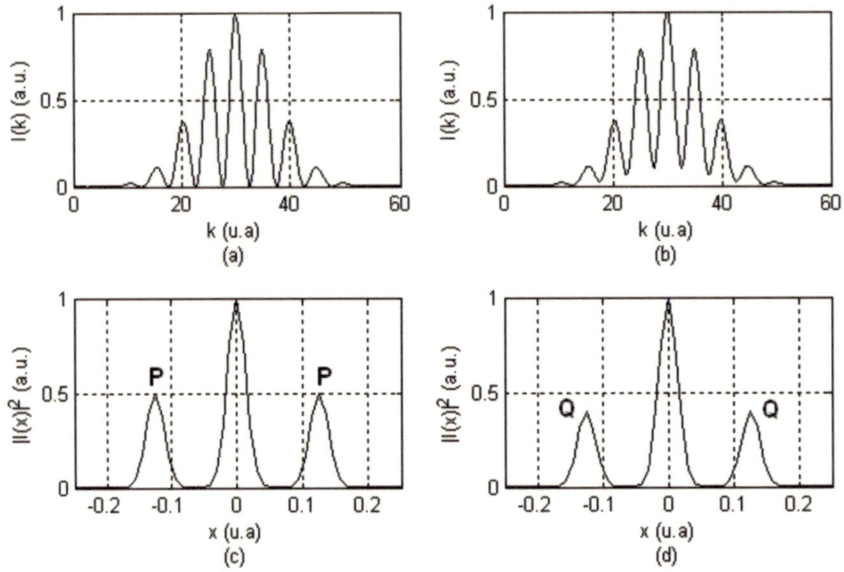

Fig. 11. a) and b) shows the total interference signal as a function of wavenumber k, for two different β relations. c) and d) shows its corresponding Fourier transform as a function of x (the conjugate variable of k) .

Figure 12 shows an example of how the position of the interference peak changes according to the increment of OPD between both arms. The peaks in order zero no contain OPD information.

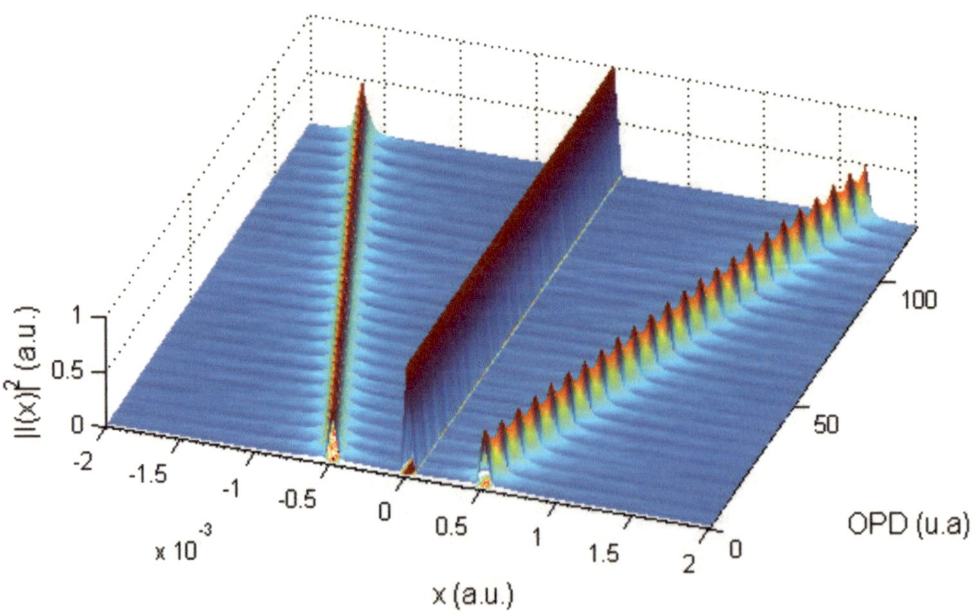

Fig. 12. Interference peak vs OPD increments.

2.3 Signal detection and parameter limits

In an ideal situation, the intensity function *I(k)* would be obtained with a "perfect" spectrometer. In a real situation, it is important to evaluate the spectrometer parameters in order to optimize the measurements and evaluate possible artifacts, errors or limitations introduced for the "real" set-up. We assume a detector model based on a grating spectrometer with a linear array sensor. There are some important issues to be considered in this evaluation, more details can be found in (Hu et al, 2007; Jeon et al, 2011).

a. The spectrometer is used for separating the different wavelengths of the light source spectrum and to focus each of them on the sensor that is considered as a linear array of N_p pixels. After its calibration it is possible to define a k axis in the array direction. The spectral range of the spectrometer (Δk_a) is given by the interval $[k_1, k_{Np}]$ where k_1 is the value assigned to the first pixel and k_{Np} to the last one. We also define the "*Range*" function as follows:

$$Range(k) = \prod\left(\frac{k-k_a}{\Delta k_a}\right) \tag{10}$$

k_a indicates the central wavelength in the spectrometer. Along this work we assume $\Pi(z)$ a *rect* function defined as 0 if $|z|>1/2$ and 1 if $|z|<1/2$.

b. The dimensions of the spot of each wavelength focused on the sensor is taken into account by the point spread function (*Psf(k)*) (Hu et al, 2007). The width, the shape and the position of the center of this spot depends on the spectrometer characteristics such as the entrance slit width, diffraction grating and focal length and eventually on the spatial profile of the light source in the spectrometer entrance (Dorrer et al, 2000). We assume a Gaussian shape (Hu et al, 1997) where its FWHM (Δk_{psf}) is a measurement of the spot size.

$$Psf(k) = psf_0.e^{-4\ln(2)\frac{(k-k_i)^2}{\Delta k^2_{psf}}} \tag{11}$$

The " ideal" detected signal is then modified by the convolution with this function:

$$I_r(k) = \int_{-\infty}^{\infty} I(k').Psf(k-k')dk' = I \otimes Psf(k) \tag{12}$$

c. The value obtained in a particular pixel "i" is the total intensity integrated over a width (Δk_{pix}), which is defined by the pixel dimensions, the grating dispersion and the focus length of the spectrometer (Leitgeb et al 2003; Hu et al 2007). This average value can be written as:

$$I_r(k_i) = \int_{-\infty}^{\infty} (I \otimes Psf)(k).Pix(k-ki)\,dk = I \otimes Psf \otimes Pix(k_i) \tag{13}$$

The convolution takes into account the process described before. A first approach to this function can be to assume a rectangular shape where its width is: $\Delta k_{pix} = \Delta k_a / N_p$ (Hu et al, 2007) (Wang et al, 2008). Here it is assumed by simplicity that the fill factor is 100%.

$$Pix(k - k_i) = \Pi(\frac{k - k_i}{\Delta k_{pix}}) \tag{14}$$

More complex dependence with k can be thought, considering additional effects such as crosstalk or fill factor (Jeon et al, 2011).

d. The measurement of the total intensity $I(k)$ obtained from the detector is a discrete collection of N_p values. Each value corresponds to the average intensity measured in one pixel, for this reason, it is convenient to define a sampling comb of delta functions as (Dorrer et al, 2000):

$$C(k) = \sum_{i=-\infty}^{+\infty} \delta(k - k_i) \tag{15}$$

Where k_i is the value assigned to the pixel with index i (Wang et al,2006); (Hu et al, 2007).

The following figure shows a schematic of the spectrometer sensor and a detail of how two wavelengths (λi and λj) have been focused on it. The picture also shows how the *"pix"* and *"psf"* function combine to define the average value in pixel j.

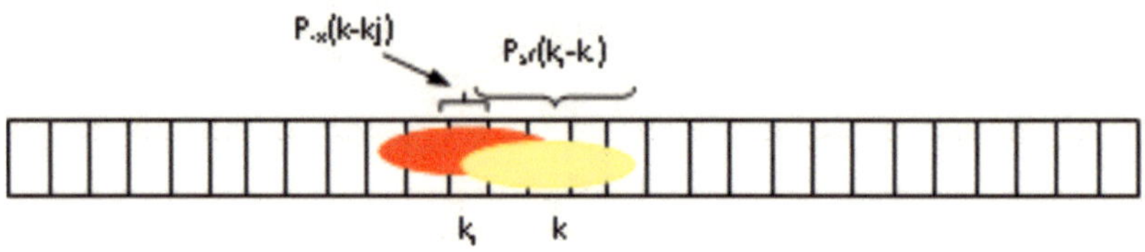

Fig. 13. Schematic of the spectrometer sensor array and the "psf" and "pixel" functions.

Then, the final expression for the signal intensity is:

$$I(k) = Range(k)(C \otimes psf \otimes pixel \otimes I)(k) \tag{16}$$

$$I(k) = \prod(k - k_a) \sum_{i=-\infty}^{\infty} \delta(k - k_i)(pixel \otimes psf \otimes I(k)) \tag{17}$$

Under these considerations the A scan signal $FI(x)$ is obtained applying the Fourier transform of equation 17. Here the x variable is the conjugated of k.

$$\begin{aligned} FI_4(x) = {} & F(\prod(k - k_a)) \\ & \otimes F\left(\sum_{i=-\infty}^{\infty} \delta(k - k_i) \right) \\ & \otimes (F(pixel(k)) F(psf(k)) F(I(k))) \end{aligned} \tag{18}$$

That can be written as:

$$FI(x) = FI_0 \operatorname{sinc}\left(\frac{\Delta k_a}{2}x\right) \otimes \sum_{n=-\infty}^{\infty} \delta(x - n\Delta x) \otimes \left[\operatorname{sinc}\left(\frac{\Delta k_{pix}}{2}x\right).\exp\left(-\frac{\Delta k^2{}_{psf}x^2}{16\ln(2)}\right)F(I(k))\right] \quad (19)$$

This expression is easily obtained from equation 18 and some mathematics relations (Papoulis, 1969). FI_o is a constant factor that also absorbs all the pure phase terms that result from the Fourier transform. The new delta comb has a period $\Delta x = 2\Pi/\Delta k_{pix}$, the interval available for signal processing avoiding aliasing.

The expression in brackets in equation 19, shows the Fourier transform of the "ideal." intensity and the modifications introduced by the Fourier transform of the *psf* and the *pixel* function.

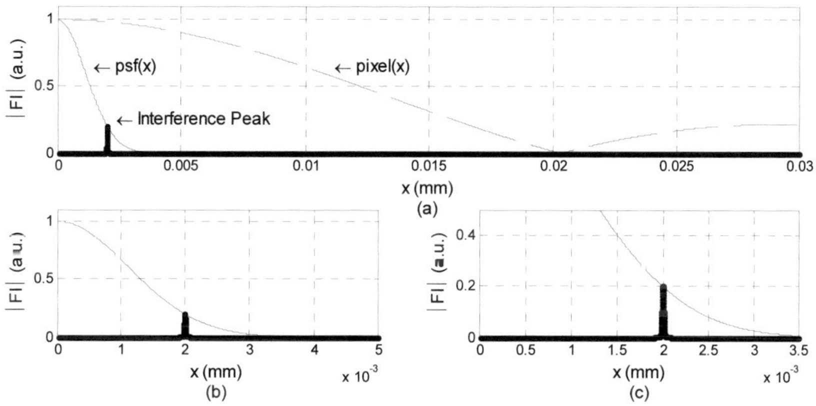

Fig. 14. a) Interference peak and the psf and pixel Fourier Transform functions. b) and c) Zoom of figure a).

This modification means a sensitivity fall-off when the OPD value increase. In the figure 14 it is shown these terms for a typical situation.

As it can be appreciated from the figure 14, the shape of both functions (the *sinc* and *psf* Fourier transform) generate a decrease in the visibility amplitude of the interference peak that goes to zero when x increase. So, there are two parameters that limit the dynamic range of the technique. One is: $x_{pix}=2\pi/\Delta k_p$, the first zero of the *sinc* function. The other is the FWHM of the Gaussian function obtained as the Fourier transform of the *psf*: $x_{psf}=4.\ln(2)/\Delta k_{psf}$ (Bajraszewski et al, 2008), (Leitgeb et al, 2003).

2.4 Signal processing

In the last section we showed some considerations in the interference image, imposed by the real characteristics of the detector. The SDLCI detection process ends with an array of N_p

data points that contain the information of the depth profile of the sample. This process usually continues with the analysis of the Digital Fast Fourier Transform (FFT) of this vector. The classical algorithm is shown in equation 20, where N_p is the I[n] length.

$$FI[q] = \sum_{n=0}^{N_p-1} I[n] W_{N_p}^{qn} \quad / \quad W_{N_p}^{qn} = e^{-j\left(\frac{2\pi}{N_p}\right)} \tag{20}$$

In figure 15 a typical example is shown. The interference signal is registered in a 640-point array (figure 15-a). Its conventional FFT is shown in figure 15-b. The interference spectrum has been obtained in a wavelength range from 780 to 880 nm. The modulation corresponds to a situation in which there is only one surface (*r1* and *s1*) in both arms of the interferometer, as presented in equation 4.

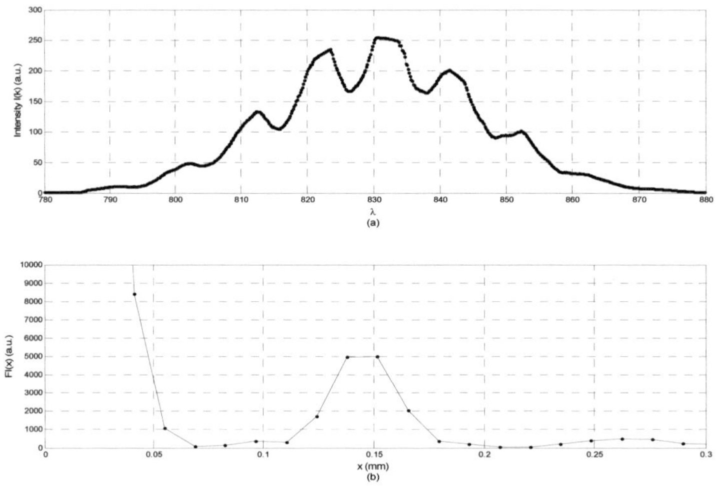

Fig. 15. a) Interference signal and b) Interference peak.

The OPD value Δx_{r1s1} in this scheme is the abscissa coordinate of the maximum point of the lateral peak in the Fourier transform curve (figure 15-b). With the conventional Cooley-Tukey FFT algorithm the sampling interval is:

$$\Delta x_{samp} = \frac{2\pi}{N_p \Delta k_{pix}} \tag{21}$$

Where N_p and Δk_{pix} are the parameters previously defined. In this step, spectral calibration is a critical process. First, a conversion of the measured spectrographs from λ-space to k-space is needed. As the spectra obtained by the spectrometer are not necessarily evenly spaced, a posterior resample to be uniformly spaced in k-axis is required. This implies a careful calibration of each pixel of the sensor. There are several works in which this point is presented (Dorrer et al, 2000; Hu et al, 2007) and different experimental and post process arrangements are proposed (Bajraszewski et all, 2008; Jeon et al 2011).

Another point that can be avoided or smoothed by signal processing is the edge effects in the Fourier transform of the *Range* function proposed in equation 10, to take into account the spectrometer range. To modify the shape of the rectangular function the windowing technique can be used (Oppenheim et al, 2000). With this tool it is possible to minimize the effects that result in spectral leakage in the *FI(x)* signal, and to increase the resolution in the OPD measurement.

There are many types of windows available in the literature. Figure 16 shows common functions (Rectangular, Bartlett, Hanning, Hamming, Blackman, Blackman-Harris and Gaussian).

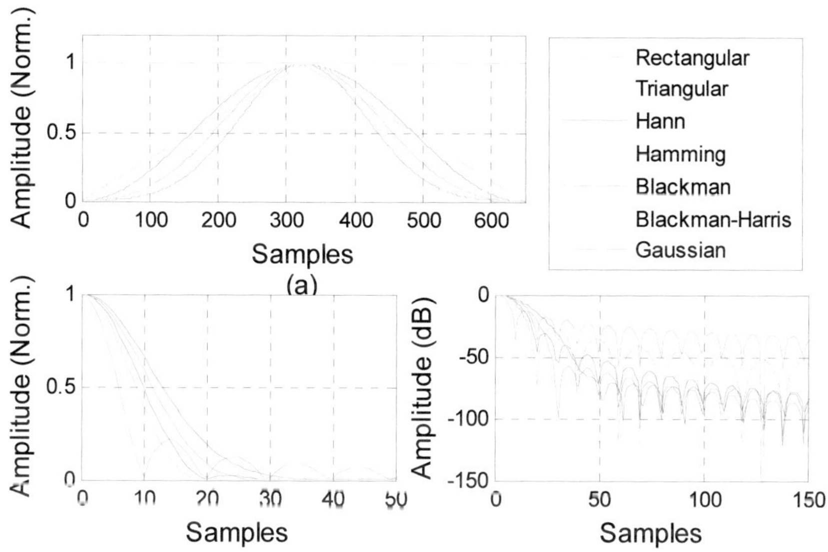

Fig. 16. Typical windows functions and b) the corresponding FFT curves. c)

A spectral analysis of the different window functions is recommended to determine the best to use. In some cases it is possible to apply more than one window simultaneously.

To increment the resolution some techniques can be used (Boaz Porat, 1997). One of the most popular is known as Zero Padding (ZP) (Dorrer et al, 2000; Yun et al, 2003). In this technique the original array *I[k]* is extended with zeros, then, the length of the new array changes to $M_p > N_p$. As a consequence, the sampling interval can be reduced by a N_p/M_p factor with the disadvantages that mean computing longer vectors and more time consuming.

An alternative technique that allows reducing the sampling interval Δx_{samp} in a similar way to ZP but in a localized region of the x axis is the Chirp Fourier Transform (CFT). This predefines the sampling interval and the region of interest on the x axis.

If a sequence I[n] with $0 \leq n \leq N_p - 1$ is used, we can write the DFT:

$$F(\theta[q]) = \sum_{n=0}^{N_p-1} I[n] e^{-j\theta[q]} \quad / \quad \theta[q] = \theta_0 + q\Delta\theta, \ 0 \leq q \leq q-1 \tag{22}$$

Where $\Delta\theta$ is the sampling resolution desired, θ_0 the initial x value and K the number of additional points. Therefore, through θ_0, $\Delta\theta$ and N_p, you can delimit the frequency range in which you want to work. The operation number required is approximately M+N.

In figure 17, the original signal FI[x], FI_{ZP}[x] and FI_{CFT}[x] are shown, the sequence I[k] has 640 points but in the FI[x] only six point defined the interference peak (figure 17-a). To increment the resolution in a factor of 5, we used ZP and CFT. The resolution level is equal in both cases but FI_{ZP}[x] has 3200 points and FI_{CFT}[x] has only 81 points as showed in figure 17-b and 17-c.

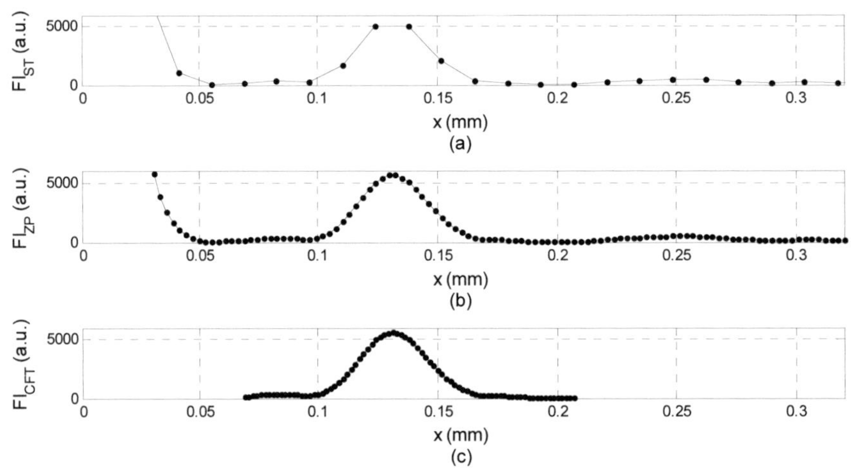

Fig. 17. (a) FI[x], (b) FI_{ZP}[x] and (c) FI_{CFT}[x].

3. Experimental results

In this section we show results obtained from an experimental configuration similar to that presented in section 2.1. The idea is to show that with a relatively simple set-up, it is possible to apply this technique in a variety of applications with interesting results. The light source is a superluminescent infrared diode, centered at 840nm, with a 20- nm bandwidth and a 5-mw output power. The interferometer is a Michelson type, in air , and the detector is a spectrometer Ocean Optics model 4000. With this configuration we get a 2-mm dynamic range and the axial resolution is lower than 10 microns.

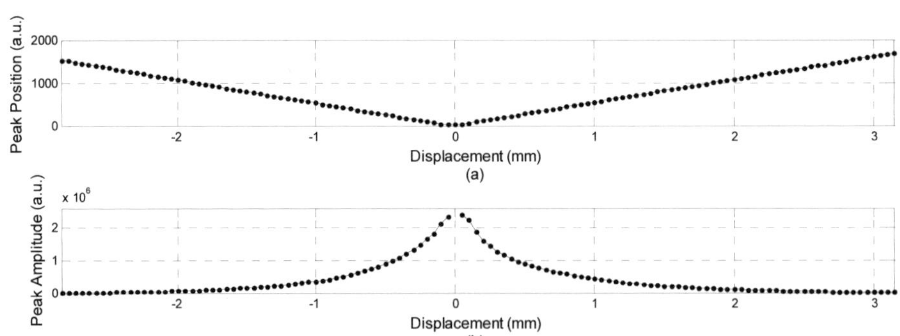

Fig. 18. a) Calibration curve and b) Amplitude variation of interference peak.

At first a calibration process is shown. As it has been mentioned before, it is necessary to link the interferometer signal with the OPD absolute values. A simple way to do this is to move the reference surface through controlled displacements (OPD) or steps, and simultaneously register the abscise coordinate of the center of the FTI peak that corresponds to each step. Figure 18-a shows the peak abscise coordinate (pixel number) versus displacement. Figure 18-b shows the peak amplitude versus the corresponding displacement.

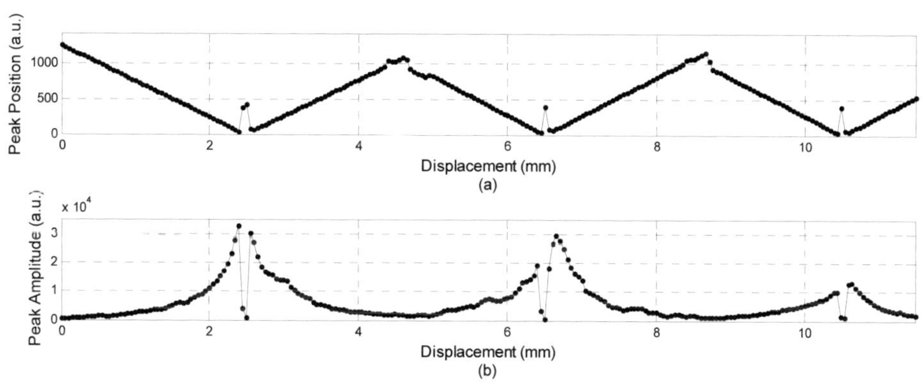

Fig. 19. a) OPD in pixel vs Displacement and b) Amplitud variation of interference peak.

The amplitude fall-off shows the consequence of the "real parameters" of the detector explained in section 2.3. For this particular set-up the *psf* function is predominant.

When the OPD value is close to the limit imposed by Nyquist, aliasing effects can affect the interference signal (Dorrer et al, 2000). If the Nyquist limit is transposed, the measurement shows no-real OPD values that can be misunderstood (figure 19).

In the next example the technique is used to obtain the profile of a metal sample. The sample is a gauge (class 2-norm ISO 3650, model M7T the C.E. Johansson Inc.); its nominal thickness is 1100 ± 0.45μm; this gauge was placed on a second gauge which was used as a reference plane, as shown in figure 20. A 4- mm length profile was obtained on the region indicated in the figure 20 (2 mm on the gauge surface and 2mm on the reference plane).

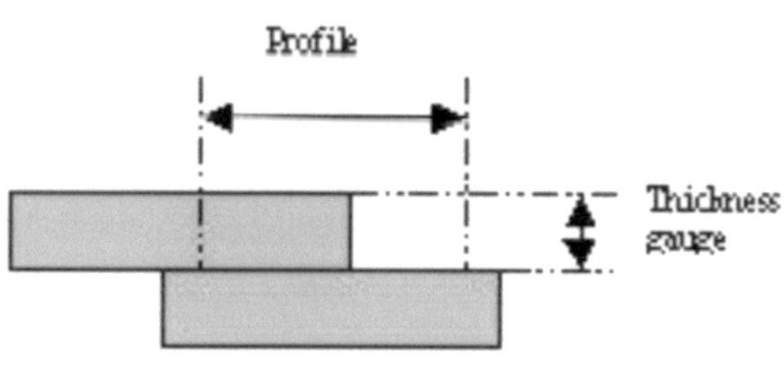

Fig. 20. Schematic of the sample.

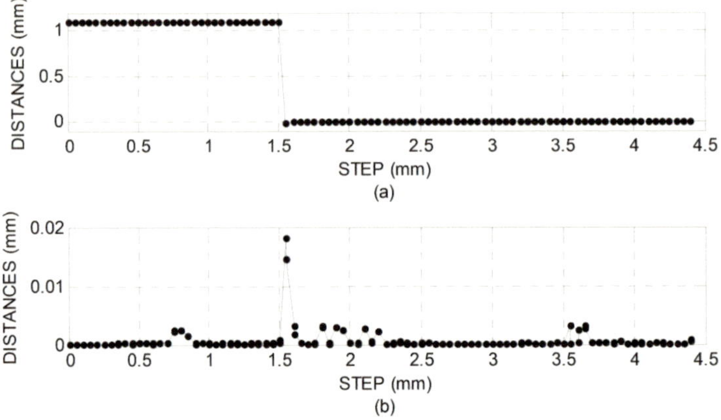

Fig. 21. a) Profile obtained by SDLCI by a sequence of points measured on the sample surface. b) Dispersion in the axial distance measurement.

The profile is obtained by measuring the axial distance in a sequence of points on the sample surface (figure 21-a). Each point is measured after a lateral displacement by steps of 50 microns. The value of gauge thickness obtained from this measurement is: 1100 ± 0.71 μm. Figure 21-b shows the dispersion in each point measured, which is much lower than the coherence length of the source, except on the borders of the sample step.

In the following example an application on 3D- surface measurement is shown. These measurements are critical to the successful manufacturing of precision parts. Components and structures ranging from submillimeter to centimeter size can be found in many fields including the automotive, aerospace, semiconductor and data storage industries.

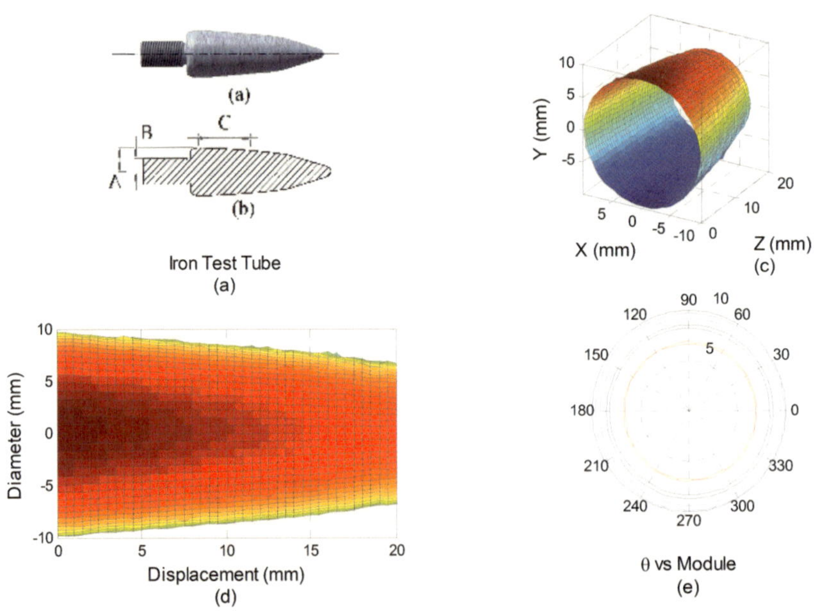

Fig. 22. b)- c) and d) are 3D scan results of a sector of the sample shown in a).

The last example is to show a different configuration in the interferometer system.

With this set-up it is possible to obtain measurement of two sample's face simultaneously employing a third reference surface, which is an advantage over other techniques as the interferometric gauge block, (Decker & Pekelsky, 1997). This is very useful for thickness measurements in opaque samples or where the refraction index is unknown. In figure 23 it is shown a basic Sagnac-Michelson interferometer set-up (Morel & Torga, 2009). This is a ring interferometer type were M_1 is the reference mirror; M_2 and M_3 are the mirrors employed to send the light to each face of the sample (S in the figure 1).

BS_1 and BS_2 are two beam splitters. The light source is a superluminescent diode and the detector system ends in a spectrometer. The reference arm ends in mirror E_1. After BS_2, the two beams are directed to each of the faces of the sample; after their reflections in the sample face and reflection E_1, we obtain three interference signals which let us know the sample thickness. We measure the interference signals in the detector system.

In order to identify the origin of each of the interference signals employed in the measurements, we define D_r as the optical path difference (OPD) between the reference arm and the beam in the ring interferometer when the sample has been removed; D_2 and D_3 represent the OPD between the reference arm and the reflections in each of the sample faces, and D_1 is the OPD between both samples faces.

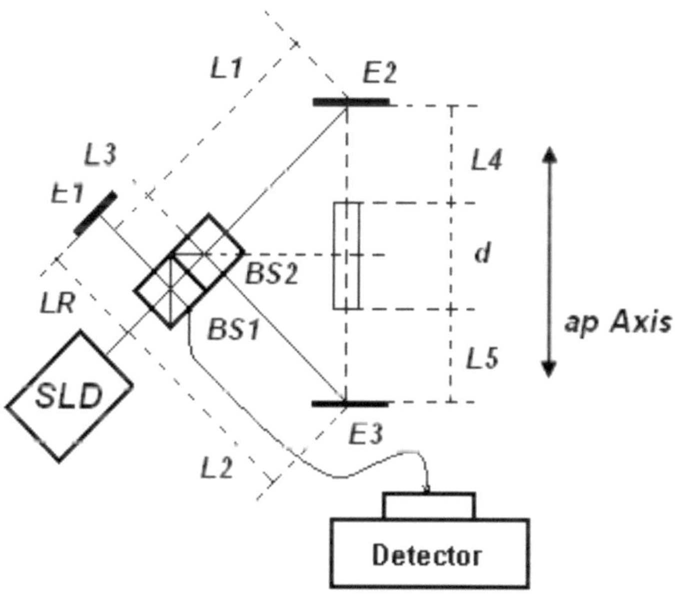

Fig. 23. Sagnac-Michelson interferometer.

The expressions for the different OPD are defined as follow:

$$D_r = (L_1 + L_2 + 2L_3 + L_4 + L_5 + d) - 2(LR) \qquad (23)$$

$$D_1 = 2(L_4 + L_1 + L_3) - 2(L_5 + L_2 + L_3) \qquad (24)$$

$$D_2 = 2(L_4 + L_1 + L_3) - 2(LR) \quad (25)$$

$$D_3 = 2(L_5 + L_2 + L_3) - 2(LR) \quad (26)$$

We showed that it is possible to get the sample thickness (d) with the following relations:

$$d = \left| \frac{(D_2 + D_3) - 2D_r}{2} \right| \quad (27)$$

As D_2 and D_3 are the OPD between the sample and a reference plane, a lateral displacement of the sample let us obtain the surface topography of each of the faces. From the same curves it is possible to improve the alignment of the sample with the reference plane (in our set-up the mirror E_1).

An example of a typical measurement obtained with this set-up is shown in figure 25. The nominal thickness of each gauge is 1.1 mm and 1.05 mm. Both are positioned so that one side has an exposed step and the other side forms a planar surface. The total lateral displacement in the experiment is about 10 mm. We called e1 to the step between the two gauges; e_3 and e_2 are the thickness of each gauge.

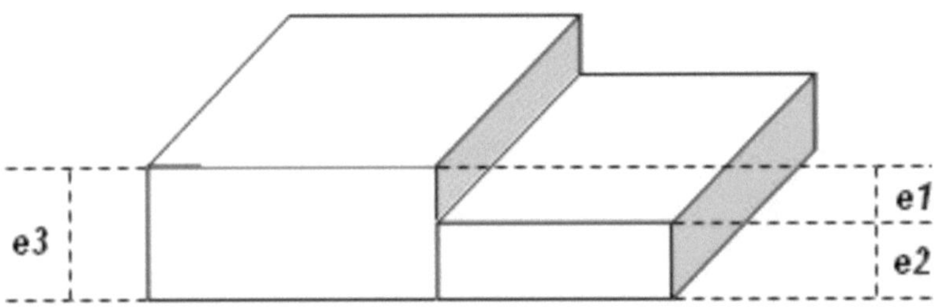

Fig. 24. Schematic of the sample.

In figure 25.a it is shown the profile obtained from the measurement of D_1. Figures 25.b and 25.c show the profiles of both faces of the sample (D_2 and D_3). From these results it is possible to obtain the height of the step (e1 in figure 24) and the thickness of both gauges (e3 and e2 in the figure 24). The slopes of these curves let us to obtain the relative alignment of each of the sample faces with a reference plane and the relative alignment between both gauges.

The values obtained for the gauges thickness are 1091.7 μm and 1047.46 μm, with an average dispersion of 1.32 μm, in very good agreement with the expected value.

There is a dark zone in the curves that appears when the sample presents abrupt changes in surface topography, so the light is scattered in high angles and the collected light intensity is under the detection sensitivity. The optic to focus light on the sample can be selected to minimize dark zone usually lowering the dynamical range of the interferometer.

This method enables simultaneous measurements of physical thickness and refractive group index without any prior knowledge on samples as showed in (Park et al, 2011).

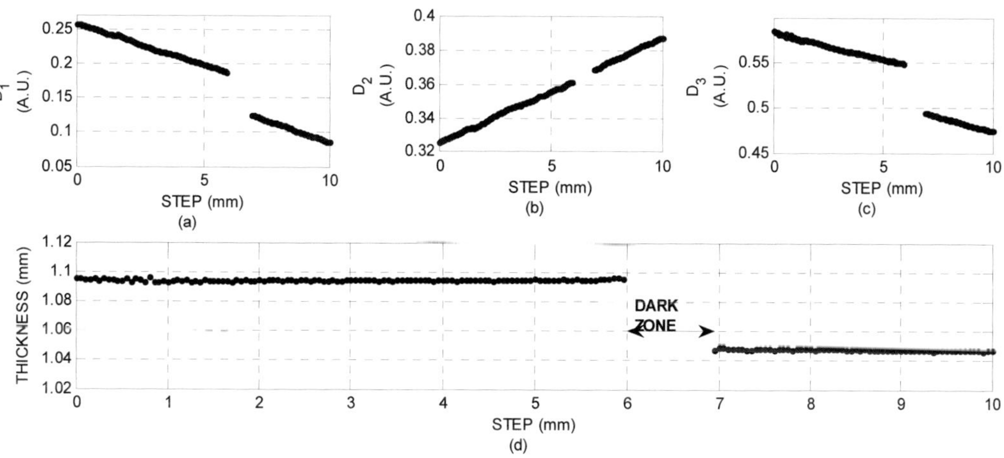

Fig. 25. Surface profiles and thickness obtained from the sample shown in figure 24.

4 Conclusions

In this chapter we present a brief introduction to the theory of low coherence interferometry and the main configurations used nowadays. Then, we focus in giving a detailed analysis of the technique "Spectral domain low coherence interferometry" that offer interesting advantages over alternative methods. We give a description of the limitations and experimental design considerations to take account in the detection of the interference intensity and in the signal processing. We also show how this knowledge is important to improve the information obtained from the images.

At the end we show some experimental results with the intention of giving examples of the enormous potential of this technique, specially in the non- destructive tests area, where we think is nowadays largely unexplored.

5. Acknowledgment

E. Morel and J. Torga are professors at the Universidad Tecnológica Nacional and are members of the Consejo Nacional de Investigaciones Científicas y Técnicas de la República Argentina.

Thanks to Hernán Miranda for his collaboration and special thanks to Pamela Morel for her support and images design.

6. References

Bajraszewski,T; Wojtkowski, M.; Szkulmowski, M.; Szkulmowska, A.; Huber, R. and Kowalczyk A., "Improved spectral optical coherence tomography using optical frequency comb" Opt. Express 16, 4163-4176 (2008). doi:10.1364/OE.16.004163.

Boaz, P.; "A course in digital signal processing" John Wiley & Sons, 1997. ISBN: 978-0-471-14961-3.

Born, M. & Wolf, E. (2002), Cambridge University Press, ISBN 13 978-0-521-64222-4, London, England.

Bruce L. Danielson and C. Y. Boisrobert, "Absolute optical ranging using low coherence interferometry," Appl. Opt. 30, 2975-2979 (1991). doi:10.1364/AO.30.002975.

Choma, M,; Sarunic, M.; Yang, C.; and Izatt, J. "Sensitivity advantage of swept source and Fourier domain optical coherence tomography," Opt. Express 11, 2183-2189 (2003). doi:10.1364/OE.11.002183.

Decker JE, Pekelsky JR. "Gauge block calibration by optical interferometry at the national research council of Canada". Measurement Science Conference Pasadena, California, 23–24 January 1997; NRC Internal Report no. 40002.

Dorrer, C.; Belabas, N.; Likforman, J.; and Joffre, M.; "Spectral resolution and sampling issues in Fourier-transform spectral interferometry," J. Opt. Soc. Am. B 17, 1795-1802 (2000). doi:10.1364/JOSAB.17.001795.

Drexler, W. & Fujimoto, J. "Optical Coherence Tomography: Technology and Applications", Springer Publishing, (2008). ISBN 978-3-540-77549-2, 1400 pages.

Fercher, A.; Hitzenberger, C.; Kamp, G.; El-Zaiat, S.; "Measurement of intraocular distances by backscattering spectral interferometry," Opt.Commun. 117, 43-48,(1995). DOI: 10.1016/0030-4018(95)00119-S.

Fujimoto, J. (2002). Optical Coherence Tomography: Introduction. In: Bouma B E & Tearney G J (eds), Handbook of Optical Coherence Tomography, Marcel Dekker, Inc., New York,USA.

Goode, B. (2009). Optical Coherence Tomography: OCT aims for industrial application, Laser Focus World, Vol 45 (9).

Goodman, J. (1984). Statistical Optics, John Wiley & Sons, ISBN 0-471-01502-4.

Häusler, G. and Lindner, M.; "Coherence radar and spectral radar - new tools for dermatological diagnosis" J. Biomed. Opt. 3, 21-31(1998). doi:10.1117/1.429899.

Huang, D.; Swanson, EA .; Lin, CP.; Schuman, JS .; Stinson, WG .; Chang, W .; Hee, MR .; Flotte, T.; Gregory, K.; Puliafito, CA. and Fuyimoto, J.; "Optical coherence tomography", Science, 22 November 1991:Vol. 254. DOI: 10.1126/ science.1957169.

Hu, Z.; Pan, Y. & Rollins, A.; "Analytical model of spectrometer-based two-beam spectral interferometry," Appl. Opt. 46, 8499-8505 (2007). doi:10.1364/AO.46.008499.

Jeon, M.; Kim, J.; Jung, U.; Lee, C.; Jung, W. and Boppart, S.A.; "Full-range k-domain linearization in spectral-domain optical coherence tomography," Appl. Opt. 50, 1158-1163 (2011). doi:10.1364/AO.50.001158.

Morel, E.N. ; Torga, J.R.; "Dimensional characterization of opaque samples with a ring interferometer", Optics and Lasers in Engineering, Volume 47, Issue 5, May 2009, Pages 607-611, ISSN 01438166.

Leitgeb, R.; Hitzenberger, C. and Fercher, A.; "Performance of fourier domain vs. time domain optical coherence tomography," Opt. Express 11, 889-894 (2003). doi:10.1364/OE.11.000889.

Oppenheim A.V., R.W. Schafer y J.R. Buck Tratamiento de señales en tiempo discreto, Prentice Hall, Madrid (2000). ISBN: 8420529877.

Papoulis, A.; (1962), "The Fourier Integral and its Applications", Mc Graw Hill, June 1, 1962, ISBN-13: 978-0070484474.

Park, S.J.; Park, K.S.; Kim, Y.H. and Lee,*B.H.;* "Simultaneous Measurements of Refractive Index and Thickness by Spectral-Domain Low Coherence Interferometry Having Dual Sample Probes", ieee Photonics Technology Letters, 1041-1135, vol. 23, no. 15, august 1, 2011. DOI: 10.1109/lpt.2011.2155642.

Schmitt, J. ;(1999). "Optical coherence tomography (OCT): a review" IEEE Journal of Selected Topics in Quantum Electronics", Vol. 5, No. 4, July/August. DOI: 10.1109/2944.796348. ISSN: 1077-260X.

Swanson, E. "OCT News". Available from: www.octnews.org.

Takada, K.; Yokohama, I.; Chida, K. & Noda J.; (1987). "New measurement system for fault location in optical waveguide devices based on an interferometric technique," Appl. Opt. 26, 1603-1606. doi:10.1364/AO.26.001603.

Vakhtin, A.; Peterson, K.; Wood, W. & Kane, D. (2003), "Differential spectral interferometry: an imaging technique for biomedical applications," Opt. Lett. 28, 1332-1334. doi:10.1364/OL.28.001332.

Wang, H.; Pan, Y.; and Rollins A.; "Extending the effective imaging range of Fourier domain optical coherence tomography using a fiber optic switch," Opt. Lett. 33, 2632-2634 (2008). doi:10.1364/OL.33.002632.

Wang, Z.; Yuan, Z.; Wang, H.; and Pan, Y.; "Increasing the imaging depth of spectral-domain OCT by using interpixel shift technique," Opt. Express 14, 7014-7023 (2006). doi:10.1364/OE.14.007014.

Wieasuer, K.; Pircher, M.; Götzinger, E.; Bauer, S.; Engelke, R.; Ahrens, G.; Grützner, G.; Hitzenberger, C. & Stifter, D. (2005), "En-face scanning optical coherence tomography with ultra-high resolution for material investigation," Opt. Express 13, 1015-1024.

Wiesner M.; Ihlemann, J.; Müller, H.; Lankenau, E. & Hüttmann, G. (2010). "Optical coherence tomography for process control of laser micromachining"; Review of Scientific Instruments, Vol. 81, Iss. 3, pg. 033705.

Wojtkowski, M. ; Leitgeb, R.; Kowalczyk, A.; Bajraszewski, T. and Fercher, A.; "In vivo human retinal imaging by Fourier domain optical coherence tomography", J. Biomed. Opt. 7, 457 (2002); doi:10.1117/1.1482379.

Wojtkowski, M.; Bajraszewski, T.; Targowski, P.; Kowalczyk, A.; "Real-time in-vivo ophthalmic imaging by ultrafast spectral interferometry" Proc. SPIE 4956, 4956-11 (2003). doi:10.1364/OE.11.000889.

Youngquist, R; Sally Carr, & Davies D. (1987 "Performance of fourier domain vs. time domain optical coherence tomography," Opt. Express 11, 889-894 (2003). doi:10.1364/OE.11.000889.

Yun, S.; Tearney, G.; Bouma, B.; Park, B. and Boer, J.; "High-speed spectral-domain optical coherence tomography at 1.3 μm wavelength," Opt. Express 11, 3598-3604 (2003). doi:10.1364/OE.11.003598.

7

Speckle Interferometry for Displacement Measurement and Hybrid Stress Analysis

Tae Hyun Baek[1] and Myung Soo Kim[2]
[1]*School of Mechanical and Automotive Engineering, Kunsan National University,*
[2]*Department of Electronic Engineering, Kunsan National University,*
Daehangno, Gunsan City,
Jeonbuk,
The Republic of Korea

1. Introduction

1.1 Application of ESPI to measurement of out-of-plane displacement in a spot welded canti-levered plate

Laser speckle interferometry can be used to detect the locations of stress concentration of an object and deformation over a whole area to be measured through the shape information in fringe patterns (Cloud, 1998; Petzing & Tyrer, 1998). ESPI can obtain interferometric fringes by a subtraction process ($I = |I_{\text{before}} - I_{\text{after}}|$) of the image data. Small displacement can be measured from such interferometric fringes corresponding to the order of the laser wavelength.

Several researches about in-plane displacement and vibration properties analyzed by ESPI have been reported and the application of ESPI is increasing (Rastogi, 2001). In this paper, the out-of-plane displacements of a partially spot welded canti-levered plate and of a normal canti-levered plate are measured and compared by application of 4-step phase shifting method to the analysis of fringe patterns in ESPI (Baek et al., 2002).

1.2 Optics of ESPI

1.2.1 ESPI for measurement of out-of-plane displacement

Figure 1-1 shows the arrangement of the ESPI optical system for measuring out-of-plane displacement. The interferometric fringe patterns caused by phase differences between the specimen and the reference plane with a PZT (piezoelectric transducer) must be analyzed because the interferometric fringe patterns contain information about out-of-plane displacement of the specimen. The phase difference is made by an optical path difference. The out-of-plane displacement, w, caused by the optical path difference is as follows:

$$w = \frac{\lambda}{4\pi}\phi \tag{1-1}$$

In Eq. (1-1), w is z-directional (out-of-plane) displacement, λ is the wavelength of the laser, and ϕ represents the phase difference between before and after out-of-plane displacement.

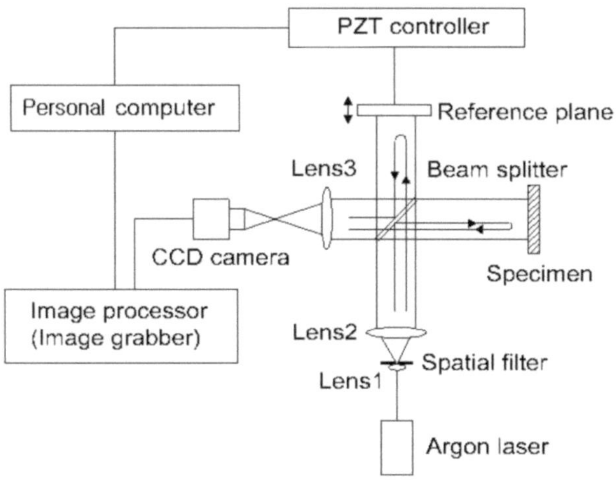

Fig. 1-1. Block diagram of ESPI optical system for measuring out-of-plane displacement.

1.2.2 4-step phase shifting method in ESPI

The 4-step phase shifting method used in this paper is employed to move the reference plane with the PZT by $\pi/2$ radians in each step and to obtain four fringe patterns with relative phase difference. The phase map is obtained from the four fringe patterns using the arc tangent function. The light intensity of the fringe pattern in ESPI, I_i, is expressed as follows:

$$I_i = I_0\{1 + m(x,y)\cos[\phi(x,y) + \alpha]\} \tag{1-2}$$

In the above equation, $I_i(x,y)$ is the measured light intensity, I_0 is the average intensity, $m(x,y)$ is the contrast, $\phi(x,y)$ is the phase difference, and α is the phase introduced by PZT. The phase map, where the magnitude and sign of displacement can be known, can be obtained as follow.

$$\phi(x,y) = \tan^{-1}\left[\frac{I_4 - I_2}{I_1 - I_3}\right] \tag{1-3}$$

In Eq. (1-3), I_1, I_2, I_3 and I_4 are the light intensities at $\alpha = 0$, $\pi/2$, π, $3\pi/2$ radians, respectively. Eq. (1-3) uses the four fringe patterns with different phases, so that this method is called the 4-step phase shifting method. The phase map obtained by this method has the phase between $-\pi$ radians and $+\pi$ radians due to the property of the arc tangent function. Thus, the phase map has the discontinuous phase at every 2π radians, but this discontinuity can be eliminated by use of the phase unwrapping process and the continuous displacement of the specimen can be obtained (Ghiglia & Pritt, 1998).

The speckle pattern contains lots of noise and the noise must be filtered out before the phase unwrapping process. The Gaussian blur process in a commercial image processing software package (Adobe Photoshop, Version 5.5) is used to eliminate the speckle noise.

1.3 Experiment

1.3.1 Specimen and experimental set-up

The specimen used in this experiment is a 2 mm-thick canti-lever made of steel plate (E=200 GPa, $\nu = 0.3$). The shape and size of the specimen is shown in Fig. 1-2. Fig. 1-2 (a) is the normal canti-levered plate that is not spot-welded, and Fig. 1-2 (b) is the canti-levered plate that is spot-welded on the rear side.

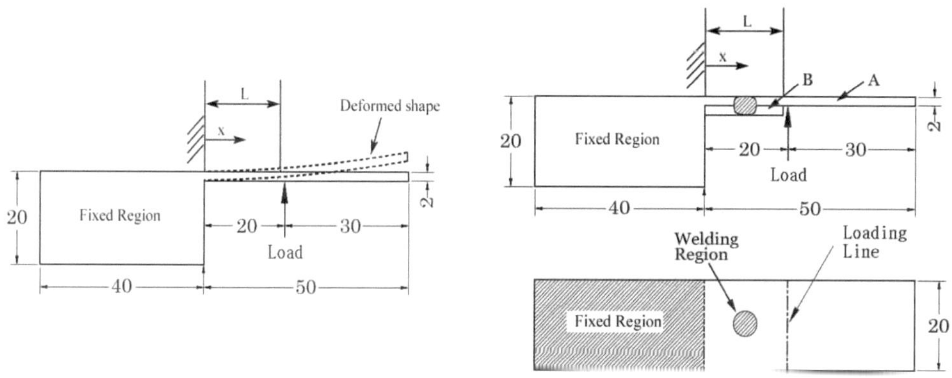

(a) Before weld (b) After weld

Fig. 1-2. Dimensions of the specimen used for measurement of out-of-plane displacement. (a) Normal canti-levered plate, (b) Spot welded canti-levered plate.

Figure 1-3 shows the alignment of the optical components in which the optical set-up for Twyman-Green interferometry is used to measure the out-of-plane displacement. The phase shifting method is used to advance the precision degree of the speckle pattern in the measurement of out-of-plane displacement, and the phase shifting is performed by the PZT that is controlled with a personal computer.

The expression for the deflection curve for a canti-lever beam subjected to a concentrated load P, as shown in Fig. 1-2 (a), is

$$\delta = \frac{P}{6EI}\left(x^3 - 3Lx^2\right) \tag{1-4}$$

In Eq. (1-4), E is the Young's elastic modulus, I is the moment of inertia, L is the distance from the fixed support to the loading point, and x is the distance from the fixed support to any arbitrary point.

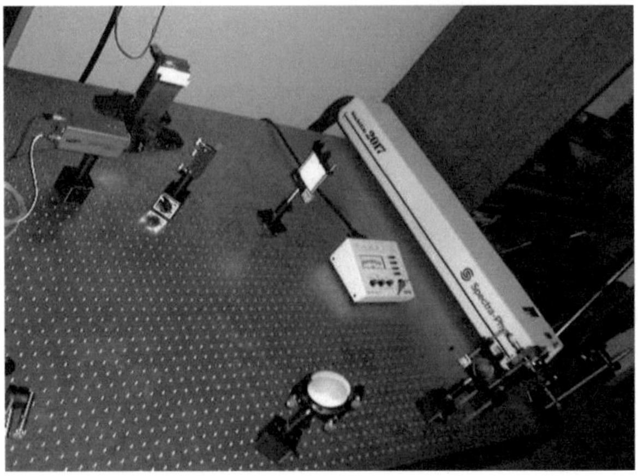

Fig. 1-3 Optical systems for ESPI measurement of out-of-plane displacement.

1.3.2 Results of experiment

Figure 1-4 shows the speckle fringe patterns of a normal canti-lever at difference phases for $\alpha = 0, \pi/2, \pi, 3\pi/2$. Figure 1-5 shows the speckle fringe patterns of a spot welded canti-levered plate at difference phases for $\alpha = 0, \pi/2, \pi, 3\pi/2$. Those fringe patterns are obtained through a subtraction of before-displacement and after-displacement results of a specimen. The different phases in Figs. 1-4 and 1-5 are created by the PZT which controls the phases of fringe pattern by $\alpha = 0, \pi/2, \pi, 3\pi/2$, as mentioned before.

As seen in Figs. 1-4 and 1-5, the change of fringe patterns is shown near the welded area of the spot welded specimen. To eliminate the noise of the high frequency component in the speckle, the Gaussian blur process is applied to the original image obtained in the experiment. Figures 1-6 (a) and (b) are the original image and the Gaussian blurred image, respectively. When the Gaussian blurred image is compared with the original image in Fig. 1-6, it is clear that the high frequency noise is eliminated.

Figure 1-7 shows the light intensities along the line A-A in Fig. 1-6 (a) and (b). There is no doubt that the high frequency noise is eliminated in the Gaussian blurred image. Figure 1-8 is the phase map calculated by Eq. (1-3) in which four Gaussian blurred images are used.

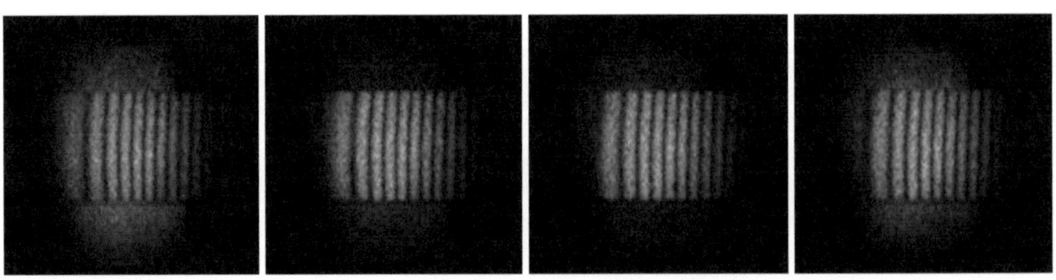

Fig. 1-4. Speckle fringe patterns of a normal canti-levered plate at difference phases for $\alpha = 0, \pi/2, \pi, 3\pi/2$ radians.

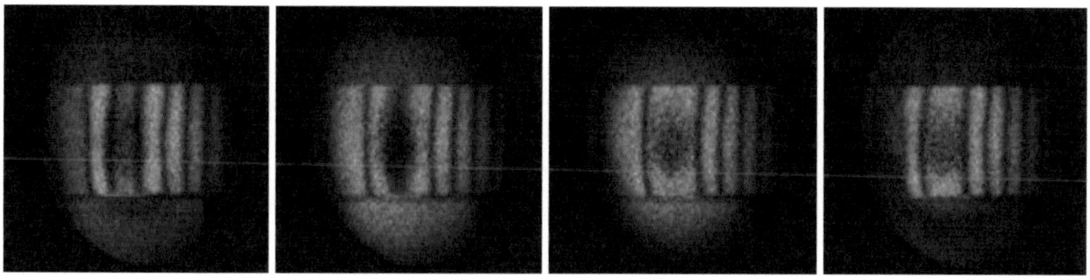

Fig. 1-5. Speckle fringe patterns of a spot welded canti-levered plate at difference phases for $\alpha = 0, \pi / 2, \pi, 3\pi / 2$ radians.

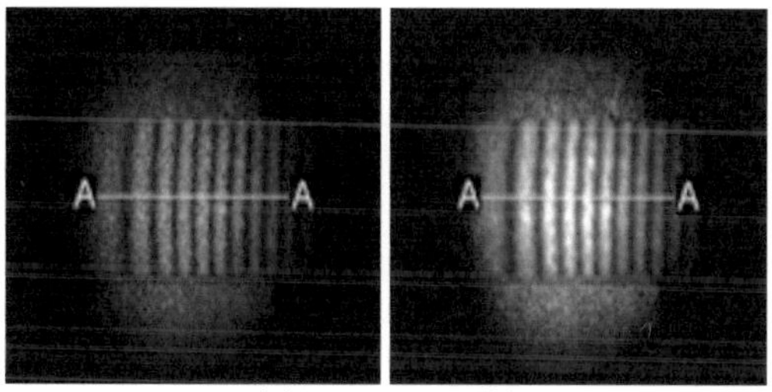

(a) (b)

Fig. 1-6. (a) Original image and (b) Gaussian blurred image obtained from Fig. 1-4.

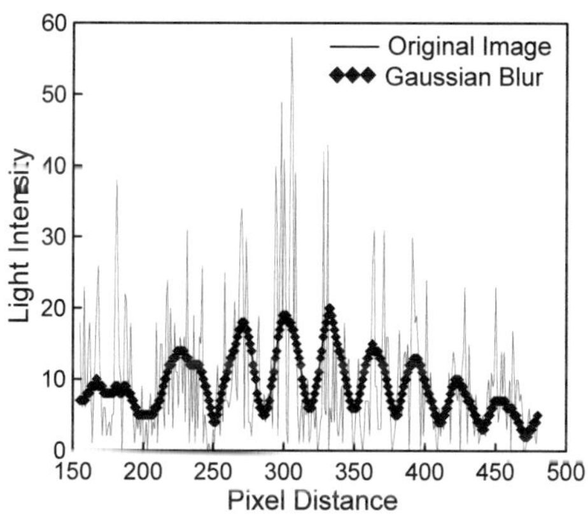

Fig. 1-7. Comparison of light intensity along line A-A of Figs. 1-6 (a) and (b).

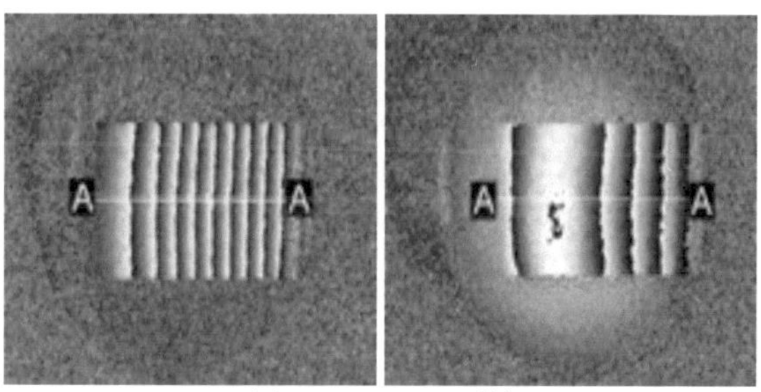

(a) (b)

Fig. 1-8. Wrapped phase images of (a) normal and (b) spot welded canti-lever plate.

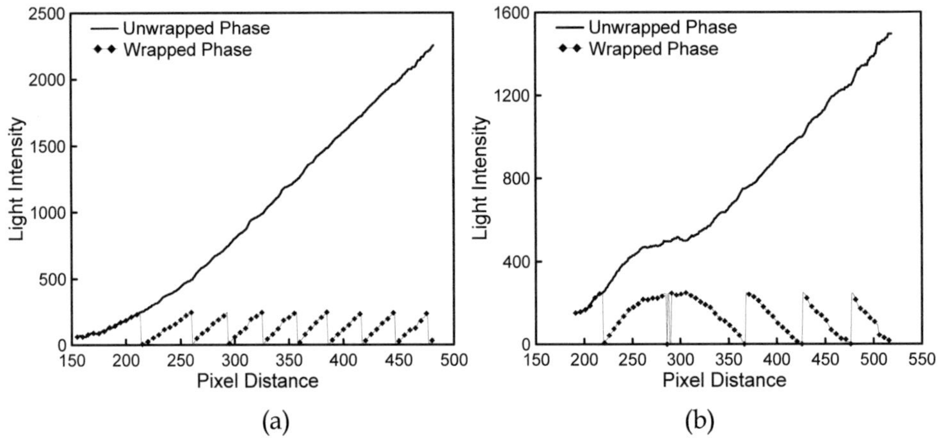

Fig. 1-9. (a) Wrapped and unwrapped phase distributions along line A-A of Fig. 1-8 (a) and along line A-A of Fig. 1-8 (b).

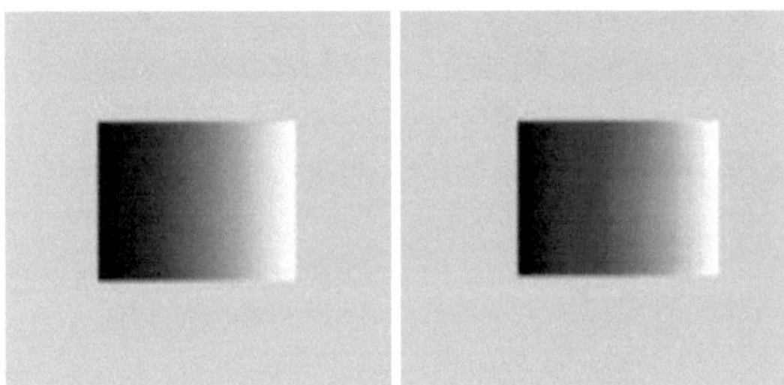

(a) (b)

Fig. 1-10. Unwrapped phase maps of Fig. 1-8 (a) and (b).

When Figs. 1-8 (a) and (b) are compared, the phase map of the normal canti-levered plate is uniform, but that of the spot welded canti-levered plate shows a phase reversal at the welded area. Figure 1-9 (a) shows wrapped and unwrapped phase distributions along line A-A of Fig. 1-8 (a) of the normal canti-levered plate. Figure 1-9 (b) shows wrapped and unwrapped phase distributions along line A-A of Fig. 1-8 (b) of the spot welded canti-levered plate. Figure 1-10 (a) and (b) are the unwrapped phase maps of Fig. 1-8 (a) and (b). Figure 1-11 is 3-D view of the unwrapped phase image of Fig. 1-10. It is clearly seen in Figs. 1-11 (a) and (b) that continuous displacement occurred in the normal canti-levered plate but the displacement at the spot welded area was hump-shaped in the spot welded canti-levered plate.

Figure 1-12 (a) shows the displacement distribution obtained from the theory and from the phase shifting method along line A-A of Fig. 1-8. It shows that the result of ESPI is almost the same as that of the theoretical calculation for the normal canti-levered plate which is not spot welded. The maximum error of 0.076 μm occurs at approximately 7.9 mm from the fixed area of the canti-levered plate. However, in general, the measured displacement by the ESPI experiment is quite close to the theoretically expected displacement. Thus, it is proved that the physical out-of-plane displacement can be directly measured by the ESPI method.

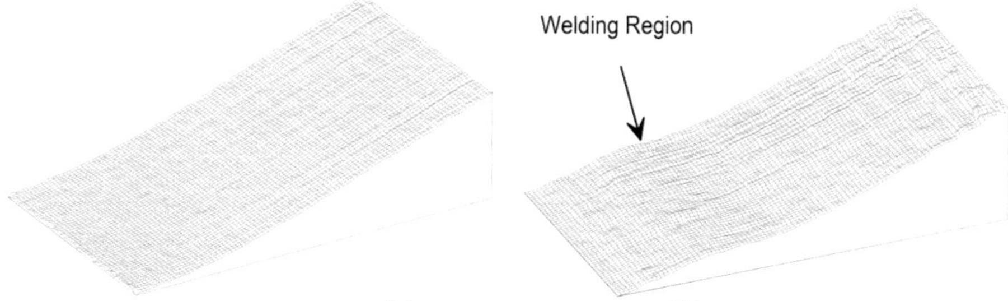

(a) (b)

Fig. 1-11. 3-D view of unwrapped phase image of (a) Fig. 11(a) and (b) Fig. 11(b).

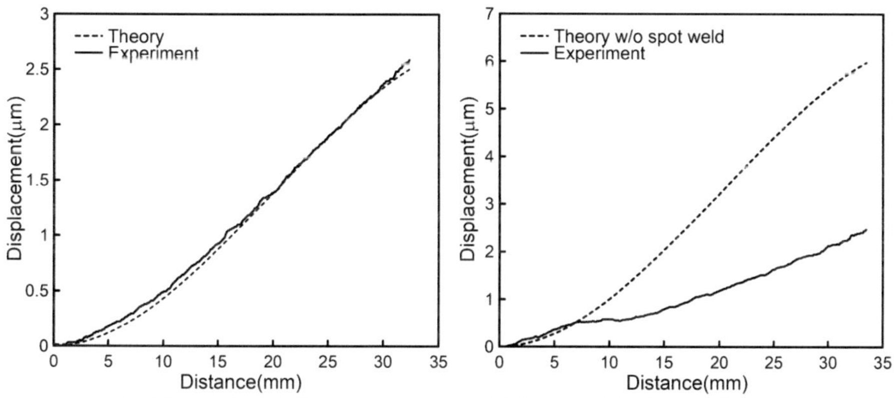

(a) Normal canti-lever (b) Spot welded canti-lever

Fig. 1-12. Displacement distribution obtained from theory and from the phase shifting method along line A-A of Fig. 1-8 (a) and (b).

Figure 1-12 (b) shows the displacement distribution obtained from the spot welded canti-levered plate. The displacements, 0.582 μm , 1.183 μm , 2.134 μm , are measured at 10 mm, 20 mm, 30 mm from the fixed area of the spot welded canti-levered plate, respectively.

As a reference, the other displacements, 1.006 μm , 3.219 μm , 5.432 μm , are estimated at 10mm, 20mm, 30mm from the fixed area of the normal canti-levered plate. Therefore, the displacement for the same load decreases as the canti levered plate is reinforced by spot-welding, and the spot welded area that is not visible can be easily detected by use of speckle interferometry.

1.4 Conclusions and discussions

The 4-step phase shifting method applied to an ESPI experiment has been used for the measurement of out-of-plane displacement in the normal canti-levered plate and the spot welded canti-levered plate. The measured displacement of the normal canti-levered plate agreed to the theoretical value within 0.076 μm . That is, it is proved that the physical out-of-plane displacement can be directly measured and precise measurement with a nanometer resolution is possible. Also, a welded area that is not visible from the surface can be detected and a small out-of-plane displacement in the welded area can be measured. The distribution of displacement shows a slight non-linearity in the period of fringe, and therefore, further work is necessary for more precise measurement. Also, noise due to speckle has to be eliminated before the phase shifting method is applied. The Gaussian blur process is used to eliminate the noise in this work.

2. Measurement for in-plane displacement of tensile plates with through-thickness circular hole and partly through-thickness circular hole by use of speckle interferometry

2.1 Introduction

Speckle interferometry is an optical technique to measure displacement of a specimen by a coherent light from a laser. There are several phase extraction methods in speckle interferometry to measure the displacement, and phase shifting method (PSM) is one of the methods (Creath, 1988). In PSM, a known phase produced by piezoelectric transducer (PZT) is added into optical beam and resultant interference patterns are processed to get information about displacement of a specimen (Kim & Baek, 2006). In this paper, in-plane displacement of a steel plate with a partly through-thickness circular hole and a steel plate of a through-thickness circular hole is measured by simple optical system of speckle interferometry with PSM. Especially, the circular hole of steel plate with a partly through-thickness circular hole is not visible because the circular hole is cut on the rear side of the plate. This means that one cannot see any deformation or defect of the specimen.

2.2 Phase shifting method in speckle interferometry

Figure 2-1 is an optical system to measure in-plane displacement of specimens by use of PSM in speckle interferometry. A specimen is placed in a loading device and tensile load is applied to the specimen by the loading device in order to make in-plane displacement on the specimen. When two optical beams from a laser illuminate a specimen in speckle interferometry as shown Fig. 2-1, the optical beams make interference fringe patterns. The interference fringe patterns, I_i, can be represented by Eq. (2-1).

$$I_i = I_0 + I_c \cos\left[\phi(x,y) + \alpha_i\right] \tag{2-1}$$

where I_0 is the average intensity of fringe pattern and I_c is the contrast of fringe pattern.

A known phase, α_i, is added into one of the optical beams through controlling PZT. $\phi(x,y)$ is the phase of fringe pattern caused by the in-plane displacement of a specimen which tensile load is applied to. When 0, $\pi/2$, π, and $3\pi/2$ radians are used for α_i, four fringe patterns are taken through CCD camera and stored in PC. When I_1, I_2, I_3, and I_4, represent the fringe patterns for the known phases of 0, $\pi/2$, π, and $3\pi/2$ radians, respectively, the phase, $\phi(x,y)$, is calculated by use of Eq. (2-2).

$$\phi(x,y) = \tan^{-1}\left[\frac{I_4 - I_2}{I_1 - I_3}\right] \tag{2-2}$$

$\phi(x,y)$ in Eq. (2-2) is so-called wrapped phase. Unwrapped phase $\phi_u(x,y)$ is obtained with unwrapping algorithm of the wrapped phase (Ghiglia & Pritt, 1998). Then, the in-plane displacement of specimen, u , is obtained from the phase $\phi_u(x,y)$ as follows (Rastogi, 2001) ;

$$u = \frac{\lambda}{4\pi \sin(\theta/2)} \phi_u(x,y) \quad (2\text{-}3)$$

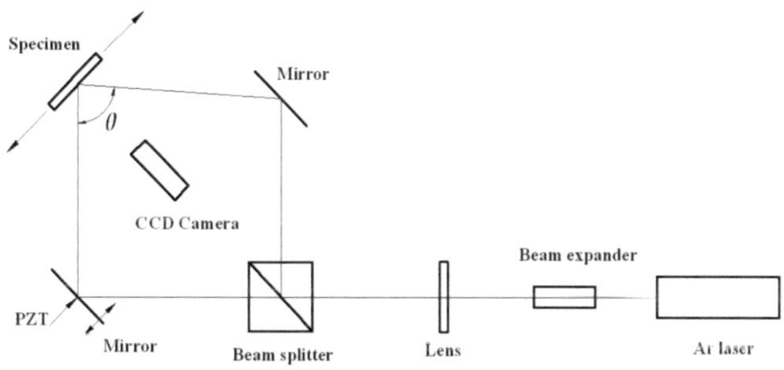

Fig. 2-1. Schematic diagram of speckle interferometry with phase-shifting method for in-plane displacement measurement (θ is the angle between two incident lights onto the specimen).

2.3 Optical experiment

The specimens used in this experiment are a rectangular steel plate with a through-thickness circular hole at the center and a rectangular steel plate with a partly through-thickness circular hole at the center on rear side as in Fig. 2-2. Both of the rectangular steel plates have the same size that is 147 mm x 27.7 mm with thickness of 1.2 mm. The diameter of through-thickness circular hole is 12 mm. Also the diameter of partly through-thickness circular hole is 12 mm but it is uniformly cut by 0.8 mm on the rear side. Figure 2-3 is the picture of optical experiment system. Ar laser with wavelength of 515 nm is used in the optical experiment.

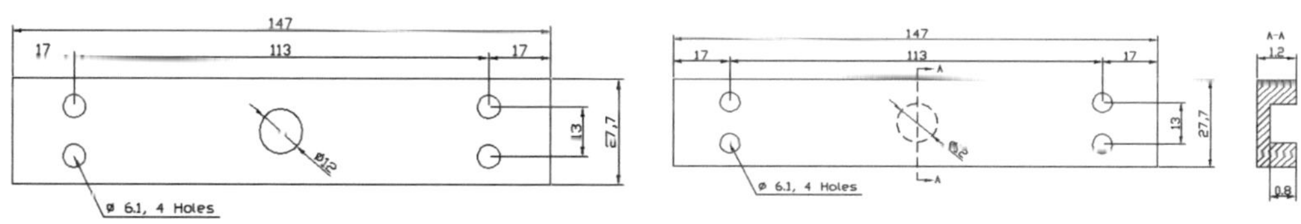

(a) (b)

Fig. 2-2. Dimensions of specimens used in-plane displacement (unit : mm) (a) rectangular steel plate with through-thickness circular hole (b) rectangular steel plate with partly through-thickness circular hole.

At first, optical experiment is performed with the steel plate with through-thickness circular hole at the center. Four fringe patterns, I_1 , I_2 , I_3 , and I_4 , are taken in the experiment and they are stored in PC. However, the four fringe patterns have lots of speckle noises, so that they are processed by use of an image processing algorithm. In this work, Gaussian blur algorithm that is available commercially in Adobe Photoshop is used to process the fringe patterns. The processed four fringe patterns are shown in Fig. 2-4. The same experimental procedures are used to get four fringe patterns for the steel plate with partly through-thickness circular hole at the center on rear side.

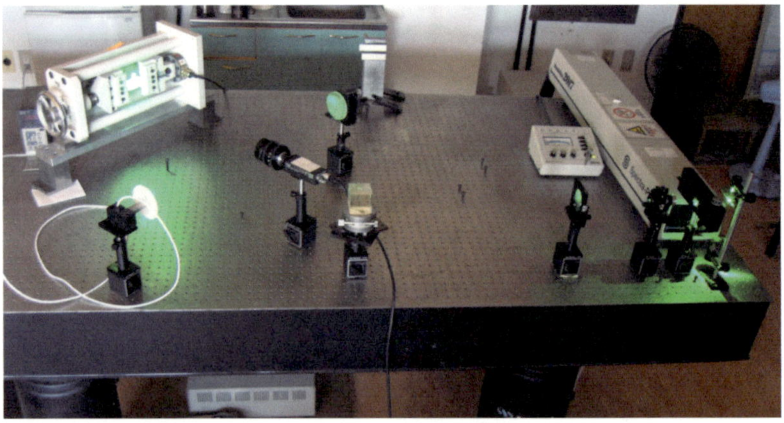

Fig. 2-3. Optical setup for measurement of in-plane displacement.

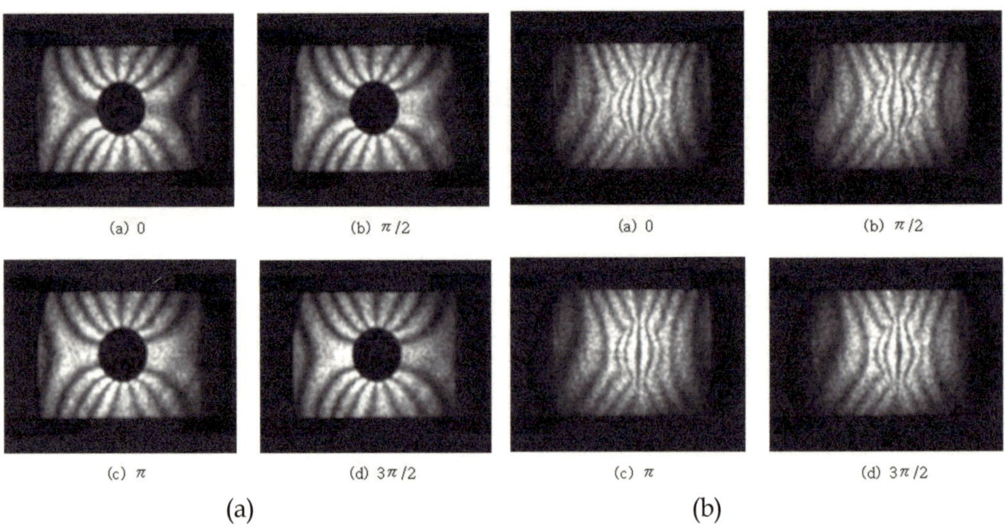

Fig. 2-4. Fringe patterns of specimen : (a) Steel plate with a through-thickness circular hole processed by Gaussian Blur (b) Steel plate with a partly through-thickness circular hole processed by Gaussian Blur.

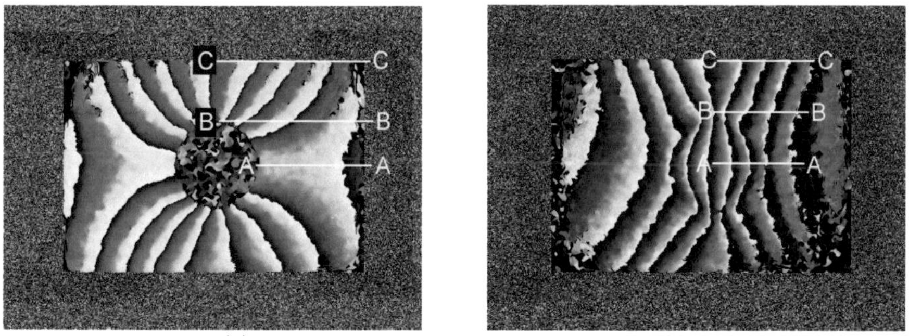

(a) (b)

Fig. 2-5. Wrapped phase maps (a) steel plate with a through-thickness circular hole, (b) steel plate with a partly through-thickness circular hole.

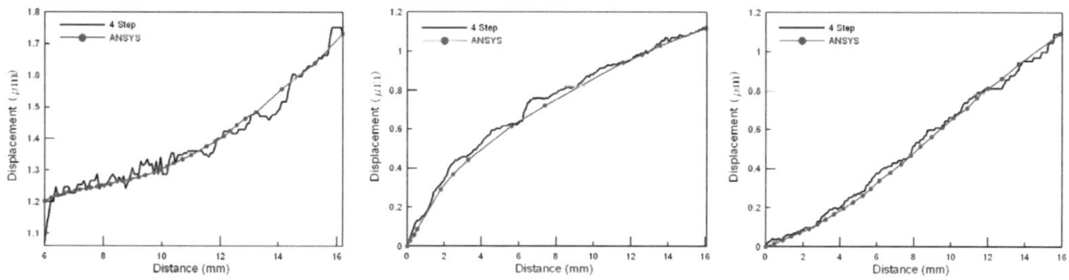

(a) A-A (b) B-B (c) C-C

Fig. 2-6. Displacement distribution along A-A, B-B, and C-C line of Fig. 2-5 (a).

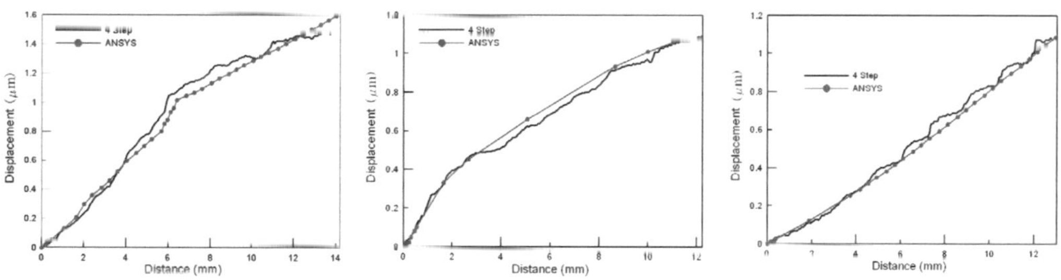

(a) A-A (b) B-B (c) C-C

Fig. 2-7. Displacement distribution along A-A, B-B, and C-C line of Fig. 2-5 (b).

Using Eq. (2-2), the wrapped phases, $\phi(x,y)$, are obtained for the two specimens and shown in Fig. 2-5 (a) and (b). Quantitative data are acquired along the lines, A-A, B-B, and C-C in Fig. 2-5. The phases, $\phi(x,y)$, along the lines are unwrapped and the in-plane displacement of the specimens is plotted in Figs. 2-6 and 2-7 by use of Eq. (2-3).

For comparative purpose, the two specimens of Fig. 2-2 are analyzed by ANSYS. Figures 2-8 (a) and (b) are the models for ANSYS. The physical properties used for the analysis are the same as the physical properties of structural steel that are E=200 GPa and $\nu = 0.3$. The ANSYS discretization for the rectangular steel plate with a through-thickness circular hole at the center used 8 node quadrilateral elements. The ANSYS discretization for the rectangular steel plate with a partly through-thickness circular hole at the center used 10 node tetrahedral elements.

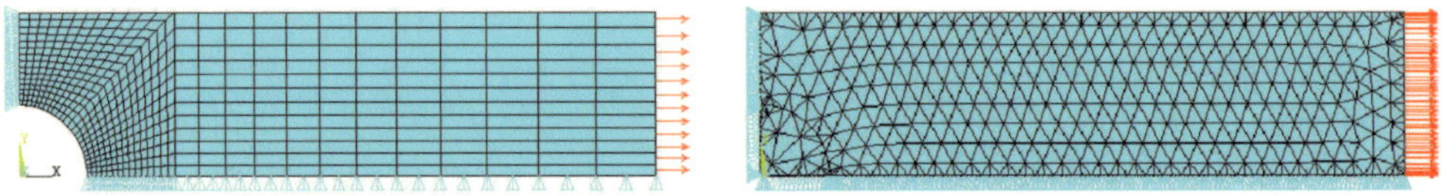

(a) (b)

Fig. 2-8. ANSYS discretization. (a) Steel plate with a circular hole in Fig. 2 (a). (b) Steel plate with partly through-thickness circular hole in Fig. 2 (b).

The results of the two specimens from ANYSYS are plotted also in Figs. 2-6 and 2-7. As shown in Figs. 2-6 and 2-7, the results of the optical experiments for speckle interferometry with 4-step phase shifting method agree with those of ANSYS.

2.4 Conclusions and discussions

In-plane displacements of two specimens are measured by simple optical system of speckle interferometry using 4-step phase shifting method; one is the rectangular steel plate with a through-thickness circular hole and the other is the rectangular steel plate with a partly through-thickness circular hole. The circular hole of the steel plate with a partly through-thickness circular hole is not visible because the circular hole is cut on the rear side of the plate, so that one cannot see any deformation or defect of the specimen. The fringe patterns acquired by optical experiment of speckle interferometry are processed by image processing algorithm of Gaussian blur in Adobe Photoshop and the in-plane displacements of the two specimens are obtained by the processed fringe patterns. Also the in-plane displacements of the two specimens are calculated by use of ANSYS.

The results of optical experiments are quite comparable to those of calculation with ANSYS. Based on the optical experiments, speckle interferometry can be applied to easily detect or measure defect or deformation in a specimen that is not visible.

3. A hybrid stress measurement using only x-displacements by Phase Shifting Method with Fourier Transform (PSM/FT) in laser speckle interferometry and least squares method

3.1 Introduction

Stress raisers have been one of the main concerns when it comes to design analysis. Due to the complexities associated with it, numerous and continuing investigation are done to develop techniques to accurately measure stress concentration around the geometric boundaries. Several methods can be found in the literature ranging from FEM, the use of photoelastic-data, hybrid method and other various numerical and experimental procedures (Lekhnitskii; Tsai & Cheron, 1968; Kobayshi, 1993; Dally & Riley, 1991; Pilkey's, 2008).

In this paper, we present stress concentration measurement method using only x-component displacement data of selected points along straight lines away from the geometric discontinuity. In conjunction with our previous studies (Baek et al., 2000; Baek & Rowlands, 1999; Baek & Kim, 2005; Baek et al., 2006), the hybrid method employing the least-squares method integrated with Laurent series representation of the stress function was used to estimate reliable edge data around the circular hole in a tensile-loaded plate from a relatively few measured x-displacement data away from the boundary. Speckle interferometry has been explored and integrated with other methods for the optical measurement of in-plane and out-of-plane displacement in a material (Schwider, 1989; Steinzig & Takahashi, 2006). Different from the previous works, this study utilized considerably a few number of in-plane micro-scale x-displacement only measured by speckle interferometry using PSM/FT for hybrid stress analysis.

The present technique employs fairly general expressions for the stress functions, and traction-free conditions which are satisfied at the geometric discontinuity using conformal mapping and analytical continuation. The approach is illustrated using the x-displacement as input data obtained from phase-shifting method in speckle interferometry.

3.2 Theoretical background

3.2.1 Basic equations

In the absence of body forces and rigid body motion, the stresses and displacements under plane and rectilinear orthotropy can be written as (Gerhardt, 1984; Rhee & Rowlands, 2002).

$$\begin{aligned} u &= 2\,\mathrm{Re}\left[p_1\Phi(\zeta_1) + p_2\Psi(\zeta_2)\right] \\ v &= 2\,\mathrm{Re}\left[q_1\Phi(\zeta_1) + q_2\Psi(\zeta_2)\right]. \end{aligned} \tag{3-1}$$

The two complex stress functions $\Phi(\zeta_1)$ and $\Psi(\zeta_2)$ are related to each other by the conformal mapping and analytic continuation. For a traction-free physical boundary, the two functions within sub-region Ω of Fig. 3-1 can be written as Laurent expansions, respectively (Gerhardt, 1984; Baek & Rowlands, 2001)

$$\Phi(\zeta_1) = \sum_{k=-m}^{m} \zeta_1^k \quad \text{and} \quad \Psi(\zeta_2) = \sum_{k=-m}^{m} \left(\overline{c_k} B \zeta_2^k + c_k C \zeta_2^k\right) \tag{3-2}$$

The coefficients of Eq. (3-2) are $c_k = a_k + i b_k$ where a_k and b_k are real numbers. In addition to satisfying the traction-free conditions on the hole boundary Γ, the stresses and displacements of Eqs. (3-1) and (3-2) associated with these stress functions $\Phi(\zeta_1)$ and $\Psi(\zeta_2)$ satisfy equilibrium and compatibility.

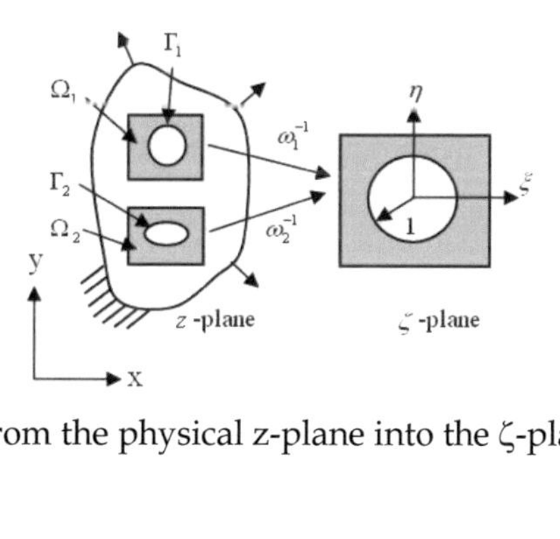

Fig. 3-1. Mapping of holes from the physical z-plane into the ζ-plane.

The primes denote differentiation with respect to the argument. Complex material parameters μ_l (l = 1, 2) are the two distinct roots of the characteristic Eq. (3-3) for an orthotropic material under plane stress (Lekhnitskii, Tsai & Cheron, 1968; Gerhardt, 1984)

$$S_{11}\mu^4 + (2S_{12} + S_{66})\mu^2 + S_{22} = 0 \tag{3-3}$$

where S_{ij} are the elastic compliances. The material properties p_l and q_l are defined as

$$\begin{aligned} p_1 &= S_{11}\mu_1^2 + S_{12},\quad p_2 = S_{11}\mu_2^2 + S_{12} \\ q_1 &= S_{12}\mu_1 + \frac{S_{22}}{\mu_1},\quad q_2 = S_{12}\mu_2 + \frac{S_{22}}{\mu_2} \end{aligned} \tag{3-4}$$

The inverse of the mapping function ω namely ω^{-1}, maps the geometry of interest from the physical z-plane into the ζ-plane ($\zeta_l = \xi + \mu_l \eta$). For orthotropic materials, the conformal transformation from the unit circle in the ζ-plane to the hole in the z-plane of radius R is shown in Fig. 3-1 and is given by

$$z_l = \omega_l(\zeta_l) = \frac{R}{2}\left[(1 - i\mu_l)\zeta_l + \frac{1 - i\mu_l}{\zeta_l}\right] \tag{3-5}$$

where $i = \sqrt{-1}$. The inverse of (3-5) is

$$\omega^{-1}(z_l) = \zeta_l = \frac{z_l \pm \sqrt{z_l^2 - R^2(1 + \mu_l^2)}}{R(1 - i\mu_l)} \tag{3-6}$$

The branch of the square root of Eq. (3-6) is chosen so that $|\zeta_l| \geq 1$ (l = 1, 2). Complex quantities B and C in Eq. (3-2) depend on material properties defined as

$$B = \frac{\bar{\mu}_2 - \bar{\mu}_1}{\mu_2 - \bar{\mu}_2},\quad C = \frac{\bar{\mu}_2 - \mu_1}{\mu_2 - \bar{\mu}_2} \tag{3-7}$$

3.2.2 Least-squares method

Combining Eqs. (3-2), (3-3) and (3-5) gives the following expressions for the displacements through regions Ω_1 and Ω_2 of Fig. 3-1. In matrix form,

$$\{d\} = [U]\{c\} \tag{3-8}$$

where $\{u\} = \{u, v\}^{\mathrm{T}}$, and $\{c\} = \{a_k, b_k\}^{\mathrm{T}}$. $[U]$ is a rectangular coefficient matrix whose size depends on the number of terms k of the power series expansions of Eq. (3-1) and given by

$$U(1, j) = 2\,\mathrm{Re}\left\{p_1\zeta_1^k + p_2(C\zeta_2^k + B\zeta_2^{-k})\right\} \tag{3-9a}$$

$$U(2, j) = 2\,\mathrm{Re}\left\{q_1\zeta_1^k + q_2(C\zeta_2^k + B\zeta_2^{-k})\right\} \tag{3-9b}$$

In Eqs. (3-9a) and (3-9b), $j = 2(k + m) + 1$ if $k < 0$ and $j = 2(k + m) - 1$ if $k > 0$.

Knowing {*d*} at various locations in Eq. (3-8) allows the best values of unknown coefficients {*c*} in a least square sense (Sanford, 1980). For *n* measured x-displacements and *m* terms, the coefficients {*c*} of Eqs. (3-8) were obtained from Eq. (3-10) by the least squares method expressed as

$$\underset{m+1}{\{c\}} = \left(\underset{[(m+1)\times 2n]}{[U]^T} \underset{[2n\times(m+1)]}{[U]} \right)^{-1} \underset{[(m+1)\times 2n]}{[U]^T} \underset{(2n\times 1)}{\{d\}} \tag{3-10}$$

3.2.3 In-plane displacement by speckle interferometry

In speckle interferometry using phase-shifting method with Fourier transformation (PSM/FT) (Morimoto & Fujisawa, 1994), interference fringe pattern is obtained by subtracting the pattern from the pre-loading and post-loading conditions of the specimen. The intensity of the fringe pattern is calculated as

$$I(x,y:\alpha) - A(x,y)\cos[\phi(x,y)-\alpha] + B(x,y) \tag{3-11}$$

where $A(x,y)$ is the amplitude of the brightness in the pattern and $B(x,y)$ is the average brightness. Piezoelectric transducer (PZT) can control an optical path length. α is the known phase which is added into one of the two optical beams by controlling the PZT. The known phase α covers the region from 0 to 2π radians at equal intervals. $\phi(x,y)$ is the phase of fringe pattern caused by the in-plane displacement of a specimen.

A phase-shifting method using Fourier transform had been well-developed for the measurement of in-plane displacement by speckle interferometry (Morimoto & Fujisawa, 1994; Kim et al., 2005). The phase ϕ can be calculated as

$$\phi(x,y) = -\tan^{-1}\left(\frac{\mathrm{Im}\left[F_\alpha(x,y:\omega_0)\right]}{\mathrm{Re}\left[F_\alpha(x,y:\omega_0)\right]} \right) \tag{3-12}$$

where

$$F_\alpha(x,y:\omega_0) = \int_{-\pi}^{\pi} \left[A\cos(\phi(x,y)-\alpha)+B\right] e^{-j\alpha} d\alpha = \pi A e^{-j\phi(x,y)} \tag{3-13}$$

is the α-directional Fourier transform of Eq. (3-11). In Eqs. (3-12) and (3-13), ω_0 is a fundamental frequency. Using the calculated phase, ϕ, the displacement of a specimen can be obtained.

3.3 Optical experiments

Speckle interferometry experiment was performed with setup as shown in Fig. 3-2 to acquire the needed data. In Fig. 3-2, LA is a laser, PA is a pin-hole assembly, CL is a collimating lens, BS is a non-polarizing beam splitter, MR1 and MR2 are mirrors. SL is a specimen installed in a tensile loading device, CCD is a CCD camera, and PC is a personal computer. PC controls the movement of PZT through the control board CNT (Kim et al., 2005).

Figure 3-3 shows the steel plate specimen (E=200 GPa, $\nu = 0.3$) which was subjected to tensile load during the procedure. The test specimen of Fig. 3 had been used for experiment for the development and application of phase-shifting method in speckle interferometry. The accuracy and reliability of the said in-plane displacement measuring method had been established (Baek et al., 2008). For this reason, it is a useful tool for the hybrid stress analysis presented herein (Baek & Rowlands, 1999; Baek & Rowlands, 2001). In this study, phase-shifting method using Fourier transform was utilized.

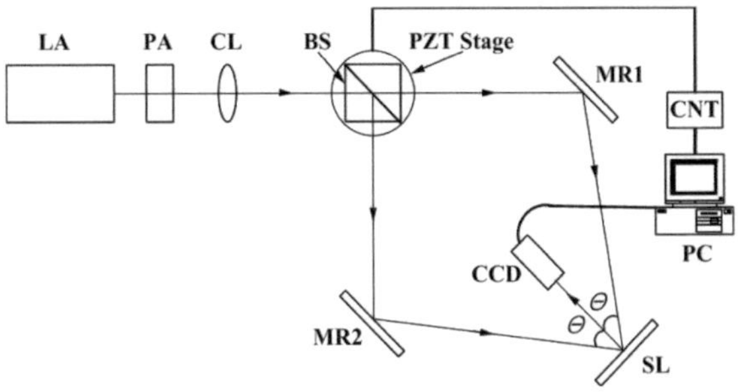

Fig. 3-2. Schematic diagram of speckle interferometry experiment for x-displacement data acquisition.

Figure 3-4 shows the picture of optical experiment system for speckle interferometry by PSM/FT. The in-plane x-displacement of the specimen, u, along the longitudinal direction is obtained through

$$u = \frac{\lambda}{4\pi \sin\theta} \phi_u \tag{3-14}$$

where u is the longitudinal displacement, λ is the wavelength of light from a laser and ϕ_u is the phase obtained from longitudinal displacement.

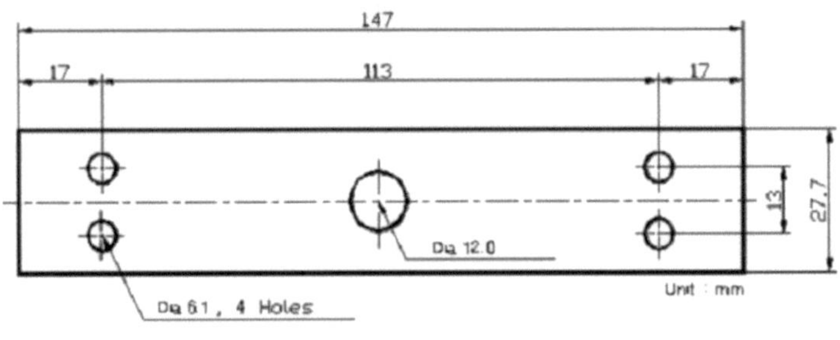

Fig. 3-3. Finite-width uni-axially tensile-loaded steel plate.

The laser used in the experiment is He-Ne laser and λ is 633 nm. The angle between incident light and vertical line to the specimen, θ as shown in Fig. 3-2, is around 26°. The specimen is installed in a loading device that applies tensile load to the specimen in order to make in-plane displacement. Fringe patterns of $I(x, y: \alpha)$ in Eq. (3-11) are taken through CCD camera in Fig. 3-2. The fringe patterns consist of 32 patterns which are sequentially phase-shifted by PZT stage and are saved in PC. Phase shifting at each step is $\pi/16$ radian.

Size of each fringe pattern is 640 x 480 with 8 bits brightness. 16 fringe patterns out of the 32 fringe patterns are shown as examples at every interval of $\pi/8$ radian for the rectangular steel plate containing a hole in Fig. 3-5 (Kim et al., 2005).

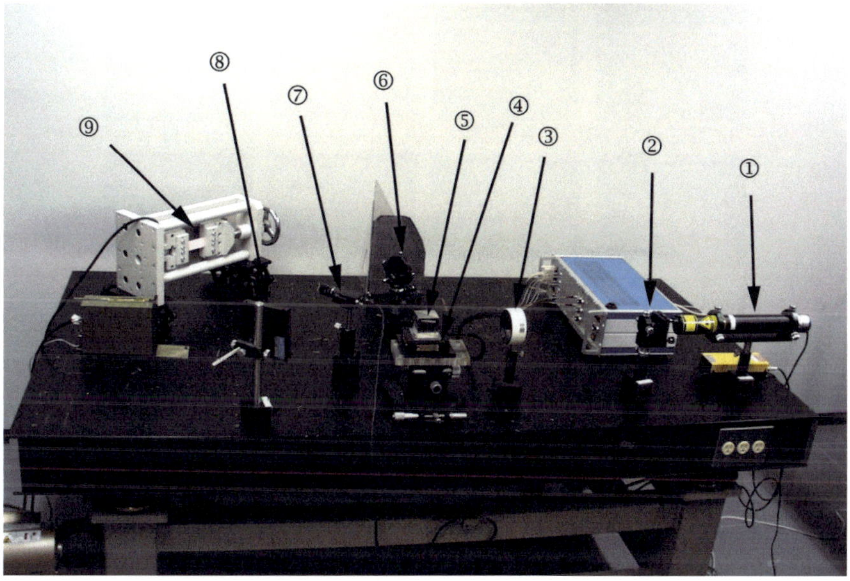

Fig. 3-4. Picture of optical experiment system for speckle interferometry by PSM/FT. (① He-Ne CW laser, ②pin-hole assembly, ③ collimating lens, ④ PZT control stage, ⑤ beam splitter, ⑥ mirror, ⑦ CCD camera, ⑧ mirror, ⑨ specimen in loading device).

The accuracy of the least-squares method for calculating displacement components $\{u,v\}^T$ as presented in the preceding section was related with the input data acquired through experiment by phase-shifting method in speckle interferometry (Baek & Rudolphi, 2010).

Percent errors for each data was calculated and shown in Table 3-1. As can be observed, the level of precision of the hybrid method is too close with the original input data as indicated by the % Errors. Thus, the method is significantly accurate and reliable for calculating such parameter.

The simplicity of the test specimen being isotropic, containing symmetrically-shaped discontinuity facilitates the ease of reliable verification. Finite element anlysis was done to establish a benchmark for comparison purposes. Figure 3-7 shows the ABAQUS discretization of the quarter steel plate. In the vicinity of the circular hole, the ABAQUS model of Fig. 3-7 utilizes elements on the edge of the hole as small as 1.5° by 0.013*r*, where *r* is the radius of the circular hole. Tangential stress around the quarter hole was determined and compared as shown in Fig. 3-8. Different values of m term in the complex stress functions was tested to see its effect, it came out that at point of high stress concentration ($\theta = 90^\circ$ in Fig. 3-8). Well-comparable results were attained at decreasing value of m with the best value equal to 1 - accurate by less than one percent error. Figure 3-9 reveals the preceding observation and shows normalized tangential stress at the edge of the quarter hole in a steel plate with $m = 1$ in the complex stress function.

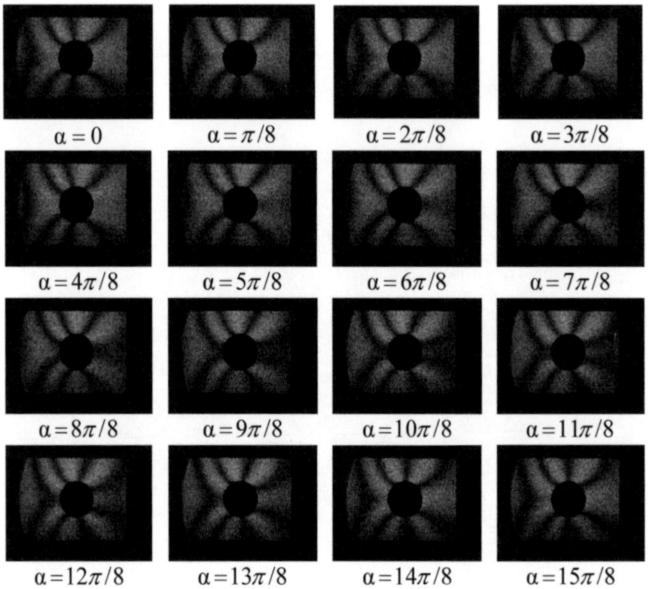

Fig. 3-5. 16 fringe patterns of rectangular steel plate containing a circular hole at every interval of $\alpha = \pi/8$ radian.

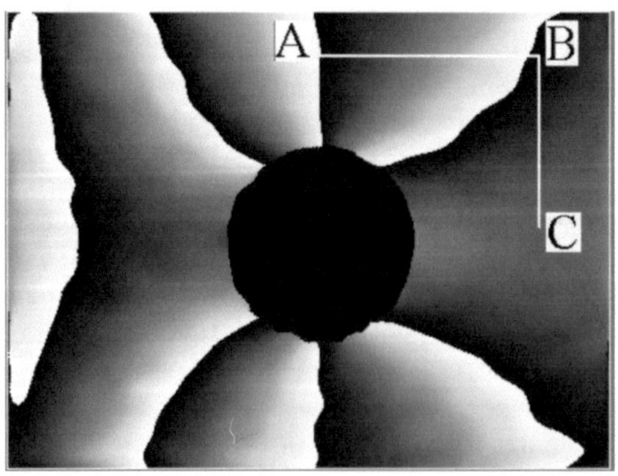

Fig. 3-6. Wrapped phase of steel plate with circular hole. Lines AB and BC are x-displacement extraction regions.

Data Point	Input: x-disp. (μm)	Calculated: x-disp. (μm)	% Error*
1	0	0	-
2	2.6032	2.5791	-0.926
3	2.8628	2.8397	-0.807
4	4.3754	4.3678	-0.174
5	5.5334	5.5499	0.298
6	6.8069	6.8555	0.714
7	7.3259	7.3421	0.221
8	7.8783	7.8483	-0.381
9	8.0719	8.0536	-0.227

*Note: $\%\text{Error} = \dfrac{\text{Calculated} - \text{Input}}{\text{Input}} \times 100(\%)$

Table 3-1. Comparison between input and calculated x-displacement for different values of $m = 1$ of the complex stress functions.

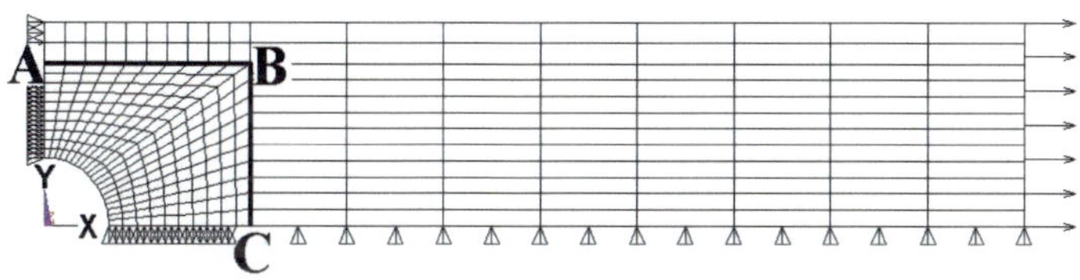

Fig. 3-7. ABAQUS discretization of the quarter steel plate (t = 1.15mm).

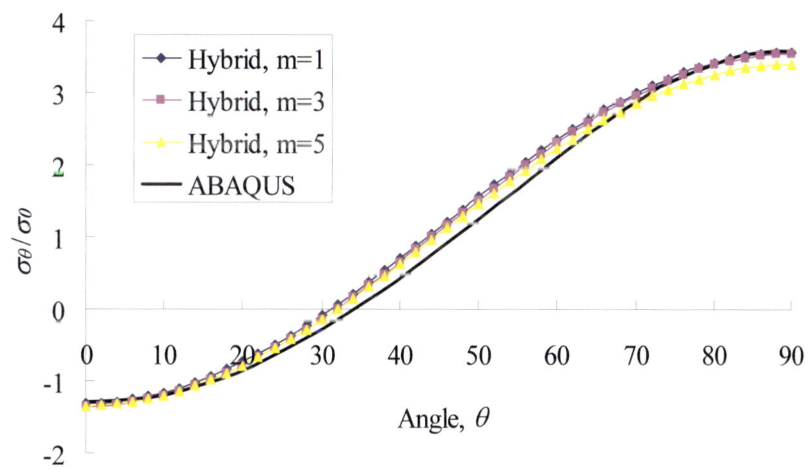

Fig. 3-8. Normalized tangential stress at the edge of the quarter hole in a steel plate with different values of m term in the complex stress function.

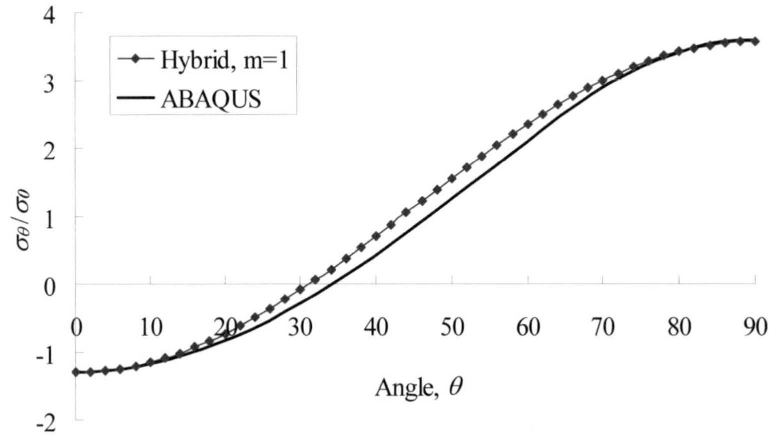

Fig. 3-9. Normalized tangential stress at the edge of the quarter hole in a steel plate with $m=1$ in the complex stress function.

3.4 Conclusions and discussions

A reliable hybrid method for characterizing stress around the circular hole in a tensile-loaded steel plate is presented. The method utilized only few micro-scale x-displacement data measured by a well-established optical technique, speckle interferometry, in conjunction with phase shifting method using Fourier transform to calculate stress components and eventually stress concentration at $\theta = 90^{\circ}$. The use of few input data may reduce experiment time and relatively increase data processing speed.

Different values of m term in the complex stress functions were tested to see its effect. In the comparison between input x-displacement data and calculated data the hybrid method is effective with an error below 1% in all values of *m*. On the other hand, it came out that at point of high stress concentration ($\theta = 90^{\circ}$), well-comparable results were attained at decreasing value of *m*. The best value of m is consistently known to be equal to 1 of which results are accurate by less than one percent error. Results showed that the method is accurate and reliable as compared with the widely-used FEM software, ABAQUS.

4. Acknowledgement

This research was supported by Basic Science Research Program through the National Research Foundation of Korea (NRF) funded by the Ministry of Education, Science and Technology (Grant number: 2010-0021248)

5. References

Cloud, G. (1998). *Optical Methods of Engineering Analysis*, Cambridge University Press, ISBN 978-0521636421, Cambridge, United Kingdom

Petzing, J. & Tyrer, J. (1998). Recent Developments and Applications in Electronic Speckle Pattern Interferometry, *Journal of Strain Analysis*, Vol.33, No.2, pp. 153-169, ISSN 0309-3247

Rastogi, P. (Editor) (2001). *Digital Speckle Pattern Interferometry and Related Techniques,* John Wiley and Sons, Ltd., ISBN 978-0471490524, New York, USA

Baek, T.; Kim, M., Na, E. & Koh, S. (2003). Application of ESPI to Measurement of Out-of-plane Displacement in a Spot Welded Canti-levered Plate, *International Journal of the Korean Society of Precision Engineering,* Vol.4, No.5, pp. 41-46

Ghiglia, D. & Pritt, M. (1998). *Two-Dimensional Phase Unwrapping,* John Wiley & Sons, Inc., ISBN 978-0471249351, New York, USA

Creath, K. (1988). Phase-measurement interferometry techniques, *Progress in Optics,* Vol.26, pp. 349-393, ISSN 0079-6638

Kim, M. & Baek, T. (2006). Measurement for in-plane displacement of tensile plates with partly through-thickness circular hole and through-thickness circular hole by use of speckle interferometry, *Key Engineering Materials,* Vols.321-323, pp. 77-80, ISSN 1162-9795

Lekhnitskii, S.; Tsai, S. & Cheron, T. (1968). *Anisotropic Plates,* Gordon and Breach

Kobayshi, A. (Editor) (1993). *Handbook on Experimental Mechanics,* 2nd Revised Edition, Society for Experimental Mechanics, VCH Publishers, Inc., ISBN 978-1560816409,

Dally, J. & Riley, W. (1991). Experimental Stress Analysis, 3rd Edition, McGraw-Hill, Inc., New York, USA

Pilkey, W. & Pilkey, D. (2008). *Peterson's Stress Concentration Factors,* John Wiley & Sons, Inc., ISBN 978-0470048245, New Jersey, USA

Baek, T.; Kim, M., Rhee, J. & Rowlands, R. (2000). Hybrid Stress Analysis of Perforated Tensile Plates using Multiplied and Sharpened Photoelastic Data and Complex-Variable Techniques, *JSME International Journal, Series A: Solid Mechanics and Material Engineering,* Vol.43, No.4, pp. 327-333, ISSN 0914-8809

Baek, T. & Rowlands, R. (1999). Experimental Determination of Stress Concentrations in Orthotropic Composites, *Journal of Strain Analysis,* Vol.34, No.2, pp. 69-81, ISSN 0309-3247

Baek, T. & Kim, M. (2005). Computer Simulation of Photoelastic Fringe Patterns for Stress Analysis, *Lecture Notes in Computer Science,* Vol.3398, pp. 214-221, ISSN 0302-9743

Baek, T. Koh, S. & Park, T. (2006). An Improved Hybrid Full-field Stress Analysis of Circularly Perforated Plate by Photoelasticity and Finite Element Analysis, *Key Engineering Materials,* Vol.326-328, pp. 1209-1212, ISSN 1162-9795

Schwider, J. (1989). Phase shifting Interferometry: reference phase error reduction, Applied Optics, Vol.28, Issue 18, pp. 3889-3892

Steinzig, M. & Takahashi, T. (2006). Residual Stress Measurement Using the Hole Drilling Method and Laser Speckle Interferometry, Part IV: Measurement Accuracy, Experimental Techniques, Vol. 27, Issue 6, pp. 59-63, ISSN 0732-8818

Gerhardt, T. (1984). A Hybrid/Finite Element Approach for Stress Analysis of Notched Anisotropic Material, *ASME Journal of Applied Mechanics,* Vol.51, No.4, pp. 804-810

Rhee, J. & Rowlands, R (2002). Moiré-Numerical Hybrid Analysis of Cracks in Orthotropic Media, *Experimental Mechanics,* Vo.42, No.3, pp. 311-317, ISSN 1741-2765

Baek, T. & Rowlands, R. (2001). Hybrid Stress Analysis of Perforated Composites using Strain Gages, *Experimental Mechanics,* Vol.41, No.2, pp. 194-202, ISSN 1741-2765

Sanford, R. (1980). Application of the Least-Squares Method to Photoelastic Analysis, Experimental Mechanics, Vol.20, No.6, pp. 192-197, ISSN 1741-2765

Morimoto, Y. & Fujisawa, M. (1994). Fringe Pattern Analysis by a Phase-Shifting Method using Fourier Transform, *Optical Engineering*, Vol.33, No.11, pp. 3709-3714

Kim, M. ; Baek, T., Morimoto, Y., & Fujigaki, M. (2005). Application of Phase-shifting Method using Fourier Transform to Measurement of In-plane Displacement by Speckle Interferometry, *Journal of the Korean Society for Nondestructive Testing*, Vol.25, No.3, pp. 171-177

Baek, T.; Chung, T. & Panganiban, H. (2008). Full-Field Stress Determination Around Circular Discontinuity in a Tensile-Loaded Plate using x-displacements Only, *Journal of Solid Mechanics and Material Engineering*, Vol.2, No.6, pp. 756-762

Baek, T. & Rowlands, R. (1999). Experimental Determination of Stress Concentrations in Orthotropic Composites, *Journal of Strain Analysis*, Vol.34, No.2, pp. 69-81, ISSN 0309-3247

Baek, T. & Rudolphi, T. (2010). A Hybrid Stress Measurement Using only x-displacement by Phase Shifting Method with Fourier Transform (PSM/FT) in Laser Speckle Interferometry and Least Squares Method, *International Journal of Precision Engineering and Manufacturing*, Vol.11, No.1, Korean Society for Precision Engineering, pp. 49-54, ISSN 0217-9849

8

Phase-Shifting Interferometry by Amplitude Modulation

Cruz Meneses-Fabian and Uriel Rivera-Ortega
Benemérita Universidad Autónoma de Puebla, Facultad de Ciencias Físico-Matemáticas, Puebla, México

1. Introduction

In optics, the superposition of two or more light beams at any point over space can produce the apparition of interference fringes. When these fringes are applied to resolve a problem in industry or they are related with some property of an investigation matter of interest in some area of physics, chemistry, biology, etc., the evaluation of them is a very necessary task. One of the most used methods for phase extraction, as a result of fringes evaluation, is based on a phase change between the interference beams by a known value, while their amplitudes are keeping constant. It is called phase-shifting interferometry, phase-sampling interferometry, or phase-stepping interferometry, which are abbreviated by "PSI" (Schwider, 1990). In this technique a set of N interferograms changed in phase are created, which are represented by a set of N equations, where each equation has three unknowns called as background light, modulation light and the object phase. These spatial unknowns are considered constant during the application of the PSI technique. Then a $N \times 3$ system is formed and therefore it can be resolved when $N \geq 3$. Many methods to introduce a constant phase have been proposed as for example by changing the optical frequency, wavelength, index of refraction, distance, optical path, for instance; but also with some properties or effect of the light such as the polarization, diffraction, Zeeman effect, Doppler effect, for instance (Schwider, 1990; Malacara, 2007). In this chapter, our major interest aims to propose a new method to generate a phase change in the interferogram based on the amplitude modulation of the electric field (Meneses-Fabian and Rivera-Ortega, 2011).

Interferometry uses the superposition principle of electromagnetic waves when certain conditions of coherence are achieved to extract information about them. If light from a source is divided into two to be superposed again at any point in space, the intensity in the superposition area varies from maxima (when two waves crests reach the same point simultaneously) to minima (when a wave trough and a crest reach the same point) ; having by this what is known as an interference pattern. Interferometry uses interferometers which are the instruments that use the interference of light to make precise measurements of surfaces, thicknesses, surface roughness, optical power, material homogeneity, distances and so on based on wavefront deformations with a high accuracy of the order of a fraction of the wavelength through interference patterns. In a two wave interferometer one wave is typically a flat wavefront known as the reference beam and the other is a distorter wavefront whose shape is to be measured, this beam is known as the probe beam.

There are several well studied and know methods to generate phase-shifts which will be briefly discussed, however the present method of PSI based on the amplitude modulation into two beams named reference beams in a scheme of a three beam interferometry will be amply discussed; this discussion will be done for a particular case where the phase difference between the reference beam is conditioned to be $\pi/2$; and for a general case where the phase difference between the reference beam should be within the range of $(0,2\pi)$ but $\neq \pi$.

We can represent n optical perturbations in a complex form with elliptical polarization and traveling on z direction as follows:

$$\mathbf{E}_k = (\mathbf{i}E_{kx} + \mathbf{j}E_{ky}e^{i\delta_k})e^{i\phi_k} \tag{1}$$

where k goes from $1 \ldots n$, ϕ_k is the phase of each wave and δ_k is the relative phase difference between the component of each wave (by simplicity the temporal and spatial dependencies have been omitted). In the particular case that $\delta_k = m\pi$ where m is an integer number, the wave will be linearly polarized; on the other hand, if $E_{kx} = E_{ky}$ and $\delta_k = (2m+1)\pi/2$ the wave will be circularly polarized.

The phenomenon of interference can occur when two or more waves overlap in space. Mathematically the resulting wave is the vector addition,

$$\mathbf{E}_T = \sum_{k=1}^{n} \mathbf{E}_k \,. \tag{2}$$

When the field is observed by a detector, the result is the average of the field energy by area unit during the integration time of the detector, that is, the irradiance, which can be demonstrated that is proportional to the squared module of the amplitude. However, it is usually accepted the approximation

$$I = |\mathbf{E}_T|^2 = \left|\sum_{k=1}^{n} E_k\right|^2 . \tag{3}$$

An interferometer is an instrument used to generate wave light interference to measure with high accuracy small deformations of the wave front. The general scheme of a two wave interferometer can be observed in Figure 1, where the electromagnetic wave $\mathbf{E}$ is typically divided in two coherent parts that is, in a wave $\mathbf{E}_1$ and $\mathbf{E}_2$, where $\mathbf{E}_1$ is the reference wave and $\mathbf{E}_2$ is the probe wave.

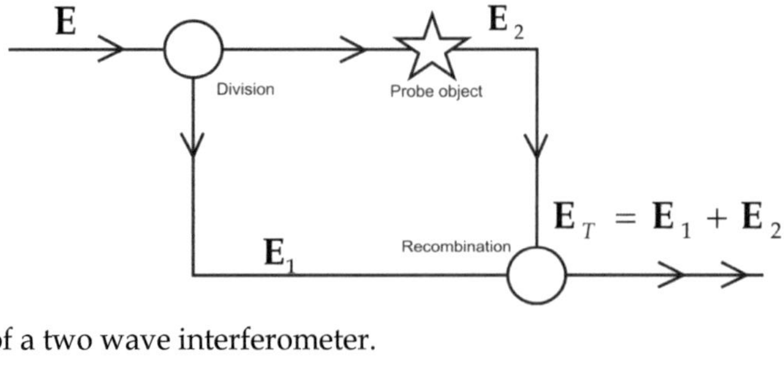

Fig. 1. Scheme of a two wave interferometer.

After the waves have travelled along two separated arms and they have accumulated phase delays, they recombine again by means of a beam splitter giving as a result a field $\mathbf{E}_T$.

For the particular case of two waves ($n=2$), and using the Eq. (3) for a vector treatment, we can find that the corresponding irradiance is

$$I = |\mathbf{E}_T|^2 = |\mathbf{E}_1 + \mathbf{E}_2|^2 = |\mathbf{E}_1|^2 + |\mathbf{E}_2|^2 + 2\mathrm{Re}\{\mathbf{E}_1^* \cdot \mathbf{E}_2\}; \tag{4}$$

It can be seen in Eq. (4) that there are three terms, by doing the math they can be expressed as follows

$$|\mathbf{E}_k|^2 = \mathbf{E}_k \cdot \mathbf{E}_k^* = \left[(\mathbf{i}E_{kx} + \mathbf{j}E_{ky}e^{-i\delta_k})e^{i\varphi_k}\right] \cdot \left[(\mathbf{i}E_{kx} + \mathbf{j}E_{ky}e^{i\delta_k})e^{-i\varphi_k}\right] = E_{kx}{}^2 + E_{ky}{}^2, \tag{5}$$

where the values for $k=1,2$; by doing this the first two terms of Eq. (4) will be obtained, the third term of this equation is obtained by

$$\mathbf{E}_1^* \cdot \mathbf{E}_2 = \left[(\mathbf{i}E_{1x} + \mathbf{j}E_{1y}e^{-i\delta_1})e^{-\phi_1}\right] \cdot \left[(\mathbf{i}E_{2x} + \mathbf{j}E_{2y}e^{i\delta_2})e^{\phi_2}\right], \tag{6}$$

therefore the resulting interference term is

$$2\mathrm{Re}\{\mathbf{E}_1^* \cdot \mathbf{E}_2\} = 2E_{1x}E_{2x}\cos\phi + 2E_{1y}E_{2y}\cos(\phi+\delta), \tag{7}$$

where $\delta = \delta_2 - \delta_1$ y $\phi = \phi_2 - \phi_1$. By taking Eqs. (5-7) a general expression for the interference of two waves is obtained

$$I = a_x + b_x\cos\phi + a_y + b_y\cos(\phi+\delta), \tag{8}$$

where $a_x = E_{1x}{}^2 + E_{2x}{}^2$, $a_y = E_{1y}{}^2 + E_{2y}{}^2$, $b_x = 2E_{1x}E_{2x}$ y $b_y = 2E_{1y}E_{2y}$. It can be observed in Eq. (8) that the interference of two elliptically polarized waves can also be generated by the addition of two interference patterns, one of them with components in x direction $(a_x + b_x\cos\phi)$ and the other with components in y $(a_y + b_y\cos(\phi+\delta))$.

On the other hand, it is known that $A\cos\phi + B\sin\phi = C\cos(\phi+\psi)$ if $C = \sqrt{A^2+B^2}$ and $\tan\psi = B/A$; by applying this identity to Eq. (8) is obtained

$$I = a + b\cos(\phi+\psi), \tag{9}$$

where a is known as the background light, b as the modulation light and ψ indicates an additional phase shifting, which can be expressed by

$$a = a_x + a_y; \quad b^2 = b_x^2 + 2b_xb_y\cos\delta + b_y^2; \quad \tan\psi = \frac{b_y\sin\delta}{b_x + b_y\cos\delta}. \tag{10}$$

It is worth noting from Eq. (10) that both the phase shifting ψ as the background a and the modulation light b depend on the components a_x, a_y, b_x, b_y and also on the phase difference between the amplitude of the waves δ. Note that a phase shifting ψ by varying b_x, b_y and δ will also generate a change in a and b.

2. Phase-shifting interferometry

The first studies in the phase shifting techniques can be found in the work of Carré 1966, but it really started with Crane 1969, Moore 1973 , Brunning *et al.* 1974 and some others. These techniques have also been applied in speckle patter interferometry (Nakadate and Saito 1985; Creath 1985; Robinson and William 1986) and also to holographic interferometry (Nakadate *et al.* 1986; Stetson and Brohinski 1988).

In phase shifting interferometry, a reference wave front is moved along its propagation direction respecting to the probe wave front changing with this the phase difference between them. It is possible to determine the phase of the probe wave by measuring the irradiance changes corresponding to each phase shifting.

2.1 Some phase-shifting methods

There are many experimental ways to generate phase shifting, the most common are:

2.1.1 Moving a mirror

This method is based on the change in the optical path of a beam by means of moving a mirror that is in the beam trajectory. This movement can be made by using a piezoelectric transducer (Soobitsky, 1988; Hayes, 1989). The phase-shifting is given by $\psi = (2\pi/\lambda)(dco)$, where dco is the optical path difference. As an illustrative example if the mirror is translated a distance of $\psi = \lambda/8$ and due to the beam travels two times the same trajectory, the value of the phase-shifting is $\psi = \pi/2$. Examples of interferometers with phase-shifting generated by a piezoelectric are: Twyman-Green, Mach-Zehnder, Fizeau, which are represented in Figure 2. The first two make a phase-shifting by means of moving a mirror that is placed in the reference beam. In the Fizeau interferometer the phase-shifting is made by the translations of the reference or probe object.

2.1.2 Rotating a phase plate

A phase-shifting can also be generated by means of rotating a retarder phase plate (Crane 1969; Okoomian 1969; Bryngdahl 1972; Sommargren 1975; Shagam y Wyant 1978; Hu 1983, Zhi 1983; Kothiyal and Delisle 1984, 1985; Salbut y Patorski 1990). As a particular case, if a circularly polarized wave passes through a half wave retarder plate rotated 45° the direction of rotation will be inverted, thus a phase-shifting of $\psi = \pi/2$ will be present.

2.1.3 Displacing a diffraction grating

This method is based on the perpendicular displacement respecting to the light beam of a diffraction grating (Suzuki y Hioki 1967; Stevenson 1970; Bryngdahl 1976; Srinivasan *et al.*1985, Meneses *et al.* 2009). If the diffraction grating is moved a small distance Δy, the changes in the phase are given by $y = (2\pi m/d)\Delta y$ where d is the period of the grating and m is the diffraction order. As an example, if the grating of Figure 3 is perpendicularly moved respecting to the optical axis, a phase-shifting will be generated.

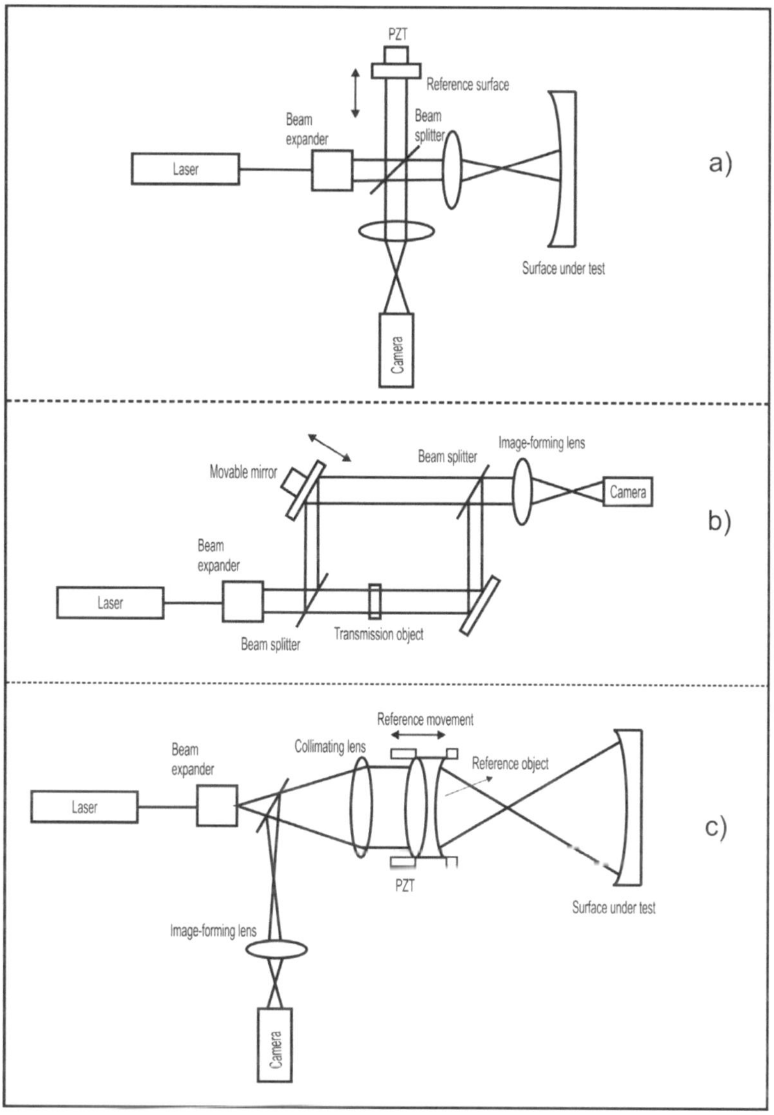

Fig. 2. Scheme of the most commonly used interferometers: a) Twyman-Green, b) Mach-Zehnder, c) Fizeau.

2.1.4 Tilting a glass plate

Another method to generate phase-shifting is by means of inserting a glass plate in the light beam (Wyant y Shagam 1978). The phase-shift ψ_0 is generated when the plate is tilted an angle ϑ respecting to the optical axis hence $\psi = (t/n)(n\cos\vartheta' - \cos\vartheta)$, where t is the thickness of the plate, n is the refraction index and $k = 2\pi/\lambda$. The angles ϑ and ϑ' are the angles formed by the normal and the light beams outside and inside the plate respectively. An special requirement is that the plate must be placed in a collimated light beam to avoid aberrations.

2.1.5 Rotating a polarizer

If in Eq. (1) $k = 1,2$, $E_{1x} = E_{1y} = E_{2x} = E_{2y} = E_0$, $\delta_1 = \pi/2$, $\delta_2 = -\pi/2$ then we have two circularly polarized waves with the same amplitude with opposite rotations, that is

$$\mathbf{E}_1 = (\mathbf{i} + e^{i\frac{\pi}{2}}\mathbf{j})E_0 e^{i\phi_1}, \tag{11}$$

$$\mathbf{E}_2 = (\mathbf{i} + e^{-i\frac{\pi}{2}}\mathbf{j})E_0 e^{i\phi_2}. \tag{12}$$

Expressing the Eqs. (11-12) with the notation for the Jones vector

$$E_1 = E_0 \begin{pmatrix} 1 \\ i \end{pmatrix} e^{i\phi_1}, \tag{13}$$

$$E_2 = E_0 \begin{pmatrix} 1 \\ -i \end{pmatrix} e^{i\phi_2}; \tag{14}$$

if we interfere those waves, then there will not be an interference term so no fringes will be observed, however if those waves are passed through a polarizer at an angle α and they interfere then

$$E_1' = \begin{bmatrix} \cos^2\alpha & \sin\alpha\cos\alpha \\ \sin\alpha\cos\alpha & \sin^2\alpha \end{bmatrix} \begin{bmatrix} 1 \\ i \end{bmatrix} E_0 e^{i\phi_1} = E_0 \begin{bmatrix} \cos\alpha \\ \sin\alpha \end{bmatrix} e^{i(\alpha+\phi_1)}, \tag{15}$$

$$E_2' = \begin{bmatrix} \cos^2\alpha & \sin\alpha\cos\alpha \\ \sin\alpha\cos\alpha & \sin^2\alpha \end{bmatrix} \begin{bmatrix} 1 \\ -i \end{bmatrix} E_0 e^{i\phi_2} = E_0 \begin{bmatrix} \cos\alpha \\ \sin\alpha \end{bmatrix} e^{-i(\alpha-\phi_2)}, \tag{16}$$

$$I = 2E_0^{\ 2}\left[1 + \cos(\phi - 2\alpha)\right], \tag{17}$$

where

$$\begin{bmatrix} \cos^2\alpha & \sin\alpha\cos\alpha \\ \sin\alpha\cos\alpha & \sin^2\alpha \end{bmatrix}, \tag{18}$$

is the Jones matrix that represents a linear polarizer at an angle α. It is observed in Eq. (17) that there is a phase–shift in the interference pattern which is twice the angle at which the polarizer is placed.

An application of phase-shifting generated by polarization is by using the scheme of Figure 3, which is based in a common path interferometer consisting of two windows in the input plane. In this interferometer is possible to have four interference patters in one shot, each pattern shifted by 90°. To do this a binary grating is used to generate the interference between the diffraction orders of two circularly polarized beams with opposite rotations; a linear polarizer at an angle α is placed in front of each diffraction order thus, the phase-shifting is obtained (Rodriguez-Zurita *et al.* 2008).

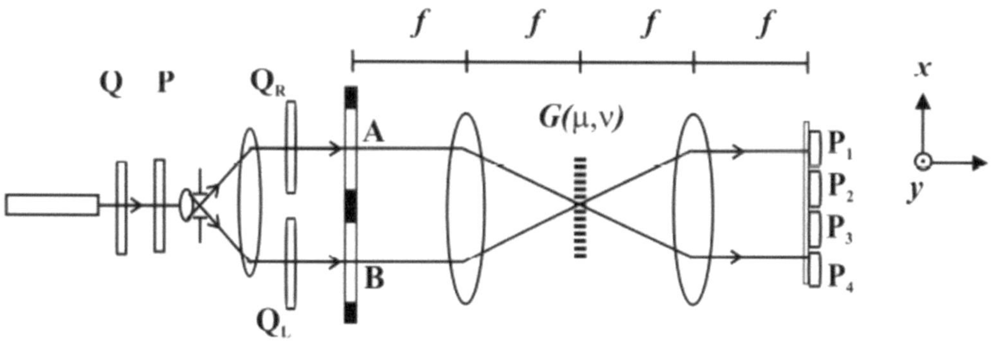

d)

Fig. 3. Scheme of interferometers where the method of phase-shifting by polarizations is used.

2.1.6 Changing the laser source

Other method to generate phase-shift is by means of changing the frequency of the laser source, there are two ways to do this; one of them is by illuminating the interferometer with a Zeeman laser. The laser frequency is divided in two orthogonally polarized output frequencies by means of a magnetic field (Burgwald and Kruger 1970). The other method is by using an unbalanced interferometer, that is an interferometer with a very long optical path difference and using a laser diode with its frequency controlled by electrical current as proposed by Ishii *et al.* (1991) and subsequently studied by Onodera e Ishii (1996). This method is based on the fact that the phase difference in an interferometer is proportional to the product of the optical path difference and the temporal frequency.

In summary, the known techniques for phase-shifting that have been mentioned are applied in two beams interferometers; however none of these techniques use a variation on the amplitude fields, that is why this option will be discussed in this chapter.

2.2 Phase extraction methods

The phase-shifting interferometry is based on the reconstruction of the phase ϕ by sampling a certain number of interference pattern which differ from each other due to different values of ψ_0. If a shift of ψ_0 is made for N steps, then N intensity values I_n will be measured, (where $n = 1,...,N$)

$$I_n = a + b\cos(\phi + \psi_{0n}). \tag{19}$$

where $\psi_{0n} = 2\pi n / N$.

Eq. (19) can be rewritten as follows

$$I_n = A + B\cos\psi_{0n} + C\sin\psi_{on}, \tag{20}$$

where

$$A = a; \quad B = b\cos\phi; \quad C = -b\sin\phi. \tag{21}$$

It can be shown that based on a least-squares fit that B and C meet the next equations in an analytical form

$$B = \frac{2}{N}\sum_{n=1}^{N} I_n \cos\psi_{0n} \; ; \quad C = \frac{2}{N}\sum_{n=1}^{N} I_n \sin\psi_{0n} \,. \tag{22}$$

A combination of Eq. (21) and Eqs. (22) can give us the basic equation of the Phase Sampling Interferometry (PSI)

$$\phi = \tan^{-1}\frac{-C}{B} = \tan^{-1}\frac{\sum I_n \sin\psi_{0n}}{\sum I_n \cos\psi_{0n}} \,. \tag{23}$$

In general, a minimum of three samples are needed to know the phase ϕ because there are three unknowns in the general interference equation Eq. (9): a, b and ϕ. However a better accuracy can be guaranteed with more than three shifts.

2.2.1 Three steps technique

Since we need a minimum of three interferograms to reconstruct the wavefront, the phase can be calculated with a phase-shift of $\pi/2$ per exposition. The three intensity measurements can be expressed as (Wyant, Koliopoulus, Bhushan and George 1984)

$$I_1 = a + b\cos\left(\phi + \frac{1}{4}\pi\right), \tag{24}$$

$$I_2 = a + b\cos\left(\phi + \frac{3}{4}\pi\right), \tag{25}$$

$$I_3 = a + b\cos\left(\phi + \frac{5}{4}\pi\right), \tag{26}$$

the phase at each point is

$$\phi = \tan^{-1}\left(\frac{I_3 - I_2}{I_1 - I_2}\right). \tag{27}$$

2.2.2 Four steps technique

A common algorithm used to calculate the phase is the four steps method (Wyant 1982). In this case the four intensity measurements can be expressed as

$$I_1 = a + b\cos\phi \,, \tag{28}$$

$$I_2 = a + b\cos\left(\phi + \frac{1}{2}\pi\right) = a - b\sin\phi \,, \tag{29}$$

$$I_3 = a + b\cos(\phi + \pi) = a - b\cos\phi \,, \tag{30}$$

$$I_4 = a + b\cos\left(\phi + \frac{3}{2}\pi\right) = a + b\sin\phi \,, \tag{31}$$

and the phase at each point is

$$\phi = \tan^{-1}\left(\frac{I_4 - I_2}{I_1 - I_3}\right) . \tag{32}$$

2.2.3 The Fourier transform method

The deformations of the wavefront in an interferogram can also be calculated by a method that uses the Fourier transform. This method was originally proposed Takeda *et al.* (1982) using the Fourier transform in one dimension along an scanned line. Later Macy (1983) extended the Takeda's method in two dimensions by adding the information of multiple scanned lines and obtaining slices of the phase in two dimensions. Bone *et al* (1986) extended the Macy's work by applying the Fourier transform in two dimensions.

Once the interferogram is obtained, its Fourier transform is calculated, by doing this an image in the Fourier space as in Figure 4a) is obtained, then one lateral spectrum is taken and filtered by means of a layer so that all the irradiance values outside the layer will be multiplied by cero, after that this spectrum is translated to the origin and its Fourier transform is obtained giving by this the resulting wavefront under test.

To describe mathematically this process, the general equation for the irradiance is taken (Eq. (9)) expressing the cosine as a complex exponential

$$I(x,y) = a(x,y) + c(x,y)e^{-i\psi} + c^*(x,y)e^{i\psi} \,, \tag{33}$$

being $c(x,y) = (1/2)be^{i\phi(x,y)}$ and the phase of the form $\psi = 2\pi\mu_0 x$, where μ_0 is known as the spatial-carrier frequency

Applying the Fourier transform to Eq. (33)

$$\tilde{I}(\mu,\upsilon) = \tilde{a}(\mu,\upsilon) + \tilde{c}(\mu + \mu_0,\upsilon) + \tilde{c}^*(\mu - \mu_0,\upsilon) \,, \tag{34}$$

where $\tilde{a}$, $\tilde{c}$ y $\tilde{c}^*$ are complex Fourier amplitudes. By means of a digital filtering one lateral spectrum is isolated by using a filtering window and then translated to the origin by doing $\mu_0 = 0$, as shown in Figure 4b).

Obtaining the inverse Fourier transform of $\tilde{c}(\mu,\upsilon)$ we have $c(x,y) = (1/2)be^{i\phi(x,y)}$, therefore the resulting phase is $\phi(x,y) = \tan^{-1}\dfrac{\mathrm{Im}\,c(x,y)}{\mathrm{Re}\,c(x,y)}$.

2.4 Unwrapping the phase

The calculated phase will present discontinuities because it is obtained by using the inverse tangent function Eq. (23). Because the inverse tangent is a multivalued function, the solution

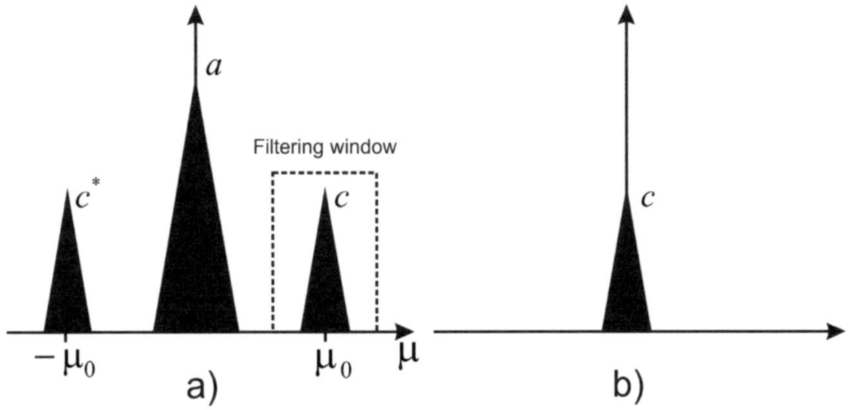

Fig. 4. a) Intensity distribution obtained with the FT of the interference pattern, b) spectrum translated to the origin.

for ϕ is a saw tooth function (Figure 5a), where the discontinuities occur every time ϕ changes by 2π . If ϕ increases, the slope of the function is positive and vice versa if the phase decreases. The final step in the process of measuring the fringe pattern is to unwrap the phase along a line or path counting the discontinuities at 2π and adding 2π each time the angle of the phase jumps from 2π to cero and subtracting 2π if the angle changes from cero to 2π . Figure 5b) shows the dates from Figure 5a) after the unwrapping. The key of a trustable unwrapping algorithm is its capacity of detecting the discontinuities with high accuracy

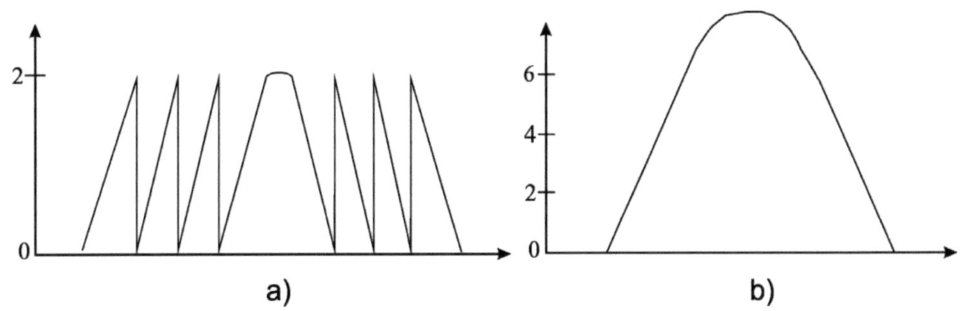

Fig. 5. a) Discontinuities of a wrapped phase, b) unwrapped phase. The basic principle of the phase unwrapping is to "integrate" the wrapped phase (in units of 2π) along a data line. The phase gradient is calculated for each pixel

$$\Delta\phi = \phi_n - \phi_{n-1} , \tag{35}$$

where n is the pixel number. If $|\Delta\phi|$ exceeds a certain threshold like π , then a discontinuity is assumed. This phase discontinuity is fixed by adding or subtracting 2π depending on the sign of $|\Delta\phi|$, Itoh (1982). The most common principle used to fix those phase discontinuities is based on the fact that the phase difference among any pair of points measured by integrating the phase along a path between these points is independent from the chosen path, provided they don't pass through a phase discontinuity.

3. PSI by amplitude modulation

As it was described in the previous section, many proposed techniques for phase-shifting interferometry (PSI) are based on the interference of two waves where just one interference term is present and the phase shift is done with a constant phase difference between them.

In the present chapter a new method for phase-shifting based on the amplitude variation of the field in a scheme of a three beam interferometer is widely discussed. In that interferometer, two beams will be considered as the reference beam and the other beam as the probe beam. The expression for the irradiance due to the interference of those beams will have three interference terms, however due to a constant phase difference of $\pi/2$ is introduced between the reference beams, one interference term will be canceled and the two remaining will be put in quadrature. Because of this and applying some trigonometric identities to the resulting pattern it is possible to model it mathematically as a two wave interferometer where the interference term will contain an additional phase that depends on the amplitude variations of the reference beam. Since the phase-shift depends on the variation of the amplitude of the fields, the visibility may not remain constant, however it will be shown that if the amplitudes are seen as an ordered pair over an arc segment in the first quadrant of a circle whose radius is the square root of the addition of the squared amplitudes it is possible to keep a constant visibility.

But it could be difficult to get experimentally a phase difference of $\pi/2$ between the reference beams, to overcome this difficulty it is necessary an analysis for a general case where the phase difference between the reference beams is arbitrary; it will be shown that despite that the conditions for the particular case are not obtained it is still possible to generate phase-shifting by means of the amplitude variations of the fields and to keep a constant visibility, those amplitudes must be seen as ordered pairs over ellipses.

3.1 Ideal case

Let's have three waves interfering at any point in space, which are linearly polarized at the same plane and traveling in the z direction

$$E_n = A_n \exp(i\phi_n), \tag{36}$$

with $n = 1,2,3$, being A_n the amplitude considered as nonnegative real, ϕ_n is the phase that contains the wavefront. According to Eq. (2) the interference of these three waves at any point in space is the addition of the three fields, being the irradiance according to Eq. (3)

$$I = |E_1 + E_2 + E_3|^2, \tag{37}$$

doing the math in Eq. (37)

$$I = A_1^2 + A_2^2 + A_3^2 + 2A_1A_2\cos(\phi_1 - \phi_2) + 2A_1A_3\cos(\phi_1 - \phi_3) + 2A_2A_3\cos(\phi_2 - \phi_3), \tag{38}$$

in which three interference terms, the background and the modulation light that is given by the addition of the intensities of each wave are present. Let's consider the first and third wave as the reference wave, and the second wave as the probe wave. By simplicity the next conditions will be chosen

$$\phi_1 = 0\,;\ \phi_2 = \phi \ \text{and}\ \phi_3 = \pi/2\,. \tag{39}$$

The waves E_1 and E_3 will be chosen as homogeneous no tilted plane waves with a phase difference between them of $\pi/2$, ϕ is the phase of the object contained in the second wave. Substituting Eq. (39) into Eq. (38) we have

$$I = A_1^{\,2} + A_2^{\,2} + A_3^{\,2} + 2A_1A_2\cos\phi + 2A_2A_3\sin\phi\ . \tag{40}$$

In this equation it can be observed that one of the three interference terms has been cancelled and the two remaining are now in quadrature. Regrouping the above equation

$$I = A_1^{\,2} + A_2^{\,2} + A_3^{\,2} + 2A_2(A_1\cos\phi + A_3\sin\phi)\,, \tag{41}$$

that can be rewritten as

$$I = A_1^{\,2} + A_2^{\,2} + A_3^{\,2} + 2A_2\sqrt{A_1^{\,2} + A_3^{\,2}}\left(\frac{A_1}{\sqrt{A_1^{\,2} + A_3^{\,2}}}\cos\phi + \frac{A_3}{\sqrt{A_1^{\,2} + A_3^{\,2}}}\sin\phi\right), \tag{42}$$

$$I = A_r^{\,2} + A_2^{\,2} + 2A_rA_2(\cos\phi\cos\psi + \sin\phi\sin\psi)\,, \tag{43}$$

where $\cos\psi = A_1\big/\sqrt{A_1^{\,2} + A_3^{\,2}}$ and $\sin\psi = A_3\big/\sqrt{A_1^{\,2} + A_3^{\,2}}$. Eq. (43) can be expressed as

$$I = A_r^{\,2} + A_2^{\,2} + 2A_rA_2\cos(\phi - \psi)\ , \tag{44}$$

which as it was indicated in Eq. (9) is the expression for a fringe pattern due to the interference of two waves, being

$$A_r^{\,2} = A_1^{\,2} + A_3^{\,2}\ ;\ \tan\psi = \frac{A_3}{A_1}\,, \tag{45}$$

where A_r is the reference amplitude and ψ is an additional phase, both of them depending on the amplitude variations of the first wave A_1 and the third wave A_3. It can be observed from this relationship that it is possible to generate a phase-shifting with the variations of those amplitudes; however because this also affects A_r, there may be a change in the visibility of the fringes, so it can be thought that it would not be possible to apply the PSI techniques for the phase extraction (because one important condition to apply the PSI phase extraction techniques is that the visibility remains constant). One way to keep A_r constant is to consider the amplitudes of waves one and three as an ordered pair (A_1, A_3) which must be contained over an arc segment of radius A_r in the first quadrant (because the amplitudes are considered positive), hence A_r will be within the range $[0, \pi/2]$, as can be seen in Figure 6. However it is possible to generate a negative amplitude modulation if the phase difference in the reference waves is π, having negative real amplitudes, by this the range of ψ will be $[0, 2\pi]$ and the phase extraction techniques in PSI could be applied without any modification.

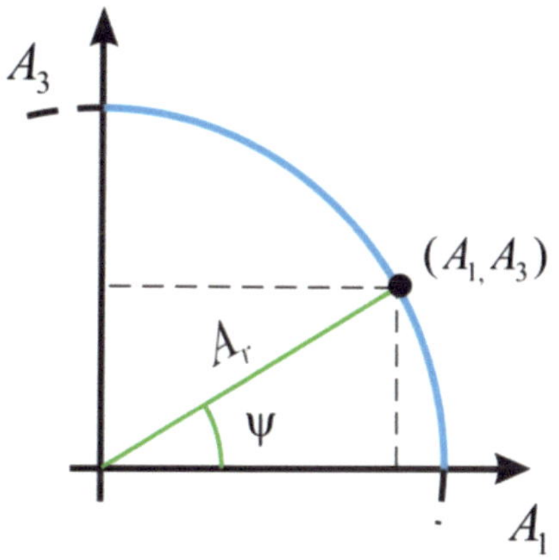

Fig. 6. Amplitudes of first and third wave in a reference system. Any point in the arc keeps A_r constant, and ψ is varied as required for PSI.

In this case the amplitudes are in quadrature, A_1 and A_3 are given by

$$A_1 = A_r \cos\psi \, ; A_3 = A_r \sin\psi \, . \tag{46}$$

In a possible experimental setup the phase difference of $\pi/2$ between the first and third wave could be achieved by means of a retarder plate of a quarter wavelenght. The amplitude variations can be done by using neutral density filters or by using the diffraction orders generated by a grating (for example a Ronchi ruling), where each order is attenuated in accord with the sinc function.

To prove the viability of the present proposal, we have carried out a numerical simulation in which for simplicity, the following considerations have been assumed

$$A_2 = A_r = 1 \, ; \; A_1, A_3 \in [0,1] \, ; \; \phi = x^2 + y^2 \, , \tag{47}$$

such a form guarantees that the interference pattern will have a maximum contrast. Therefore, the three fields could take the form

$$E_1 = A_1 \, ; E_2 = \exp(i\phi) \; E_3 = iA_3 \, , \tag{48}$$

being the interference patter

$$I = 2 + 2\cos(\phi - \psi) \, . \tag{49}$$

Figure 7a) shows the values of the amplitudes needed for a phase-shifting of three steps $N = 3$. The left graphic indicates three points over the arc $(1,0)$, $(\sqrt{2}/2, \sqrt{2}/2)$, $(0,1)$ which are the amplitude variations to obtain the phase steps $\psi = 0, \pi/4, \pi/2$ respectively, while A_r remains constant. Figure 7b) shows the interferograms modeled by Eq. (49) for the phase steps indicated in Figure 7a). A bar diagram above each interferogram indicates the amplitude levels of the waves needed for each phase-shift. In a very similar way Figure 8 show the phase-shifting for the case of four steps $N = 4$.

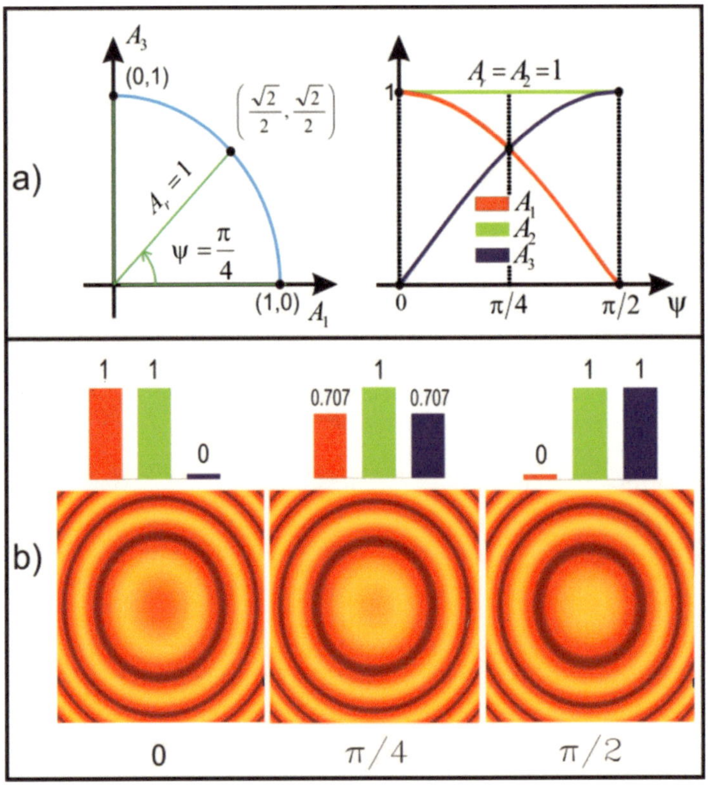

Fig. 7. Amplitude variations for a PSI of three steps: a) a point on the arc yields a phase-shift while the reference amplitude is kept constant; b) interferogram shifted in phase by the amplitude variations indicated in a).

The columns of Figure 9a) show simulated interference pattern for different phase-shifts, which are obtained when the amplitude values A_1 and A_3 are over straight lines that form an angle ψ respecting to the axis A_1, Figure 10b); by doing this, the visibility of the patterns remains constant, but to get a maximum visibility the value of the amplitude A_2 must be equal to the value of A_r, which can be observed in the first row of Figure 9a), where the values of A_1 and A_3 are over an arc segment of radius $A_r = 1$, Figure 9b). If we have different values of ψ but its corresponding amplitudes are not over the same arc segment, the interference patterns will not have a constant visibility, what can be seen in Figure 9a) if for each value of ψ we take a different row (indicated by different symbols).

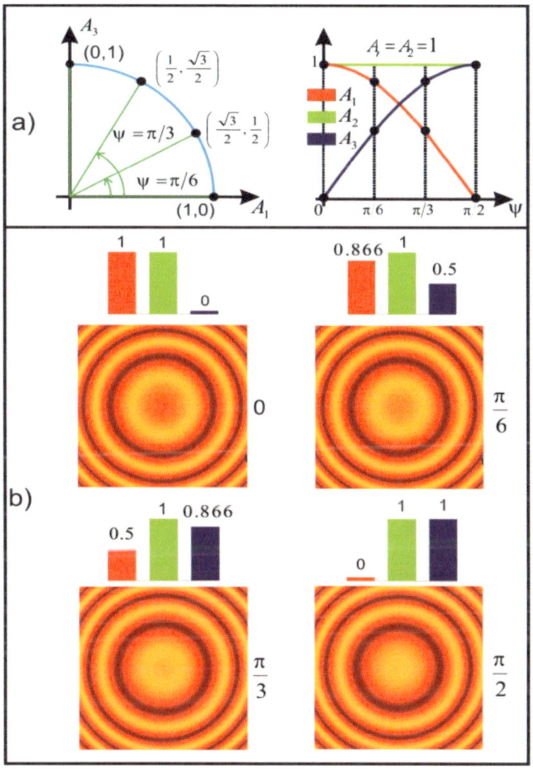

Fig. 8. Amplitude variations for a PSI of four steps: a) a point over the arc yields a phase-shift while the reference amplitude is kept constant; b) interferograms shifted in phase by the amplitude variations indicated in a).

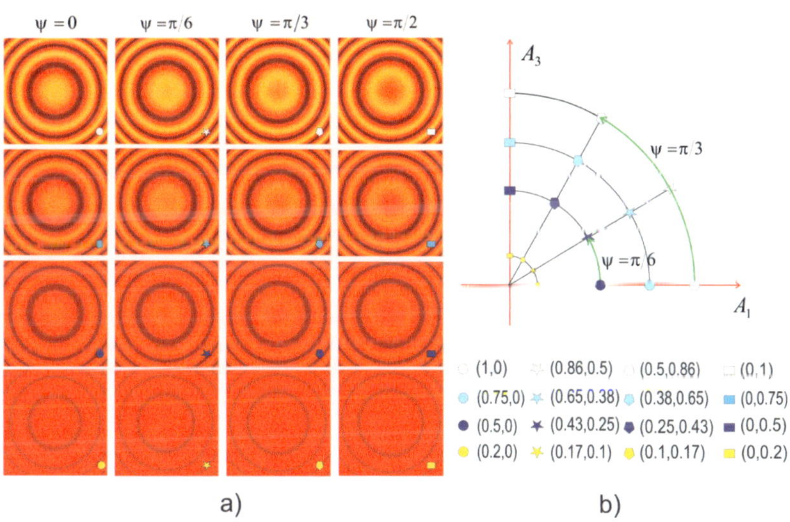

Fig. 9. Interference patterns with different visibility a) Phase-shifted interferograms due to the amplitudes shown in b); b) points over arcs and straight lines that give phase-shift keeping in some cases a constant visibility.

3.2 General case

As demonstrated in the previous section, it is possible to have a new method of PSI by means of the field amplitude variations based on the scheme of a three beam interferometer modeled as a two beam interferometer, where the reference beam and a constant phase term (used to generate the phase-shift) were given in function of the two reference beams. Due to a phase difference of $\pi/2$ between the reference beams one of the interference terms was canceled, leaving the two remaining in quadrature.

However under experimental conditions it is not always possible to obtain a phase difference of $\pi/2$ between the reference beams. Despite of this it will be shown that it is still possible to generate phase-shift by means of the amplitude variation of the fields, where now to keep a constant visibility the amplitudes must be seen as ordered pairs over an ellipse instead of a circle, extending with this the range of the phase-shifting until $[0,\pi]$ instead of $[0,\pi/2]$.

Let's consider again three linearly polarized waves at the same plane traveling on z direction as shown in Eq. (36), whose irradiance can be expressed as in Eq. (37)

$$I = A_1^2 + A_2^2 + A_3^2 + 2A_1A_2\cos(\phi_1 - \phi_2) + 2A_1A_3\cos(\phi_1 - \phi_3) + 2A_2A_3\cos(\phi_2 - \phi_3), \quad (50)$$

where the intensities of the three waves are present in the background light and also in the three interference terms. For the general case, the phases that will be considered are

$$\phi_1 \neq 0;\ \phi_2 = \phi_1 + \Delta\phi_1 + \phi;\ \text{and}\ \phi_3 = \phi_1 + \Delta\phi_2, \quad (51)$$

$\Delta\phi_1$ is a constant phase difference between the first and second wave, $\Delta\phi_2$ is a constant phase difference between the first and third wave; therefore between the second and third wave will exist a phase difference of $\Delta\phi_2 - \Delta\phi_1$. In summary, in the absence of a phase object it will be considered a phase difference between each pair of waves. It is important to notice that when $\Delta\phi_1 = 0$ and $\Delta\phi_2 = \pi/2$ the general case studied in section 3.1 will be obtained. Substituting Eq. (51) into Eq. (50) we have

$$I = A_1^2 + A_2^2 + A_3^2 + 2A_1A_2\cos(\phi + \Delta\phi_1) + 2A_1A_3\cos(\Delta\phi_2) + 2A_2A_3\cos(\phi + \Delta\phi_1 - \Delta\phi_2), \quad (52)$$

$$I = A_1^2 + A_2^2 + A_3^2 + 2A_1A_3\cos\Delta\phi_2 + 2A_2\left[A_1\cos(\phi + \Delta\phi_1) + A_3\cos(\phi + \Delta\phi_1 - \Delta\phi_2)\right], \quad (53)$$

$$I = A_1^2 + A_2^2 + A_3^2 + 2A_1A_3\cos\Delta\phi_2 + 2A_2\left[A_1\cos(\phi + \Delta\phi_1) + \right.$$

$$\left. + A_3\cos\Delta\phi_2\cos(\phi + \Delta\phi_1) + A_3\sin\Delta\phi_2\sin(\phi + \Delta\phi_1)\right], \quad (54)$$

$$I = A_1^2 + A_2^2 + A_3^2 + 2A_1A_3\cos\Delta\phi_2 + 2A_2\left[(A_1 + A_3\cos\Delta\phi_2)\right.$$

$$\left.\cos(\phi + \Delta\phi_1) + A_3\sin\Delta\phi_2\sin(\phi + \Delta\phi_1)\right], \quad (55)$$

$$I = A_2^2 + A_r^2 + 2A_2A_r\left[\cos\psi\cos(\phi + \Delta\phi_1) + \sin\psi\sin(\phi + \Delta\phi_1)\right], \quad (56)$$

where

$$\cos\psi = \frac{A_1 + A_3\cos\Delta\phi_2}{A_r}\ ;\ \sin\psi = \frac{A_3\sin\Delta\phi_2}{A_r}\,, \tag{57}$$

$$I = A_2^{\ 2} + A_r^{\ 2} + 2A_2A_r\cos(\phi + \Delta\phi_1 - \psi)\,. \tag{58}$$

It has been deduced the general expression of a fringe pattern of two waves where A_r is equivalent to the reference amplitude and ψ is an additional phase, both given by

$$A_r^{\ 2} = A_1^{\ 2} + A_3^{\ 2} + 2A_1A_3\cos\Delta\phi_2\ ; \tag{59}$$

$$\tan\psi = \frac{A_3\sin\Delta\phi_2}{A_1 + A_3\cos\Delta\phi_2}\,, \tag{60}$$

it can be seen that A_r and ψ depend on the amplitude variations of A_1 and A_3 as well as the phase difference between the first and third wave $\Delta\phi_2$.

To apply the PSI techniques, the visibility in Eq. (58) must remain constant for each phase-shift; this can be achieved if the amplitude A_r in la Eq. (59) remains constant while ψ varies with the changes of A_1 and A_3. To define the behavior of the amplitudes which satisfy the condition to keep A_r constant, Eq. (59) will be rewritten as

$$\sin^2\Delta\phi_2 = \frac{A_1^{\ 2}}{A_r^{\ 2}}\sin^2\Delta\phi_2 + \frac{A_3^{\ 2}}{A_r^{\ 2}}\sin^2\Delta\phi_2 + 2\frac{A_1A_3}{A_r^{\ 2}}\sin^2\Delta\phi_2\cos\Delta\phi_2\,, \tag{61}$$

which can take the form of the equation of an ellipse

$$\sin^2\alpha = \frac{x^2}{a^2} + \frac{y^2}{b^2} - 2\frac{xy}{ab}\cos\alpha\,, \tag{62}$$

such that

$$\tan 2\upsilon = \frac{2ab}{a^2 - b^2}\cos\alpha\,, \tag{63}$$

where υ is the inclination angle of the ellipse, a is the maximum value of the ellipse over the x axis and b is the maximum value over the y axis.

Being in our case $a = -b$, so Eq. (61) can be written as in Eq.(62), where

$$a = \frac{A_r}{\sin\Delta\phi_2}\ ; \tag{64}$$

substituting the values of a, b and $\alpha = \Delta\phi_2$ in Eq. (60) a family of ellipses inclined to an angle υ will be obtained

$$\tan 2\upsilon = -\frac{\cos\Delta\phi_2}{0} = \begin{cases} \text{indet} & \Delta\phi_2 = \pi/2 + n\pi \\ -\infty & \Delta\phi_2 \in [0,\pi/2)\cup(3\pi/2,2\pi], \\ +\infty & \Delta\phi_2 \in (\pi/2,3\pi/2) \end{cases} \tag{65}$$

therefore υ can take the next values

$$\upsilon = \begin{cases} \text{indet}, & \Delta\phi_2 = \pi/2 + m\pi \\ -\pi/4, & \Delta\phi_2 \in [0,\pi/2)\cup(3\pi/2,2\pi]. \\ \pi/4, & \Delta\phi_2 \in (\pi/2,3\pi/2) \end{cases} \tag{66}$$

The amplitudes in Eq. (61) can take the next parametric form

$$A_1 = \frac{A_r}{\sin\Delta\phi_2}\sin(\Delta\phi_2 - \psi)\,;\; A_3 = \frac{A_r}{\sin\Delta\phi_2}\sin\psi\,, \tag{67}$$

whose parameter is the phase-shift ψ within a valid range of

$$\psi \in \begin{cases} [0,\Delta\phi_2], & \Delta\phi_2 \in (0,\pi) \\ [\Delta\phi_2 - \pi,\pi], & \Delta\phi_2 \in (\pi,2\pi)' \end{cases} \tag{68}$$

that is obtained considering that the amplitudes A_1 and A_3 are positive, therefore they must be in the first quadrant and that $\Delta\phi_2 \neq m\pi$ being m an integer number. However it is possible to have a negative modulation in the amplitude if a phase difference of π in the reference waves is properly implemented, thus it is possible to have real negative amplitudes, hence ψ will be within the range of $[0,2\pi]$, and the known PSI techniques could be applied without modifications.

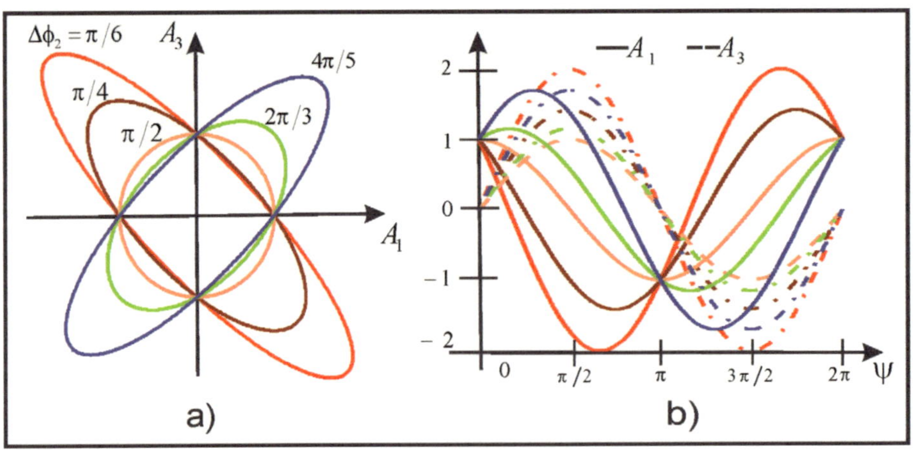

Fig. 10. Amplitudes of the wave one and three put in a reference system for several values of $\Delta\phi_2$. a) family of ellipses; any point on any ellipse keeps A_r constant while ψ is varied as it is required for PSI, b) amplitudes in a parametric form respecting to ψ corresponding to the ellipses in a).

It can be observed form Eq. (68) that $\Delta\phi_2$ determines directly the range of the phase-shifting. When a certain phase-shift is needed, the amplitudes A_1 and A_3 will be given by Eq. (67) which comply the conditions for a constant visibility and they could be seen as points over the arc of an ellipse at the first quadrant as it is shown for several values of $\Delta\phi_2$ in Figure 10.

A numerical simulation will be shown in order to prove the viability of the proposal. For simplicity, the next considerations have been taken

$$A_2 = A_r = 1\,;\ \Delta\phi_1 = \phi_1 = 0\,;\ \Delta\phi_2 = 2\pi/3\,;\ \phi = x^2 + y^2\,, \tag{69}$$

therefore the three fields can be expressed as

$$E_1 = A_1\,;\ E_2 = \exp(i\phi)\,;\ E_3 = \exp(i\,2\pi/3)\,, \tag{70}$$

the interference pattern can be written as

$$I = 2 + 2\cos(\phi - \psi)\,, \tag{71}$$

such a form that the interference pattern will have a maximum contrast, where the amplitudes A_1, A_3 and the additional phase ψ are given by

$$1 = A_1^{\,2} + A_3^{\,2} - A_1 A_3\,, \tag{72}$$

$$\tan\psi = \frac{\sqrt{3}A_3}{2A_1 - A_3}\,. \tag{73}$$

Substituting Eq. (69) in Eq. (67) is gotten

$$A_1 = \cos\psi + \frac{1}{\sqrt{3}}\sin\psi\,;\ A_3 = \frac{2}{\sqrt{3}}\sin\psi\,. \tag{74}$$

Figure 12a) shows the ellipse at the first quadrant for this particular case, which is inclined to $\pi/4$ with an ellipticity of $e = \sqrt{3}$. Figure 12b) shows the amplitudes corresponding to the phase shift within the range $\psi \in [0, 2\pi/3]$.

Figure 12 shows the interferograms for $N = 9$, being the phase step of $\Delta\psi = \pi/12$; the phase-steps $\psi_k = k\pi/12$ with $k = 0,1,\ldots,8$ were generated by the amplitudes shown in Figure 12b), and these are indicated with points over the arc corresponding to the ellipse in Figure 12a).

This analysis has the advantage that even if there is a phase difference of $\Delta\phi_2$ between the reference waves it is still possible to generate phase-shifting with the proposed method, besides that it will not be necessary to use an optical device (as phase retarders) to generate $\Delta\phi_2$. The variations of the amplitudes can be done by using neutral filters or also by using the diffraction orders produced by a grating as for example a Ronchi ruling, where each diffraction order is attenuated according to the sinc function.

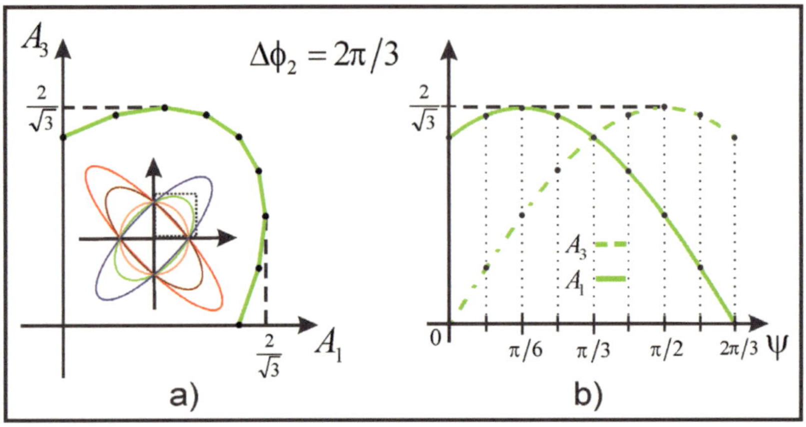

Fig. 11. Amplitude variations for a PSI of 9 steps when $\Delta\phi_2 = 2\pi/3$: a) a point over the arc yields a phase-shifting while the reference amplitude is kept constant; b) amplitudes (A_1, A_3) for the phase steps indicated in a).

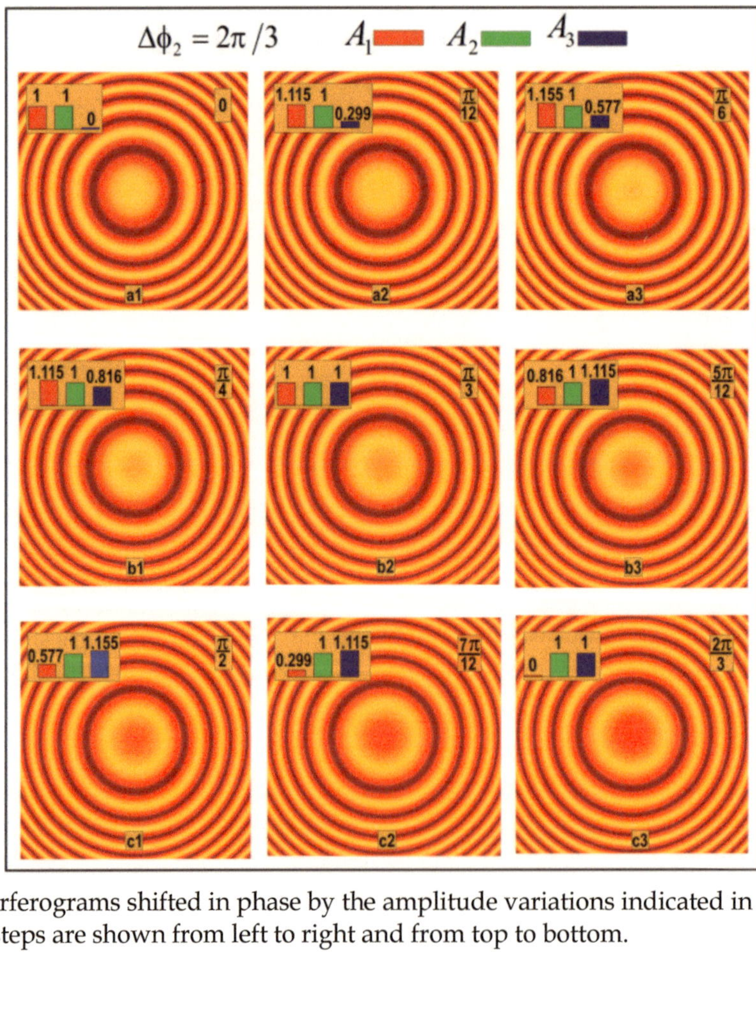

Fig. 12. Interferograms shifted in phase by the amplitude variations indicated in Figure 11. The phase steps are shown from left to right and from top to bottom.

4. Conclusions

In this chapter have been discussed the phenomenon of optical interference for two waves elliptically polarized, the phase shifting interferometry and the commonly used methods to generate that shifting, as well as the methods for phase extraction. It has been demonstrated with a numerical analysis and a computer simulation the viability of a new method of phase-shifting based on the amplitude variation of two fields considered as the reference beams in a scheme of a three beam interferometer, for which two cases were analyzed:

A particular case was considered when the phase difference between the reference waves is $\pi/2$, hence one of the three interference terms is cancelled while the two remaining are put in quadrature. To get a constant visibility, the amplitude of the reference waves must be over an arc segment in the first quadrant of a circle whose radius is A_r. Due to the amplitudes are considered to be real positive, the phase-shifting will be within the range of $[0,\pi/2]$. But, theoretically this range could be extended until $[0,2\pi]$ if the amplitudes are modulated moreover in their negative part, what can be done by an appropriate phase change in the reference waves by π radians.

In a more general case, the phase differences $\Delta\phi_2$ between the reference waves was considered to be arbitrary and within the range $\Delta\phi_2 \in [0,2\pi]$. In this study it was shown that despite of this the phase-shifting by amplitude modulation is also possible, and it includes the particular case given when $\Delta\phi_2 = \pi/2$. Besides in order to keep a constant visibility during the PSI application, the amplitudes must be over an ellipse instead of a circumference, which is inclined at $\pm\pi/4$ if $\Delta\phi_2 \in (\pi/2, 3\pi/2)$ or $\Delta\phi_2 \in [0,\pi/2) \cup (3\pi/2, 2\pi]$, respectively, also it was found that $\Delta\phi_2$ directly influences in both the ellipticity and the phase-stepping range when the amplitudes are modulated no-negative and can also reach a range until of $\psi \in [0,\pi]$, however when the negative part is taken in account, the range can reach until $\psi \in [0,2\pi]$.

5. Acknowledgment

This work was partially supported by Vicerrectoría de Investigación y Estudios de Posgrado of Benemérita Universidad Autónoma de Puebla under grant MEFC and by Programa de Mejoramiento del Profesorado (México) under grant PROMEP/103.5/09/4544.

6. References

Bone D. J., Bachor H.-A., and Sandeman R. J., (1986). Fringe-Pattern Analysis Using a 2-D Fourier Transform, *Appl. Opt.*, Vol. 25, pp. 1653-1660

Born M. and Wolf E.; Cambridge University Press (1993). *Principles of Optics*

Bruning J. H., Herriott D. R., Gallagher J. E., Rosenfield D.P., White A. D., and Brangaccio D. J., (1974). Digital Wavefront Measuring Interferometer for Testing Surfaces and Lenses, *Appl. Opt.*, Vol. 3, pp. 2693-2703

Bryngdahl O., (1972). Polarization Type Interference Fringe Shifter, *J. Opt. Soc. Am.*, Vol. 62, pp. 462-464

Bryngdahl O., (1976). Heterodyne Shear Interferometers Using Diffractive Filters with Rotational Symmetry, *Opt. Commun.*, Vol. 17, pp. 43

Burqwald G. M. and Kruger W. P., (1970). An instant-On Laser for Distant Measurement, *Hewlett Packard J.*, Vol. 21, pp. 14

Cai L. Z., Liu Q., and Yang X. L., (2004). Generalized Phase-Shifting Interferometry with Arbitrary Unknown Phase Steps for Diffraction Objects, *Opt Lett.*, Vol. 29 pp. 183-185

Carré, P., (1966). Instalation et Utilisation du Comparateur Photoelectrique et Interferentiel du Bureau International des Poids et Measures, *Metrología*, Vol. 2, pp. 13-23

Crane R., (1969). Interference Phase Measurement, *Appl. Opt.*, Vol. 8, pp. 538-542

Creath K., (1985). Phase-Shifting Speckle Interferometry, *Appl. Opt.*, Vol. 24, pp. 3053-3058

Gasvik K. J.; John Wiley & Sons Ltd (1996). *Optical Metrology*

Goodwin E. P. and Wyant J. C., SPIE Press, Washington, (2006). *Field guide to Interferometric Optical Testing*

Hech E.; Addison-Wesley (1972). *Optics*

Hu H. Z., (1983). Polarization Heterodyne Interferometry Using Simple Rotating Analyzer, Theory and Error Analysis, *Appl. Opt.*, Vol. 22, pp. 2052-2056

Ishii Y., Chen J., and Murata K., (1987). Digital Phase Measuring Interferometry with a Tunable Laser .Diode, *Opt Lett.*, Vol. 12, pp. 233-235

J. R. P Angel and P. L. Wizinowich, (1988). A Method of Phase Shifting in the Presence of Vibration, *Eur. Southern Obs. Conf Proc.*, Vol. 1, pp. 561-567

Itoh K., (1982). Analysis of Phase Unwrapping Algorithm, *Appl. Opt.*, Vol. 21, pp. 2470

Kothiyal M. P. and Delisle C., (1984). Optical Frequency Shifter for Heterodyne Interferometry Using Counterrotating Wave Plates, *Opt. Lett.*, Vol 9, 319-321

Macy William. W., Jr., (1983). Two-Dimensional Fringe Pattern Analysis, *Appl. Opt.*, Vol. 22, pp. 3898-3901

Malacara D., Servín M. and Malacara Z., "Interferogram Analysis for Optical Testing," (Marcel Dekker, New York, 1998), pp. 247-248.

Malacara D.; Wiley (2007). *Optical Shop Testing*

Martínez A., Sánchez M. de M. and Moreno I., (2007). Phasor Analysis of Binary Diffraction Gratings With Different Fill Factors, *Eur. J. Phys.*, Vol. 28, pp. 805-816

Mendoza-Santoyo F., Kerr D., and Tyrer J. R., (1988). Interferometric Fringe Analysis Using a Single Phase Step Technique, *Appl. Opt.*, Vol. 27, pp. 4362-4364

Meneses-Fabian Cruz and Rivera-Ortega U., (2011). Phase-shifting interferometry by wave amplitude modulation, *Opt. Lett.* Vol. 36, pp. 2417-2419

Meneses-Fabian C., Rodriguez-Zurita Gustavo, Encarnacion-Gutierrez Maria-del-Carmen, Toto-Arellano Noel I., (2009). Phase-shifting interferometry with four interferograms using linear polarization modulation and a Ronchi rating displaced by only a small unknown amount, *Optics Communications,* Vol. 282, pp. 3063–3068

Nakadate S., Saito H. and Nakajima T., (1986). Vibration Measurement Using Phase-Shifting Stroboscopic Holographic Interferometry, *Opt. Acta,* Vol. 33, pp. 1295-1309

Nakadate S. and Saito J., (1985). Fringe Scanning Speckle Pattern Interferometry, *Appl. Opt.*, Vol. 24, pp. 2172-2180

Novak M, Millerd J., Brock N., North-Morris M., Hayes J., and Wyant J., (2005). Analysis of a micropolarizer array-based simultaneous phase-shifting interferometer, *Appl. Opt.*, Vol. 44, pp. 6861- 6868

Okoomian H. J., (1969). A Two Beam Polarization Technique to Measure Optical Phase, *Appl. Opt.*, Vol. 8, pp. 2363-2365

Onodera R. and Ishii Y., (1996). Phase–Extraction Analysis of Laser-Diode Phase Shifting Interferometry That is Insensitive to Changes in Laser Power, *J. Opt. Soc. Am.*, Vol. 13, pp. 139-146

Rastogi K.; Artech House, Inc (1997). *Optical Measurement Techniques and Applications*

Robinson D. and Williams D., (1986). Digital Phase Stepping Speckle Interferometry, *Opt. Commun.*, Vol. 57, pp. 26-30

Rodriguez-Zurita G., Meneses-Fabian C., Toto-Arellano N. I., Vázquez-Castillo J., and Robledo-Sánchez C., (2008). One-shot phase-shifting phase-grating interferometry with modulation of polarization: case of four interferograms, *Opt. Express,* Vol 16, pp. 7806-7817

Ronchi V., (1964). Forty Years of History of a Grating Inteferometer, *Appl. Opt.* Vol 3, pp. 437-451

Schwider J., "Advanced Evaluation Techniques in Interferometry," Progress in Optics. Vol. XXVIII, E. ..Wolf, ed., (Elsevier Science, 1990), pp. 274-276

Servin M., Marroquin J. L., and Cuevas F. J., (1997). Demodulation of a Single Interferogram by Use of a Two-Dimensional Regularized Phase-Tracking Technique, *Appl. Opt.*, Vol 36, pp. 4540-4548

Shagam R. N. and Wyant J. C., (1978). Optical Frequency Shifter for Heterodyne Interferometers Using Multiple Rotating Polarization Retarders, *Appl. Opt.*, Vol. 17, pp. 3034-3035

Sommargreen G. E., (1975). Up/Down Frequency Shifter for Optical Heterodyne Interferometry, *J. Opt Soc. Am.*, Vol. 65, pp. 960-961

Soobitsky J. A., *Piezoelectric Micromotion Actuator*, U.S. Patent No. 4, 577, 131 1986

Srinivasan V., Liu H. C., and Halioua M., (1985). Automated Phase-Measuring Profilometry: A Phase Mapping Approach, *Appl. Opt.*, Vol. 24, pp. 185-188

Stetson K. A. and Brohinsky W. R., (1988). Fringe-Shifting Technique for Numerical Analysis of Time Average Holograms of Vibrating Objects, *J. Opt. Soc. Am. A*, Vol. 5, pp. 1472-1476

Stevenson W. H., (1970). Optical Frequency Shifting by Means of a Rotating Diffraction Grating, *Appl. Opt.*, Vol 9, pp. 649-652

Susuki. T. and Hioki R., (1967). Translation of Light Frequency by a Moving Grating, J. *Opt. Soc. Am.*, Vol. 57, pp. 1551-1551

Takeda M., Ina H., and Kobayashi S., (1982). Fourier-Transform Method of Fringe-Pattern Analysis for Computer-Based Topography and Interferometry, *J. Opt Soc. Am*, Vol. 72, pp. 156-160

Villa J., I. De la Rosa, and. Miramontes G., (2005). Phase Recovery from a Single Fringe Pattern Using an Orientational Vector-Field-Regularized Estimator, *Opt. Soc. Am.*, Vol. 22, pp. 2766-2773

Wizinowich P. L., (1990). Phase Shifting Interferometry in the Presence of Vibration: A New Algorithm and System, *Appl. Opt.*, Vol. 29, pp. 3271-3279

Wyant J. C. and Shagam R. N., (1978). Use of Electronic Phase Measurement Techniques in Optical Testing, *Proc ICO-11, Madrid,* pp. 659-662

Meng X.F., Cai L.Z., Wang Y.R., Yang X.L., Xu X.F., Dong G.Y., Shen X.X., Cheng X.C., (2008). Wavefront reconstruction by two-step generalized phase-shifting interferometry, *Opt. Commun.*, Vol. 281, pp. 5701-5705

9

N-Shots 2N-Phase-Steps Binary Grating Interferometry

Cruz Meneses-Fabian, Gustavo Rodriguez-Zurita and Noel-Ivan Toto-Arellano
Facultad de Ciencias Físico-Matemáticas, Benemérita Universidad Autónoma de Puebla México

1. Introduction

In physical optics, the interference effect consists in superposing two or more optical fields in a region of space, which mathematically results in the vector sum of them, it is certain because of in physical optics the superposition principle is valid. So, when this sum is observed with some optical detector such as a human eye or a CCD camera, the irradiance of the total field is obtained, and it can be understood as the sum of the irradiance from each individual field, known as background light, plus an interference additional term per each pair of fields, which consists of a cosine of the phase difference between the two waves and of a factor given mainly by the product of waves amplitudes known as a modulation light. The total effect shows brilliant and obscure zones known as interference fringes, also called as a fringe pattern, or an interferogram (Born & Wolf, 1993). This effect was first reported by Thomas Young in 1801 (Young, 1804; Shamos, 1959), more late it also was observed by Newton, Fizeau, Michelson, etc. (Hecht, 2002). They designed many optical arranges now known as interferometers in order to interfere two o more waves trying to meet the optimal conditions to have a maximum quality in the fringes. Many studies have demonstrated that the fringe quality, (better known as visibility of interference pattern), as well as the shape and the number of fringes is depending of each parameter in the optical field such as the amplitude, the polarization state, the wavelength, the frequency, the coherence degree, the phase and because of it so many applications among physics and other sciences like biology, medicine, astronomy, etc., or also in engineering have been extensively made (Kreis, 2005). However, an intermediate step between the fringe pattern and the direct application in some topic of interest as the evaluation and processing of this pattern must be realized. In this regarding, many proposal have been amply discussed, for example in interferometry of two waves when the phase difference is the variable to be calculated, one of the techniques more widely used consists of performing consecutively shifts of constant phase between the waves that interfere. Then, for each phase-step a new interferogram is gotten and therefore for N steps, N interferograms are obtained. Mathematically, a Nx3 system of equations is formed since the with phase-step the object phase, the background light, and the modulation of light are considered unknown with respect to position and constant with respect to time, is known as phase-shifting interferometry (PSI) (Creath, 1993; Malacara, 2007; Schwider, 1990). The phase shift is introduced by a shifting device, which can be done with Zeeman effect shifters, acousto-optical modulators, rotating polarizers or translating gratings, among other possibilities. Spatial techniques also have been introduced and widely studied. These consist of introducing a spatial phase variation into interferogram (Takeda, et. al., 1982). Typically, this variation is a linear function, but when the fringes forms closed loops this carrier is not appropriate, and to overcome this shortage a quadratic phase has been proposal (Malacara, et. al., 1998). Others methods for phase extraction have also been proposed (Moore & Mendoza-Santoyo, 1995; Peng, et. al., 1995).

Based on the interference of two monochromatic and coherent waves, the present chapter speaks about the phase shifting interferometry. The principal idea consists of obtaining two interferograms shifted in phase by 180° in a single-shot. By performing an arbitrary phase-step, another two interferograms shifted 180° in phase are captured in a second shot. This way, four interferograms result shifted in phase in two shots with two phase steps, considering the first phase step equal to zero. The arbitrary phase-step is considered to be between 0° and 180° and it will be measured under the concept of generalized phase-shifting interferometry (GPSI) (Xu, et. al., 2008). Therefore, in general with N-phase-steps $2N$ interferograms will be captured in N shots only. Fringe patterns are obtained from an interferometer build using a $4f$ optical correlator of double Fourier transform, where, at the input plane two apertures are considered. One of them is crossed by a reference beam and the other is considered as a probe window where a phase object is placed. These windows are considered as an input transmittance function. In the Fourier plane, a binary grating (Ronchi ruling) with certain period is placed as a spatial filter function. Then with the appropriate conditions of the wavelength, the grating period, and the focal length of the lenses, at the image plane the interference of the fields in each window is achieved and replicated around diffraction order. Then, by using orthogonal linear polarization in the windows, it is possible to demonstrate that a phase shifting of 180° can be obtained by observing the superposition with adequate transmittance angle of another linear polarizer. An arbitrary phase-shifting is later obtained with a grating displacement.

In summary, a method to reduce the number of captures needed in phase-shifting interferometry is proposed on the basis of grating interferometry and modulation of linear polarization. In this chapter, the case of four interferograms is considered. A common-path interferometer is used with two windows in the object plane and a Ronchi grating as the pupil, thus forming several replicated images of each window over the image plane. The replicated images, under proper matching conditions, superimpose in such a way so that they produce interference patterns. Orders 0 and +1 and -1 and 0 form useful patterns to extract the optical phase differences associated to the windows. A phase of π is introduced between these orders using linear polarizing filters placed in the windows and also in the replicated windows, so two π -shifted patterns can be captured in one shot. An unknown translation is then applied to the grating in order to produce another shift in the each pattern. A second and final shot captures these last patterns. The actual grating displacement and the phase shift can be determined according to the method proposed by (Kreis, 1986) before applying proper phase-shifting techniques to finally calculate the phase difference distribution between windows. Along this chapter a theoretical model is amply discussed and it is verified with both a numeral simulation and experimental results.

2. Theoretical analysis

Phase-shifting interferometry retrieves phase distributions from a certain number N of interferograms (Creath, 1993; Malacara, 2007). Each interferogram I_k must result from phase displacements by certain phase amounts φ_k $(k=0\ldots N-1)$ in order to form a solvable system of equations (Malacara, et. al., 1998; Creath, 1993; Millerd & Brock, 2003). Because one of these phase amounts can be taken as reference, say $\varphi_0=0$, it is possible to use the corresponding phase shifts, each denoted by $\alpha_{k+1}=\varphi_{k+1}-\varphi_k$. Among several possibilities, the case of $N=4$ interferograms and $N-1=3$ equal shifts $\alpha_1=\alpha_2=\alpha_3=\alpha=90°$, has been demonstrated to be very useful (Schreiber & Bruning, 2007;), especially for well contrasted interferograms (Schwider, 1990). To obtain phase-shifted interferograms, a number of procedures have been demonstrated but many of them needs of N shots to capture all of those interferograms. Thus, a simplification is desirable in order to reduce the time of capture. Single-shot interferometers capturing all needed interferograms simultaneously are good examples of this effort (Barrientos-Garcia, et. al., 1999; Novak, 2005). On the other side, phase shifts can be induced by mechanical shifts of a proper element, as a piezoelectric stack (Bruning, et. al., 1974) or a grating (Schwider, 1990). Modulation of polarization is another useful technique (Creath, 1993). In grating interferometry, a grating can be transversally displaced by a quarter of a period to obtain shifts of $\alpha=90°$ (Meneses-Fabian, et. al., 2006), for example. But in order to obtain several values, the same number of displacements is required (Hariharan, et. al., 1987; Hu, et. al., 2008; Novák, et. al., 2008). Besides, when using gratings as phase shifters, the grating displacement must be carried out with sufficient precision. The higher the grating frequency, the smaller the grating displacement required. Thus, the use of high frequency rulings could compromise the precision of the phase shift.

In this chapter, a method to reduce the number of captures from 2N to N is proposed by means of common-path grating interferometry (Meneses-Fabian, et. al., 2006) in conjunction with crossed linear polarization filters for modulation of polarization (Nomura, et. al., 2006) and grating displacements. A common-path phase-shifting interferometer can be constructed with two-window in the object plane of a $4f$ Fourier-transform system and a grating as its pupil (Arrizon & De-La-Llave, 2004). One-shot phase-shifting interferometers have already been proposed with phase-gratings and elliptical polarization (Rodriguez-Zurita, et. al., 2008a; Kreis, 1986). However, a more common Ronchi grating can be used for the case of $N=4$ because the diffraction efficiencies for diffraction orders ± 1 are sufficiently good enough to display adequate interferograms. Interference of first-neighboring orders can be obtained in the image plane when proper matching conditions are fulfilled and the phase shifting can be performed with grating displacement driven by an actuator (Meneses-Fabian, et. al., 2006). Because several diffraction orders are to be found in the image plane, some additional shifts can be induced by use of linear polarization instead of circular or elliptical polarization. The calculation of the values of the shifts induced by a grating displacement can be carried out with a method developed by Thomas Kreis (Kreis, 1986), thus alleviating the need of a more detailed calibration. This method is particularly useful for the case of $N=4$, where the proposed simplification reduces the number of required grating displacements to only one. This particular value does not need to be known beforehand because it can be calculated directly by the above mentioned method of Kreis to find a solution for the optical phase. Then, the number of shots required to capture the four interferograms would result in only two. We restricted ourselves to the case of $N=4$ precisely but with α_i not necessarily of the same value. Experimental results are presented.

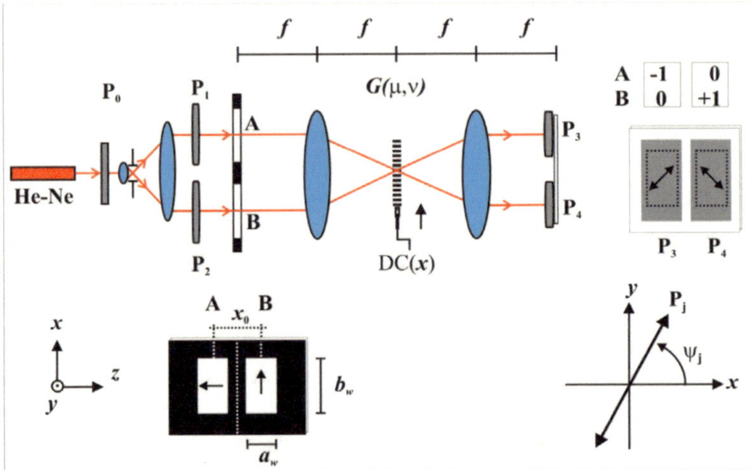

Fig. 1. Experimental setup. Pj , j = 0…4: linear polarizers. A, B: rectangular windows in object plane. $G(\mu,\nu)$: Ronchi grating (μ,ν the spatial frequencies escalated by λf). DC(x): actuator. f: focal length. Side views of windows (lower left) and polarizers with $\psi = \psi_j$ in image plane (lower right) are also sketched.

2.1 Theoretical background

Fig.1 shows the experimental setup. It comprises a 4*f* Fourier-transform system under monochromatic illumination at wavelength λ. The transforming lenses have a focal length of *f*. Linear polarizer P_0 have its transmission axis at 45° and the linear polarizers P_1 and P_2, over windows A and B, have its transmission axis at 0° and 90° respectively. The object plane (input plane) consists of two similar rectangular windows A and B, each of sides a_w and b_w. The windows centers are separated by the distance x_0. In general, an amplitude distribution of the form $A(x,y)$ can be considered in the window A as a reference wave, while $B(x,y)\exp\left[i\phi(x,y)\right]$ can represent the amplitude distribution in the window B (a test object, for instance). Then, the input transmittance can be expressed by.

$$\mathbf{t}(x,y) = w(x+\frac{1}{2}x_0,y)A\left(x+\frac{1}{2}x_0,y\right)\mathbf{J}_A + w(x-\frac{1}{2}x_0,y)B\left(x-\frac{1}{2}x_0,y\right)\exp\left[i\phi\left(x-\frac{1}{2}x_0,y\right)\right]\mathbf{J}_B \quad (1)$$

where $\mathbf{J}_A = \begin{pmatrix}1\\0\end{pmatrix}$, $\mathbf{J}_B = \begin{pmatrix}0\\1\end{pmatrix}$ are Jones vectors corresponding to orthogonal linear polarization states and the window function is written as $w(x,y) = rect(x/a_w)\cdot rect(y/b_w)$, where $rect(x/a_w)$ is the rectangle function on x direction of width a_w and $rect(y/b_w)$ is the rectangle function on y direction of width b_w.

A binary absorptive grating $G(\mu,\nu)$ with spatial period u_d and bright-band width u_w is placed in the frequency plane (Fourier plane, Fig.1). An actuator (DC) can translate the grating longitudinally through a given distance u_0. The grating (a Ronchi grating) can be written.

$$G(\mu,\nu) = rect\left(\frac{\mu-\mu_0}{\mu_w}\right) \otimes \sum_{n=-\infty}^{\infty} \delta(\mu - n\mu_d), \tag{2}$$

with the spatial frequency coordinates are given by $(\mu,\nu) = (u/\lambda f, v/\lambda f)$, where (u,v) are the actual spatial coordinates, and $\mu_k = u_k / \lambda f$ with $k = 0, w, d$ as a label for displacement, bright-band width and grating period, respectively. The symbol $\otimes$ means convolution.

In the image plane, the amplitude distribution can be written as the convolution between the amplitude of the object and the impulse response of the system, i.e.,

$$\mathbf{t}_O(x,y) = \mathbf{t}(x,y) \otimes \Im^{-1}\{G(\mu,\nu)\}, \tag{3}$$

with $\Im^{-1}$ the inverse Fourier-transform operation assumed to be performed by the second transforming lens as a convention taken in this work in accordance with an inversion in the image coordinates. Using Eq. (2), the convolution results in

$$\mathbf{t}_O(x,y) = \sum_{n=-\infty}^{\infty} C_n \left\{ w_A\left[x - \left(\frac{n}{N_0} - \frac{1}{2}\right)x_0, y\right]\mathbf{J}_A + w_B\left[x - \left(\frac{n}{N_0} + \frac{1}{2}\right)x_0, y\right] \exp\left[i\phi\left(x - \left(\frac{n}{N_0} + \frac{1}{2}\right)x_0, y\right)\right]\mathbf{J}_B \right\}, \tag{4}$$

where $w_A(x,y) = w(x,y)A(x,y)$ and $w_B(x,y) = w(x,y)B(x,y)$ have been defined and $\Im^{-1}\{G(\mu,\nu)\} = \sum_{n=-\infty}^{\infty} C_n \delta(x - n/\mu_d)$ has been substituted with $C_n = \frac{1}{2}\mathrm{sinc}(n/2)\exp(i2\pi n u_0/u_d)$ for $\mu_w = \frac{1}{2}\mu_d$, and assuming that x_0 equals some multiple integer N_0 of the period, in other words, $x_0 = N_0/\mu_d = N_0(\lambda f/u_d)$, (diffraction orders matching condition). According to Eq. (4), the amplitude in the image plane consists of a row of copies of the entrance transmittance, each copy separated by $1/\mu_d = \lambda f/u_d$ from the first neighbours. By adjusting the distance x_0 between windows such that $N_0/\mu_d = x_0$ and also assuring that the inequality $a_w \le \lambda f/u_d$ is satisfied, the field amplitude $\mathbf{t}_w(x,y)$ of window $w\left[x - (n/N_0 - 1/2)x_0, y\right]$ (as observed through a linear polarizer with transmission axis at angle ψ_n with respect to the horizontal) can be described as

$$\mathbf{t}_w(x,y) = \left\{C_n A(x,y)\cos(\psi_n) + C_{n-N_0} B(x,y)\sin(\psi_n)\exp\left[i\phi(x,y)\right]\right\}\mathbf{J}_\psi \tag{5}$$

where $\cos(\psi_n)\mathbf{J}_\psi = \mathbf{J}_\psi^L \mathbf{J}_A$, $\sin(\psi_n)\mathbf{J}_\psi = \mathbf{J}_\psi^L \mathbf{J}_B$, and

$$\mathbf{J}_\psi^L = \begin{pmatrix} \cos^2(\psi_n) & \sin(\psi_n)\cos(\psi_n) \\ \sin(\psi_n)\cos(\psi_n) & \sin^2(\psi_n) \end{pmatrix}; \quad \mathbf{J}_\psi = \begin{pmatrix} \cos(\psi_n) \\ \sin(\psi_n) \end{pmatrix}. \tag{6}$$

In particular, for the case $N_0 = 1$ and knowing that the irradiance results proportional to the square modulus of the field amplitude, $I(x,y) = |\mathbf{t}_w(x,y)|^2$, the corresponding interference pattern is given by

$$I(x,y) = a_n(x,y) + b_n(x,y)\cos\left[\phi(x,y) - 2\pi\frac{u_0}{u_d}\right], \tag{7}$$

where $a_n(x,y)$ is known as background light, $b_n(x,y)$ is known as a modulation light and they are given by

$$a_n(x,y)=\frac{1}{4}\mathrm{sinc}^2\left(\frac{1}{2}n\right)\cos^2\psi_n I_A(x,y)+\frac{1}{4}\mathrm{sinc}^2\left(\frac{1}{2}(n-1)\right)\sin^2\psi_n I_B(x,y), \tag{8a}$$

$$b_n(x,y)=\frac{1}{4}\mathrm{sinc}\left(\frac{1}{2}n\right)\mathrm{sinc}\left[\frac{1}{2}(n-1)\right]\sin(2\psi_n)\sqrt{I_A(x,y)I_B(x,y)}, \tag{8b}$$

with $I_A(x,y)=|A(x,y)|^2$ and $I_B(x,y)=|B(x,y)|^2$. The pattern in Eq. (7) results to be modulated by the functions sin and sinc. Note that $a_n(x,y)$ and $b_n(x,y)$ would be independent of position if the illumination is uniform in each window at the input plane. Otherwise, $a_n(x,y)$ and $b_n(x,y)$ can be smooth functions of the position and, in such a case they must give rise to corresponding spectra of a given extension, however small.

In the practice, $n=0,1$ are two cases of interest, for $n=0$ the window of observation would be $w[x+x_0/2,y]$ and Eq. (8) could be reduced to

$$a_0(x,y)=\frac{1}{4}\cos^2\psi_0 I_A(x,y)+\frac{1}{\pi^2}\sin^2\psi_0 I_B(x,y), \tag{9a}$$

$$b_0(x,y)=\frac{1}{2\pi}\sin(2\psi_0)\sqrt{I_A(x,y)I_B(x,y)}, \tag{9b}$$

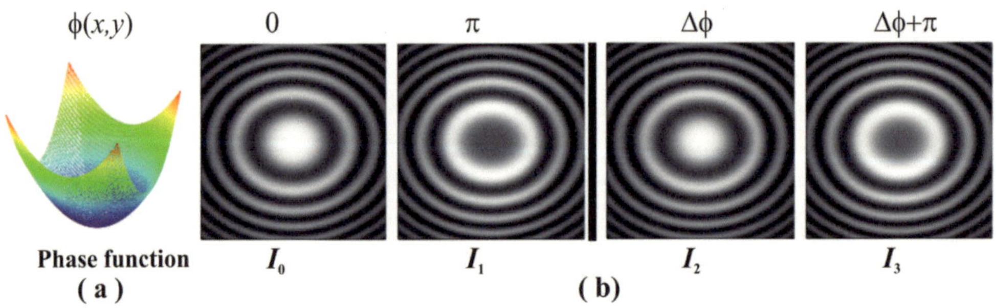

Fig. 2. Numerical simulation of phase shifting interferometry for four unequal phase-steps, which are obtained with two camera shots, that is with $N=2$, then $2N=4$ interferograms changed in phase are obtained: (a) phase function considered in this simulation and (b) for interferograms changed in phase by $\varphi=0,\Delta\phi,\pi,\Delta\phi+\pi$.

and for $n=1$ the window of observation would be $w[x-x_0/2,y]$ and Eq. (8) could be reduced to

$$a_1(x,y)=\frac{1}{\pi^2}\cos^2\psi_1 I_A(x,y)+\frac{1}{4}\sin^2\psi_1 I_B(x,y), \tag{10a}$$

$$b_1(x,y)=\frac{1}{2\pi}\sin(2\psi_1)\sqrt{I_A(x,y)I_B(x,y)}. \tag{10b}$$

Comparing Eq. (9) and Eq. (10) is easy to note that $a_0(x,y)$ and $a_1(x,y)$ in general are different and only they are equals in two cases:

case 1: when $I_A(x,y) = I_B(x,y)$, then from Eq. (9a) and Eq. (10a) must be complied

$$\frac{1}{4}\cos^2\psi_0 + \frac{1}{\pi^2}\sin^2\psi_0 = \frac{1}{\pi^2}\cos^2\psi_1 + \frac{1}{4}\sin^2\psi_1, \tag{11}$$

which conduce to

$$\cos^2\psi_0 + \cos^2\psi_1 = 1, \tag{12}$$

obtaining so

$$\cos\psi_0 = \pm\sin\psi_1, \tag{13}$$

and finally can be established

$$\psi_0 = \begin{cases} \pi/2 - \psi_1 \\ \pi/2 + \psi_1 \end{cases}, \tag{14}$$

which means that the polarizer placed at angle ψ_0 over the window $w[x + x_0/2, y]$ and the polarizer placed at angle ψ_1 over the window $w[x - x_0/2, y]$ could be at complementary angles, when $\psi_0 = \pi/2 - \psi_1$, or well could be at 90° each other, when $\psi_0 = \pi/2 + \psi_1$. For example: if $\psi_1 = \pm 45°$, then $\psi_0 = 45°$, or well $\psi_0 = -45°$.

For the modulation light, in order to do $b_0(x,y)$ equal to $b_1(x,y)$ or $-b_1(x,y)$, from Eq. (9b) and Eq. (10b) it must be complied

$$\sin(2\psi_0) = \pm\sin(2\psi_1), \tag{15}$$

whereof can be deduced

$$\psi_0 = \begin{cases} \pm\psi_1 \\ \pm\pi/2 \mp \psi_1 \end{cases}, \tag{16}$$

a particular case of $\psi_0 = \pm\psi_1$ in Eq. (16) would be only coincident with Eq. (14) for $\psi_1 = \pm 45°$ as it was written in the above example, but the case $\psi_0 = \pi/2 - \psi_1$ is totally coincident with Eqs. (14), so it can be used to apply to PSI technique.

case 2: when $I_A(x,y) \neq I_B(x,y)$, from Eq. (9a) and Eq. (10a) it must be complied

$$\frac{1}{4}\cos^2\psi_0 I_A(x,y) + \frac{1}{\pi^2}\sin^2\psi_0 I_B(x,y) = \frac{1}{\pi^2}\cos^2\psi_1 I_A(x,y) + \frac{1}{4}\sin^2\psi_1 I_B(x,y), \tag{17}$$

as a particular case, we can suppose

$$\frac{1}{4}\cos^2\psi_0 = \frac{1}{\pi^2}\cos^2\psi_1\text{; and } \frac{1}{\pi^2}\sin^2\psi_0 = \frac{1}{4}\sin^2\psi_1, \tag{18}$$

whereof leads to plan a system of equations, where $\cos^2\psi_0$ and $\cos^2\psi_1$ would be the unknowns,

$$\begin{aligned}\pi^2\cos^2\psi_0 - 4\cos^2\psi_1 &= 0\\ -4\cos^2\psi_0 + \pi^2\cos^2\psi_1 &= \pi^2 - 4\end{aligned}, \tag{19}$$

being possible to found

$$\cos^2\psi_0 = \frac{4}{\pi^2+4};\ \text{and}\ \cos^2\psi_1 = \frac{\pi^2}{\pi^2+4}, \tag{20}$$

from Eq. (20) the following relation can be deduced

$$\cos^2\psi_0 = \sin^2\psi_1, \tag{21}$$

then, $\psi_0 = \pi/2 - \psi_1$, which are complementary angles, as in the *case 1* illustrated with Eq. (14). Finally, taking in account Eq. (20) the values of the polarizer angles can be computed,

$$\psi_0 = \pm 57.51°;\ \text{and}\ \psi_1 = \pm 32.49°. \tag{22}$$

these values for transmission angles of the polarizers also must comply $b_0(x,y) = b_1(x,y)$ or $b_0(x,y) = -b_1(x,y)$ due to they are a particular case of Eq. (16).

In order to achieve the PSI technique, selecting P_3 at angle $\psi_0 = 45°$ for $n = 0$ and P_4 at angle $\psi_1 = -45°$ for $n = 1$ with no grating displacement, $u_0 = 0$, with idea to implemented *case 1* [Meneses-Fabian, et. al., 2009], this results $a_0(x,y) = a_1(x,y) = a(x,y)$ and also $b_0(x,y) = -b_1(x,y) = b(x,y)$, and of this manner two complementary patterns are first obtained at the image plane within replication regions given by $w[x + x_0/2, y]$ and $w[x - x_0/2, y]$. The possible changing in the fringe modulations was narrowly reduced after a normalization procedure, since the interferograms can be converted in digital patterns with the same modulation. Such patterns can be written as

$$I_0(x,y) = a(x,y) + b(x,y)\cos[\phi(x,y)], \tag{23a}$$

$$I_1(x,y) = a(x,y) - b(x,y)\cos[\phi(x,y)]. \tag{23b}$$

They correspond to patterns with a phase shift of $\alpha_1 = \pi$. Secondly, by performing an arbitrary translation of value $0 < u_0 < u_d/2$, the introduced phase shift is less than π radians. Then, another two interferograms result. Each one can be expressed as follows

$$I_2(x,y) = a(x,y) + b(x,y)\cos[\phi(x,y) - \Delta\phi], \tag{23c}$$

$$I_3(x,y) = a(x,y) - b(x,y)\cos[\phi(x,y) - \Delta\phi]. \tag{23d}$$

The corresponding phase shift for them is $\alpha_3 = \pi$. Considering patterns I_1 and I_2 they differ by a phase shift of $\alpha_2 = \Delta\phi - \pi$, where $\Delta\phi = 2\pi \cdot u_0 / u_d$. With this procedure, four interferograms with phase displacements of $\varphi_0 = 0$, $\varphi_1 = \pi$, $\varphi_2 = \Delta\phi$, $\varphi_3 = \Delta\phi + \pi$ can be obtained using only an unknown grating shift and, thus, two camera shots. The desired phase distribution can be calculated from

$$\phi_w(x,y) = \arctan\left\{ \frac{I_2(x,y) - I_3(x,y) - [I_0(x,y) - I_1(x,y)]\cos(\Delta\phi)}{[I_0(x,y) - I_1(x,y)]\sin(\Delta\phi)} \right\}, \tag{24}$$

where ϕ_w denotes the wrapped phase to be unwrapped further. From Eq. (24), for the case of $\Delta\phi = 90°$, the well-known formula for four shifts can be obtained (Schwider, 1990). Eq. (25) requires, of course, the knowledge of the value $\Delta\phi$ to be useful. In order to calculate $\Delta\phi$ from the same captured interferograms, the procedure suggested by (Kreis, 1986) can be applied. This procedure is based on the Fourier transform analysis of fringes and a variant of it is proposed in the following sections to conceal it with the desired phase extraction.

For the case 2, the polarizer angle must be placed to $\psi_0 = 57.51°$ and $\psi_1 = -32.49°$ in order to keep constant the visibility in the interference pattern as it was made in *case 1*, this manner the same structure of Eq. (23) could be obtained and the solution for the wrapped phase is obtained with the same Eq. (24). Note that the *case 2* is more general, so also the polarizer angle $\psi = \pm 57.51°$ is also valid for the case 1.

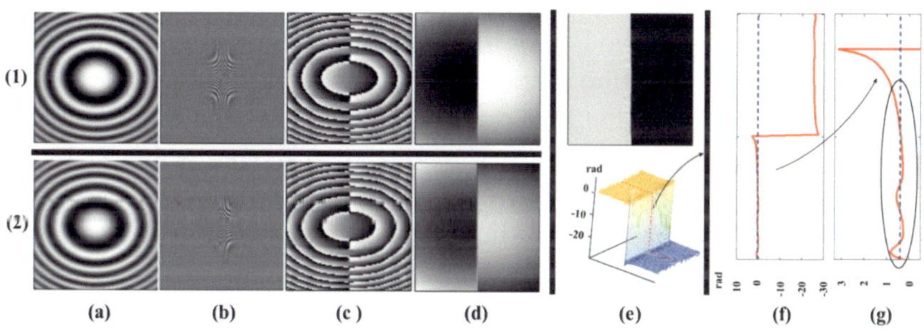

Fig. 3. Numerical simulations for determination of $\Delta\phi$ by using the modified Kreis method: (a) Patters indicated in Eq. (25), (b) Fourier transform from Eq. (25a) and spectrum filtered, (c) wrapped phases, (d) unwrapped phases, (e) phase difference of the unwrapped phases in (b1) and (b2), (f) a data line from (e), (g) subsample of the data line in (f).

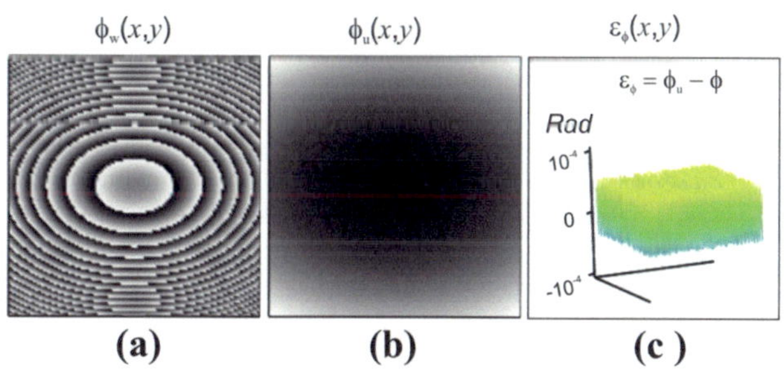

Fig. 4. Numerical simulations of phase extraction: (a) wrapped phase obtained with Eq. (24), (b) unwrapped phase, and (c) phase difference between the calculated phase in (b) and the proposal phase in Fig. (2.a) as a measurement of the error.

2.2 Determination of $\Delta\phi$

Subtraction of Eq. (23b) from Eq. (23a) and Eq. (23d) from Eq. (23c) gives

$$g_1(x,y) = I_0(x,y) - I_1(x,y) = 2b(x,y)\cos[\phi(x,y)], \tag{25a}$$

$$g_2(x,y) = I_2(x,y) - I_3(x,y) = 2b(x,y)\cos[\phi(x,y) - \Delta\phi], \tag{25b}$$

procedure which eliminates $a(x,y)$. In Eq. (25), some dependence on position has been considered to include effects such as non uniform illumination, non linear detection or imperfections in the optical components. These subtractions avoid to apply the Fourier transformation and the spatial filtering usually performed for the same purpose of eliminating $a(x,y)$ (Kreis, 1986). It is remarked that the Fourier transform procedure to eliminate $a(x,y)$ introduces an error due to the fact that, in general, the spectra from $a(x,y)$ and $b(x,y)\cos\phi(x,y)$ can be found mixed one with each other over the Fourier plane. Therefore, filtering out the $a(x,y)$-spectrum around the zero frequency excludes also low frequencies from $b(x,y)\cos\phi(x,y)$ and, as a consequence, there is a corresponding loss of information related to $\phi(x,y)$ and $\Delta\phi$. To calculate $\Delta\phi$, the method introduced by (Kreis, 1986) is employed (an alternative can be seen in (Meng, et. al., 2008). An advantage of the variant that is proposed in this work consists of the elimination of $a(x,y)$ by subtracting two patterns. This way, there is no loss of information due to frequency suppression in the Fourier plane, as is the case of the method as proposed by (Kreis, 1990). The proposed technique is illustrated in the example of the Fig.2. In addition, the technique is valid even when the phase function is more complex. The 3-D plot at the left is the phase distribution $\phi(x,y)$, while the other plots are phase-shifted interferograms calculated from $\phi(x,y)$. Interferograms I_0 and I_1 are mutually shifted by π radians, as well as I_2 and I_3, but between I_0 and I_2, and I_1 and I_3 is the same arbitrary phase of $\Delta\phi = \pi/7 \approx 0.39269908$ radians. This situation illustrates the kind of results that can be obtained with the setup of Fig.1. Then, the problem is to find $\Delta\phi$ from the four interference patterns assuming that its value is not known. The solution is illustrated in Fig.3. The plots included are shown as an array of rows (seven letters) and columns (two numbers).

According to Eqs. (25.a) and (25.b), Figs.3-a1 and 3-a2 show the subtractions $g_1(x,y)$ and $g_2(x,y)$ respectively, where the four irradiances I_k were taken from the same interferograms of Fig. 2. Next, Fig. 3-b1 shows the Fourier spectrum of 3-a1 only ($g_1(x,y)$), while Fig. 3-b2 depicts its resulting filtered spectrum in accordance with the method of

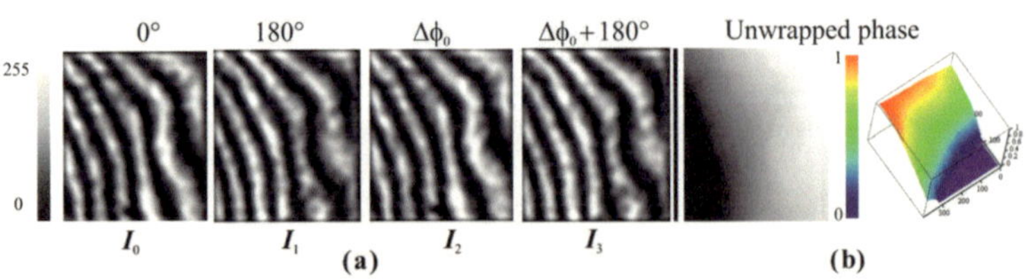

Fig. 5. Tilted wavefront for testing. (a) Interference patterns. (b) Unwrapped phase. Phase shift measured according to the modified Kreis method: $\Delta\phi = 14.4° = 0.2513274$ radians.

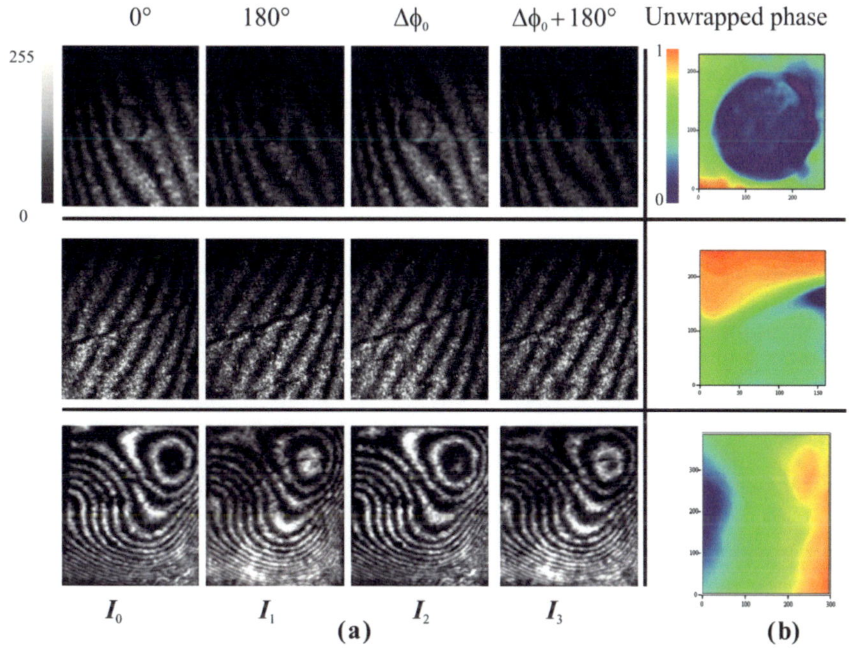

Fig. 6. Test objects. (a) Interference patterns (b) Unwrapped phase. Upper row: phase dot, $\Delta\phi = 50.007° = 0.874$ radians. Central row: phase step, $\Delta\phi = 25.726° = 0.449$ radians. Lower row: still oil, $\Delta\phi = 60.7° = 1.059$ radians.

Kreis. The used filter is a unit step, so the suppression of the left half of the spectrum is achieved. The following stage of the procedure to find $\Delta\phi$ consists of extracting the wrapped phase from the inverse Fourier transform of the already filtered $g_1(x,y)$, which is shown in Fig. 3-c1. Fig. 3-c2 shows the same procedure as applied to Fig. 3-2a (i.e., $g_2(x,y)$). At this stage, the phases of the inverse Fourier transform of each filtered interference pattern $g_1(x,y)$ and $g_2(x,y)$ are obtained. Therefore, these phases are wrapped, modified phases whose respective numerical integration results in two unwrapped, modified phases (Figs. 3-d1 and 3-d2). These unwrapped phases result in monotonous functions which are different from the desired phase ϕ , but their difference in each point (x,y) gives the modified phase difference $\Delta\phi'$ (Figs. 3-e). This modified phase difference has to be constant for all interference pattern points, so an average over some range can be sufficient to calculate $\Delta\phi$ with a good approximation. Fig. 3-f shows a line of Fig. 3-c1, and Fig. 3-g shows a section of the Fig. 3-f, where the used region to measure phase difference is indicated with an elliptic trace. The dots show the ideal phase difference $\Delta\phi$, while the red line shows the modified phase difference $\Delta\phi'$. The resulting value over the entire lower region of Fig. 3-e2 was of $\overline{\Delta\phi'} = 0.39273364$, where the bar means average, so its difference with respect to the initial induced phase $\Delta\phi$ is of the order of 3.4556×10^{-5} rads. Taken $\overline{\Delta\phi'}$ as $\Delta\phi$, the wrapped phase distribution ϕ_w can be determined with Eq. (24) and the desired phase distribution ϕ can be identified with ϕ_u, the phase calculated from ϕ_w with standard unwrapping algorithm [Malacara, et. al., 1998]. The results of these stages are shown in Fig. 4.

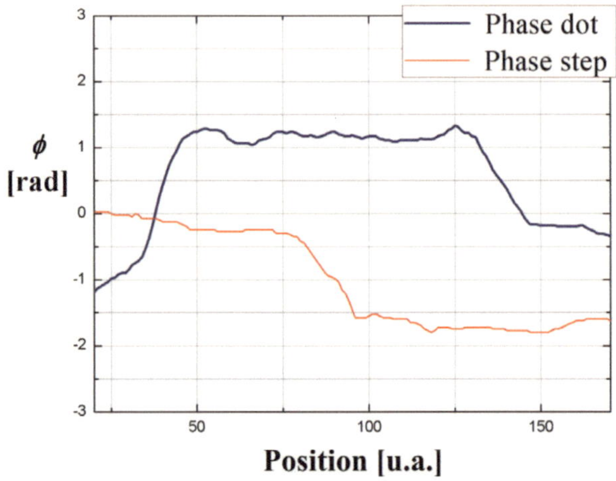

Fig. 7. Typical slice profiles of the phase dot and the phase step along arbitrary directions.

3. Experimental results

The experimental setup follows closely the sketch of Fig.1, with $f = 479mm$, a laser He-Ne emitting at $\lambda = 632.8nm$, with a linear polarization at 45° by using P_0. $a_w = 10mm$, $b_w = 13mm$, and $x_0 = 12.45mm$, so the condition $a_w < x_0$ is fulfilled. The period of the Ronchi grating was $u_d = 25.4\mu m$. The grating was mounted on an actuator (Newport CMA-25CC). Note that the diffraction-order matching condition $x_0 = N_0(\lambda f/u_d)$ for $N_0 = 1$ is satisfied. The CCD camera (COHU 4815) is adjusted to capture the images of two interference patterns ($w[x+x_0/2,y]$ and $w[x-x_0/2,y]$) simultaneously. The fringe modulations of this interferogram pair are mutually complementary. By one actuator displacement, another pair of complementary interferograms can be obtained. This is shown in Fig. 5, where the calculated value of the introduced phase was of $\Delta\phi = 14.4° = 0.2513274$ radians. The unwrapped phase is also shown. Each captured interferogram was subject to the same processing so as to get images with gray levels ranging from 0 to 255 before the use of Eq. (8). The sets of interferograms from three more objects are shown in Fig. 6. Two of these objects were prepared by evaporating magnesium fluoride (MgF2) on glass substrates (a phase dot, upper row, and a phase step, center row). The third object was oil deposited on a glass plate (bottom row). Each object was placed separately in one of the windows using the interferometer of Fig.1. Each set of four interferograms were obtained with a grating displacement between two of them as described. En each example, the phase-shifts induced by polarization can be visually identified from the complementary modulation contrasts of each pair. The calculated unwrapped phase for each object is shown in the rightmost column. Two typical profiles for some unwrapped phase are shown in Fig. (7). The resulting unwrapping phases do not display discontinuities, as is the case when the wrong $\Delta\phi$ is taken in Eq. (8).

4. Conclusion

The proposed phase-shifting interferometer is able to capture four useful interferograms with only one grating displacement and two shots. This grating displacement only requires being smaller than a quarter of period, but this condition is possible to verify by observing that the fringes do not shift enough to adopt a complementary fringe modulation. Because in this proposed method the phase shifts are achieved either by modulation of polarization or by grating displacement, not all of the shifts result necessarily of the same value. The requirements of the arrangement are not very restrictive because it uses only very basic optical components, such as linear polarizers at angles of 0°, $\pm 45°$, or 90°. A Ronchi grating from 500 to 1500 lines per inch can be used. As remarked before, the higher the grating frequency, the larger the distance between windows and more space to place samples becomes available, but the grating displacement has to be smaller. This last feature does not represent an impediment in this method because the actuator responsible for the displacement does not need of a calibration within a certain range, neither a precise displacement at a given prescribed value. Only one unknown displacement is needed and its value can be calculated each time it is employed. As for the fringe modulation, it is close to unity when using the 0 and ±1 diffraction orders for a typical Ronchi grating. Also, the two used patterns have the same fringe modulation because, as long as the two diffraction orders which superpose are of equal value, the involved amplitudes are the same. These features to extract static phase distributions make this proposal competitive as compared to the existing ones. Although in this chapter was discussed the PSI method for four steps, this proposal can be extended for $N \geq 3$, with which $2N$ interferograms changed in phase would be obtained with N camera shots.

5. Acknowledgment

This work was partially supported from VIEP-BUAP under grant MEFC and PROMEP under grant PROMEP/103.5/09/4544) are greatly appreciated.

6. References

Arrizon, V. & De-La-Llave, D. (2004) Common-path interferometry with one-dimensional periodic filters, *Optics Letters*, Vol. 29, 141-143

Barrientos-García B.; Moore A. J.; Perez-Lopez C.; Wang L. & Tshudi T. (1999). Transient Deformation Measurement with Electronic Speckle Pattern Interferometry by Use of a Holographic Optical Element for Spatial Phase Stepping, *Applied Optics*. Vol. 38, 5944-5947, ISSN 1559-128X

Born, M. and Wolf, E.; Cambridge University Press (1993). *Principles of Optics*

Bruning, J. H.; Herriott, D. R.; Gallagher, J. E.; Rosenfeld, D. P.; White, A. D. & Brangaccio, D. J. (1974) Digital wavefront measurement interferometer for testing optical surfaces and lenses, *Applied Optics*, Vol. 13, pp. 2693-2703.

Creath K., "Phase-measurement interferometry techniques," Progress in Optics, Vol. XXVI, E. Wolf, ed., (Elsevier Science, 1993), pp. 349-393.

Hariharan, P.; Oreb, B. F.; Eiju, T. (1987) Digital phase-shifting interferometry: a simple error-compensating phase calculation algorithm, *Applied Optics*, Vol. 26, 2504-2506

Hech E.; Addison-Wesley (1972). *Optics*

Kreis, T. (1986). Digital holographic interference-phase measurement using the Fourier-transform method, *Journal of the Optical Society of America A*, Vol. 3, 847-855.

Kreis, T.; WILEY-VCH Verlag GmbH & Co. KGaA (2005). *Handbook of Holographic interferometry, optical and digital methods*

Malacara, D.; Servín, M.; & Malacara, Z.; Marcel Dekker, New York (1998). *Interferogram Analysis for Optical Testing*

Malacara D.; Wiley (2007). *Optical Shop Testing*

Meneses-Fabian, C.; Rodriguez-Zurita, G.; Vazquez-Castillo, J. F.; Robledo-Sanchez C. & Arrizon, V. (2006) Common-path phase-shifting interferometry with binary pattern, *Optics Communications*, Vol. 264, 13-17.

Meneses-Fabian, C.; Rodriguez-Zurita, G.; Encarnacion-Gutierrez, Ma. C. & Toto-Arellano N-I. (2009). Phase-shifting interferometry with four interferograms using linear polarization modulation and a Ronchi grating displaced by only a small unknown amount, *Optics Communications*, Vol. 282, 3063-3068.

Meng, X. F.; Cai, L. Z.; Wang, Y. R.; Yang, X. L.; Xu, X. F.; Dong, G. Y.; Shen, X. X. & Cheng, X. C. (2008). *Optics Communications*, Vol. 281 5701–5705.

Millerd J. E. & Brock N. J. (2003). Methods and apparatus for splitting, imaging, and measuring wavefronts in interferometry, U. S. Patent 20030053071A1.

Moore, A. J. & Mendoza-Santoyo, F., (1995). Phase demodulation in the space domain without a fringe carrier, *Optical Laser Engineering*, Vol. 23, pp. 319–330.

Nomura,T.; Murata, S.; Nitanai, E. & Numata, T. (2006). Phase-shifting digital holography with a phase difference between orthogonal polarizations, *Applied Optics*, Vol. 45 4873-4877.

Novák, M.; Millerd, J.; Brock, N.; North-Morris, M.; Hayes, J. & Wyant, J. (2005). Analysis of a micropolarizer array-based simultaneous phase-shifting interferometer, *Applied Optics*, vol. 44, pp. 6861-6868, ISSN 1559-128X.

Novák, J.; Novák, P. & Mikš, A. (2008) Multi-step phase-shifting algorithms insensitive to linear phase shift errors, *Optics Communications*, Vol. 281, 5302–5309

Peng, X.; Zhou, S. M. & Gao Z. (1995). An automatic demodulation technique for a non-linear carrier fringe pattern, *Optik*, Vol. 100, pp. 11–14.

Rodriguez-Zurita, G.; Meneses-Fabian, C.; Toto-Arellano, N. I.; Vazquez-Castillo, J. & Robledo-Sanchez C. (2008a), One-shot phase-shifting phase-grating interferometry with modulation of polarization: case of four interferograms, *Optics Express*, Vol. 16 9806-9817

Rodriguez-Zurita, G.; Toto-Arellano, N. I.; Meneses-Fabian, C. & Vazquez Castillo, J. (2008b) One-shot phase-shifting interferometry: five, seven, and nine interferograms, *Optics Letters*, Vol. 33, 2788-2790

Schreiber, H. & Bruning, J.H., (2007). Phase shifting interferometry, Chapter 14, in: *Optical Shop Testing*, D. Malacara Ed., Wiley & Sons, New York, 547-655.

Schwider, J. "Advanced Evaluation Techniques in Interferometry," Progress in Optics. Vol. XXVIII, E. Wolf, ed., (Elsevier Science, 1990), pp. 274-276.

Shamos, M. (1959). Great experiments in Physics, pp. 96-101, Holt Reinhart and Winston, New York

Takeda M., Ina H., and Kobayashi S., (1982). Fourier-Transform Method of Fringe-Pattern Analysis for Computer-Based Topography and Interferometry, *Journal of the Optical Society of America A*, Vol. 72, pp. 156-160

Xu, X. F.; Cai, L. Z.; Wang, Y. R.; Meng, X. F. & Sun, W. J. (2008). Simple direct extraction of unknown phase shift and wavefront reconstruction in generalized phase-shifting interferometry: algorithm and experiments, *Optics Letters*, Vol. 33, 776-778.

Young, T. (1804). Experimental demonstration of the general law of the interference of the light, *Philosophical transactions of the Royal Society of London*, Vol. 94, 2

10

Path Length Resolved Dynamic Light Scattering Measurements with Suppressed Influence of Optical Properties Using Phase Modulated Low Coherence Interferometry

Babu Varghese and Wiendelt Steenbergen
Biomedical Photonic Imaging Group, MIRA Institute for Biomedical Technology and Technical Medicine, University of Twente, Enschede, The Netherlands

1. Introduction

In optical Doppler measurements, the path length of the light is unknown. This complicates the noninvasive diagnosis of tissue with light. For example, in laser Doppler blood flowmetry, the coherent light delivered into the tissue interacts with static as well as moving scatterers, e.g. red blood cells and it records values averaged over different and basically unknown path lengths. One of the important limitations of this technique is the dependence of the perfusion signal on the optical properties of the tissue, i.e. absorption coefficient, scattering coefficient and anisotropy factor. These dependences result from the varying optical path length of detected photons; the longer the optical path length the greater is the probability for Doppler scattering events to occur, thus yielding an overestimation of the blood perfusion, compared to the short path length situation [1].

If red blood cells in vascular blood can be regarded as independent scatterers, the average number of collisions between photons and red blood cells which is used to determine the concentration of blood cells moving in the tissue is given by

$$\overline{m} = \Sigma_{sc}(rbc) * [RBC] * L \quad (1)$$

Where Σ_{sc} is the scattering cross section of RBCs, [RBC] is the number of moving RBCs in 1 mm^3 of tissue and L is the mean path length of the detected light. For a homogeneous tissue of fixed concentration and blood volume, the value of $\overline{m}$ is proportional to the average path length of the detected light. The average path lengths will be different for different tissue types due to the changes in tissue optical properties in terms of absorption and scattering and thus laser Doppler flowmetry provides only a relative measure of the perfusion level. Therefore, development of techniques for monitoring Doppler shifts with path length information would result in more-quantitative and more reliable tissue perfusion information.

To facilitate quantitative path length resolved dynamic light scattering measurements with suppressed influence of optical properties we have developed an improved method based on phase modulated low coherence interferometry [2-3]. In this chapter, we aim at describing the state-of-art of this novel interferometric technique and we show that we can measure dynamic properties of particles, independent of the optical properties of the surrounding tissue matrices. Furthermore, we demonstrate the feasibility of phase modulated low coherence interferometry in measuring *in vivo* optical path lengths and path length resolved Doppler shifts.

2. Review of coherence domain path length resolved approaches in laser Doppler flowmetry

To obtain path length distributions with widths of a few millimeters, several successful approaches based on low coherence interferometric methods were reported. In low coherence interferometry, a user-positioned coherence gate selects the light that has traveled a known optical path length in the medium to interfere with reference light. Dougherty et al. presented a new approach based on coherence modulation of semiconductor lasers using the variable coherence properties of the semiconductor laser [4]. In this technique, they exploited variations in effective coherence length properties of certain types of laser diodes by regulating the input drive current to these devices. For a long coherence length, all photons interfere, while for a short coherence length only photons with almost the same path length will interfere. This will relatively suppress the deep photons, since the (few) deep photons will only interfere with the few deep photons but not with the (many) shallow photons. However, these methods still give no control over the optical path length traveled by the detected light. McKinney et al. [5] and Haberland et al. [6] used a wavelength modulated continuous wave source as a variable-coherence source for measuring path length distributions. The frequency of the modulation used was much faster than the integration time of the detection and the authors demonstrated that the speckle contrast ratio measured in that way was linked to the photons path-length distribution [5]. Haberland et al. [6] used such a wavelength modulated source to demonstrate that chirp optical coherence tomography (OCT) can be an alternative to short coherence tomography with the advantage of a simplified optical set-up. However they reduced their investigations to unscattered light. Later, Tualle et al. [7] reported the development of a low cost interferometric set-up to record the scattered light by the use of a wavelength modulated continuous wave source. The principle of this technique relies on the facts that the shape of the time-resolved signal corresponds to the path-length distribution of the scattered light inside the turbid medium and the path-length differences can be measured using an interferometer. They showed that the study of the speckle pattern fluctuations within the modulation period can provide much more information, and that this information can be used to completely reconstruct the scattered light path-length distribution, or equivalently to perform time-resolved measurements. However, the slow rate that they used for the wavelength modulation limited their experiments to static scattering media.

With a fiber-optic low coherence Michelson interferometer, Bizheva et al. demonstrated that particle dynamics of highly scattering media can be imaged and quantified in the single scattering regime with dynamic low coherence tomography (LCI) by examining the intensity fluctuations of the backscattered light and extracting information from the photocurrent power spectrum [8]. Later, they showed that dynamic LCI permits path-length-resolved measurements of particle dynamics in highly scattering media with the ability to separate singly scattered, multiply scattered, and diffusive light and the results were compared with the predictions of the dynamic light scattering (DLS) and diffusive wave spectroscopy (DWS) theories in the single scattered and diffusion regimes, respectively. They showed the dependence of detection of multiply scattered light on the geometry of the detection optics and on the anisotropy of the scattering [8]. Even though these studies modeled the single scattering and the diffusive regimes of light fluctuations, they did not model the transition between the two regimes. With a free beam Michelson interferometer, Wax et al. applied path-length resolved DLS spectroscopy and a theoretical model was developed to predict this transition regime across the full range of path lengths from single scattering through diffusive transport [9]. By comparing the trends in the measured power spectra for various-sized microspheres with a theoretical treatment that decomposes the total power spectrum by the number of scattering events, they correlated the detection of multiply scattered light with scattering anisotropy. We reported the development of fiber-optic Mach-Zehnder interferometer for path length resolved measurements with two spatially separated fibers for illumination and detection, as used in conventional laser Doppler perfusion monitors [10]. Low coherence interferometry with phase modulation of the reference beam has been adopted by Ishii et al. in single scattering spectroscopy to analyze the characteristics of extremely dense colloidal suspensions [11]. Doppler optical coherence tomography based on low-coherence single-mode fiber optic Michelson interferometry has been proposed for path length resolved measurements adopting on axis back reflection and confocal detection of singly scattered photons [12]. Here two embodiments were reported due to the possibility of performing interferometric measurements either in the time domain or in the Fourier domain. In time domain OCT the path length of the reference arm is varied in time. In frequency domain OCT the broadband interference is acquired with spectrally separated detectors either by encoding the optical frequency in time with a spectrally scanning source (Swept source OCT) or with a dispersive detector, like a grating and a linear detector array (Spectral Domain or Fourier Domain OCT) [13-14]. In these techniques, optical path length distributions can be obtained about photons which where Doppler shifted by the medium. The photons that has been scattered by static structures can also be made to contribute to the interferometric signal by modulating the phase in the reference path.

3. Phase modulated low coherence Mach-Zehnder interferometry

In our research, we have developed a new bio-optical technique for path length resolved laser Doppler perfusion monitoring, by combining the principles of coherence gated interferometry and laser Doppler blood flowmetry [2-3]. The method is based on a phase modulated fiber optic low coherence Mach-Zehnder interferometer, in which the limited temporal coherence acts as a band pass filter in selecting the photons that have traveled a specific path length. We use a fiber-optic Mach-Zehnder interferometer (Fig.1) with a superluminescent diode (λ_c=832nm, $\Delta\lambda_{FWHM}$ =17 nm, L_C =18 µm) as the light source. A single mode fiber-optic coupler with a splitting ratio of 90:10 is used to create a reference arm (10%) and a sample arm (90%). Single mode fibers (mode field diameter=5.3 mm, NA=0.14) are used for illumination, while multimode graded-index fibers (core diameter =100 mm, NA=0.29) are used for detection, providing a large detection window. The path length of the reference arm is varied by reflection of the light in a translatable retroreflector and the position of the retroreflector is adjusted to yield an optical path length equal to the optical path length of a certain part of the photons in the sample arm. The reference beam is polarized using a linear polarizer and the phase is sinusoidally modulated at 6 kHz using an electro optic broadband phase modulator with a peak optical phase shift of 2.04 radians applied to the modulator. The AC photocurrent is measured with a 12 bit analogue to digital converter sampling at 40 kHz. The coherence length of the light source, and the intermodal dispersion in the detection fiber, define the path-length resolution of the measurement.

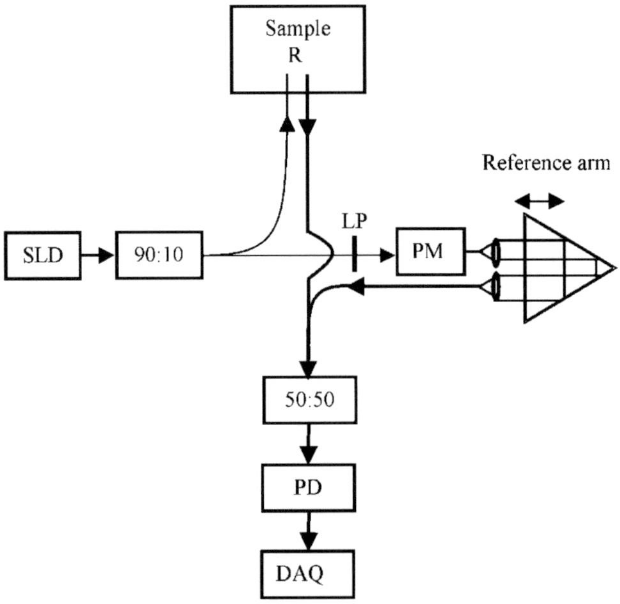

Fig. 1. Schematic of the fiber optic Mach-Zehnder interferometer. (Figure reprinted from Fig. 2 in Ref. [2] Copyright (2008) with permission from SPIE).

In phase modulated low coherence interferometry, for sufficiently small phase modulation angles, the power spectrum measured when the path length difference between the reference light and the scattered light is within the coherence length of light source, is a mixture of homodyne interference of light remitted by the sample, and heterodyne interference between sample light and unmodulated reference light, and a heterodyne spectral component around the phase modulation frequency resulting from the interference between sample light and modulated reference light. Path length resolved optical properties of the media are measured from the heterodyne peak appearing at the modulation frequency. This can be understood from the phasor description of the interfering fields, as shown in fig. 2. Here the reference wave and two Doppler shifted sample waves are represented by phasors R, S_1 and S_2, respectively. The angle ϕ is the peak phase deviation due to phase modulation in the reference light, which is depicted by a sinusoidal oscillation of the reference phasor between the extreme phasors R_1 and R_2 on either side of the average phasor R. For small values of ϕ, the oscillation of the reference phasor between R_1 and R_2 is equivalent to the summation of the average reference phasor R, and two phasors M_1 and M_2 of equal length that rotate in opposite directions with a constant angular speed equal to the phase modulation frequency ω_m. The initial phase of M_1 and M_2 should be chosen such that their sum phasor M is perpendicular to R. The amplitude of M_1 and M_2 is $OM_1=OM_2=1/2.OR.\tan\phi/2$ to achieve the desired phase modulation angle. The sample waves S_1 and S_2 with Doppler shifts ω_{D1} and ω_{D2} interfere with both M_1 and M_2. Since only positive frequencies show up in the power spectrum interference peaks are expected at $\omega_m - \omega_{Di}$ and $\omega_m + \omega_{Di}$ $(i=1,2)$. This interference of sample light with reference light will be called 'heterodyne'. In practice from a turbid sample waves are obtained with a distribution of Doppler shifts, leading to a similar distribution of spectral components centered around ω_m. Hence the shape of the peak around ω_m corresponds to the Doppler shift distribution. The component of the reference light represented by the average phasor (OR) that still is at the original light source frequency will also interfere with the scattered light from the sample and thus, apart from the peak around the phase modulation frequency, a heterodyne component will also occur at low frequencies. Finally, the sample phasors S_1 and S_2, will mutually interfere to generate beats at frequency $\omega_{D1} - \omega_{D2}$, a component that we call 'homodyne'. Hence, the spectrum at low frequencies is a mixture of homodyne interference of light remitted by the sample, and heterodyne interference between sample light and unmodulated reference light, while the spectral component around the phase modulation frequency is the pure Doppler shift distribution. The power spectrum measured for widely different optical path lengths in the sample and the reference arm, contains the ordinary homodyne signal due to the mutual interference of scattered light over almost equal optical path lengths in the sample.

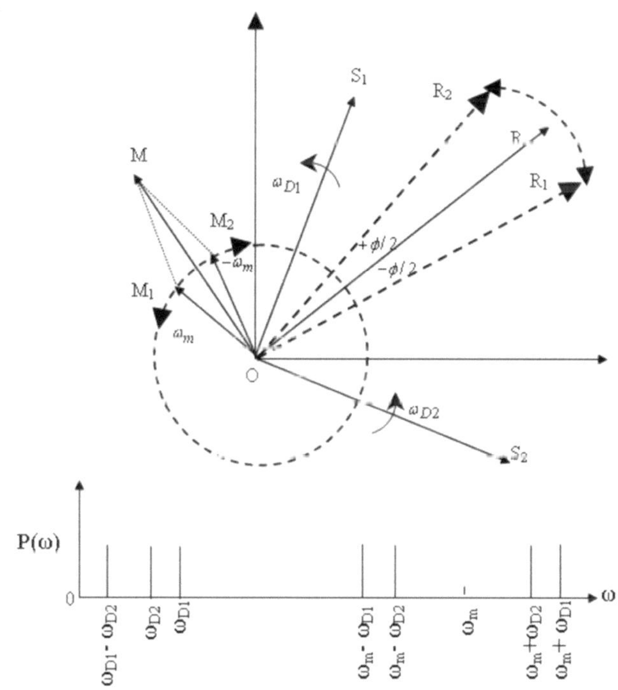

Fig. 2. Phasor diagram for the interfering fields (top) and the resulting power spectrum (bottom). (Figure reprinted from Fig. 1 in Ref. [2] Copyright (2008) with permission from SPIE).

The fundamental output quantity of a laser Doppler perfusion monitor is the first moment of the power spectrum $P(\omega)$ of the detector signal; in general, the ith moment is being defined as

$$M_i = \int_a^b P(\omega)\omega^i d\omega \tag{2}$$

Here a and b are device dependent low and high cut-off frequencies. With i=0, a quantity is obtained which is proportional to the concentration of moving red blood cells, while i=1 describes red blood cell flux, which is the product of concentration and the root mean square of the red cell velocity, at least for low blood concentrations [15]. In our instrument, for large phase modulation angles ($\Delta\phi$ =2.04 radians) the power spectra contain interference peaks at both the phase modulation frequency and higher harmonics (Fig.3). Optical path length distributions are obtained by adding the areas of all interference peaks (after subtraction of the background noise, and within a bandwidth of ± 2 kHz around all center frequencies) in the power spectrum [16]. The area of the Doppler broadened peak, excluding the statically scattered light contribution at the interference peaks, forms an estimation of the amount of Doppler shifted light at that specific optical path length. The average Doppler shift corresponding to the Doppler shifted light is calculated from the weighted first moments (M_1/M_0) of the heterodyne peak at the modulation frequency, after correction for the sample signal and for the reference arm noise (in a bandwidth of 50 Hz-2 kHz close to the phase modulation frequency and its higher harmonics, indicated by a and b in Eq.2)

$$M_i = \sum_{j=1}^{3} \int_{j\omega_m+a}^{j\omega_m+b} P(\omega)(\omega - j\omega_m)^i d\omega \tag{3}$$

To determine the parameters path length resolved, we measured the power spectra with the Mach-Zehnder low coherence setup. First, the background noise from the power spectrum around the modulation frequency (ω_m = 6 kHz) is subtracted. The calculated $M_0(0,b,\omega_m)$ of the broadened interference peak (until b=2 kHz from the phase modulation peak) is proportional to the total number of detected photons for that given (by the reference arm) path length. The full width at half maximum (FWHM) of the interference signal in a statically scattering medium has a value δ_s (δ_s = 50-60 Hz in our system) whereas in the case of dynamic media a Doppler broadened spectral peak around the phase modulation frequency is formed [2]. The area of the Doppler broadened peak, excluding the statically scattered light contribution at the interference peaks, forms an estimation of the amount of the Doppler shifted light at that specific optical path length. For a given optical path length, the fraction of Doppler shifted photons f_D is then given by $f_D = M_0(\delta_s,b,\omega_m)/ M_0(0,b,\omega_m)$. Here we regard the Doppler fraction f_D as a measure of the concentration of particles moving in the static matrix. The average speed of the moving particles is represented by the average Doppler shift of the Doppler shifted fraction of the detected light, which in terms of equation (1) is $\langle\omega_D\rangle = M_1(\delta_s,b,\omega_m)/M_0(\delta_s,b,\omega_m)$ with the frequency ω_D in Hz.

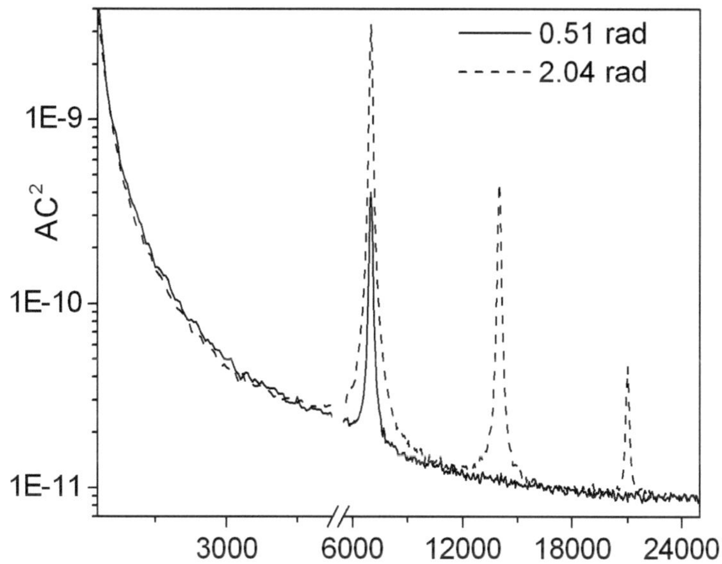

Fig. 3. Power spectra measured for water suspension of Polystyrene microspheres for two different peak optical phase shifts (0.51 and 2.04 radians), with the position of the retroreflector corresponding to an optical path length difference of 1.3 mm. (Figure reprinted from Fig. 1 in Ref. [16] Copyright (2008) with permission from Elsevier B.V.).

4. Path length resolved dynamic light scattering measurements with suppressed influence of optical absorption properties of surrounding tissue matrices

To study the effect of absorption on Doppler shift, measurements were performed on three samples with identical scattering properties but increasing absorption levels [2]. The media were an aqueous suspension of 25% of Intralipid 20% [17] and the same suspensions with absorption coefficients of 0.50 mm^{-1} and 0.85 mm^{-1}.

The estimations of path length distributions of photons in the aqueous Intralipid suspension (μ_a = 0.001 mm^{-1}) and for identical suspensions with different absorption levels (0.50 mm^{-1} and 0.85 mm^{-1}) are shown in fig.4. The estimation of the optical path length distribution is obtained for increasing absorption levels. The minimum path length is the same for all absorptions and is related to the fiber distance of 500 micrometer. At this path length all the distributions start to increase independent of the absorption. As the photons with longer path length have a greater probability to be absorbed in an increasingly absorbing medium, M_0 decreases with the absorption. Hence the path length distribution narrows and shows a decrease in the average intensity as the absorption coefficient increases. Lambert-Beer's law can describe the effect of absorption on the path length distribution. According to the law of Lambert-Beer, the light intensity I_0 in an absorbing medium decays exponentially as $I(L)= I_0$

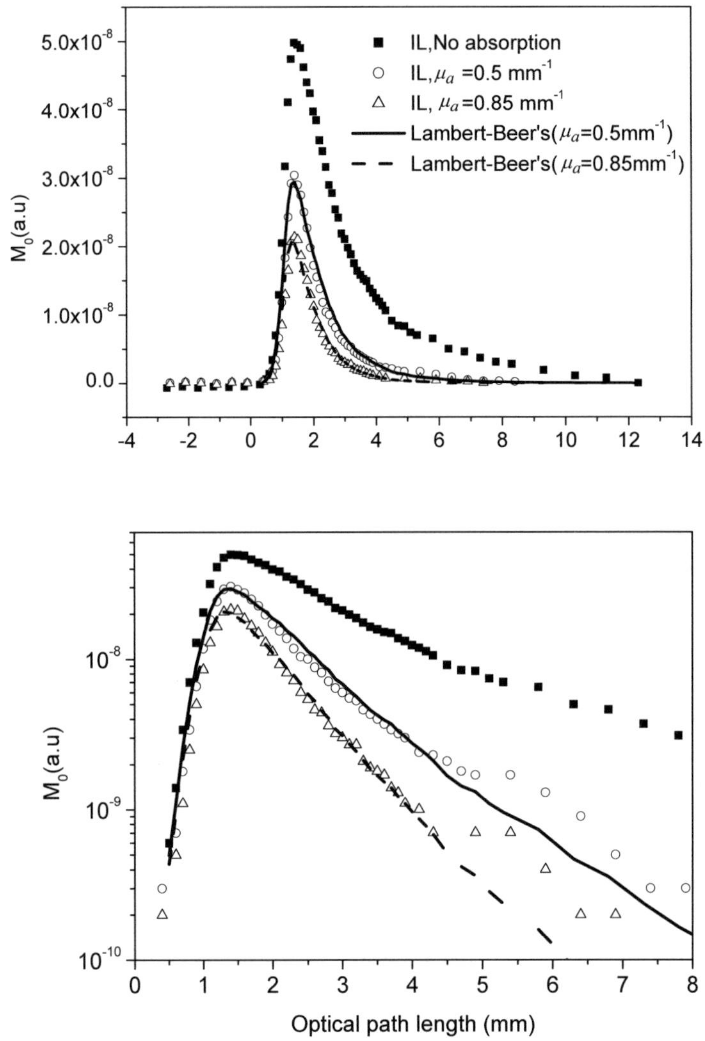

Fig. 4. Optical path length distributions estimated from the zero order moment of the phase modulation peak for an aqueous Intralipid suspension (μa = 0.001 mm^{-1}) and for identical suspensions with two different absorption coefficients (0.50 mm^{-1} and 0.85 mm^{-1}), but equal reduced scattering coefficient (linear and logarithmic scales). The lines result from the application of Lambert-Beer's law on the experimental dataset for zero absorption. (Figure reprinted from Fig. 4 in Ref. [2] Copyright (2008) with permission from SPIE).

$exp(-L.\ \mu_a/n)$ with L the optical path length, and μ_a and n the absorption coefficient and the refractive index of the medium. To validate that the results shown in figure 4 represent the true optical path length distributions, we verify whether the path length distribution of the original Intralipid and the same suspensions with high absorption coefficients are mutually related by Lambert-Beer's law. The path length distributions of original Intralipid multiplied by the exponential decay function $exp(-L.\ \mu_a/n)$ and the experimental data are shown in figure 4 in linear and logarithmic scales. There is a good agreement between the experimental data and the calculated values (for n = 1.33, μ_a = 0.50 and 0.85 mm^{-1}) on the basis of Lambert-Beer's law up to an optical path length of 4.5 mm, which proves that path length distributions have been correctly measured.

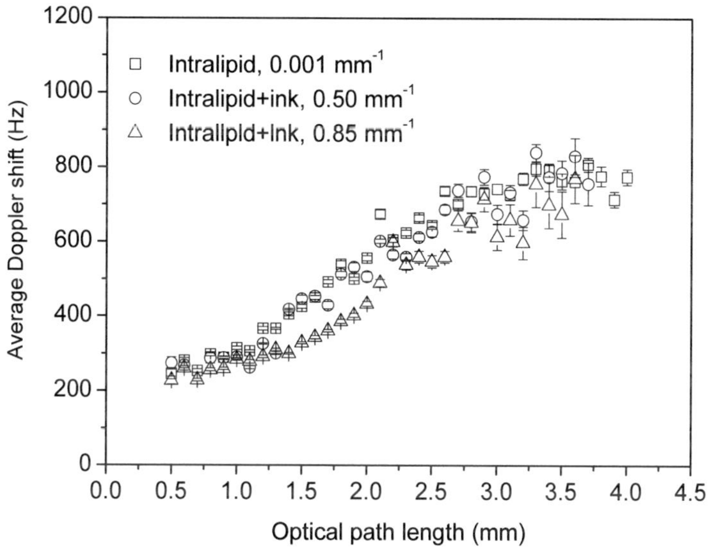

Fig. 5. The average Doppler shift extracted from the phase modulation peak, as a function of the optical path length for an aqueous Intralipid suspension with different absorption coefficients. (Figure reprinted from Fig. 5 in Ref. [2] Copyright (2008) with permission from SPIE).

The average Doppler shift, measured from the width of Doppler broadened phase modulation interference peaks is represented in fig.5 as a function of the optical path length. The average Doppler shift increases with the optical path length, which can be expected from the increase in the number of scattering events with the optical path length. For a given medium with a constant scattering coefficient but absorption coefficients μ_a = 0.001 and 0.50 mm^{-1} the Doppler broadening of path length resolved heterodyne spectra is shown to be independent of the absorption level, for a given optical path length. Therefore, our results indicate that for absorption levels realistic for tissue, our method enables Doppler measurements independent of the absorption level of the medium in which the moving particles are embedded.

5. Path length resolved dynamic light scattering measurements with suppressed influence of scattering properties of surrounding tissue matrices

To study the effect of scattering of surrounding tissue matrices on Doppler shift, mixed static-dynamic scattering phantoms were prepared with aqueous suspensions of polystyrene microspheres of ∅4.7 μm and ∅0.20 μm respectively [19]. Three scattering phantoms with the same concentration of particles ∅0.20 μm ($g=0.18$, $\mu_{s'}=0.55$ mm^{-1}, $\mu_a=0.001$ mm^{-1}) were prepared and two scattering levels of the static medium were realized ($\mu s'=1.4$, 0.8, 0.4 mm^{-1}). For estimating the flux of particles moving inside static matrices in absolute terms, we related the outcomes of our measurements to the optical and dynamical properties of the dynamic part of the medium. Here we focus on the concentration of moving particles, which may be retrieved from models which relate the measured Doppler fraction f_D to the contribution of the dynamic part of the medium to the total scattering coefficient of the entire medium. We will consider a simple exponential decay model and compare it with the gold standard provided by the Monte Carlo simulation technique. In the exponential decay model we assume that the fraction of *unshifted* light decays exponentially with the traveled optical path length l_{opt}. Consequently, the fraction of Doppler shifted photons will be given by $f_D=1-exp(-\mu_{s,dyn}\ell_{opt}/n)$, with $\mu_{s,dyn}$ the scattering coefficient of the ensemble of moving particles. Monte Carlo simulations were performed with the algorithm and software as described by De Mul [18]. The single mode fiber used in the experiment was modeled as a point source. Photon detection was performed in a ring with inner and outer radius of 0.25 and 0.35 mm (in agreement with the core diameter and position of the real detection fiber), concentric to the light beam for illumination. The simulated numerical apertures for illumination and detection were identical to the experimental values. The three mixed static-dynamic phantoms were exactly mimicked, with the scattering phase functions being calculated using Mie's theory. Photons which were scattered by the ∅0.20 μm particles were given a Doppler label. For each medium and each path length, 20000 photons were detected.

Figure 6 shows the fraction of Doppler shifted photons f_D as a function of the optical path length, for the three media. As expected, the measured Doppler fraction increases with the optical path length and the confounding influence of the surrounding static matrices is suppressed. Furthermore, fig. 6 shows the results of Monte Carlo simulations and for the exponential decay model $f_D=1-exp(-\mu_{s,dyn}\ell_{opt}/n)$. The models in general predict higher values of the Doppler fraction than the experimental values. Furthermore, the experimental and Monte Carlo results show biphasic behaviour, with a different trend for optical path lengths below and larger than 2 mm. For a given optical path length, f_D is independent of the influence of static matrices, in particular for optical path lengths larger than 2 mm. Furthermore f_D increases with optical path length with a trend that can be depicted by the simple exponential decay model. However, the theoretically predicted Doppler fractions are higher than the experimental values. Nevertheless, if in this model we define scattering coefficient $\mu_{s,dyn}$ as a fitting parameter, it appears that the exponential decay model properly fits the observations for $\ell_{opt}>2$mm, with $\mu_{s,dyn}=0.55$ mm^{-1}. For $\ell_{opt}>2$mm, the predicted Doppler fractions by Monte Carlo simulations are similar for the three media and are in good agreement with the experimental results. Figure 7 shows the average Doppler shift generated by the moving particles $<\omega_D>=M_1(\delta_s,b,\omega_m)/M_0(\delta_s,b,\omega_m)$ as a function of optical path length. For optical path lengths larger than 2 mm, $<\omega_D>$ increases linearly with optical path

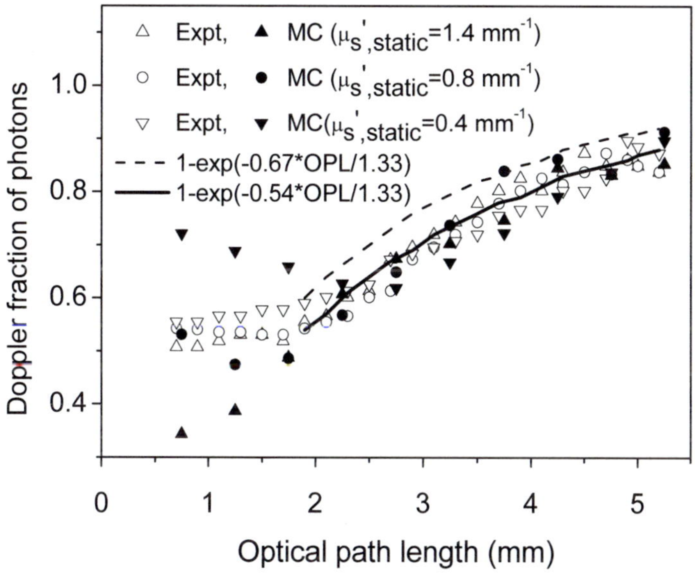

Fig. 6. The fraction of Doppler shifted photons as a function of optical path length, as a result of experiments (open markers), Monte Carlo simulations (filled markers), and experimental decay models, one with a theoretical decay rate (dashed line), and one with the best fit to the experimental results (thick line). (Figure reprinted from Fig. 2 in Ref. [19] Copyright (2010) with permission from OSA).

length as expected theoretically and experimentally [3]. However, $<\omega_D>$ also shows a different behaviour for optical path lengths smaller than 2mm. The average Doppler shift decreases with increasing scattering of the static material.

The overall dependence on the static matrix optical properties on the Doppler shift is small, as depicted in figure 7. In the case of a higher scattering coefficient (μ_s'=1.4mm^{-1}), for optical path lengths between 2.5 and 3.5 mm, the Doppler broadening is lower in comparison with those obtained for the lower scattering levels. We may express the overall dependence of the measured concentration, represented by f_D, and the particle velocity, represented by M_1/M_0, by their average value. This yields average Doppler fractions $<f_D>$ of 0.692, 0.685 and 0.694, and average Doppler shifts $<M_1/M_0>$, of 447.7, 447.9 and 442.2 Hz, for μ_s'=0.4, 0.8 and 1.4 mm^{-1}, respectively. Measurements at a single optical path length may be more suitable in practice. For single path lengths, figures 6 and 7 feature maximum variations of 10% for both f_D and M_1/M_0. These results clearly illustrate that the average Doppler shift measured with the low coherence interferometer, averaged over all optical path lengths, is much less sensitive to the influence of the scattering properties of the static medium.

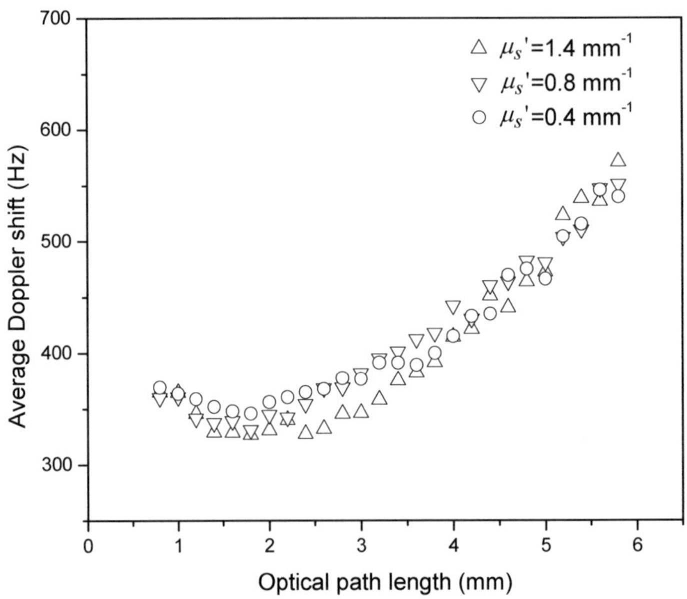

Fig. 7. The weighted Doppler shift measured as a function of optical path length in the medium. (Figure reprinted from Fig. 3 in Ref. [20] Copyright (2010) with permission from OSA).

6. Path length resolved optical Doppler perfusion monitoring

To assess the feasibility of the technique for path length resolved optical Doppler perfusion monitoring, measurements were performed on the skin of the dorsal side of the right forearm of a healthy human volunteer (Skin type- Type II) in the sitting position [20]. A probe holder (PH 08) was attached to the skin with a double-sided adhesive tape. The subject rested approximately 10 minutes prior to the measurements. Skin sites were avoided with visible large superficial blood vessels, hair and pigment variations.

The intensity of Doppler shifted and nonshifted photons measured in skin as a function of optical path length are shown in Fig. 8. The fraction of Doppler shifted photons and nonshifted photons averaged over the entire optical path length measured from the respective areas of the optical path lengths are 22 and 78%, respectively. As shown in Fig. 8, the weighted first moment M_1/M_0 of the Doppler shifted light, which represents the average Doppler shift, increased with the optical path length due to the greater probability of interaction of photons with moving scatterers for large optical path lengths. Further *in vivo* studies were performed to measure the variations in perfusion to external stimuli, inter-and intra-individual variations in optical path lengths and path length resolved Doppler shifts, and to compare these results with the perfusion signal measured with a conventional LDPM [21].

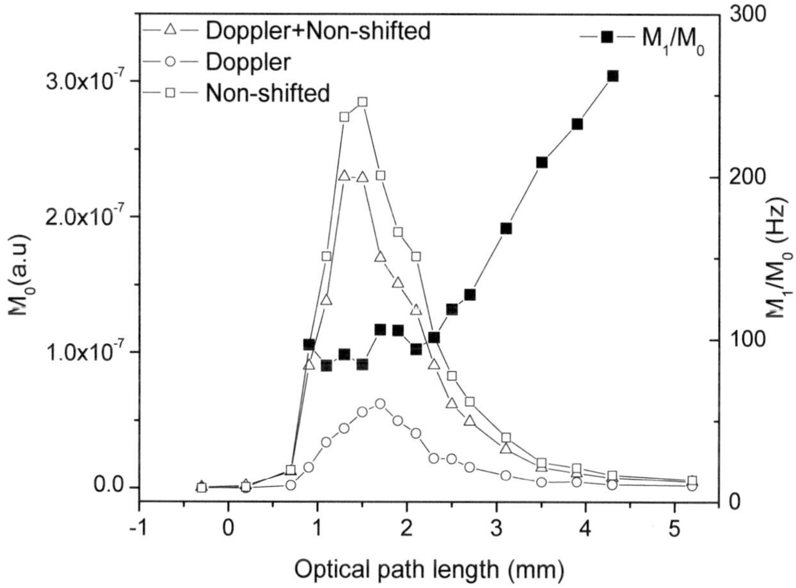

Fig. 8. Intensity of Doppler-shifted, nonshifted photons and the average Doppler shift as a function of optical path length measured in skin (Figure reprinted from Fig. 1 in Ref. [8] Copyright (2008) with permission from SPIE).

Here we have presented optical path length distributions and path length resolved Doppler shifts of multiply scattered light, extracted from the spectral peak that was generated by phase modulation of the reference arm in a low coherence Mach Zehnder interferometer. As such, these data can also be obtained without modulation, but then we only can obtain information about photons which where Doppler shifted by the medium. Hence, phase modulation will enable us to measure path length distributions of static, and mixed static and dynamic media. A second advantage of using phase modulation is that the information can be shifted to higher frequencies, where often the noise level is lower and its spectrum is more flat than for low frequencies [2]. The path length resolved perfusion measurements presented here may overcome the inherent limitation of conventional LDPM that restrict its clinical usefulness, where the perfusion signal depends on an unknown photon path length. This will enable to correctly interpret or counter-act the inter-and intra-individual variations in the LDF readings introduced by the variance in tissue optical properties. This approach enables to discriminate between the Doppler-shifted photons resulting from interaction with the moving red blood cells and the non-shifted light scattered only by the surrounding static tissue matrices [20]. Another important feature of this approach is the tunable depth resolved perfusion information that can be achieved. By changing the optical path length in the reference arm, the photons migrated deeper into the tissue can be made to interfere with the reference light and thus enable to discriminate between the perfusion signal from superficial and deeper layers of tissue. Determination of superficial burn depth may be an important application of our technique [19]. However, further developments and fundamental research are required in developing this into a tool that is suitable for use in a clinical environment, with acceptable measurement times and suitable patient interfaces.

7. Conclusions

To summarize, we have developed a new bio-optical method "Path length resolved optical Doppler perfusion monitoring," to determine path length distributions of multiple scattered light in static and dynamic turbid media using phase modulated coherence gated interferometry. We have shown that path length-resolved dynamic light scattering can measure the dynamic properties of a medium independent of its optical absorption properties, at least when absorption levels are applied in the range found for biological tissues. Furthermore, we showed that our method enables optical Doppler or dynamic light scattering measurements of dynamic media embedded in a static medium, with suppressed dependence of the effect of the scattering coefficient of the static matrix in which the moving particles are embedded. Also, we have presented the first path length resolved Doppler measurements of multiply scattered light from human skin. The results presented here show that this approach has potential applications in discriminating between statically and dynamically scattered light in the perfusion signal. In general, path length resolved dynamic light scattering, of which the basic technique is presented in this work, may overcome the influence of photon path lengths on the measured perfusion signal in laser Doppler techniques and makes it possible to perform depth resolved perfusion measurements with suppression of the confounding influence of optical properties in the tissue matrix.

8. References

[1] P. Shepherd, and P. Å. Öberg, Laser-Doppler Blood Flowmetry (Kluwer Academic, Boston, 1990).

[2] B. Varghese, V. Rajan, T. G. Van Leeuwen, and W. Steenbergen, "Path-length-resolved measurements of multiple scattered photons in static and dynamic turbid media using phase-modulated low-coherence interferometry," J. Biomed. Opt. 12(2), 024020 (2007).

[3] B. Varghese, V. Rajan, T. G. Van Leeuwen, and W. Steenbergen, "Quantification of optical Doppler broadening and optical path lengths of multiply scattered light by phase modulated low coherence interferometry," Opt. Express, 15 (15), 9157-9165 (2007).

[4] G. Dougherty, "A laser Doppler flowmeter using variable coherence to effect depth discrimination," Rev. Sci. Instrum. 63, 3220–3221(1992).

[5] J.D. McKinney, M.A. Webster, K.J. Webb and A.M. Weiner, Characterization and imaging in optically scattering media by use of laser speckle and a variable-coherence source. Opt. Lett. 25, 4–6 (2000).

[6] U.H.P. Haberland, V. Blazek and H.J. Schmitt, Chirp optical coherence tomography of layered scattering media. J. Biomed. Opt. 3, 259–266 (1998).

[7] J. M. Tualle, E. Tinet, S. Avrillier, "A new and easy way to perform time resolved measurements of the light scattered by turbid medium", Opt. Comm., 189, 211-220 (2001).

[8] K. K. Bizheva, A. M. Siegel, and D. A. Boas, "Path-length-resolved dynamic light scattering in highly scattering random media: The transition to diffusing wave spectroscopy," Phys. Rev. E, 58 (1998).

[9] A. Wax, C. Yang, R. R. Dasari, and M. S. Feld, "Path-length-resolved dynamic light scattering: modeling the transition from single to diffusive scattering," Appl. Opt., 40, 4222-4227 (2001).

[10] A. L. Petoukhova, W. Steenbergen, and F.F.M.de Mul, "Path-length distribution and path-length resolved Doppler measurements of multiply scattered photons by use of low-coherence interferometer," Opt. Lett.26, 1492-1494 (2001).

[11] K. Ishii, R. Yoshida, and T. Iwai, "Single-scattering spectroscopy for extremely dense colloidal suspensions by use of a low-coherence interferometer," Opt. Lett. 30, 555-557 (2005).

[12] D. Huang, E. A. Swanson, C. P. Lin, J. S. Schuman, W. G. Stinson, W. Chang, M. R. Hee, T. Flotte, K. Gregory, C. A. Puliafito, and J. G. Fujimoto, "Optical coherence tomography", Science 254, 1178 (1991).

[13] J. M. Schmitt, "Optical Coherence Tomography (OCT): A Review", IEEE Selected Topics in Quantum Electronics, 5(4), 1205-1215 (1999)..

[14] A. F. Fercher, C. K. Hitzenberger, C. K. Kamp and S. Y. El-Zayat, "Measurement of intraocular distances by backscattering spectral interferometry," Opt. Comm., 117, 43-48, (1995).

[15] G. E. Nilsson, "Signal processor for laser Doppler tissue flow. meters," Med. Biol. Eng. Comput, 22, 343-348 (1984).

[16] B. Varghese, V. Rajan, T. G. Van Leeuwen, and W. Steenbergen, "High angle phase modulated low coherence interferometry for path length resolved Doppler measurements of multiply scattered light," Opt. Commun. 281(3), 494–498 (2008).

[17] S. T. Flock, S. L. Jacques, B. C. Wilson, W. M. Star, and M. J. C. van Gemert, "Optical properties of intralipid: a phantom medium for light propagation studies," Lasers Surgery Med. 12, 510-519 (1992).

[18] F.F.M. De Mul, "Monte-Carlo simulation of Light transport in Turbid Media", in: Handbook of Coherent Domain Optical Methods, Biomedical Diagnostics, Environment and Material Science, Tuchin, Valery V. (Ed.), 2004, Kluwer Publishers, 465-533 (2004).

[19] B. Varghese, V. Rajan, T. G. Van Leeuwen, and W. Steenbergen, "Measurement of particle flux in a static matrix with suppressed influence of optical properties, using low coherence interferometry," Opt. Express 18, 2849-2857 (2010).

[20] B. Varghese, V. Rajan, T. G. van Leeuwen and W. Steenbergen, "Path length resolved optical Doppler perfusion monitoring," J. Biomed. Opt. Lett., 12(6):060508 (2007).

[21] B. Varghese, V. Rajan, T. G. van Leeuwen and W. Steenbergen, "In vivo optical path lengths and path length resolved doppler shifts of multiply scattered light," Lasers in Surgery and Medicine, 42(9), 692-700(2010).

11

Interferometric Measurement in Shock Tube Experiments

Masanori Ota, Shinsuke Udagawa, Tatsuro Inage and Kazuo Maeno
Graduate School of Engineering, Chiba University
Japan

1. Introduction

This chapter describes applications of interferometry to the shock tube experiments. The first topic is Laser Interferometric Computed Tomography (LICT) technique to realize the three-dimensional (3D) density measurement of high-speed and unsteady flow field behind shock waves discharging from nozzles. The second topic is measurement of propagating shock wave in micro-scale shock tube by interferometic approach. Micro-scale shock tube is being researched in several fields of science recently and micro-scale shock wave has possibilities of applications for various fields - medical, engineering, ...etc. Clarifying the characteristics of micro-scale shock tube to generate the micro-scale shock wave is very important step for the application.

2. Laser Interferometric Computed Tomography (LICT) technique

The purpose of this investigation is to develop Laser Interferometric Computed Tomography (LICT) technique to observe high-speed, unsteady and three-dimensional (3-D) flow field that includes shock wave, and to clarify 3-D flow phenomena induced by shock waves. In our previous study, 3-D complex flow discharged from a square nozzle and a pair of circular nozzles was measured by LICT tchnique (Maeno et al., 2005; Ota et al., 2005; Honma et al., 2003a, 2003b). The shock Mach number at the exits of the nozzles were both higher and lower than 2.0. As a result, various phenomena of 3-D flow field were clarified by several imaging technique such as pseudo-color images, pseudo-schlieren images by pseudo-schlieren technique, 3-D isopycnic images, etc. Three-dimensional and complex flow phenomena behind shock wave were elucidated precisely and reported (Honma et al., 2003a, Maeno et al., 2005), therefore this chapter reports measurement results mainly.

2.1 Experimental apparatus

Diaphragmless shock tube is employed to produce a shock wave with good reproducibility. Figure 1 illustrates a schematic diagram of LICT experimental apparatus and observation system. The observation system consists of a CCD camera, a Mach-Zehnder interferometer, a pulsed nitrogen laser, a delay/pulse generator, an oscilloscope, and a personal computer.

The shock wave is generated by a diaphragmless shock tube driver in the low-pressure tube of 3.1 meters in length, and its inner cross section is 40 mm x 40 mm square. A rotating plug is installed at the end of the low-pressure tube. The duct is open to the low-pressure test section.

To obtain the 3-D image of flow field, we need multidirectional projection data for a reproducible flow. A set of experiments has been performed for several rotation angles at the combination of fixed initial gas conditions for the high-pressure chamber and the low-pressure tube. Figure 2 shows coordinate system of rotating plug relative to the light pass s. We define x and y axes as shown in Fig.2, where these axes rotate with rotating plug. The z is central axis of rotating plug and is perpendicular to x and y axes. Rotation angle θ can be controlled from outside the shock tube with introduced rotation driving equipment. The experiment is performed for 19 rotation angles between 0 degree and 90 degrees at five-degree intervals while the light path s is fixed, taking benefit of the two-axis symmetrical characteristics of the flow field. The 3-D density distribution is reconstructed from a set of projection data for the same M_i and z_s. The Mach number of the incident shock wave M_i is calculated by pressure jump across the shock wave at the pressure transducer installed at 61mm ahead of the inlet of the rotating plug as shown in Fig. 3. In this paper M_i is fixed to 2.0. Mach-Zehnder interferometer is conventional one as shown in Fig. 4

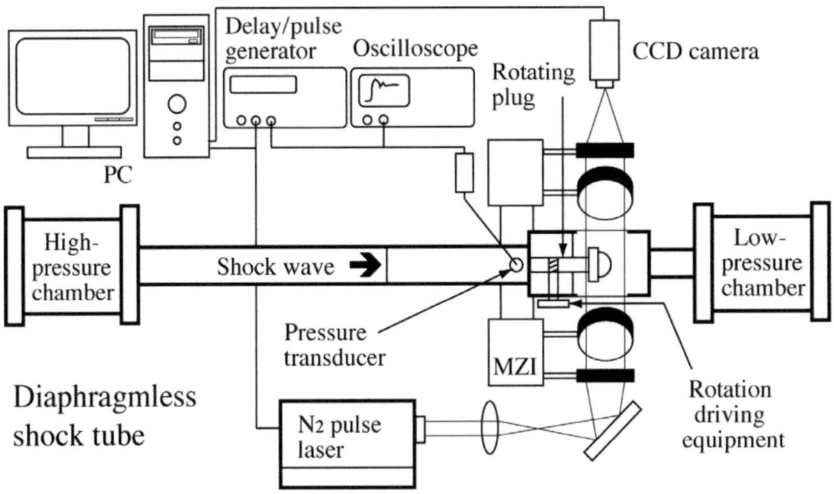

Fig. 1. Schematic diagram of experimental apparatus.

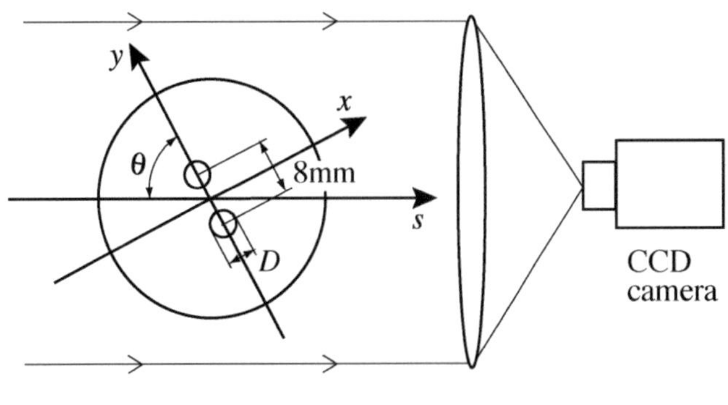

Fig. 2. Coordinate system of the rotating plug.

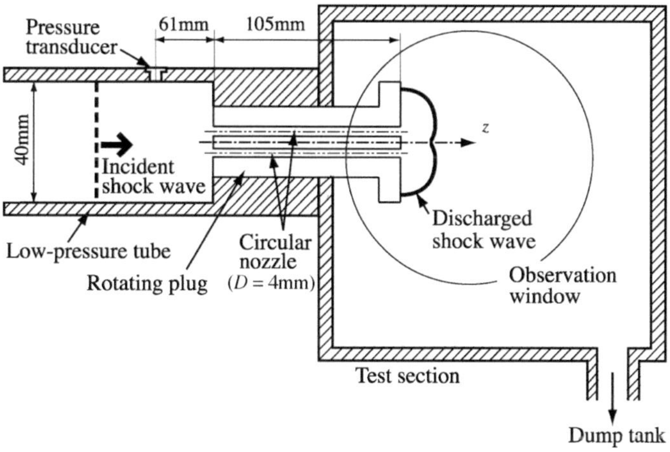

Fig. 3. Layout of rotating plug.

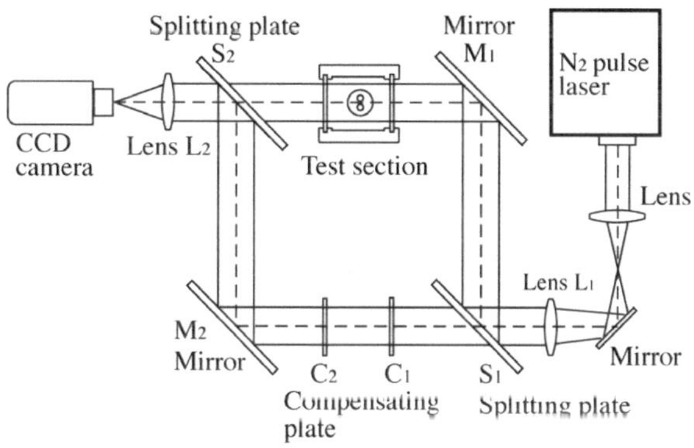

Fig. 4. Mach-Zehnder interferometer.

2.2 Projection data for LICT measurement

Figure 5 shows a finite-fringe interferogram at rotation angle $\theta = 90^o$ taken by CCD camera with Mach Zehnder interferometer illustrated in Fig. 4. To obtain higher accuracy, we have taken interferogram of the lower half part of whole flow field, taking benefit of the two-axis symmetrical characteristics of the flow field. Thick blanked line indicates the central axis of rotating plug, and thin blanked line indicates the central axis of circular nozzles. The distance between these two central axes is 4mm. In this figure z_s is a frontal position of the primary shock wave, D is a diameter of circular nozzle (4mm), and z_s/D is the normalized frontal position of the primary shock wave. In this paper we discuss the case data of z_s/D = 2.50.

In LICT measurement, three-dimensional density distribution is reconstructed from multidirectional projection data. Projection data are calculated from Eq. 1, where ΔH is displacement of fringe pattern, Δh is interval of fringe pattern, λ is wavelength of observation light, and K is Gladstone-Dale index. ΔH and Δh in the right hand side in Eq. 1 are calculated from finite-fringe interferogram as shown in Fig. 5. Firstly, we calculate Δh at A section of no flow area. Secondly, we calculate ΔH from displacement of fringe pattern comparing with A section. We obtain the integrated value of density change along the light pass from Eq. 1. Figure 6 shows the calculated projection data at B section of rotation angle $\theta = 90^{\circ}$. The horizontal axis indicates the vertical position of the finite-fringe image shown in Fig. 5, the origin of the coordinates is located in left top position of this image. The vertical axis of Fig.6 indicates the calculated integrated value of density change and its value is normalized with the initial density (ρ_0) at A section. We repeat this process for all projection angles at one cross section to obtain the multidirectional projection data at one cross section. Then 2-D density distribution is reconstructed from these projection data with appropriate reconstruction algorithm. Finally, 3-D density distribution is obtained as collection of reconstructed 2-D density distribution.

$$\int_0^d \{\rho(x,y,z) - \rho_0\}\, ds = \frac{\Delta H}{\Delta h} \frac{\lambda}{K} \tag{1}$$

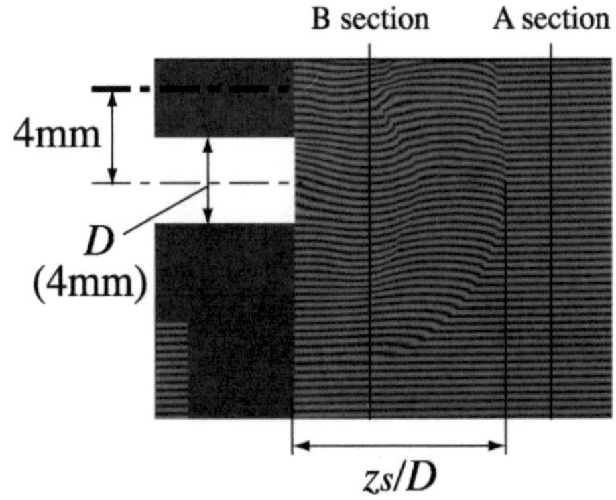

Fig. 5. Finite-fringe interferogram at rotation angle $\theta = 90^{\circ}$.

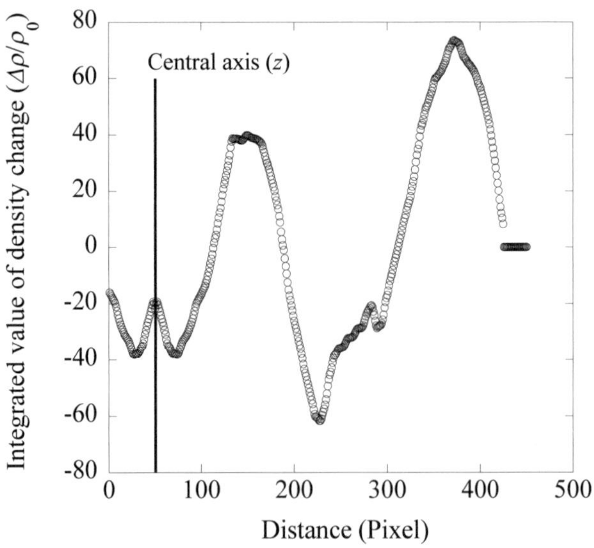

Fig. 6. Calculated projection data at B section in Fig. 5.

2.3 Results and discussion

Figure 7 shows the pseudo-color images of normalized density distribution and the pseudo-schlieren images indicating the magnitude of the density gradient $\nabla|\rho/\rho_0|$ at y-z cross-section. In pseudo-color image the density level is shown in visible-light color spectrum. The black line with white blank indicates the rotating plug with circular nozzles, where white blank corresponds to a pair of circular nozzles. The experimental condition is $M_i = 2.0$, $z_s/D = 2.50$ as mentioned above. The smoothing filter is applied to the reconstructed 3-D density distribution with FBP algorithm before the calculation of the density gradient, for reducing the noise in the resultant pseudo-schlieren image. In pseudo-color image two vortex rings around the discharged flow from two circular open cylinders are illustrated. The primary shock wave (PSW), secondary shock wave (SSW), contact surface (CS1, CS2), and transmitted shock wave (TSW) are exhibited clearly in pseudo-schlieren image.

Pseudo-color image of normalized density distribution at x-y cross-section is shown in Fig. 8. Position of the x-y cross section is indicated with normalized distance z/D where z is the distance between the rotating plug's wall and the x-y cross section. In these images cross sectional shape of PSW, TSW and two vortex rings around the discharged flow is exhibited. Figure 9 shows the pseudo-schlieren images at x-y cross-sections. In these images PSW, the shock-vortex interaction between TSW and vortex, shock-shock interaction between TSW and PSW are exhibited clearly.

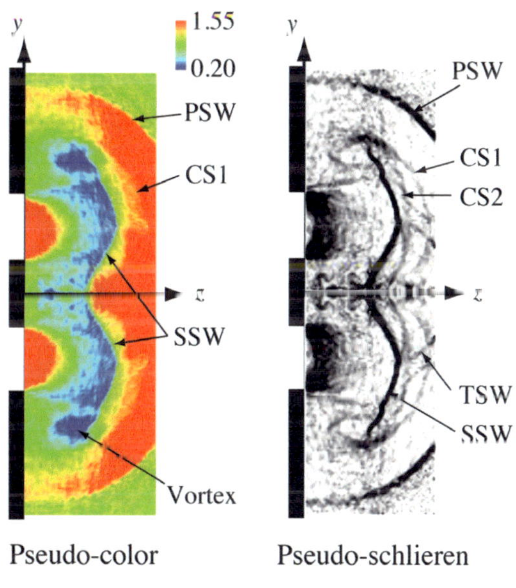

Fig. 7. Pseudo-color (left) and pseudo-schlieren (right) image at y-z cross-section.

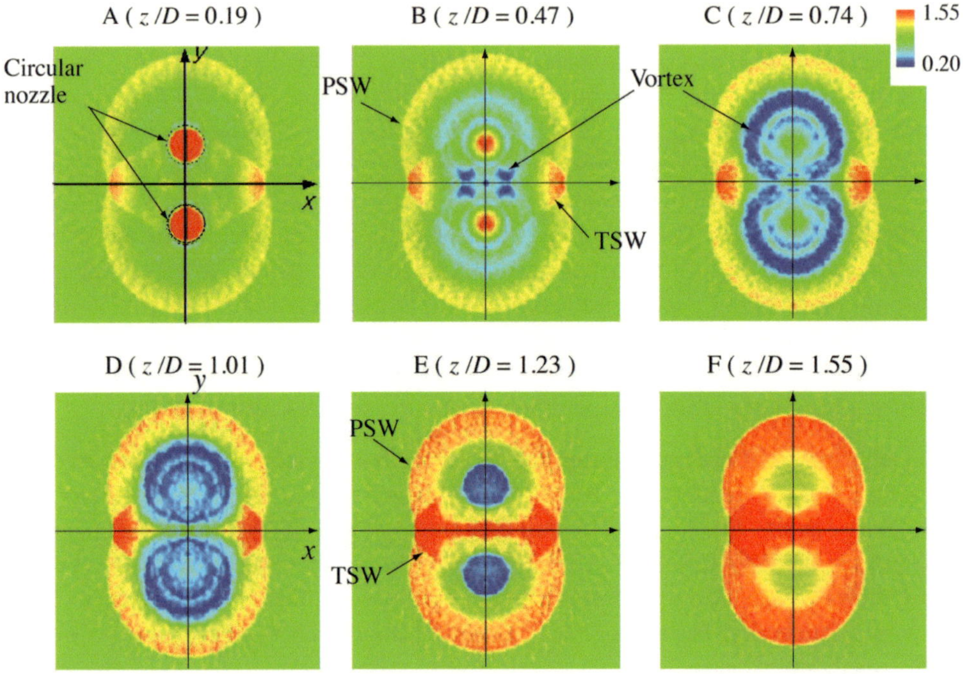

Fig. 8. Pseudo-color images of density distribution at *x-y* cross-sections.

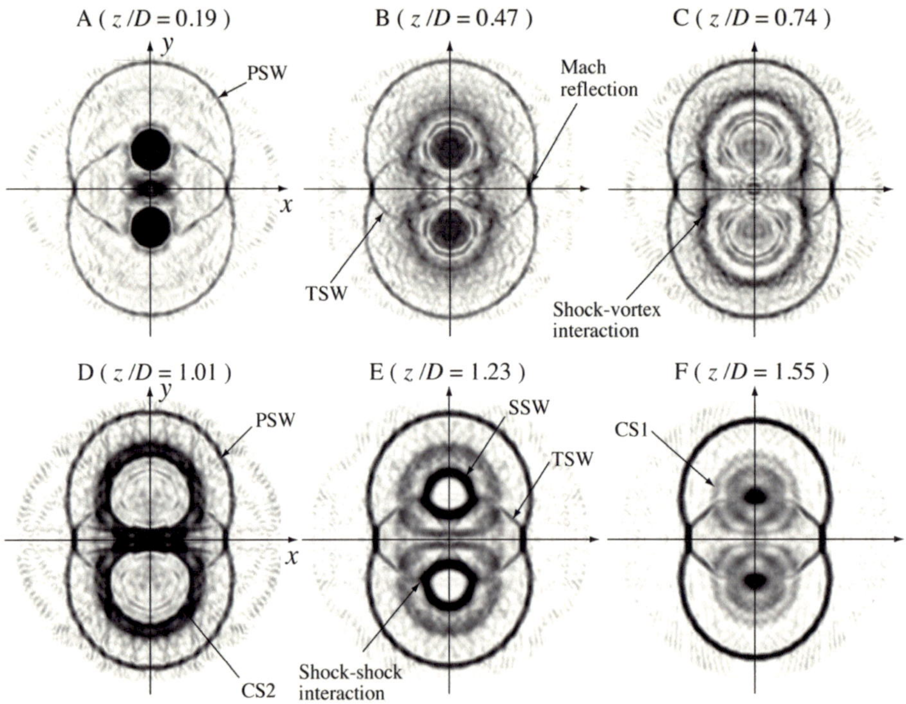

Fig. 9. Pseudo-schlieren images of density gradient ($\nabla|\rho / \rho_0|$) at *x-y* cross-sections.

3. LICT Measurement of flow field around the object

Three-dimensional observation of flow field around object will be important to apply CT measurement to more general case. Therefore LICT technique has been applied to the flow field around a circular cylinder as shown in Fig. 10. Diameter of a cylinder is 4mm, length 10mm and it was installed at intervals of 8mm from a cylindrical nozzle. In this case the problem is reconstruction has to be done from incomplete projection data. The object in the observation area blocks off the observation light for interferometry as shown in Fig. 11, the calculated projection data also contain the blank part which corresponds to the position of the circular cylinder as indicated in Fig. 12. This section describes the reconstruction of flow field around object with ART algorithm and results.

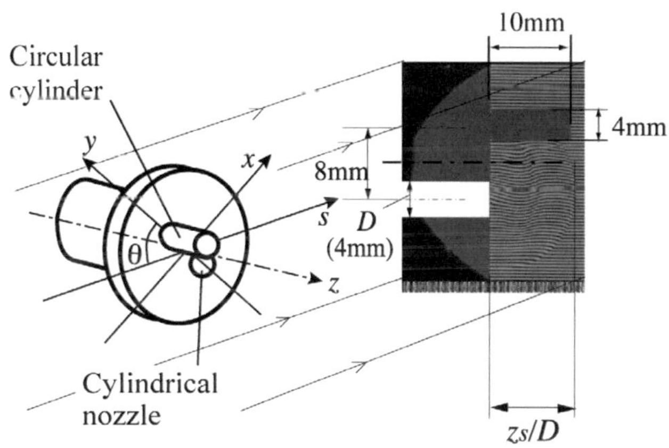

Fig. 10. Coordinate system of rotating plug and finite-fringe interferogram.

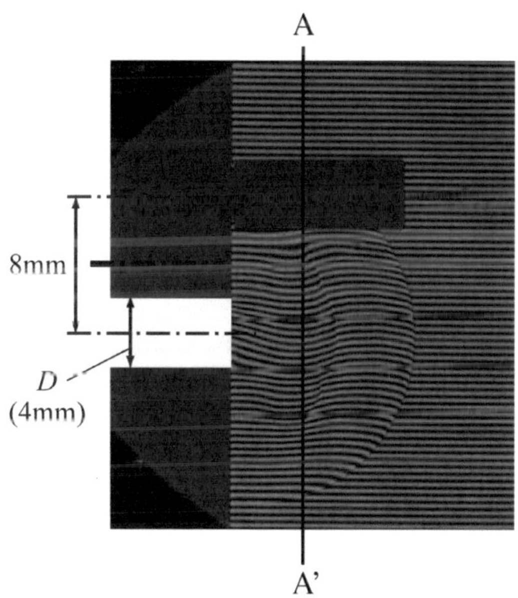

Fig. 11. Finite-fringe interferogram at rotation angle $\theta = 90^{o}$.

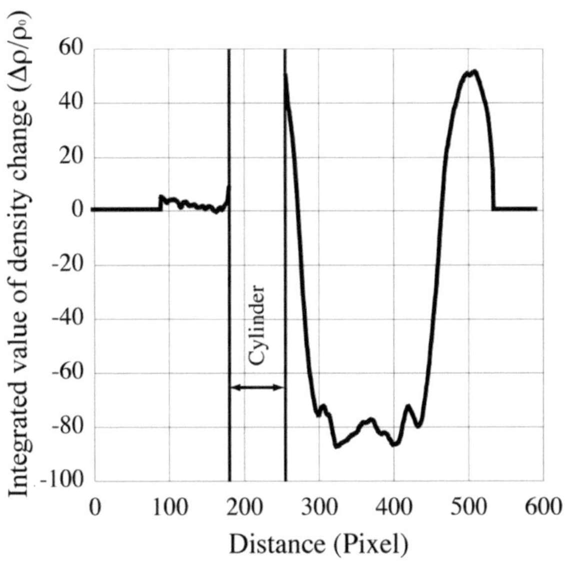

Fig. 12. Calculated projection data at A-A′ section in Fig. 11.

3.1 Reconstruction algorithm

In previous section, 3-D density distribution was reconstructed by FBP (Filtered Back Projection) with Shepp and Logan type filter (Shepp & Logan, 1974). FBP is the most common technique for the tomographic reconstruction. In this chapter the density distribution of unsteady flow field around a circular cylinder is reconstructed by ART (Algebraic Reconstruction Technique). This technique is one of the iterative reconstruction method and consists of assuming that the cross section consists of an array of unknowns, and then setting up algebraic equations for the unknowns in terms of the measured projection data (Kak & Slaney, 1988). ART is much simpler than FBP method which is the transform-based method and we have used in our previous study. For FBP a large number of projections is required for higher accuracy in reconstructed image, in the situation where it is not possible to obtain these projections the reconstructed image is suffer from many streaky noise. The reconstruction from incomplete projection data is more amenable to solution by ART. Figure 13 shows comparison between FBP and ART reconstruction. Left figure is pseudo-color image at the same plane illustrated in Fig. 7 and right figure is pseudo-schlieren image. Upper half image is reconstructed by FBP and lower half is by ART. In pseudo-color image, reconstructed density distribution by ART is clearer than FBP result. The smoothing filter is applied to the reconstructed density distribution before the calculation of density gradient for FBP result as mentioned above. On the other hand, no smoothing filter is applied for ART result however density gradient is captured distinctly. Figure 14 is comparison between FBP and ART illustrated by pseudo-color and pseudo-schlieren image at x-y cross section (normalized position z/D = 0.74). In pseudo-color image some radial noises from the center can be seen in FBP result (upper), however these noises are reduced in ART result (lower) drastically. Though no smoothing filter is applied, flow phenomena are clearly seen with ART result in pseudo-shclieren image. In our experiments density distribution is reconstructed from 19 projections that is based on our experiences.

Larger projection number will better to obtain higher accuracy, however it will cause inefficiency of experiments and data processing. Figure 13 and 14 show that ART is effective for reconstruction from incomplete projection data when LICT measurement is applied to the measurement of flow field around a circular cylinder. In this chapter blanked part in projection data where a circular cylinder is captured in projection plane does not contribute to reconstruction. Projection number is 19 rotation angles from 0° to 90° at 5° intervals as previous case.

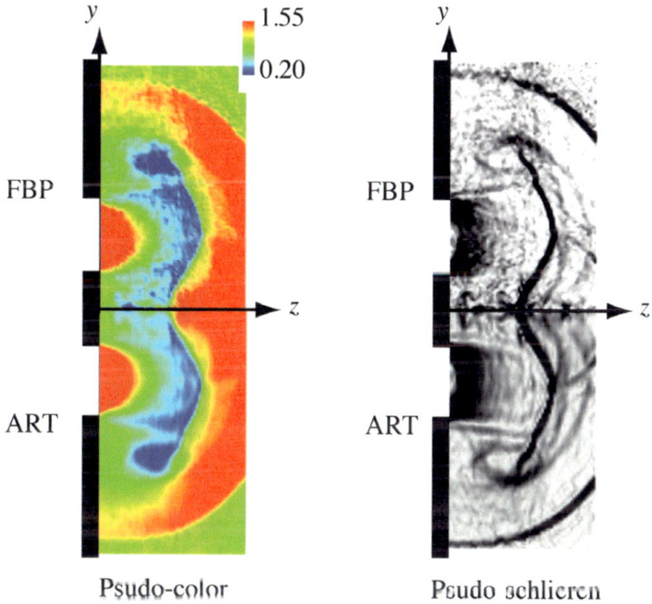

Fig. 13. Comparison between FBP and ART at *y*-*z* cross-section.

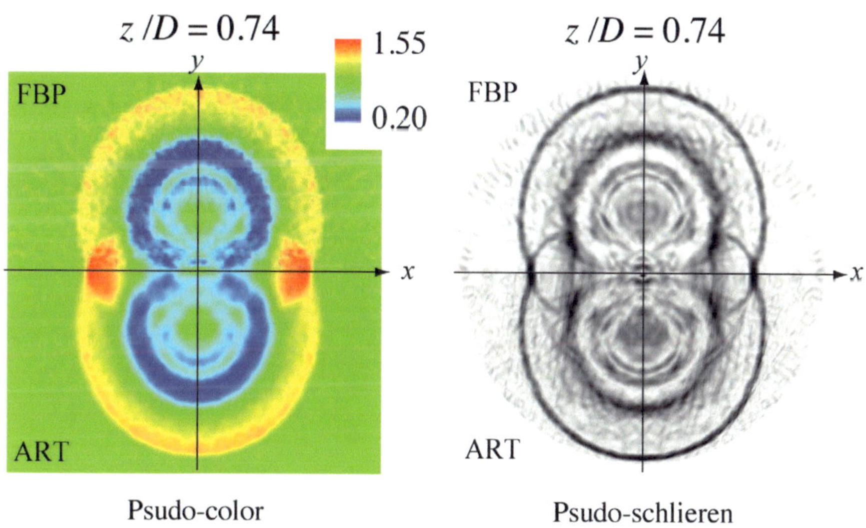

Fig. 14. Comparison between FBP and ART at *x*-*y* cross-section.

3.2 Results and discussion

The resultant image in y-z cross section is illustrated in Fig. 15. The left is pseudo-color image of normalized density distribution and right is pseudo-schlieren image. The vertical thick line with white blank indicates rotating plug's wall and white blank indicates a cylindrical nozzle. The position of a circular cylinder is indicated with two horizontal blanked lines. The vortex around discharging flow from a cylindrical nozzle is identified in pseudo-color image and primary shock wave (PSW), secondary shock wave (SSW), contact surface (CS) and reflected shock wave (RSW) from circular cylinder is clearly seen in pseudo-schlieren image.

Figure 16 illustrates pseudo-color images of normalized density distribution in x-y cross section. Six cross sections (position A~F) that are parallel to rotating plug's wall are indicated. The normalized position of cross section (z/D) is shown at upper side of each image. The position of a circular cylinder is indicated with blanked circle. In position A~C, the cross sectional shape of vortex around the discharging flow from a cylindrical nozzle is captured. The reflected shock wave (RSW) from circular cylinder is seen in position D~F. Four slanting noise from a circular cylinder is appeared in position E and F, this is influence of reconstruction from incomplete projection data.

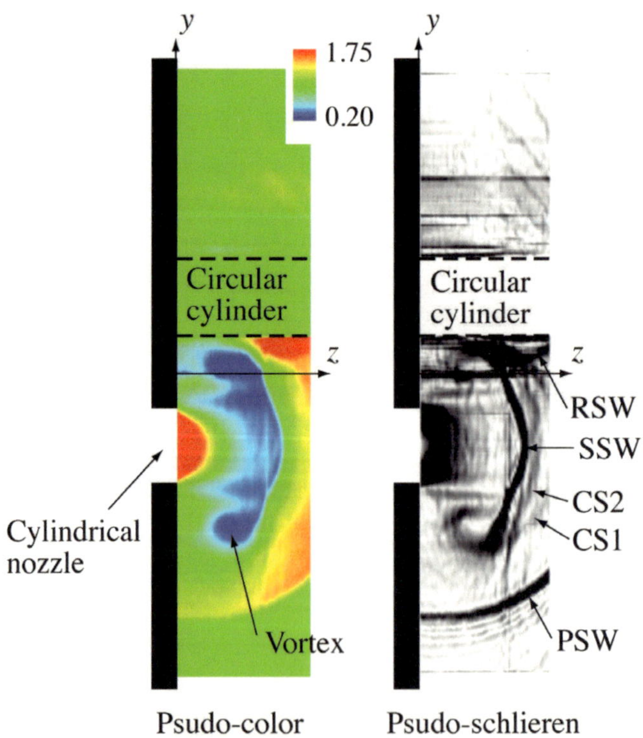

Fig. 15. Pseudo-color (left) and pseudo-schlieren (right) image at y-z cross-section.

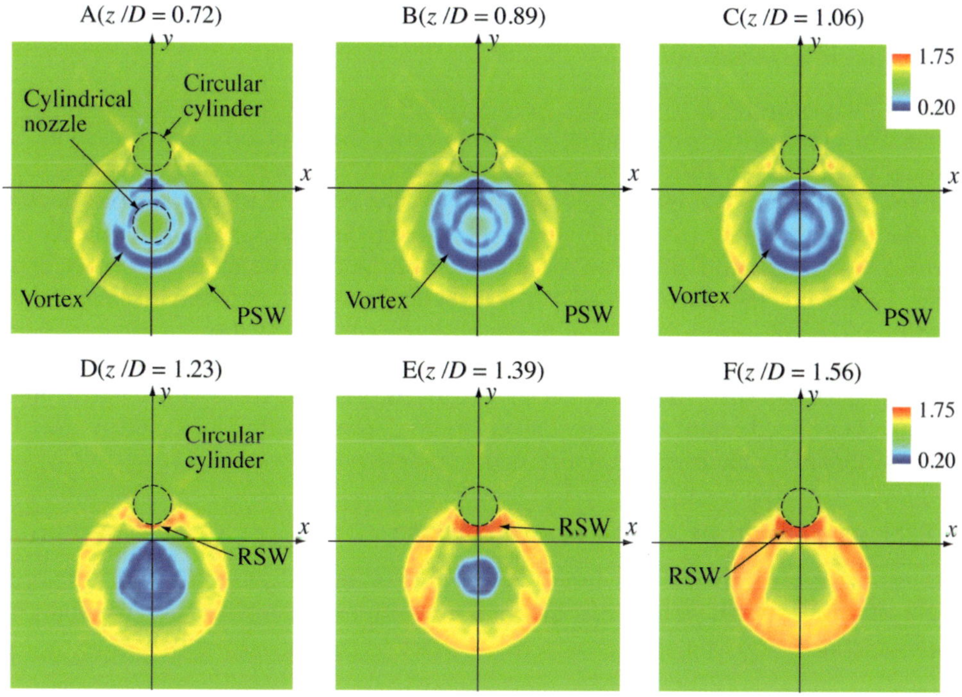

Fig. 16. Pseudo-color images of density distribution at x-y cross-sections.

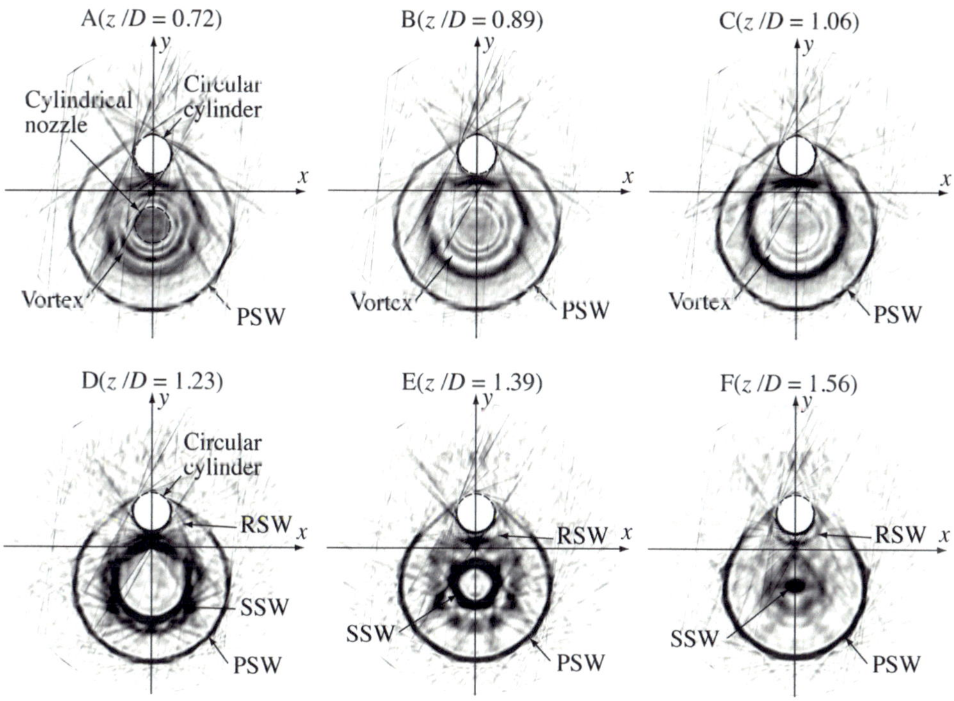

Fig. 17. Pseudo-schlieren images of density gradient ($\nabla|\rho / \rho_0|$) at x-y cross-sections.

4. Velocity measurement of shock wave in micro-scale shock tube by laser differential interferometer

Recently, the micro-shock waves have attracted attention of researchers in several fields of science. The shock wave propagating in 5.3mm inner diameter tube was measured using pressure transducers by Brouillette. As a consequence, it is experimentally clarified that the shear stress and the heat transfer between a test gas and a wall lead to significant deviations from the normal theory, especially in a small diameter shock tube (Brouillette, 2003). However, it is predicted that the pressure transducer becomes disturbance in case of using smaller tube from the difference of the representative scales between the tube and the transducer. Thus, it is very important to establish the contactless measurement method for the shock wave propagating in small diameter tubes. In this study, we measured the velocities of shock wave and the density ratios across the shock wave, generated by originally developed diaphragmless driver section, propagating in 2 and 3 mm inner diameter tubes by using laser differential interferometer.

4.1 Diaghragmless driver section with two pistons

Figure 18 shows a schematic drawing of the diahpragmless driver section, we developed; it consists of a main piston instead of a diaphragm, a sub piston, a buffer section, and a high-pressure section.

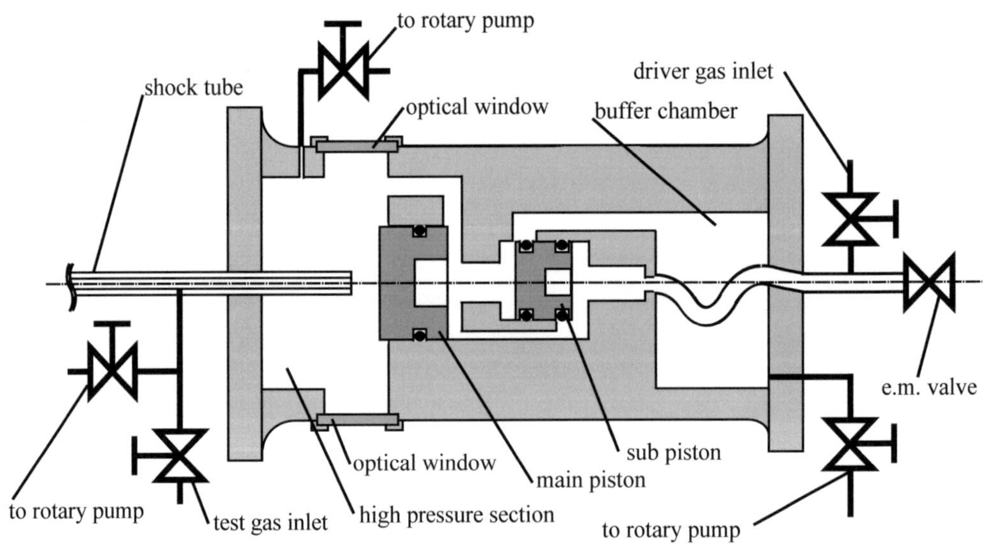

Fig. 18. Schematic drawing of the diaphragmless driver section with two pistons.

First, the buffer chamber and the region behind the sub piston are evacuated by a rotary pump, then the sub piston moves to the right side by the pressure difference. The pressure difference occurs between the high-pressure section and the region behind the main piston by the connection between the buffer chamber and the region behind the main piston after the sub piston movement. After movement of the pistons, the high-pressure section and the shock tube are evacuated by a rotary pump. The driver gas is introduced from the driver gas inlet after evacuation, and then the sub piston moves to the left side by the inlet gas pressure. The buffer chamber and the region behind the main piston are separated by the movement of a sub piston as shown in the left side of Fig. 19. Subsequently, the main piston moves to the left side by that the driver gas flows into the region behind the main piston. The high-pressure chamber and the shock tube are also separated by movement of the main piston. Test gas is introduced from the test gas inlet after the introduction of the driver gas. The buffer chamber maintains a low-pressure state, less than 0.1 kPa. The driver gas pressures p_4 is 0.9 MPa and the test gas pressure p_1 is maintained at atmospheric pressure. The sub piston moves to the right side by controlling the electromagnetic valve, and then the main piston rapidly moves to the left side by the connection of the buffer chamber and the region behind the main piston. The shock tube and the high-pressure chamber are connected by rapid movement of a main piston as shown in the right side of Fig. 19.

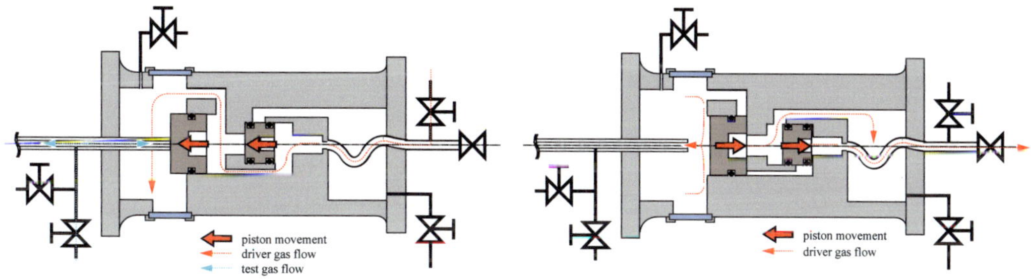

Fig. 19. Schematic drawing of the operating procedure of the diaphragmless driver section.

4.2 Measurement system

Generally the direct measurement of shock wave, by using pressure transducer is very difficult especially in the shock wave propagating in small diameter tube, caused by the difference of the representative scales between the tube and the transducer. Thus, contactless measurement by using laser interferometry is very important and useful.

4.2.1 Laser differential interferometer

Figure 20 shows the laser differential interferometer used in this experiment. The laser differential interfeormeter is the polarization phase difference interferometer by using three Wollaston prisms, developed by Smeets (Smeets, 1972; Smeets, 1977). The bright and dark, observed in the interferometric fringe, is measured by photo detectors.

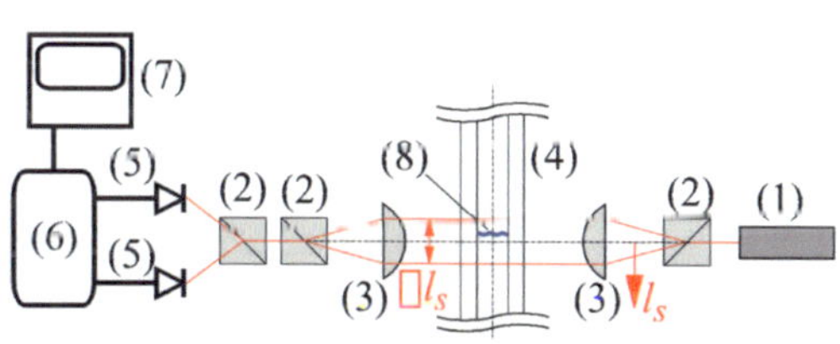

(1) He-Ne laser (5) Photo detector
(2) Wollaston prism (6) Differential amplifier circuit
(3) Convex lens (7) Oscilloscope
(4) Shock tube (8) Shock wave

Fig. 20. Schematic drawing of the laser differential interferometer.

The laser beam from He-Ne laser head is separated into two parallel and orthogonal polarizing components after passing through the first Wollaston prism and a convex lens. The two-polarized and parallel beams are converged on the second Wollaston prism by a second convex lens after passing through the shock tube. Where, there is no interferometric fringe from the combined beams caused by the discrepancy between the polarization planes of both beams. Here, the shock wave propagates in the shock tube in a direction from the bottom to top in Fig. 20, and it is considered that the shock wave arrives between two beams. The density in front of the shock wave and the density behind the shock wave are ρ_1 and ρ_2, respectively. Two beams passing through the same optical path as light vectors are E_y and E_z, respectively as shown in Fig. 21.

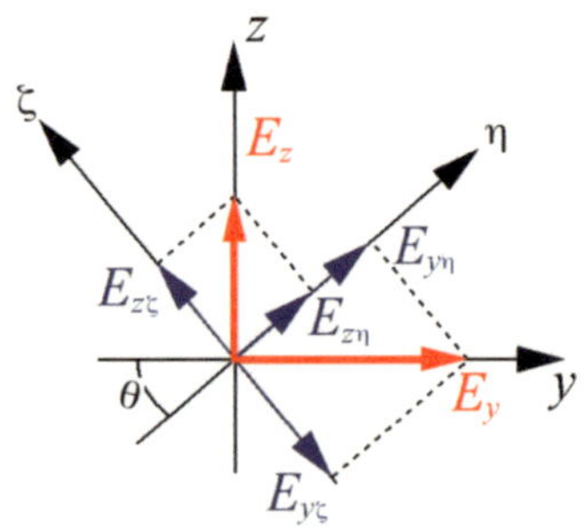

Fig. 21. Schematic drawing of the optical vectors

The light vectors E_y and E_z are denoted as follows:

$$E_y = A_y \cos\left(\frac{2\pi}{\lambda} x\right), \quad E_z = A_z \cos\left\{\frac{2\pi}{\lambda}(x + \Delta\phi)\right\} \tag{2}$$

Here A_y and A_z are the constants for a laser light intensity. x, λ, and $\Delta\phi$ are the distance of the traveling direction of laser, the wavelength of laser, and the optical path difference between both beams, respectively. The optical path difference is occurred by the density difference across the shock wave. The third Wollaston prism should be settled as rotated 45 degrees for each polarization planes. The light vector E_y is separated into $E_{y\eta}$ and $E_{y\zeta}$, respectively. Similarly, The light vector E_z is separated into $E_{z\eta}$ and $E_{z\zeta}$, respectively. Here, η and ζ are the axes inclined 45 degree from y and z axes. After passing through the second Wollaston prism, the agreement of polarization plane causes interference at η and ζ directions, respectively. The intensities of interfering lights for η and ζ directions are denoted as follows:

$$I_\eta = \left\langle \left| E_{y\eta} + E_{z\eta} \right|^2 \right\rangle = A^2 \cos^2\left(\frac{\pi}{\lambda}\Delta\phi\right), \quad I_\zeta = \left\langle \left| E_{y\zeta} + E_{z\zeta} \right|^2 \right\rangle = A^2 \sin^2\left(\frac{\pi}{\lambda}\Delta\phi\right) \tag{3}$$

Here, $A_y = A_z = A$. The intensities of two polarized interfering beams I_η and I_ζ are converted into the voltage signals V_η and V_ζ by the photo detectors as follows:

$$V_\eta = kI_\eta = kA^2 \cos^2\left(\frac{\pi}{\lambda}\Delta\phi\right), \quad V_\zeta = kI_\zeta = kA^2 \sin^2\left(\frac{\pi}{\lambda}\Delta\phi\right) \tag{4}$$

The voltage signal V is stored to an oscilloscope. Here, k is the constant defined from the characteristics of the photo detector. The voltage signal $V = V_\eta - V_\varsigma$ is amplified by the differential amplifier circuit.

$$V=\alpha\left(V_\eta-V_\varsigma\right)=\alpha kA^2\cos\left(\frac{\pi}{\lambda}\Delta\phi\right)=V_0\cos\left(\frac{\pi}{\lambda}\Delta\phi\right) \tag{5}$$

Here, α is the gain of the amplifier. Figure 22 shows the relation between the voltage signals and the optical path difference between both beams. The gain of the amplifier α is denoted as 2 in Fig. 22. Here, the refractive index of the medium is expressed as n and the length of the medium is obtained as the inner diameter of the shock tube d, the optical path difference between both beams $\Delta\phi$ is obtained as follows:

$$\Delta\phi=\left(n_2-n_1\right)d \tag{6}$$

Here, n_2 and n_1 are the refractive index of the gas behind the shock wave and the gas in front of the shock wave, respectively. The relation between the refractive index n and the density ρ is expressed by Gladstone-Dale's formula as follows:

$$n-1=K\frac{\rho}{\rho_0} \tag{7}$$

Here, ρ_0 and K are the density at the normal condition and the non-dimensional Gladstone-Dale constant, respectively. The relation between the voltage signal V and the density difference ρ_2-ρ_1 is obtained from Eq. 5, 6 and 7 as follows:

$$V=V_0\cos\left\{\frac{2\pi dK}{\lambda\rho_0}\left(\rho_2-\rho_1\right)\right\} \tag{8}$$

Here, the voltage signal V can be linearly-approximated to ΔV at V=0 neighborhood as follows, as shown in Fig. 22:

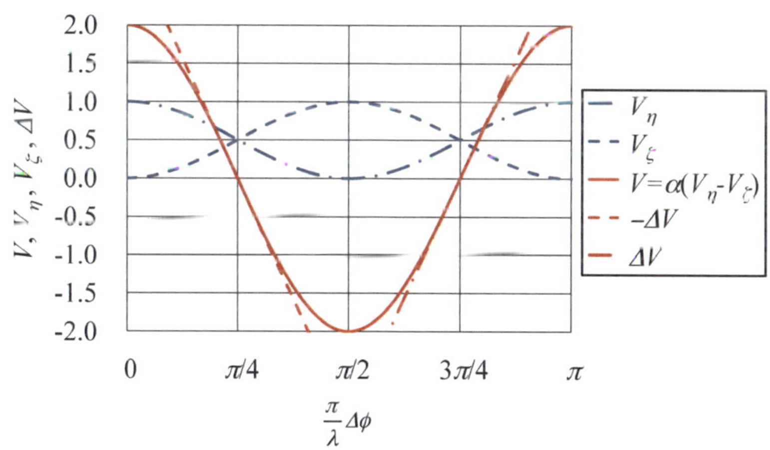

Fig. 22. The interfering signals from the laser differential interferometer.

$$\pm\Delta V=2V_0\frac{\pi dK}{\lambda\rho_0}(\rho_2\text{-}\rho_1) \tag{9}$$

Additionally, equation 9 can be rewritten by considering that the value required in this experiment is the absolute value of the voltage signal V for the density difference ρ_2-ρ_1 as follows:

$$|\pm\Delta V|=\Delta V=2V_0\frac{\pi dK}{\lambda\rho_0}(\rho_2\text{-}\rho_1) \tag{10}$$

Moreover, the above Eq. 10 can be deformed by using density ratio across the shock wave ρ_2/ρ_1 as follows:

$$\frac{\rho_2}{\rho_1}=1+\frac{\Delta V}{2V_0}=\frac{\lambda}{\pi dK}\frac{\rho_0}{\rho_1} \tag{11}$$

4.2.2 Shock wave measurement system

Figure 23 shows the shock wave measurement system used in this experiment. The glass tube, connected to the high pressure section, has the length l = 1000 mm. The position of laser differential interferometer ls is changed from 200 to 800 mm. The inner diameter of the tube d is used as 2 and 3 mm in this experiment. The pressure transducer is settled at the end of the tube to detect the reflected shock wave. The signals obtained from the interferometer and the pressure transducer are stored to an oscilloscope. The initial pressure ratio p_4/p_1 maintains constant as 9, in driven pressure p_1 = 0.1 MPa. The driver and driven gases are helium and air, respectively.

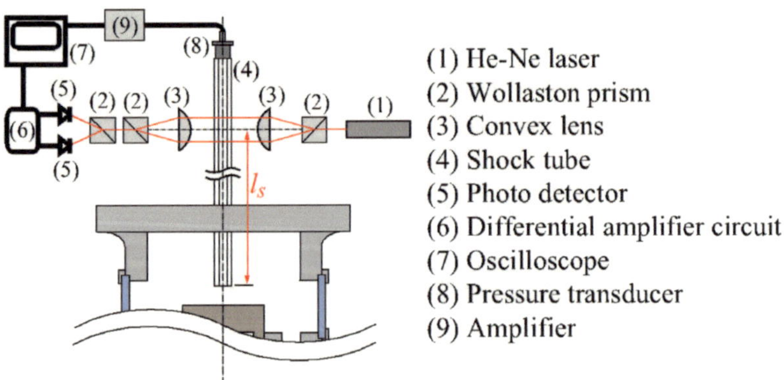

Fig. 23. The interfering signals from the laser differential interferometer.

4.3 Results and discussion

The interferometric signal of the shock wave, generated by using the diaphragmless driver section as mentioned above, propagating in the small diameter tubes is obtained from the laser differential interferometer.

4.3.1 Shock wave detection by the interferometer

Figure 24 shows the time variation of typical signal trace from the interferometer. The horizontal and vertical axes indicate the time and the voltage. The signal rising and decay in Fig. 24 show that the shock wave passes through the first beam and the second beam of the interferometer. The signal is obtained at l_s=800mm in d=2mm tube. The Mach number of shock wave M_s can be calculated by using the following equation;

$$M_s = \frac{\Delta l_s}{\Delta \tau \cdot a_1} \tag{12}$$

Here, Δls is the distance of the two beams of the interferometer, τ is the time difference between the signal rising and decay as shown in Fig. 24, and a_1 is the sound velocity of the driven gas. The Mach number is obtained as 1.53 in this case by calculating from Eq. 12.

4.3.2 Density ratio across the shock wave ρ_2/ρ_1

The density ratio across the shock wave ρ_2/ρ_1 can be calculated from Eq. 11 by using measured ΔV in Fig. 24. Figure 25 shows the relation between the density ratio across the shock wave and the Mach number. The horizontal and the vertical axes indicate the Mach number Ms and the density ratio across the shock wave ρ_2/ρ_1. The black line shows the theoretical value obtained from the following;

$$\frac{\rho_2}{\rho_1} = \frac{(\gamma_1+1)M_s^2}{(\gamma_1-1)M_s^2+2} \tag{13}$$

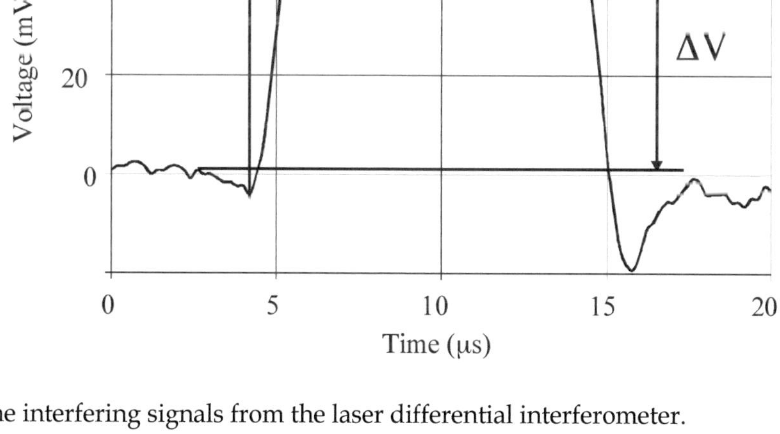

Fig. 24. The interfering signals from the laser differential interferometer.

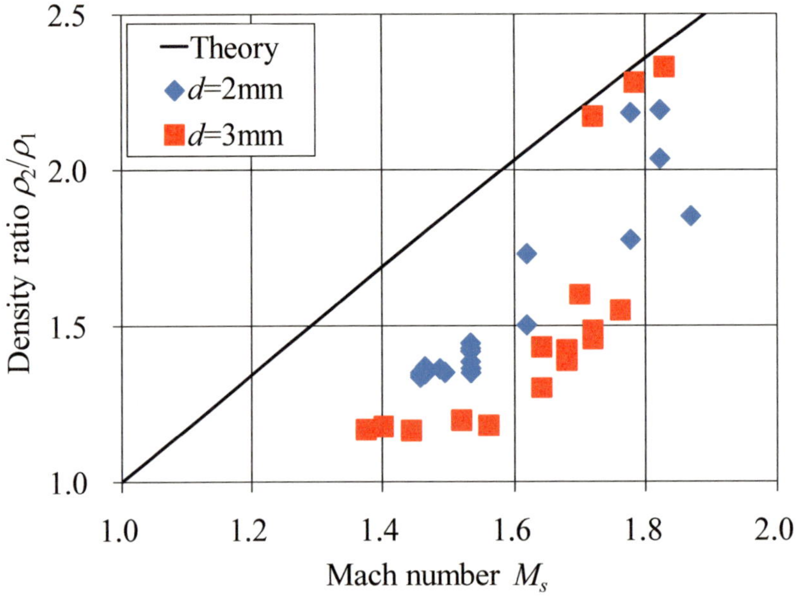

Fig. 25. Relation between ρ_2/ρ_1 and M_s.

The measured density ratio is lower than the theory. It is considered that the Gladston-Dale constant at the normal condition is used for the calculation, nevertheless the temperature behind the shock wave becomes higher. Additionally, the influence of the inner diameter of the shock tube *d* can be seen slightly.

5. Conclusion

This chapter described two topics related to interferometric measurement in shock tube experiment. LICT technique was applied to the measurement of high-speed, unsteady and 3D flow field induced by discharging shock waves, and laser differential interferometer was applied to velocity and density measurement in micro-scale shock tube.

For LICT measurement, 3D flow phenomena behind discharging shock wave from two parallel and circular nozzles were captured in detail. Various phenomena were clarified - shock-shock interaction, shock-vortex interaction, SSW, CS1, CS2, …etc. LICT measurement was also applied to flow field around a circular cylinder to aim at extending the tomographic measurement to more general case. Density distribution was reconstructed from incomplete projection data containing blanked part by ART algorithm. Three-dimensional flow phenomena can be captured from resultant images. However some improvements in reconstruction will be necessary to clarify complex flow phenomena.

For velocity measurement of shock wave in micro-scale shock tube, we measured the velocities of the shock wave and the density ratios across the shock wave propagating in the small diameter tubes, generated by the diaphragmless driver section, by using laser differential interferometer. From our results the following points can be concluded. Firstly, the relation between the Mach number and the density ratio across the shock wave is obtained. The measured density ratio is lower than the theory. It is considered that the Gladstone-Dale constant at the normal condition is used for the calculation, nevertheless the temperature behind the shock wave becomes higher. Additionally, the influence of the inner diameter of the shock tube d can be seen slightly. Secondly the detailed experiments are required to estimate the influence of the shock wave attenuation by using longer shock tube.

6. Acknowledgment

Authors greatly appreciate Hiroki Honma of Professor Emeritus of Chiba University and Professor Walter Garen of Hochschule Emden/Leer, University of Applied Sciences for their valuable advice and discussion.

7. References

Brouillette, M. (2003). Shock waves at microscales, *Shock waves,* Vol.13, No.1, (April 2003), pp. 3-12, DOI 10.1007/s00193-003-0191-4.

Honma, H., Ishihara, M., Yoshimura, T., Maeno, K. and Morioka, T. (2003a), Interferometric CT measurement of three-dimensional flow phenomena on shock waves and vortices discharged from open ends, *Shock Waves* (2003a), 13, pp.179-190.

Honma, H., Maeno, K., Morioka, T. and Kaneta, T. (2003b), Some topics on three-dimensional features of shock waves and vortices discharged from open ends, *Proc. of the 5th International Workshop on Shock/Vortex Interaction* Hosted by National Cheng Kung University, The Kaohsiung Grand Hotel Kaohsiung, October 27-31.

Kak A. C., Slaney M. (1988), Principles of Computerized Tomographic Imaging, *IEEE,* New York.

Maeno, K., Kaneta, T., Yoshimura, T., Morioka, T. and Honma, H. (2005), Pseudo-schlieren CT measurement of three-dimensional flow phenomena on shock waves and vortices discharged from open ends, *Shock Waves,* Vol.14, No.4, pp.239-249.

Ota, M., Koga, T. and Maeno, K. (2005), Interferometric CT Measurement and Novel Expression Method of Discharged Flow Field with Unsteady Shock Waves, *Japanese Journal of Applied Physics,* Vol. 44, No. 42 pp. L1293-1294.

Ota M., Koga, T. and Maeno, K. (2006), Laser interferometric CT measurement of the unsteady super sonic shock-vortex flow field dischargeing from two parallel and cylindrical nozzles, *Measurement Science and Technology,* Vol. 17, pp. 2066-2071.

Shepp, A. L. and Logan, F. B. (1974), The Fourier Reconstruction of a Head Section, *IEEE Trans. Nucl, Sci.,* Ns-21, pp.21-43.

Smeets, G. (1972). Laser interferometer for high sensitivity measurements on transient phase objects, *IEEE transactions on Aerospace and Electronic Systems,* Vol.AES-8, No.2, (March 1972), pp. 186-190, DOI 10.1109/TAES.1972.309488.

Smeets, G. (1977). Flow diagnostics by laser interferometry, *IEEE transactions on Aerospace and Electronic Systems,* Vol.AES-13, No.2, (March 1977), pp. 82-89, DOI 10.1109/TAES.1977.308441.

Part 2

Recent Interferometry Applications

12

Interferometry for Fusion Devices

Luis Esteban and Miguel Sánchez
CIEMAT-EURATOM Association
Spain

1. Introduction

Interferometers are optical arrangements capable of measuring changes in the optical paths by measuring the phase differences. They allow to perform a wide range of measurements, including the measurement of the line-electron density of the fusion plasmas Innocente & Tudisco (2006); Innocente et al. (1997); Joffrin & Riva (2003); Kornejew et al. (2006); Sanchez et al. (2008); Zeeland et al. (2006):

$$\int_0^{L_p} n_e \mathrm{d}l, \tag{1}$$

where n_e is the electron density and L_p is the integration path in the plasma.

The interferometric measurements are obtained through the observation of the interference between two or more beams, generally obtained by splitting an incident beam.

The solution to the electromagnetic free space wave equation can be expressed as:

$$\mathbf{E}(\mathbf{r},t) = \mathbf{E}(\mathbf{r})\mathrm{e}^{\mathrm{i}(\mathbf{k}\cdot\mathbf{r}-!\mathrm{t}+`)}, \tag{2}$$

$\mathbf{E}(\mathbf{r})$ is the amplitude vector, ω is the angular frequency, and $\mathbf{k}$ is the wave number vector:

$$k = \frac{2\pi}{\lambda} \cdot \mathbf{e}_k, \tag{3}$$

where λ is the wavelength in the medium, and $\mathbf{e}_k$ is a unitary vector in the direction of the propagation.

In general, interferometers use at least two beams, probing and reference, following different optical paths, L_1 and L_2, which finally interfere in an square law detector[1]. The optical paths are given by:

$$L = \int_{path(t)} N \mathrm{d}l, \tag{4}$$

where the integration is taken along a geometric path with a fixed origin (beam splitter) and end (detector), N is the refractive index.

Considering that these waves have angular frequencies ω_1 and ω_2, and that they are plane, monochromatic, and coherent (from the same source), in this case the following approximation to the electric fields is considered:

$$u_1 = u_{01}\mathrm{e}^{\mathrm{i}(\mathrm{k}_{01}\cdot\mathrm{L}_1(\mathrm{t})-!_1\mathrm{t}+`_{01})}, \tag{5}$$

[1] An square law detector measures intensities

$$u_2 = u_{02} e^{\mathrm{i}(\mathrm{k}_{02} \cdot \mathrm{L}_2(\mathrm{t}) - !_2 \mathrm{t} + `_{02})}. \quad (6)$$

At the detector, the interference signal has the form:

$$I = (u_{01} + u_{02})^2 = I_1 + I_2 + 2\sqrt{I_1 I_2} \cos\left((\omega_1 - \omega_2)t + k_0 \cdot (L_1(t) - L_2(t))\right), \quad (7)$$

where I is the intensity modulated at frequency $\Omega = \omega_1 - \omega_2$, and I_1 and I_2 are the respective intensities of the two waves. Since $k_{01} \simeq k_{02}$, the average between the two values $k_0 = (k_{01} + k_{02})/2$, has been considered in Eq. 7. Using k_0, the effective wavelength can be expressed as:

$$\lambda_{eff} = \frac{2\lambda_1 \lambda_2}{\lambda_1 + \lambda_2}. \quad (8)$$

The alternate current from the detector is:

$$i = i_0 \cos\left(\Omega t + \phi(t)\right), \quad (9)$$

the phase $\phi(t)$ corresponds to the optical path length differences.

Usually, both angular frequencies are equal (i.e., $\Omega = 0$), in this case, the interferometer is called homodyne. If $\Delta\omega \neq 0$ (the measuring and reference beams operate at slightly different frequencies), the interferometer is called heterodyne. The frequencies are typically shifted using Acousto-Optic-Modulators (AOM) for the infrared Ohtsuka & Sasaki (1974). However, other techniques can be adopted for this purpose, in agreement with the used wavelengths, for instance rotating gratings in the case of far-infrared wavelengths.

The phase $\phi(t)$ carries the information related to the difference in the optical path-lengths between the two beams, due to the mechanical movements or changes in the refractive index. The initial phases of the probing and the reference beams, θ_{01} and θ_{02}, are integration constants, which for convenience are considered zero.

With respect to the optical setup, two configurations that are commonly used for plasma refractive index measurements are the Michelson and the Mach-Zehnder schemes. However, in practice, the setups used in fusion are a mixture of these two optical configurations. Both are two-beam interferometers; in the case of the Michelson configuration, the incoming and outgoing beams, travel in the same direction and in opposite senses, while in the Mach-Zehnder one the propagation follows only one sense. The difference between both is shown in Figure 1.

This Chapter is structured as follows. In Section 2 the relation between the line-integrated electron density and the phase shift suffered by the waves that cross the plasma is deduced for a cold plasma model. In Subsection 2.2 extends the cold plasma model for weakly relativistic plasmas, such as those expected in the ITER tokamak. In Section 3 the possible wavelengths to be used in these type of diagnostics are discussed. Section 4 outlines the main characteristics of the middle IR interferometers at TJ-II and at W7-X. Finally the basic expressions of Gaussian beam propagation are developed in Section 6. In addition, several phase measuring algorithms are listed, and the interpolation phase measuring algorithm is described in detail. The main conclusions of the chapter are summarized in Section 8.

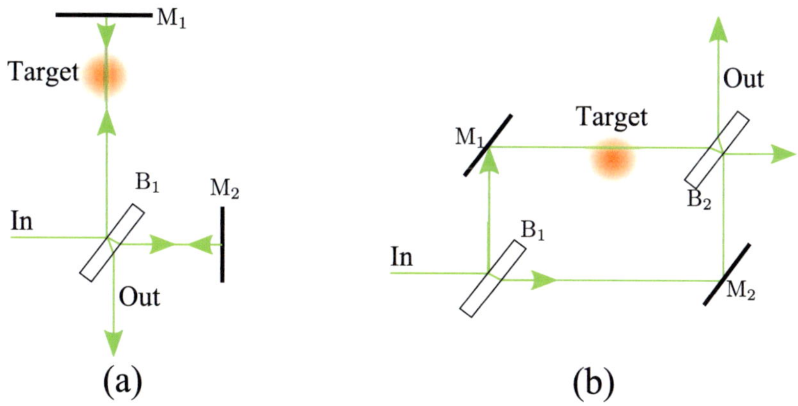

Fig. 1. a) A Michelson interferometer; the light is divided into two beams in the beam-splitter B_1; Both beams are reflected back at mirrors M_1 and M_2 and recombined in the output. b) A Mach-Zehnder interferometer. The incoming beam is divided in beam splitter B_1 into reference and measuring beams, which are reflected at mirrors M_1 and M_2, and finally recombined at beam splitter B_2.

2. Propagation of electromagnetic waves in plasmas

The expression to compute the line integrated electron density can be deduced from the Appleton-Hartree relation for a cold plasma. This expression would be valid for the TJ-II and the W7-X plasmas. However, for plasmas with an electron temperature of 20 eV, such as those expected at the ITER tokamak, a relativistic correction must be included in the line integrated electron density. In this Section the main expressions, for cold plasmas, Subsection 2.1 and for weakly relativistic plasmas, Subsection 2.2, are summarized. A detailed deduction can be found in Hartfuss et al. (1997); Hojo & Kawahata (2009); Hutchinson (1990).

2.1 Propagation of electromagnetic waves in cold plasmas

For the measurement of the electron density, the relation between this quantity and the changes in the refractive index of the plasma must be found.

Adopting the cold plasma approximation, the refractive index of the plasma is given by the Appleton-Hartree relation, Hartfuss et al. (1997); Hutchinson (1990):

$$N^2 = 1 - \frac{X(1-X)}{1 - X - \frac{1}{2}Y^2\sin^2\theta \pm \sqrt{\left[\left(\frac{1}{2}Y^2\sin^2\theta\right)^2 + (1-X)^2Y^2\cos^2\theta\right]}}. \quad (10)$$

If the magnetic field is negligible, $Y \to 0$, the refractive index becomes:

$$N^2 = 1 - X, \quad (11)$$

where:

$$X = \omega_p^2/\omega^2, \quad (12)$$

being ω_p the plasma frequency, and ω and the frequency of the radiation. The refractive index becomes:

$$N^2 = 1 - \frac{\omega_p^2}{\omega^2} = 1 - \frac{n_e}{n_c}, \quad (13)$$

n_e is the electron density and $n_{co} = \omega^2 \mathrm{m_e} \epsilon_0 / e^2$ is the cutoff plasma density[2].

The phase shift suffered by an electromagnetic wave is given by Hutchinson (1990):

$$\phi = \int_{plasma} k(z)\mathrm{d}z. \tag{14}$$

In the particular case of interferometry, the measuring beam will travel through the plasma and also through vacuum. The vacuum contribution to the k vector can be removed comparing the phases of both measuring and reference beams Hartfuss et al. (1997); Hutchinson (1990):

$$\phi = \int_{plasma} (N(z) - 1)\frac{\omega}{c}\mathrm{d}z. \tag{15}$$

ω is the radiation frequency and c is the speed of light.

Substituting Eq. 13 in Eq. 15, the phase shift becomes:

$$\phi = \frac{\omega}{c}\int_{plasma}\left\{\sqrt{\left(1 - \frac{n_e(z)}{n_c}\right)} - 1\right\}\mathrm{d}z. \tag{16}$$

If $n_e \ll n_{co}$, the refractive index, N can be approximated by the first two terms of its Taylor expansion:

$$N \simeq 1 - \frac{1}{2}\frac{n_e}{n_c}, \tag{17}$$

the phase difference, ϕ becomes:

$$\phi = \mathrm{r_e}\lambda \int_{\mathrm{plasma}} n_e(z)\mathrm{d}z. \tag{18}$$

This integral receives the name o line integrated electron density, LID and for simplicity is denoted as $\int n_e(z)\mathrm{d}z$ Hartfuss et al. (1997). The LID measurement is computed after measuring the phase differences of the output signals, $\Delta\phi$ these phase differences are calculated around the carrying frequency Ω. The path across the plasma receives the name of chord length.

2.2 Corrections for weakly relativistic plasmas

In ITER, it is expected an electron temperature in the order of 20 keV, which implies that the cold plasma approximation is no longer valid. A correction factor β, must be included for the computation of the refractive index, Hojo & Kawahata (2009). Under these conditions, considering the dispersion relation for the ordinary mode, the refractive index would now become:

$$N = \sqrt{1 - \frac{1}{\beta}\frac{\omega_p^2}{\omega^2}}, \tag{19}$$

[2] The cutoff density establishes the minimum frequency of the waves that will propagate through the plasma. Below this value the propagation modes are evanescent

where a relativistic correction β, has been introduced:

$$\beta = \frac{3K_2(\rho)}{\rho^2 \int_0^\infty (p^4/\gamma^2) e^{-\rho\gamma} dp}, \tag{20}$$

$p = |\mathbf{p}|/(m_e c)$ is the normalized momentum, $\gamma = \sqrt{(1+p^2)}$, $\rho = m_e c^2/T_e$ and $K_2(\rho)$ is the modified Bessel function. Considering a weakly relativistic plasma ($\gamma < c$), $e^{-\rho\gamma}$ can be approximated by its Taylor expansion as:

$$e^{-\rho\gamma} \simeq e^{-\rho} e^{-\rho \frac{p^2}{2}} \left(1 + \rho \frac{p^4}{8}\right), \tag{21}$$

and the correction factor β can be simplified as Hojo & Kawahata (2009):

$$\beta \simeq 1 + \frac{5}{2\rho}, \tag{22}$$

if $n \ll n_{co}$ the phase difference can be finally written as:

$$\Delta\phi = \frac{\omega}{2cn} \frac{2\rho}{5+2\rho} \int_{z_1}^{z_2} n(z) dz. \tag{23}$$

3. Wavelength choice

In practice, these phase differences will not only occur due to plasma density variations, but also due to the mechanical vibrations and thermal drifts.

$$\phi = \phi_p + \frac{2\pi\Delta l}{\lambda} \tag{24}$$

Mechanical vibrations, in one-color arrangements are one of the limiting factors for the wavelength choice. As it can be seen in the former equation, these effects become particularly important for low wavelengths. However, they can be suppressed using two-color setups. The upper limit is imposed by the cutoff density, n_{CO}, since when the probing wavelength is long enough it will not propagate through the plasma (evanescent modes). The probing wavelength must satisfy the next relation to ensure that it will propagate through the plasma:

$$\lambda < \frac{2\pi c}{e} \sqrt{\frac{m_e \epsilon_0}{n_{peak}}}, \tag{25}$$

where n_{peak} is the maximum electron plasma density and e and m_e are the electron charge and the electron mass respectively.

In interferometers operating in the middle infrared range, mechanical vibrations are higher than variations caused by plasma density fluctuations. This circumstance forces to use a second interferometer sharing the same optical path as the first one, so this contribution can be canceled in the final compensated signal[3]. These type of arrangements receive the name

[3] The compensated signal is obtained after subtracting the optical paths of both interferometers in order to cancel mechanical vibrations

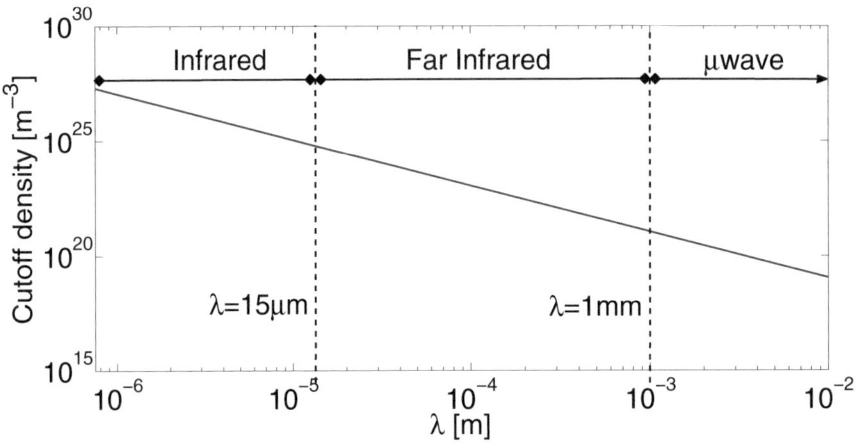

Fig. 2. Cutoff density versus wavelength from the middle infrared range to the μwave range. With shorter wavelengths higher densities can be measured. However, the vibration problem becomes more significant and forces to use two color arrangement.

of two-color interferometers. From Eq. 18 the optical path length difference from one of the interferometers is calculated as:

$$\lambda_1 \Delta\phi_1 = 2\pi \mathrm{r_e} \lambda_1^2 \int_{\mathrm{plasma}} n_e(l)\mathrm{d}l + \lambda_1 k_{01} \int_{\mathrm{out}} N\mathrm{d}l, \tag{26}$$

The second term of the equation takes into account the optical path length outside the plasma. The final line integrated density is computed subtracting the optical path-length differences between both interferometers if the second term is canceled[4], Hartfuss et al. (1997); Hutchinson (1990):

$$\int n_e \mathrm{d}l = \frac{\Delta\phi_1\lambda_1 - \Delta\phi_2\lambda_2}{2\pi p \cdot \mathrm{r}_e(\lambda_1^2 - \lambda_2^2)}, \tag{27}$$

p is an integer that corresponds to the number of times that the beams cross the plasma, r_e is the classical electron radius, λ_1 and λ_2 are the long and short wavelengths and $\Delta\phi_1$ and $\Delta\phi_2$ are the phase differences between the probing and reference signals of the two interferometers. Calculating the phase differences and hence the optical path length differences $\lambda_1\Delta\phi_1$ and $\lambda_2\Delta\phi_2$, the line integrated density is obtained from Eq. 27. An appropriate phase measuring algorithm is needed for this aim. Estimating the plasma chord-length (path-length for the line-integral), $\mathrm{L_{chord}}$, the average density is evaluated as $< n_e >= \int n_e\mathrm{d}l / L_{\mathrm{chord}}$. The plasma chord-length depends on the plasma configuration, and is estimated using the spatial profile measured by the Thomson Scattering system, Herranz et al. (2008; 2000).

Another limiting factor, is that in practice plasmas are not uniform. A density gradient normally exists in the radial direction of the plasma, this will cause a deflection of the output beam with respect to the incident one, Figure 3. If the deflected angle, θ is large enough a miss alignment may occur and by the interference signal can be lost at the detector.

In general, the deflection angle is given by Hutchinson (1990):

$$\theta = \frac{\lambda}{2\pi}\frac{\mathrm{d}\phi}{\mathrm{d}y} = \frac{\mathrm{d}}{\mathrm{d}y}\int N\mathrm{d}l. \tag{28}$$

[4] The non-cancelation of the second term results in systematic error in the final electron density

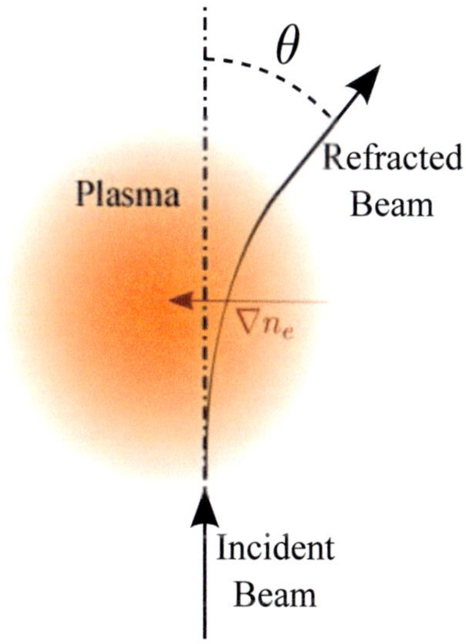

Fig. 3. Refraction suffered by a beam of light when crossing a non-uniform plasma.

Substituting Eq. 17 into the former equation, the deflection angle becomes:

$$\theta = \frac{\lambda^2 r_e}{2\pi} \int \nabla n_e \mathrm{d}l. \tag{29}$$

The deflection has a quadratic dependence with the wavelength, so for low wavelengths this effect is to be lower and for the range of interest in this chapter, $1.064 - 10.591$ μm, it is negligible. For example, if a parabolic density profile is considered, the maximum deflexion angle would be:

$$\theta = \frac{n_{peak}\lambda^2 e^2}{4\pi^2 c^2 \epsilon_0}, \tag{30}$$

where n_{peak} is the maximum electron density reached. Considering a peak density of 10^{20} m^{-3} the deflexion suffered by a 10.591 μm beam would be of 10^{-5} rad, which is several orders of magnitude below the typical alignment errors. However, for larger wavelengths, for instance in the far IR region this effect can be significant and should be taken into account. In Table ?? a summary of different wavelengths used in interferometers operating in TJ-II, W7-X, DIII-D, JET, LHD, Tore Supra and ITER fusion machines is shown Braithwaite et al. (1989); Brower et al. (2006); Gil et al. (2001); Kawahata et al. (1997); Kornejew et al. (2006); Zeeland et al. (2006).

4. Overview of the TJ-II and the W7-X interferometers

TJ II is a medium scale flexible Heliac type stellarator Alejaldre (2005) installed in the Laboratorio Nacional de Fusión of CIEMAT, Spain for the study of helical axis plasmas in a wide range of parameters Alejaldre (1989). TJ-II device has a major radius of 1.5 m, a minor of 0.28 m and the magnetic field generated in the center of the plasma is of 1 T. The plasma is heated using an ECRH[6] system based on two gyrotrons working at 53.2 GHz and delivering

[6] Electron Cyclotron Resonance Heating

Fusion Device	λ_1	λ_2	Peak density
TJ-II	$CO_2 - 10.6\ \mu m$	NdYAG $- 1.064\ \mu m$	$\sim 8 \cdot 10^{19}\ m^{-3}$
W7-X[5]	$CO_2 - 10.6\ \mu m$	CO $- 5.295\ \mu m$	$> 10^{20}\ m^{-3}$
DIII-D	$CO_2 - 10.6\ \mu m$	HeNe $- 0.633\ \mu m$	$> 10^{20}\ m^{-3}$
JET	DCN $- 195\ \mu m$	$CH_3OH - 118.8\ \mu m$	$> 10^{20}\ m^{-3}$
LHD	$CH_3OH - 118.8\ \mu m$	-	$> 10^{20}\ m^{-3}$
TORE SUPRA	DCN $- 195\ \mu m$	$CH_3OH - 118.8\ \mu m$	$> 10^{20}\ m^{-3}$
ITER	$CO_2 - 10.6\ \mu m$	CO $- 5.295\ \mu m$	$> 10^{22}\ m^{-3}$

Table 1. Optical configurations for interferometers operating in different fusion devices

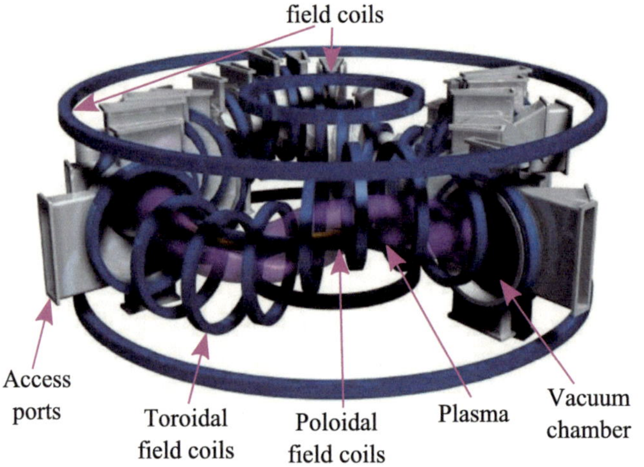

Fig. 4. Aerial and lateral view of the TJ-II device it has four periods, 32 toroidal coils, 2 poloidal coils and 2 vertical coils.

each up to 200 kW of power to the plasma, and NBI[7]. The maximum electronic densities reached are in the order of $8 \cdot 10^{19} m^{-3}$. To measure in this density range an heterodyne interferometer operating in the middle IR range is installed.

In IPP Greifswald the W7-X stellarator is currently in an advanced stage of construction. W7-X will be the biggest stellarator in the world. The major radius of W7-X will be of 5.5 m and the minor of 0.53 m with a maximum magnetic field in the axis of 3 T. W7-X will have ECRH heating systems in continuous mode (30 MW of power will be reached using this procedure), NBI and ICRH[8] (pulsed). W7-X device is an evolution of the W7-AS, being the first fusion device of the stellarator type with fully optimized, modular super-conducting coils aiming on high performance plasma under steady state conditions.

A key diagnostic that is being developed is a one channel two color heterodyne IR interferometer dedicated for density control and to control the Thomson scattering system Kornejew et al. (2006). It is planned to build it using CO_2 (10.591 μm) for the long wavelength and CO (5.295 μm) for the short one. The choice of the short wavelength has the purpose of

[7] Neutral Beam Injection
[8] Ion Cyclotron Resonance Heating

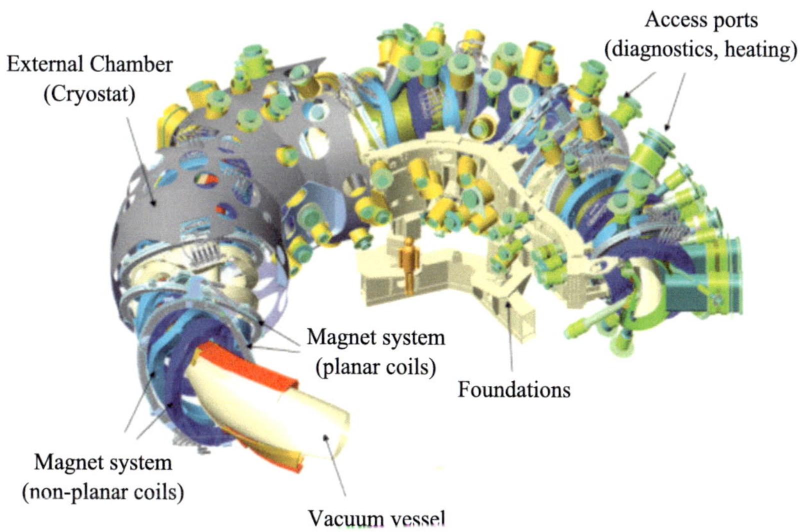

Fig. 5. Detail of the W7-X Stellarator.

minimizing the thermo-optical effect in the vessel windows Sánchez & Sánchez (2005). The configuration CO_2-CO permits the use of one single detector for both wavelengths, however a correct frequency diplexion strategy must be fulfilled. There only exists one interferometer fusion diagnostic with a similar detection scheme Innocente & Tudisco (2006).

Both the IR interferometer at TJ-II and the test prototype currently installed at IPP-Greifswald use an hybrid Michelson Mach-Zhender configuration, with two colors to compensate mechanical vibrations and heterodyne modulation. For the shake of simplicity in Figure 6 a two color Mach-Zhender arrangement operating as a fusion diagnostic is illustrated. One arm of the interferometer crosses the plasma and the other arm that travels the same optical path is used as a reference. In the probing arm (the one that crosses the plasma) a frequency displacement ω_1 is introduced using Acousto-optical modulation. When the probing beam reaches an square law detector it interferes with the reference beam and an output signal is generated. The frequency of this signal is that of the displacement introduced, ω_1. The second interferometer is used to cancel mechanical vibrations and shares the same optical path as the first one. The final line-integrated density is obtained by subtracting the measures of both devices.

In the TJ-II infrared interferometer always important signal processing tasks have been carried out by traditional analog systems. The speed and accuracy needed for this application suggests to replace these type of systems by digital electronics. Taking into account the frequencies handled the only possible solution is the use of FPGAs. All the signal processing has been implemented in FPGA type devices.

5. Detection schemes at TJ-II and W7-X

In the TJ-II Stellarator the configuration CO_2 [10.591 μm] – Nd : YAG [1.064 μm] forces to use two detectors, Figure 7. The two interference signals are sampled using two ADCs and the phases are compared with the phases of two reference signals extracted directly from the

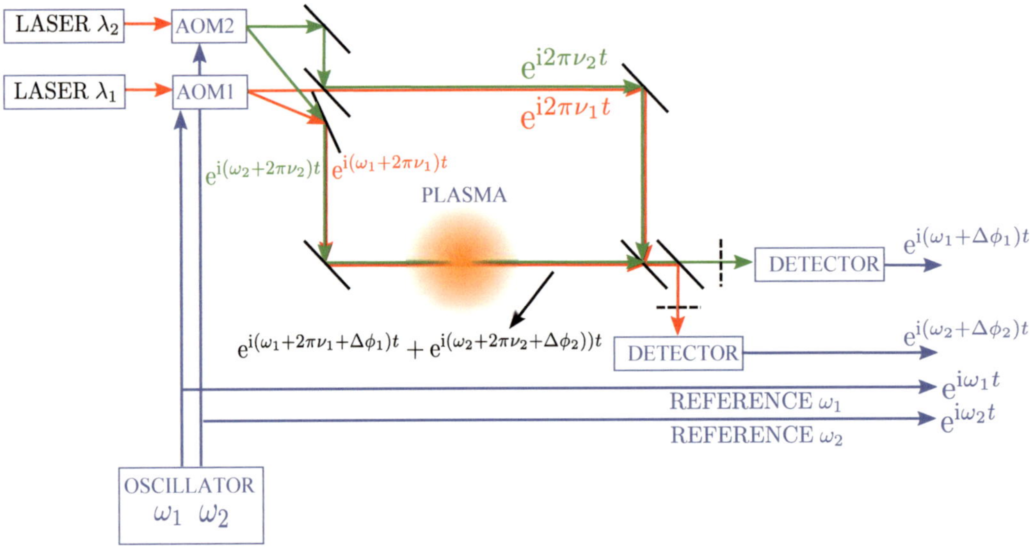

Fig. 6. Two color infrared heterodyne Mach-Zehnder interferometer employing a classical two detector scheme.

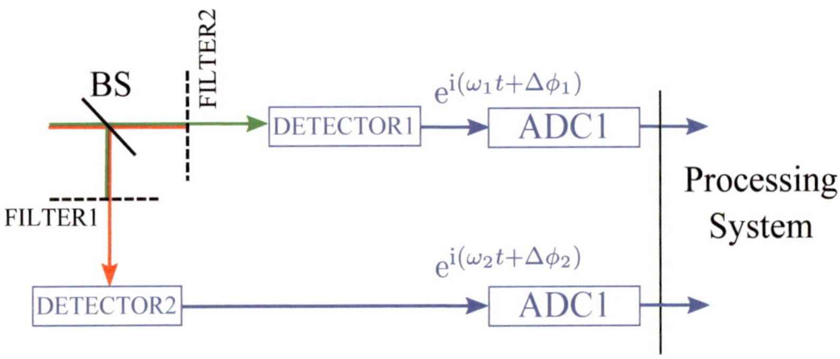

Fig. 7. Detection scheme based in two detectors, a beam splitter and two optical filters are needed to isolate the correspondent signals.

oscillator that drives the AOMs. In principle no electronic input split preprocessing filter is needed, however, in practice these are useful to increase the signal-to-noise ratio (SNR) of the input interference signals.

The combinations, CO_2 [10.591 μm] – CO [5.295 μm] and CO_2 [12.1 μm] – CO_2 [9 μm] are the ones planned to be used in the W7-X Stellarator and in the ITER Tokamak, respectively. In both cases the wavelengths can be detected by a single detector, Figure 8. Provided that the signals can be split using electronic filters the number of optical components can be significantly reduced. Taking into account the port restriction access in W7-X this would be a great advantage.

In the particular case of the W7-X interferometer, several detectors from Vigo system show the same response for both CO_2 [10.591 μm] and CO [5.295 μm] wavelengths. In Table **??** the

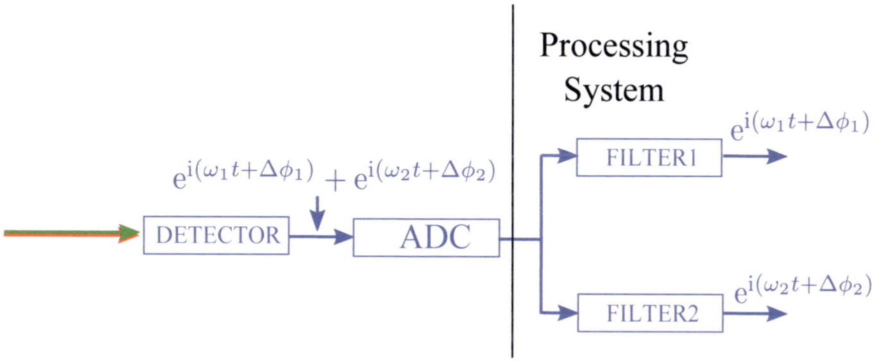

Fig. 8. Detection scheme based in two detectors, a beam splitter and two optical filters are needed to isolate the correspondent signals.

Detector	Detectivity at 5.296 [μm]	Detectivity at 10.591 [μm]
PVM-2TE-10.6	$0.8 \cdot 10^8$	$1.0 \cdot 10^8$
PCI-2TE-10.6	$2.0 \cdot 10^9$	$1.8 \cdot 10^9$
PVI-2TE-5	$5.0 \cdot 10^{11}$	–

Table 2. Detectivity for several Vigo Systems detectors at the wavelengths 5.296 μm and 10.591 μm. PV stands for photo-voltaic and PC for photo-conductive. The final choice has been the PCI-2TE-10.6.

detectivity[9] of these detectors is shown, being the photoconductive solution (PCI-2TE-10.6) the best.

The line-electron density measurement at W7-X will require a precision of 10^{18} m^{-2}, as it is described in Esteban, Sánchez, Sánchez, Kornejew, Hirsch, López, Fernández & Nieto-Taladriz (2010) the electronic filters needed to split both frequency components (CO_2 and CO interference signals), must have an attenuation in the rejection band of 43 dB. This task has been solved using digital filters Esteban, Sánchez, Sánchez, Kornejew, Hirsch, López, Fernández & Nieto-Taladriz (2010) implemented in hardware as a better alternative than the analog diplexer used in the Frascati Tokamak Innocente & Tudisco (2006). The intermediate modulating frequencies are of 25 MHz for the CO and of 40 MHz for the CO_2 and have been chosen in such a way that the useful bandwidth of the interferometer is not limited by the filters Kornejew et al. (2006); Sanchez et al. (2008).

6. Beam propagation

The predominant propagation modes of the lasers used are the TEM_{00} the beams can be considered to be gaussian. The propagation of gaussian beams is well described in the classic literature Kogelnik & Li (1966); Self (1983), in this section the basic relations are summarized.

The beam intensity is given by:

$$I(r,z) = \frac{2}{\pi\omega^2} e^{-2\left(\frac{r}{\omega}\right)^2}. \tag{31}$$

[9] The detectivity is defined as the signal-to-noise ratio (SNR) at the detector output normalized to a 1 W radiant power, a 1 cm^2 detector optical area and a 1 HZ bandwidth

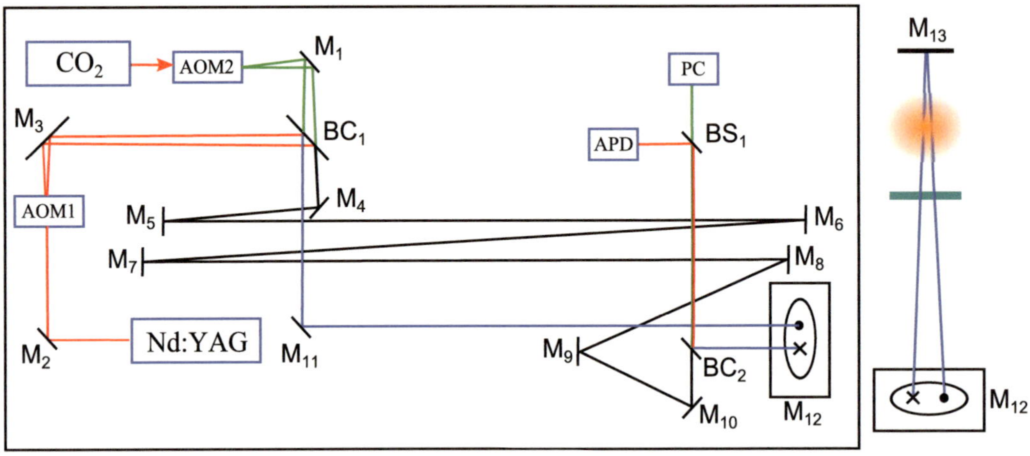

Fig. 9. Schematic layout of the one channel infrared interferometer of TJ-II, The green and red lines correspond to the CO_2 and Nd:YAG beams. The black line after the first beam combiner (BC_1) represents the path for both reference beams while the blue line represents the path for the probing beams. Both interference signals are detected in a photoconductor detector (PC), CO_2 and in an Avalanche Photo detector (APD), Nd:YAG. M_{12} is a periscope that launches both probing beams into the plasma, they cross a ZnSe window, the plasma and are reflected back to the periscope. The beam expander located at the output of the CO_2 sets both beam waists, Nd:YAG and CO_2 in the same position.

being, r the radius and ω the beam radius:

$$\omega(z) = \omega_0 \left[1 + \left(\frac{z}{z_R}\right)^2\right], \tag{32}$$

ω_0 is the beam waist and z_R is the Rayleigh range that separates the near and far fields:

$$z_R = \frac{\pi \omega_0^2}{\lambda}. \tag{33}$$

The have angle beam divergence is given by:

$$\theta = M^2 \frac{\lambda}{\pi \omega_0}, \tag{34}$$

where M^2 is the beam quality factor or the beam propagation factor.

In Figure 9 the schematic layout of the TJ-II infrared interferometer is illustrated and in Figure **??** an unfolded virtual representation has been adopted for simplicity, in which the gaussian, Nd:YAG and CO_2 beams, travel in one direction and equal distances as they do in the interferometer, in this particular case 15 m until the beams reach the detectors.

Table **??** summarizes the main propagation characteristics of both lasers. At this point the Nd : YAG beam has a diameter of 5 cm and the CO_2 of 10 cm, the beam waists of both beams are located in the same positions so the curvature radius:

$$R(z) = z \left[1 + \left(\frac{z_R}{z}\right)^2\right], \tag{35}$$

Laser	Power [W]	θ [mrad]	ω_0 [mm]	M^2
$Nd : YAG - 1.064\ \mu m$	0.3	4	0.2	< 1.2
$CO_2 - 10.591\ \mu m$	8	5.5	1.5	< 1.2

Table 3. Propagation characteristics of the Nd:YAG and CO_2 beams, the curvature radius of both beams must be the same for equal axial positions. Both optical paths must share the same optical axis.

is the same at the detectors.

The power received by the detector is calculated integrating the radiation intensity, Eq. 31 over the detector area. If the detectors are located at a position z, such that the condition, $z \gg z_0$ is met, then the beam radius $\omega(z)$ can be approximated by:

$$\omega \simeq \omega_0 \frac{z}{z_R} = \frac{\lambda z}{\pi \omega_0}, \tag{36}$$

and the intensity, for one beam becomes:

$$I(r,z) = \frac{2\pi\omega_0}{\lambda^2 z^2} e^{-2\left(\frac{r\pi\omega_0}{\lambda z}\right)^2}. \tag{37}$$

Considering that both reference and probing beams reach the detector with intensities, I_1 and I_2 and knowing that the total interference intensity at one detector is:

$$I = I_1 + I_2 + 2\sqrt{I_1 I_2}\cos(2\pi\nu + \Delta\phi), \tag{38}$$

the AC component of the interference, $\sqrt{I_1 I_2}$ has been plotted together with the reference and probing intensities as it can be seen in Figure **??** the intensity of the interference is maximum when both probing and measuring intensities are equal, $I_1 = I_2$. This condition can be met by adjusting properly the AOMs.

As a final remark it should be notices that the beam waists of both colors must be located at the same geometrical positions this issue is specially important in multichannel interferometers. In this sense, three approaches can be considered to obtain the different chords:

1. Expanded beam,
2. Using independent modules,
3. Using beam splitters,

from the experience gain at TJ-II it can be said that the first is likely to be very sensitive to mechanical movements, in any case it should be ensures a high pointing stability of the beams.

7. Phase measurement

The line-electron density computation, indicated in Eq. 27 implies the use of an appropriate phase measuring algorithm. Several solutions have been developed in this sense. Some procedures are based on counting the cycles of a high speed clock in synchronism with the zero crossings of the probing signal (2004). These approaches, typically require a clock frequency higher than the frequency of the probing signals and are likely to suffer fringe jumps and synchronism problems if they are not carefully designed. Another method suitable for measuring phase shifts lower than one period of the input signal consist in decomposing

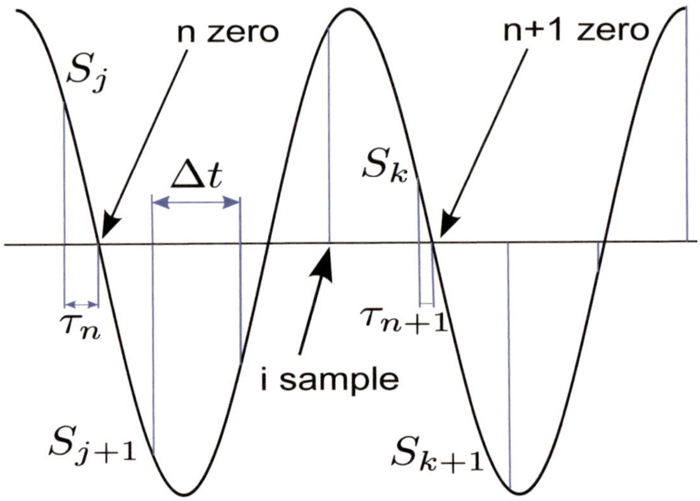

Fig. 10. Ilustration of the parameters involved in the interpolation algorithm. In this example just the negative edge zero crossings are considered.

the signals in its In-Phase & Quadrature components (IQ scheme). As the phase shifts in IR-interferometer operating as fusion diagnostic are of several periods the use of this technique requires of an additional fringe counting mechanism. Pulse width modulation based schemes are also used for phase measurements in interferometry Choi et al. (2005); Mlynek et al. (2010) Another method based on Analog-to-Digital conversion, ADC, zero crossing detection and linear interpolation Zubarev & Khilchenko (2003) was used in the TJ-II interferometer with good results Sanchez et al. (2004) this has been the deciding factor to implement this phase measuring algorithm in the real time measuring FPGA-based system. Below the IQ scheme and the interpolation algorithm are explained in detail.

7.1 Interpolation algorithm

During several TJ-II campaigns an interpolation algorithm has been used to compute the phase differences of the interferometric signals Sanchez et al. (2004). This algorithm counts both the zero crossings of the signals n, and the periods of time between zero crossings, Figure 10. Counting the number of samples in a period and adding it to the fraction of sample correspondent to each zero crossing, the phase at each sampling time can be obtained easily. The sample time fractions are computed interpolating linearly two consecutive samples around a zero cross, each value is given by Sanchez et al. (2004); Zubarev & Khilchenko (2003):

$$\tau_n = \frac{|S_j|}{|S_j| + |S_{j+1}|}\Delta t, \tag{39}$$

j is the number of zero crossings, Δt is the sample period and $|S_j|$ and $|S_{j+1}|$ are the absolute values of leading and trailing samples of a zero crossing. The total time elapsed at the j zero crossing is given by:

$$t_n = j\Delta t + \tau_n. \tag{40}$$

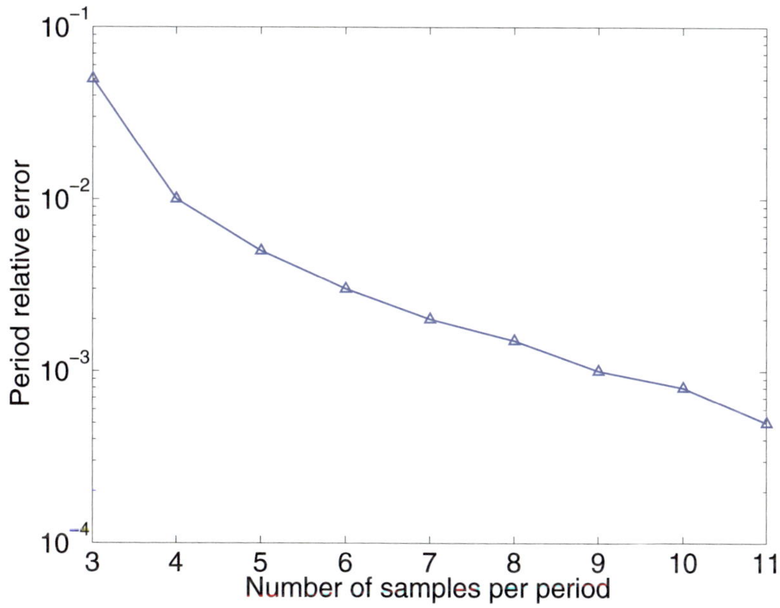

Fig. 11. Interpolation error according to the number of samples per period of the signals, as it can be seen in the figure with less than four samples per period the error increases abruptly.

Using the same procedure, t_{n+1} can be calculated. The calculation of the phase value correspondent at sample time i is:

$$\varphi_i = 2\pi n + \frac{2\pi[(i-j)\Delta t - t_n]}{t_{n+1} - t_n}. \tag{41}$$

Where $t_{n+1} - t_n$ is the current period. in Figure 11 the error introduced by the algorithm for different numbers of samples per period is illustrated

Precisions of 1/400 of CO_2 fringe have been achieved with this algorithm . This precision is equivalent to an error in the line integrated density of $\pm 10^{17} m^{-3}$. The algorithm was implemented in an off-line processing system and has been routinely working since autumn of 2003. Currently the FPGA-based processing is installed in the TJ-II IR-interferometer, allowing the readout of control signals and the processing of the electron density signal in real time Esteban et al. (2011); Esteban, Sánchez, López, Nieto-Taladriz & Sánchez (2010).

8. Conclusions

Several factors affect the propagation of electromagnetic waves in the plasmas and they affect the phase shift suffered by these waves. In most cases, the cold plasma approximation is valid. However, for electron temperatures in the order of tens of keV, relativistic effects become important, and a correction factor must be included in the equation that relates the phase shifts with the line integrated electron density.

The cutoff plasma frequency, mechanical vibrations and refraction are limiting parameters for the probing wavelengths. For high plasma densities, these limiting factors force to use short wavelengths, that are specially sensitive to mechanical vibrations. These mechanical vibrations are canceled using another interferometer with a shorter wavelength, and subtracting both results.

The optical setup of both, the two-color one-channel interferometer at the TJ-II and the two color interferometer planned to be installed at the W7-X, have been described. They both operate in the middle infrared range to measure high density plasmas, and use heterodyne frequency displacement to cope with the phase sensitivity problem. In the particular case of the W7-X interferometer, the possibility of using a single detector is opened, which represents a challenge in the design of a frequency diplexor. The basic principles of gaussian beam propagation have been summarized, and illustrated for the TJ-II one-channel heterodyne interferometer. Also, an estimation of the signal-to-noise ratio at the output of both detectors has been calculated for the TJ-II interferometer. In addition several phase measuring algorithms for these type of interferometers have been listed, and the linear interpolation based algorithm used at the the TJ-II interferometer and tested at the W7-X one has been described.

9. References

(2004). *Measurement Science and Technology* 15: 2341–2348.

Alejaldre, C. (1989). Tj-ii project: A flexible heliac stellarator, *Fusion Technology* 17.

Alejaldre, C. (2005). Tj-ii the flexible heliac, *Europhysics News* pp. 232–233.

Braithwaite, G., Gottardi, N., Magyar, G., O'Rourke, J., Ryan, J. & Véron, D. (1989). Jet polari-interferometer, *Review of Scientific Instruments* 60(9): 2825–2834.
URL: *http://link.aip.org/link/?RSI/60/2825/1*

Brower, D. L., Deng, B. H. & Ding, W. X. (2006). Divertor interferometer diagnostic for iter, *Review of Scientific Instruments* 77(10): 10E911.
URL: *http://link.aip.org/link/?RSI/77/10E911/1*

Choi, H., Park, K. & La, J. (2005). Novel phase measurement technique of the heterodyne laser interferometer, *Review of Scientific Instruments* 76(9): 093105.
URL: *http://link.aip.org/link/?RSI/76/093105/1*

Esteban, L., Sánchez, M., López, J. A., Nieto-Taladriz, O., Kornejew, P. & Hirsch, M. (2011). Development of efficient fpga-based multi-channel phase meters for ir-interferometers, *IEEE Transactions on Nuclear Science* 58: 1562 – 1569.

Esteban, L., Sánchez, M., López, J. A., Nieto-Taladriz, O. & Sánchez, J. (2010). Continuous plasma density measurement in tj-ii infrared interferometer-advanced signal processing based on fpgas, *Fusion Engineering and Design* .

Esteban, L., Sánchez, M., Sánchez, J., Kornejew, P., Hirsch, M., López, J. A., Fernández, A. & Nieto-Taladriz, O. (2010). Continous phase measurement in the w7-x ir-interferometer by means of an fpga and high speed adcs, *Fusion Science and Technology* .

Gil, C., Elbeze, D. & Portafaix, C. (2001). Development of the ir interfero-polarimeter for long pulse operation at tore supra, *Fusion Engineering and Design* 56-57: 969 – 973.
URL: *http://www.sciencedirect.com/science/article/B6V3C-44JDTMG-5G/2/6ec2412225de281427fb48803f2d8b58*

Hartfuss, H. J., Geist, T. & Hirsch, M. (1997). Heterodyne methods in millimetre wave plasma diagnostics with applications to ece, interferometry and reflectometry, *Plasma Physics*

and Controlled Fusion 39(11): 1693.
URL: *http://stacks.iop.org/0741-3335/39/i=11/a=001*

Herranz, J., Castejón, F. & Pastor, I. (2008). Herranz, castejón, and pastor reply:, *Phys. Rev. Lett.* 101(13): 139504.

Herranz, J., Pastor, I., Castejón, F., de la Luna, E., García-Cortés, I., Barth, C. J., Ascasíbar, E., Sánchez, J. & Tribaldos, V. (2000). Profile structures of tj-ii stellarator plasmas, *Phys. Rev. Lett.* 85(22): 4715–4718.

Hojo, H. & Kawahata, K. (2009). Interferometry for weakly relativistic plasmas, *Plasma and Fusion Research* 4: 010.

Hutchinson, I. H. (1990). *Principles of plasma diagnostics*, Cambridge University Press.

Innocente, A. C. P. & Tudisco, O. (2006). Two-color medium-infrared scanning interferometer for the frascati tokamak upgrade fusion test device, *Applied Optics* 45(36): 9105–9114.

Innocente, P., Martini, S., Canton, A. & Tasinato, L. (1997). Upgrade of the rfx co[sub 2] interferometer using in-vessel optics for extended edge resolution, *Review of Scientific Instruments* 68(1): 694–697.
URL: *http://link.aip.org/link/?RSI/68/694/1*

Joffrin, P. I. D. M. E. & Riva, M. (2003). Real-time fringe correction algorithm for the jet interferometer, *Rev. Sci. Instrum.* 74: 3645–3653.

Kawahata, K., Ejiri, A., Tanaka, K., Ito, Y. & Okajima, S. (1997). Design and construction of a far infrared laser interferometer for the lhd, *Fusion Engineering and Design* 34-35: 393 – 397. Fusion Plasma Diagnostics.
URL: *http://www.sciencedirect.com/science/article/B6V3C-3SN5TJ5-2H/2/2f9f024cce400 6745e3f824b0609ad87*

Kogelnik, H. & Li, T. (1966). Laser beams and resonators, *Appl. Opt.* 5(10): 1550–1567.
URL: *http://ao.osa.org/abstract.cfm?URI=ao-5-10-1550*

Kornejew, P., Hirsch, M., Bindemann, T., Dreier, A. D. H. & Hartfuss, H. J. (2006). Design of multichannel laser interferometry for w7-x, *Rev. Sci. Instrum.* 77: 10F120.

Mlynek, A., Schramm, G., Eixenberger, H., Sips, G., McCormick, K., Zilker, M., Behler, K., Eheberg, J. & Team, A. U. (2010). Design of a digital multiradian phase detector and its application in fusion plasma interferometry, *Review of Scientific Instruments* 81(3): 033507.
URL: *http://link.aip.org/link/?RSI/81/033507/1*

Ohtsuka, Y. & Sasaki, I. (1974). Laser heterodyne measurement of small arbitrary displacements, *Optics Comunications* 10(4): 362–365.

Sanchez, M., Esteban, L., Kornejew, P. & Hirsch, M. (2008). Admissible crosstalk limits in a two colour interferometer for plasma density diagnostics. a reduction algorithm., *AIP Conference Proceedings* 993(1): 187–190.
URL: *http://link.aip.org/link/?APC/993/187/1*

Sanchez, M., Sanchez, J., Estrada, T., Sanchez, E., Acedo, P. & Lamela, H. (2004). High resolution co[sub 2] interferometry on the tj-ii stellarator by using an adc-based phase meter, *Review of Scientific Instruments* 75(10): 3414–3416.
URL: *http://link.aip.org/link/?RSI/75/3414/1*

Self, S. A. (1983). Focusing of spherical gaussian beams, *Applied Optics* 22(5): 658–661.

Sánchez, M. & Sánchez, J. (2005). Thermo-optical effect in zinc selenide windows for two-color interferometer for fusion plasma diagnostics, *Rev. Sci. Instrum.* 76: 046104–046104.

Zeeland, M. A. V., Boivin, R. L., Carlstrom, T. N., Deterly, T. & Finkenthal, D. K. (2006). Fiber optic two-color vibration compensated interferometer for plasma density

measurements, *Review of Scientific Instruments* 77(10): 10F325.
URL: *http://link.aip.org/link/?RSI/77/10F325/1*

Zubarev, P. V. & Khilchenko, A. D. (2003). Precision phase discriminator for the heterodyne interferometric plasma-density measurement technique, *Instruments and Experimental Techniques* 46: 171–176. 10.1023/A:1023661531511.
URL: *http://dx.doi.org/10.1023/A:1023661531511*

13

Simultaneous Phase Shifting Shearing Interferometry for Measurement of Static and Dynamic Phase Objects

Noel-Ivan Toto-Arellano[1], David-Ignacio Serrano-García[1], Amalia Martínez-García[1] and Gustavo Rodríguez-Zurita[2]
[1]*Centro de Investigaciones en Óptica A.C. León, Gto.,*
[2]*Benemérita Universidad Autónoma de Puebla,*
México

1. Introduction

Through the use of non-destructive optical techniques to measure the surface with high accuracy, two shearing interferometers were implemented to analyze the slope of phase objects using simultaneous phase-shifting shearing interferometry [Toto et al., 2010]. These optical configurations allow obtaining n-shearograms simultaneously to retrieve the optical phase data map with higher accuracy [Rodriguez et al., 2009; Toto et al 2009].

Several optical systems have been developed to retrieve the optical phase data in a single capture employing polarization, for example, through the use of micro-polarizer arrays [Novak et al.,2005], a point diffraction interferometer[Neal, 2003], a two-beam grating interferometer [Rodriguez, 2008] and a liquid-crystal spatial modulator [Mercer and Creath, 1996], among others. These systems have been employed in several fields of application, such as optical metrology[Wyant, 2004; Cheng and Wyant, 1984], holography[Yamaguchi and Zhang, 1997; Nomura, 2006], optical tomography[Meneses, 2006], ESPI [Chen, 2010], etc. In this chapter, we present theoretical analyses and experimental results obtained by implementing several shearing simultaneous phase-shifting interferometers capable of retrieving the directional derivatives in the x-direction, or adapting the systems to obtain radial derivatives [Serrano et al., 2011], allowing direct slope measurements; In each case, the optical phase data are processed by means of the capture of n-interferograms in a single shot. To achieve this, we propose an alternative to the interferometers developed for phase shifting that use micro polarizer arrays, replacing them with a 4-f system with high frequency Ronchi gratings or Phase gratings coupled at the end of the interferometer.

In order to obtain the partial derivative of the phase object, a Cyclic Shear Interferometer (CSI) of lateral shear with crossed circular polarization has been implemented; the CSI is coupled to a 4-f system with high frequency Ronchi gratings, a pair of crossed gratings $G(\mu,\upsilon)$ placed as the system's pupil with a spatial period d. The high frequency Bi-Ronchi gratings generate replicas of the patterns required to process the derivative of the phase. The processing of the interferograms is performed using conventional methods of phase extraction with the algorithm of four shifts. For the case when the object presents radial symmetry, it is more convenient to use a radial-shear interferometer. Applications of radial-shear interferometers have been previously reported in optical testing [Malacara, 1974; Steel 1965], aspherical surface measuring [Hariharan, 1984], corneal topographic inspection [Kowalik et al., 2002], adaptive optics [Shirai et al., 2001], wavefront sensing [Geary, 2000; Hutchin 1985] and beam characterization [Hernandez, 2000; Liu, 2007], among others, for these cases, we propose the use of a Mach-Zehnder Radial-Shear Interferometer (MZRI), consisting of one telescope system on each arm.

The advantage of both systems is its ability to obtain the derivative of the object wavefront in real time, which can be associated with the partial or radial slope of the phase object, as will be seen in the following sections; additionally, the phase gratings used in these systems can be replaced by high frequency Bi-Ronchi gratings. Experimental results are presented for phase objects.

2. Lateral and radial shearing interferometers

Obtaining the derivative of the wavefront brings numerous advantages due to the high sensitivity against abrupt phase changes. In some cases, if the derivatives in the x and y directions are known, we can calculate the gradient information of the object, but this implies the design of optical systems that generate shearing in both directions, or we can obtain shearing in each direction separately. If the object under study presents radial symmetry, it is more convenient to use a radial-shear interferometer. In this section, we present in detail two interferometric systems used to generate directional derivatives and the slope associated, using simultaneous phase shifting interferometry.

2.1 Cyclic path interferometer

Figure 1(a) shows the experimental set-up to generate the derivative of the incident wave front in the x-direction. It consists of a cyclic path interferometer where the illumination comes from a polarized Verdi laser underneath, operating at λ= 532 nm. The system is illuminated with polarized light at 45° with respect to the x-axis (using a quarter-wave retarding plate, Q, and a linear polarizer, P).

The PBS generates two beams with crossed linear polarization states, which become circular after going through a wave retardation plate of $\lambda/4$ (Q'). Thus, the output of the CSI consists of two sheared wavefronts, each one with mutually orthogonal circular polarizations(Left and Right). The interference pattern generated has circular crossed polarizations, and the shear can be varied by moving mirror M'. At the output of the CSI, a 4-f system has been implemented, with two crossed high frequency Ronchi gratings placed on the Fourier plane [Toto et al., 2010](The Ronchi gratings can be substituted by Phase grids, taking into account the proper considerations at the moment of selecting the interference replicas to calculate the phase data map). For this case, shear $\Delta x = x_0$ has a smaller value than size **a** of the beam's cross section, which is smaller than order separation F_0 (Two neighboring diffraction orders have a distance of $F_0 \equiv \lambda f / d$ on the image plane for a grating); thus, two beams with shear Δx enter the 4-f system close to the optical axes. At the image plane, the superposition of the two beam replicas with mutual shear Δx would appear around each diffraction order. Since the interferograms retain their polarization characteristics, polarizing filters are placed in each generated replica in order to generate the phase shifts required to obtain the phase. Fig. 1(b) shows the two configurations of the polarizing arrangement proposed to generate the phase shifts in a single shot, used to process the optical phase. Some replicas of the interference patterns have different intensities; however, there are at least four patterns with similar intensities in each configuration. As will be shown in the following sections, the replicas of the pattern generated by the CSI are centered on each diffraction order [Rodriguez, 2009].

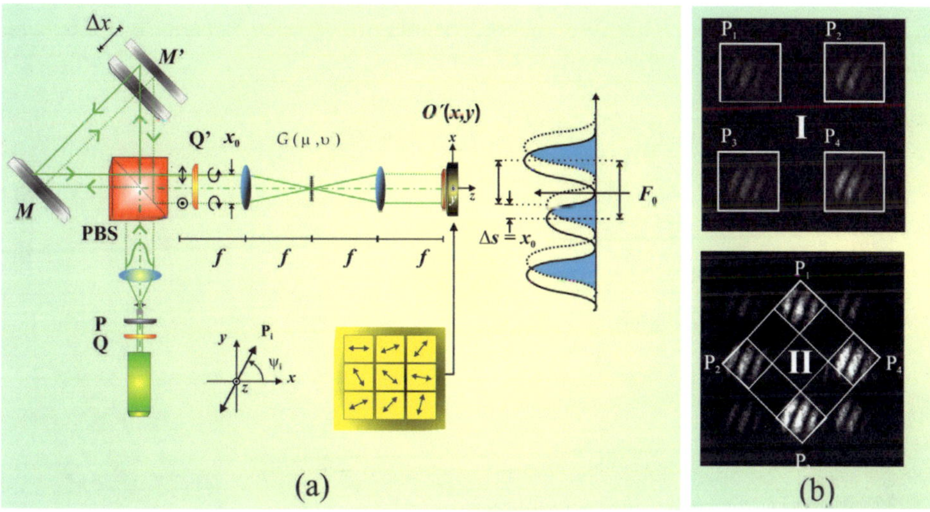

Fig. 1. Simultaneous phase shifting cyclic path interferometer. (a) CSI with variable shear coupled to 4-f system with high frequency Bi-Ronchi grating. Q, Q': Quarter wave retarders. Δx: linear shear. x_0: beam separation. $F_0 = \lambda f / d$ is the order of separation. (b) Arrangement of polarizing filters. P_i: polarizing filters, ψ_i: transmission angle of polarization.

2.2 Mach Zehnder interferometer

In several cases, such as in ophthalmic lens analysis, radial symmetry is encountered, and it is strongly necessary to obtain the radial derivative of the incident wave front in order to analyze the radial variations of the objects. For that reason, we propose the use of another optical system to obtain the radial derivative and associate it with its corresponding radial slope. Fig. 2 shows the experimental setup for the measurement of the radial slope of transparent objects. The interferometric system uses a *He-Ne* laser operating at $\lambda = 632.8$ nm. Polarized light at 45° enters the interferometer by using a quarter-wave retarder plate Q_0 and a linear polarizer P_0. The Mach-Zehnder radial-shear interferometer (MZRI) presented comprises two telescope systems, (S1, S2) [Lago and de la Fuente R. ,2008] one on each arm with vertical and horizontal linear polarized light, respectively. Their variants are widely used as part of different applications, so they can be considered to be representative cases of potential adaptations for simultaneous phase shifting interferometry. A 4-f system is coupled at the end of the interferometer; this system consists of two similar achromatic lenses of focal length $f \approx 20$ cm and a phase grid G(μ,v) placed as the system's pupil with spatial period *d*=110 *ln/mm*. In the phase grid used, $\mu = u/\lambda f$ and $\upsilon = \mathrm{v}/\lambda f$ are the frequency coordinates (*u,v*) scaled to wavelength λ and focal length *f*. The output of the MZRI consists of two versions of the same wavefront, but slightly enlarged in relation to the copy, and each one with mutually orthogonal linear polarizations (vertical and horizontal) too. A wave retarder plate of λ/4 (Q_1) is used to obtain cross circular polarization for each radial-shear wavefront (left and right, J_L and J_R) with equal amplitudes[Toto et al., 2009].

A linear polarized filter array is placed on the image plane, where each one is centered on its respective replicas of the interference pattern, as it is shown in Fig. 2 of the polarized linear array. The cross sections of the two sheared beams are $a_1 = 7.0\ mm$ and $a_2 = 8.6\ mm$, and the relative magnification of the pupils is $M_a = 1.23$.

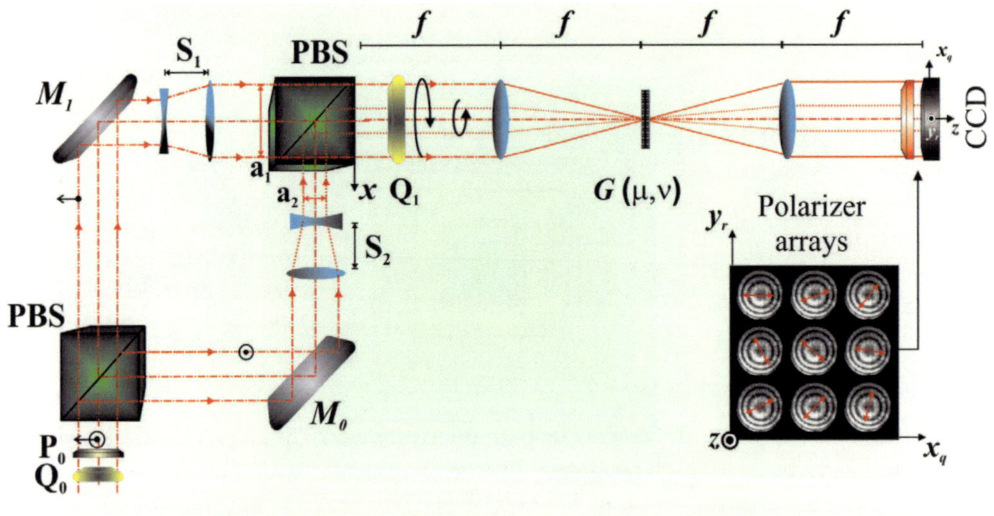

Fig. 2. Simultaneous phase shifting Mach-Zehnder radial-shear interferometer. P_i: Polarizers; Q_i: QWP operating at 632.8 *nm*; PBS: Polarizing Beam splitter; M_i: Mirrors; G(μ,v): Phase grid. Translation of coordinates around the order position: $x_q = x - qF_0$ and $y_r = y - rF_0$.

In the present section, we analyze the general case of the interference patterns obtained by the shearing interferometers presented, as well as the modulation by polarization generated in each replica in both cases (lateral or radial shear) by the Jones's Matrix representation.

3. Polarizing phase-shifting lateral shear interferometry

At the output of the CSI, two images are generated, each with circular polarization; the two fields generated for the system are described by

$$O(x+\Delta x, y) = \frac{1}{\sqrt{2}}\begin{pmatrix}1\\ i\end{pmatrix} e^{i\phi(x+\Delta x, y)} \quad O(x,y) = \frac{1}{\sqrt{2}}\begin{pmatrix}1\\ -i\end{pmatrix} e^{i\phi(x,y)} \tag{1}$$

Eq. (1) represents cross circular polarization states of same beams with lateral shear Δx. When each field is observed through a linear polarizing filter whose transmission axis is at an angle ψ, the new polarization states are

$$O'(x,y) = P_\psi \cdot O(x+\Delta x, y) \quad , \quad O''(x,y) = P_\psi \cdot O(x,y) \ , \tag{2}$$

where P_ψ is the matrix for a linear polarizer with its axis of transmission at angle ψ, given by

$$P_\psi = \begin{pmatrix} \cos^2\psi & \sin\psi\cos\psi \\ \sin\psi\cos\psi & \sin^2\psi \end{pmatrix}. \tag{3}$$

When the two fields interfere, the resulting irradiance can be written as

$$\begin{aligned} I(x,y) &= \left\| O'(x,y) + O''(x,y) \right\|^2 \\ I(x,y)_i &= 2 + 2\cos\left\{ 2\psi_i - \left[\phi(x+\Delta x, y) - \phi(x,y) \right] \right\}. \\ I(x,y)_i &= 2 \cdot \left\{ 1 + \cos\left[2\psi_i - \Delta x \frac{\partial}{\partial x}\phi(x,y) \right] \right\} \end{aligned} \tag{4}$$

When $i = 1 \ldots 4$, the relative phase can be calculated as [Sharma, 1984; Barrientos, 1999]

$$\frac{\partial}{\partial x}\phi(x,y) = \tan^{-1}\left(\frac{I_1 - I_3}{I_2 - I_4} \right), \tag{5}$$

where I_i are the intensity measurements captured in a single shot, with the values of ψ given by $\psi_1 = 0 \quad \psi_2 = 46.577° \quad \psi_3 = 92.989° \quad \psi_4 = 136.42°$. Hence, the slope is:

$$\frac{\partial w(x,y)}{\partial x} = \frac{1}{c_1} \cdot \tan^{-1}\left(\frac{I_1 - I_3}{I_2 - I_4} \right), \tag{6}$$

with $c_1 = \lambda / 4\pi \cdot \Delta x$.

4. Polarizing phase-shifting radial shear interferometry

Assuming each beam section at the end of the MZRI as $O(r) = circ[r] \cdot \exp\{ i\phi(r) \}$ and $O(r + \Delta r) = circ[r + \Delta r] \cdot \exp\{ i\phi(r + \Delta r) \}$, the field entering the 4-*f* system is given by

$$O_1(r) = \begin{pmatrix} 1 \\ i \end{pmatrix} \cdot O(r) + \begin{pmatrix} 1 \\ -i \end{pmatrix} \cdot O(r + \Delta r) \ , \tag{7}$$

representing cross circular polarization states of the same beam radially sheared. When each field is observed through a linear polarizing filter whose transmission axis is at angle ψ, the new polarization states are:

$$O_2(r) = \cdot P_{\psi}\left[\begin{pmatrix}1\\ i\end{pmatrix}\cdot O(r) + \begin{pmatrix}1\\ -i\end{pmatrix}\cdot O(r+\Delta r)\right] \quad . \tag{8}$$

As it was shown in section 3, we can show that the interference pattern can be represented as:

$$I(r)_i = 2\cdot\left\{1+\cos\left[2\psi_i - \Delta r\frac{\partial}{\partial r}\phi(r)\right]\right. \tag{9}$$

and considering also that $i = 1\ldots4$, the relative phase data map can be calculated as

$$\frac{\partial}{\partial r}\phi(r) = \tan^{-1}\left(\frac{I_1 - I_3}{I_2 - I_4}\right), \tag{10}$$

where I_i are the intensity measurements captured in a single shot. In this particular case, the radial slope can be obtained as:

$$\frac{\partial w(r)}{\partial r} = \frac{1}{c_1}\cdot\tan^{-1}\left(\frac{I_1 - I_3}{I_2 - I_4}\right) \tag{11}$$

5. Grating interferometry: pattern replication and modulation of polarization

5.1 Interferometer with Bi-Ronchi gratings

In the one-dimensional case, with μ denoting the object space, the amplitude transmission of a Ronchi grating [Ronchi ,1964.] can be written as the periodic function

$$G(\mu) = rect\left[\frac{\mu}{a_w}\right] * \sum_{n=-N}^{N}\delta\left[\frac{\mu - n\cdot d}{a_w}\right], \tag{12}$$

with $d = a_w + b_w$ being the grating period, a_w and b_w being the widths of the clear and dark bands respectively, $\delta(\mu)$ denoting the Dirac delta function, and $*$ the convolution. $\mu = u/\lambda f$ is the frequency coordinate scaled to the relevant wavelength λ and the focal length f. The Fourier transform of Eq. (12) with $N \to \infty$ is given by

$$\begin{aligned}\tilde{G}(x) &= \frac{a_w}{d}\operatorname{sinc}(a_w x)\sum_{n=-\infty}^{\infty}\delta\left(x - \frac{n}{d}\right)\\ &= \frac{a_w}{d}\sum_{n=-\infty}^{\infty}\operatorname{sinc}\left(a_w\frac{n}{d}\right)\delta\left(x-\frac{n}{d}\right) = \sum_{n=-\infty}^{\infty}C_n\cdot\delta\left(x-\frac{n}{d}\right)\end{aligned} \tag{13}$$

where C_n results as the n-Fourier complex coefficient of $G(\mu)$. In general, when Bi-Ronchi gratings[Cordero, 1998] are placed on the Fourier plane, they can be considered to be two cross amplitude Ronchi gratings,

$$G(\mu,\nu) = G(\mu)\cdot G(\nu) = \sum_{n=-N}^{N} rect\left[\frac{\mu - n\cdot d_\mu}{a_{w\mu}}\right]\cdot \sum_{l=-L}^{L} rect\left[\frac{\nu - l\cdot d_\upsilon}{a_{w\mu}}\right] \tag{14}$$

were N, L are the numbers of components of the grating, d_μ, d_υ are the respective periods along directions "μ" and "ν", and $a_{w\mu}$, $a_{w\nu}$ are the widths in clear strips along each one of both directions. The gratings studied have equal periods in both directions; then, $d_\mu = d_\upsilon = d$, as well as $a_{w\mu} = a_{w\upsilon} = a_w$. Due to the properties of the Fourier transform, the corresponding spectrum is:

$$\begin{aligned}\tilde{G}(x,y) &= \tilde{G}(x)\cdot\tilde{G}(y)\\ &= \frac{a_w{}^2}{d^2}\sum_{n=-N}^{N}\text{sinc}\left(\frac{a_w}{d}\cdot x\right)\cdot\delta\left(x-\frac{n}{d}\right)\sum_{l=-L}^{L}\text{sinc}\left(\frac{a_w}{d}y\right)\cdot\delta\left(y-\frac{l}{d}\right)\\ &= \sum_{n=-N}^{N} C_{xn}\cdot\delta\left(x-\frac{n}{d}\right)\sum_{l=-L}^{L} C_{yl}\cdot\delta\left(y-\frac{l}{d}\right)\end{aligned} \tag{15}$$

For the experimental case where gratings (100 ln/mm) have the same period and $d = 0.01mm$, the diffraction patterns generated using Eq. (15) are shown in Fig. 3. Fig. 3(a) shows the simulated diffraction pattern generated by the Bi-Ronchi grating. In Fig. 3(b), it can be seen that the amplitude spectra for each axis show more than three orders of diffraction; however, only orders 0 and ±1 allow one to obtain replicas of the interference pattern with comparable intensities, the fringe modulation it is close to unity.

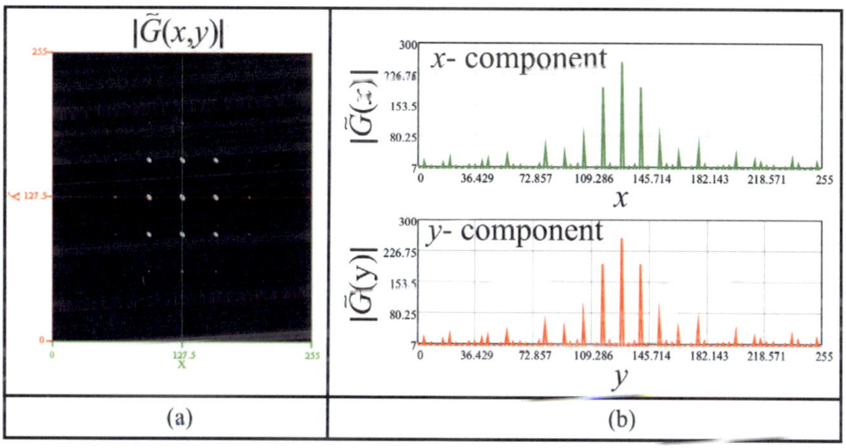

Fig. 3. Simulated diffraction pattern generated by high frequency Bi-Ronchi grating. (a) Diffraction spectra. (B) x-y components of diffraction spectra of each grating component.

5.1.1 Interference pattern replication

The 4-f system with high frequency Bi-Ronchi grating only multiplexes the interference pattern generated at the output of CSI; the shifts in the interference pattern can be generated operating linear polarizers placed at an angle ψ. Considering Eq. (1), the interference pattern $I'(x,y)$ on the image plane of the system will be:

$$
\begin{aligned}
I'(x,y) &= \left| \vec{P}_{\psi} \left[\vec{J}_L O(x+\Delta x, y) + \vec{J}_R O(x+, y) \right] * \tilde{G}(x,y) \right|^2 \\
&= \left| I(x,y) * \tilde{G}(x,y) \right|^2
\end{aligned}
, \qquad (16)
$$

where each sheared wavefront has a circular polarization, $\vec{J}_L$ (left) and $\vec{J}_R$ (right) respectively, and $I(x,y)$, the interference pattern generated by the CSI, having the form

$$
I = 1 + \cos\left[2 \cdot \psi - \Delta x \frac{\partial}{\partial x} \phi(x,y) \right]; \qquad (17)
$$

so, the interference pattern on the image plane, centered on each diffraction order, will be:

$$
\begin{aligned}
I'(x,y) &= \sum_{n=-N}^{N} \sum_{l=-L}^{L} C_{yl}^{2} \cdot C_{xn}^{2} \cdot I\left(x - \frac{n}{d_x}, y - \frac{l}{d_y} \right) \\
&= \sum_{n=-N}^{N} \sum_{l=-L}^{L} C_{yl}^{2} \cdot C_{xn}^{2} \cdot \left\{ 1 + \cos\left[2 \cdot \psi - \Delta x \frac{\partial}{\partial x} \phi\left(x - \frac{n}{d_x}, y - \frac{n}{d_y} \right) \right] \right\}
\end{aligned}
\qquad (18)
$$

Eq. (18) shows that the results are replicas of the pattern generated by the CSI displaced to the nth and l^{th} orders of diffraction; each replica of the main pattern maintains an intensity modulated by Fourier Coefficients C_n and C_l corresponding to each direction. It should be noted that the contrast of the fringes depends on the degree of circular polarization of the interfering beams [Rodriguez, 2008]. In this case, the Ronchi gratings only multiplex the pattern, so the amplitudes of the orders only increase or decrease the intensity of the interference patterns, as it is shown in Fig. 4. Figure 4 shows the experimental results obtained with a Bi-Ronchi grating. The interference patterns used are shown enclosed in a circle, independent phase shifting, can be generated operating linear polarizers. Fig. 4(a) shows the diffraction orders' overlap, indicated by a subscript. Fig. 4(b) shows the diffraction orders generated by the grating; it can be seen that they have at least 9 diffraction orders with comparable intensities. This result is due to the modulation generated by coefficients C_n and C_l (see Eq. 18). Fig. 4(c) shows the replicas of the interference patterns centered on each diffraction order.

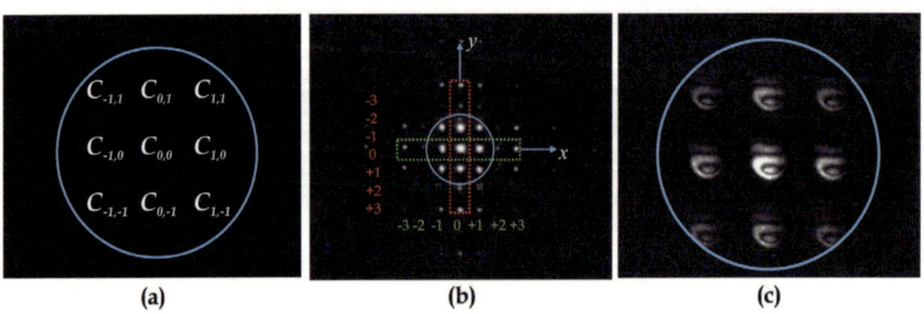

Fig. 4. Replicas of the interference patterns. (a) Fourier coefficients for each replica of the interference pattern. (b) Experimental diffraction orders generated by high frequency Bi-Ronchi gratings (100 ln/mm). (c) Replicas of the interference patterns centered on each diffraction order (Circle).

5.2 Interferometer with phase grids

Previous works have shown that when the interferometer is coupled to a 4-f system using phase grids, it is possible to obtain replicas of the interference pattern, and each one of these replicas can be modulated by polarization [Rodriguez 2008; Toto, 2008; Serrano, 2011]. A sinusoidal phase grid generated by the multiplication of two sinusoidal phase gratings, whose respective grating vectors are crossed, generates diffraction orders modulated by Bessel functions J_q and J_r [Rodriguez, 2008;Toto, 2008], and the image of the interference pattern generated by the interferometer is found centered around each of them. If we assume, for simplicity, that each beam section is of the form $w(x,y) = O(x)\cdot O(y)\cdot \exp\{i\phi(x,y)\}$, the transmittance in front of the entrance lens is given by

$$\vec{t}_i(x,y) = \vec{J}_L \cdot w(x-\frac{\Delta x}{2},y) + \vec{J}_R w(x+\frac{\Delta x}{2},y) \tag{19}$$

The displacement Δx between the wavefronts can be adjusted by translating the mirror M of the CSI. To draw theoretical conclusions, we assume that the phase grid is made up of two phase gratings (with 2πAg the grating's phase amplitude) and with orthogonal gratings vectors. Taking the rulings of one grating along the " μ " direction, and the rulings of the second grating along the " ν " direction, the resulting Fourier transform of the centered phase grid can be written as

$$\tilde{G}_2(x,y) = \sum_{q=-\infty}^{q=\infty}\sum_{r=-\infty}^{r=\infty} J_q(2\pi A_g) J_r(2\pi A_g)\,\delta(x-qF_0, y-rF_0) \tag{20}$$

where the order separation is $F_0 = \lambda f/d$, and J_q, J_r denote the Bessel function of the first kind and integer order q,r [Goodman, 1988]. On the image plane of the CSI, the amplitude can be written as

$$\begin{aligned}\vec{t}_o(x,y) &= \vec{t}_i(x,y) * \tilde{G}_2(x,y)\\ &= \sum_{q=-\infty}^{\infty}\sum_{r=-\infty}^{\infty} J_q(2\pi A_g) J_r(2\pi A_g)\left\{\vec{J}_L\ w(x-qF_0-\frac{\Delta x}{2}, y-rF_0) + \vec{J}_R \cdot w(x-qF_0+\frac{\Delta x}{2}, y-rF_0)\right\}\end{aligned} \tag{21}$$

with $(*)$ denoting convolution. On the image plane of the system, a series of replicated beams can be observed. Because of their polarization, an interference pattern can be detected when a linear polarizing filter is placed before the detector. Using Jones calculation with

$$\vec{J}_L' = \mathbf{J}_\psi^L \vec{J}_L,\ \vec{J}_R' = \mathbf{J}_\psi^L \vec{J}_R,\ \mathbf{J}_\psi^L = \begin{pmatrix} \cos^2\psi & \sin\psi\cos\psi \\ \sin\psi\cos\psi & \sin^2\psi \end{pmatrix}, \tag{22}$$

the pattern irradiance turns out to be proportional to the squared modulus of Eq. (21) in the general case. Considering the shear $\Delta x = x_0$ of a value smaller than the side a of the beam section, which in turn is smaller than the order separation F_0 (Fig. 1), two beams with shear x_0 enter the 4-f system close to the optical axes. At the image plane, the superposition of the sheared beams would appear around each diffraction order, each superposition isolated from the others. The sheared beams around a given order qr will have counter rotating circular polarization. A diffraction order qr passes through a filter designed to block out all of the remaining orders. Under these conditions, by detecting the irradiance with a linear polarizing filter with the transmission axis at an angle ψ, only the contribution of an isolated term of order qr can be considered, and its irradiance would be proportional to

$$\begin{aligned} I(x,y) &= \left(J_q J_r\right)^2 + \left(J_q J_r\right)^2 + 2J_q J_q J_r J_r \cdot \cos\left[\xi(\psi) - \Delta\phi(x_q, y_r)\right] \\ &= 2J_q^2 \cdot J_r^2\left(1 + \cos\left[\xi(\psi) - \Delta\phi(x_q, y_r)\right]\right) \end{aligned} \tag{23}$$

where a translation of coordinates was used around the order position ($x_q = x - qF_0$ and $y_r = y - rF_0$), and $\Delta\phi(x,y) \equiv \phi(x + x_0/2, y) - \phi(x - x_0/2, y)$. The pattern would present a phase shift $\xi(\psi) = 2\psi$ and fringe modulation of unity. This configuration uses the CSI as a beam divider and to produce the shear.

Figure 5 shows the diffraction pattern generated by a phase grid. Fig. 5(a) shows the diffraction orders' overlap, indicated by a subscript. Fig. 5(b) shows the diffraction orders generated by the grating; it can be seen that there are at least 9 diffraction orders with comparable intensities. This result is due to the modulation generated by coefficients $J_{n,l}$. Fig. 5(c) shows the replicas of the interference patterns centered on each diffraction order.

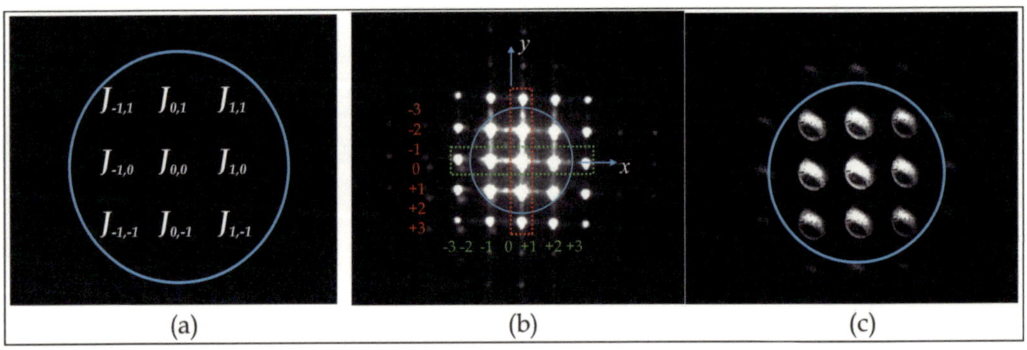

Fig. 5. Replicas of the interference patterns. (a) Fourier coefficients for each replica of the interference pattern. (b) Experimental diffraction orders generated by high frequency Bi-Ronchi gratings (100 ln/mm). (c) Replicas of the interference patterns centered on each diffraction order (Circle).

6. Slope measurements of a phase object

In general, if an object without absorption (transparent) is placed on one arm of the interferometer, it is expressed as:

$$O(x, y) = e^{i \cdot \phi(x,y)} \tag{24}$$

where $\phi(x,y)$ is a real function; this object is known as a phase object. For this analysis, we consider the special case where $|\phi(x,y)| < 1$ for both $|\phi(x,y)|^2 << 1$, resulting in the kind of transparent objects known as phase objects. Accepting the approach

$$\begin{aligned} O(x, y) &= e^{i \cdot \phi(x,y)} \\ O(x, y) &= 1 + i \cdot \phi(x, y) \end{aligned} \tag{25}$$

and taking the partial derivative of Eq. (26) with respect to x, we obtain

$$\frac{\partial}{\partial x} O(x,y) = c_0 \cdot \frac{\partial}{\partial x} \phi(x,y), \tag{26}$$

with $c_0 = i$ being a constant. It can be noted that the derivative of the phase object is proportional to the derivative of the phase, while additionally maintaining a constant phase relationship $\pi/2$ because $i = e^{i \cdot \pi/2}$. It is well known that when an object is illuminated by a single collimated beam, the relationship between the phase difference and the change in displacement [Rastogi, 1996; Ng , 1995; Bhaduri, 2006] is obtained from:

$$\frac{\partial \phi(x,y)}{\partial x} = \frac{2\pi}{\lambda} \cdot \left[\sin\theta \frac{\partial du}{\partial x} + (1 + \cos\theta) \frac{\partial w}{\partial x} \right] \Delta x \tag{27}$$

where u and w specify the displacement components. The fringe pattern has contributions from the deformation $\partial u / \partial x$ and the slope $\partial w / \partial x$. However, when the object is illuminated in the normal direction (θ=0°), the fringe pattern represents the x-partial slope, $\partial w / \partial x$. This is:

$$\frac{\partial \phi(x,y)}{\partial x} = \frac{4\pi}{\lambda} \cdot \left[\frac{\partial w(x,y)}{\partial x} \right] \Delta x, \tag{28}$$

giving us the slope of a phase object as

$$\frac{\partial w(x,y)}{\partial x} = \frac{c_1}{c_0} \cdot \frac{\partial}{\partial x} O(x,y) \tag{29}$$

with $c_1 = \lambda / 4\pi \cdot \Delta x$.

For the case of the radial slope, Eq. (29) can be used only with the consideration of the radial dependency.

6.1 Static distributions

For the case of measurements of the slope in the x-direction, experimental results are presented in Figures 6 and 7 for a microscope slide and an acetate film, respectively. The polarizer arrays used for these two proposed cases are presented in section 2.1 (See Fig. 1). In Fig. 6 (Case: configuration **I**), the upper row shows a microscope slide blocking half of the beam, with an inclination of 5 ° to axis z, resulting in a modification of the slope of the fringes. The lower row shows the results obtained with a water drop flowing into a microscope slide (since it modifies the surface of slide, the phase introduced by the drop is equivalent to obtaining a deformation in the z direction, out of plane). Fig. 7 shows the case of Configuration II of the polarizer array, where the microscope slide has been placed with a minimum inclination (above); as a result, the unwrapped phase presents the slope obtained. The second case shows water deposited on a microscope slide presenting variations in the surface. It can be seen that the patterns used with polarizer arrangement **I** have an apparent decrease of the contrast of the fringes; this is because in both cases, the samples had residual dust particles, and the water was not entirely transparent, so there was minimal absorption of the medium.

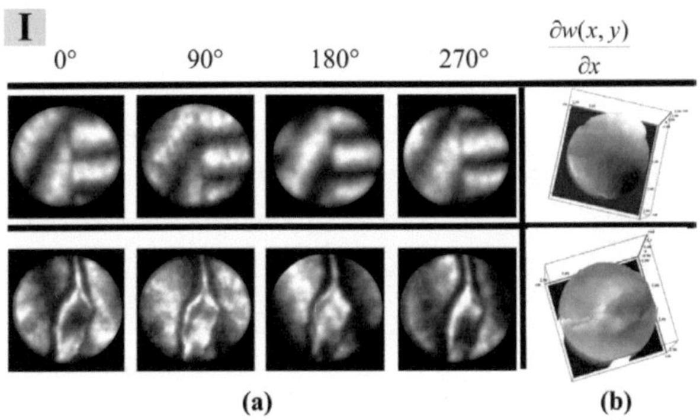

Fig. 6. Microscope slide (Configuration **I** of polarizer's array). The relative shifts corresponding to polarizing filter angles in degrees were
$\psi_1 = 0 \quad \psi_2 = 46.577° \quad \psi_3 = 92.989° \quad \psi_4 = 136.42°$ (a) Interference patterns obtained in a single shot (b) Retrieved Slope.

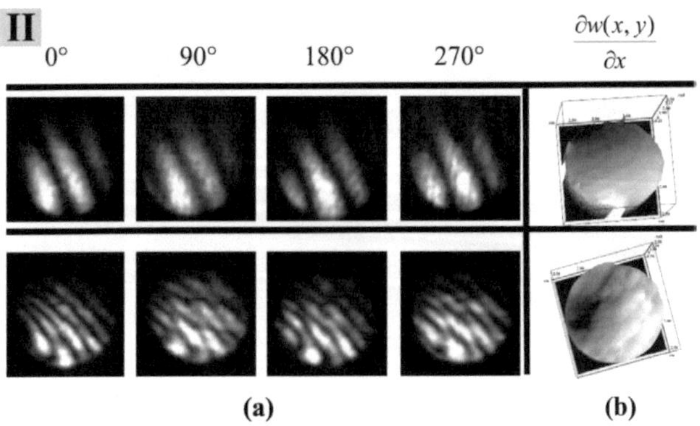

Fig. 7. Microscope slide (Configuration **II** of polarizer's array). The relative shifts corresponring to polarizing filter angles in degrees were
$\psi_1 = 0 \quad \psi_2 = 46.577° \quad \psi_3 = 92.989° \quad \psi_4 = 136.42°$ (a) Interference patterns obtained in a single shot (b) Retrieved Slope.

The experimental results for an acetate sample (average thickness of about 100 microns), on which a deformation in the *z* direction was caused, are shown in Fig. 8, where the patterns obtained in simultaneous capture and the slope associated with the derivative of the phase for configuration **I** of the polarizer array are presented. In both cases, it can be seen that the fringe pattern changes to generate a deformation around a point on the acetate.

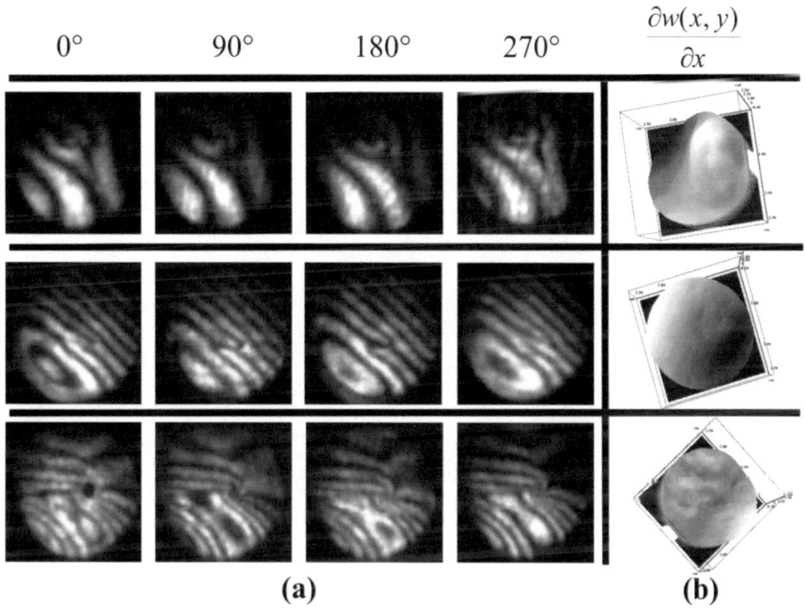

Fig. 8. Acetate. The relative shifts corresponding to polarizing filter angles in degrees were $\psi_1 = 0$ $\psi_2 = 46.577°$ $\psi_3 = 92.989°$ $\psi_4 = 136.42°$ (a) Interference patterns obtained in a single shot (b) Retrieved Slope.

Experimental results are presented using the optical system described in section 2.2 (Fig. 2). In order to extract the radial slope of a phase object, it is only analyzed using four patterns obtained in a single shot, and applying the classical method of phase extraction. Fig. 9 shows experimental results corresponding to several samples under study and their associated radial slopes. The upper row in Fig. 9 presents the radial slope of an ophthalmic lens presenting a cross section of 5 mm, the center row shows the deformed interference patterns of one drop of immersion oil placed on a microscope slide, and the lower row presents the deformation caused by lubricating oil over a microscope slide. The fringe pattern shows stress associated with the stabilization on the slide (perpendicular lines in the concentric pattern). This variation can be observed due to the radial slope variation.

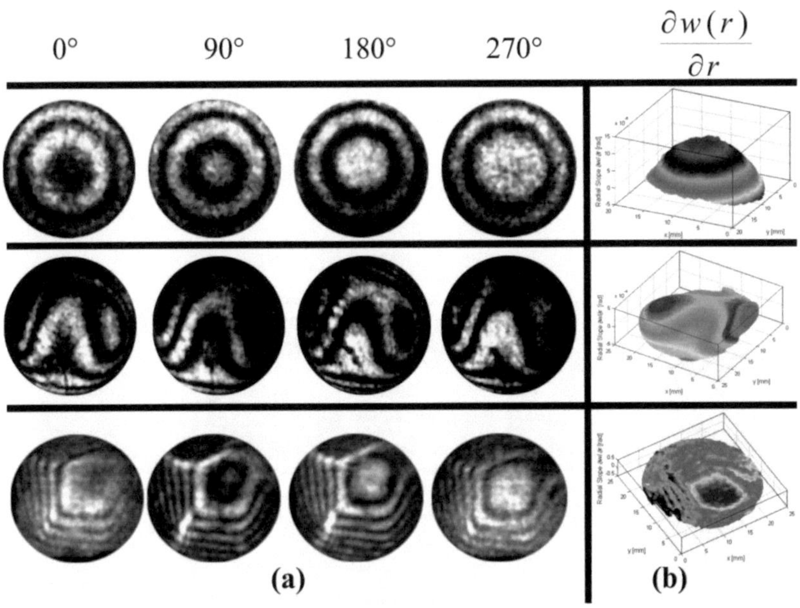

Fig. 9. Radial slope measurements. (a) Interference pattern captured in a single shot and its respective (b) radial slope retrieved.

6.2 Dynamic distributions

A dynamic phase object is shown in Fig. 10 where a transparent bubble is crossing the field of the CCD camera. The corresponding interferograms for successive captures are presented. These results show the curvature of the semi-sphere of the phase object producing a characteristic distortion of a spherical wavefront in the primary interference pattern (circle). In the animation, it is possible to see the transition that takes place when the bubble bursts and the interference patterns recover their initial profile (dot circle). The slope was calculated obtaining the phase unwrapping, according to Eq. (27). Before calculating the unwrapped phase, each interferogram was normalized and filtered using a low-pass filter.

For the case of radial symmetry, a dynamic event (4-D) is presented. Evaporation of lubricating oil placed on a microscope slide by using a tin soldiering iron is presented in Figure 11. It shows the change in the interference pattern (upper row) and its associated radial slope (lower row); the radial slope transition that occurs when the oil is evaporated can be seen. It is important to show that the results are obtained using an optical table without pneumatic suspension, and that the polarized array used is formed by conventional polarized film placed at different angles, having the advantage of not using a micro-polarized array.

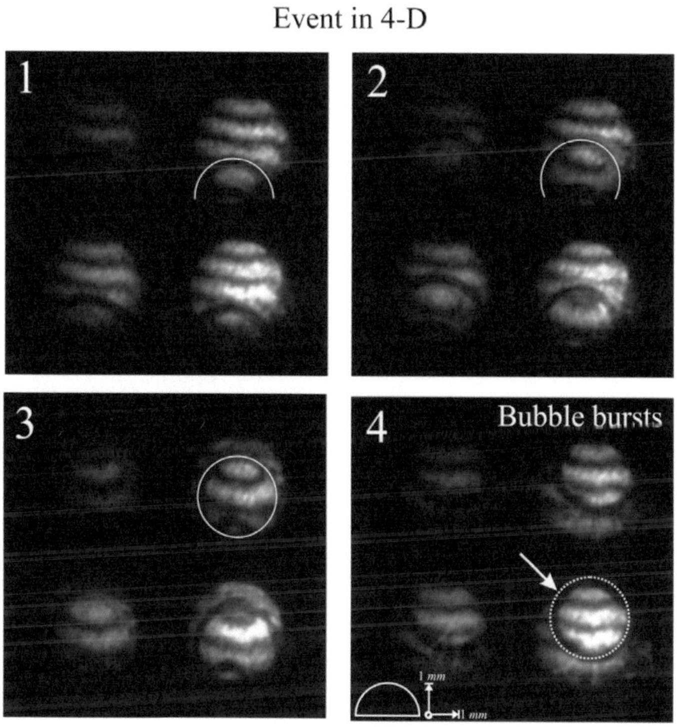

Fig. 10. Dynamic phase object. Transparent bubble moving on a microscope slide. Representative Frames.

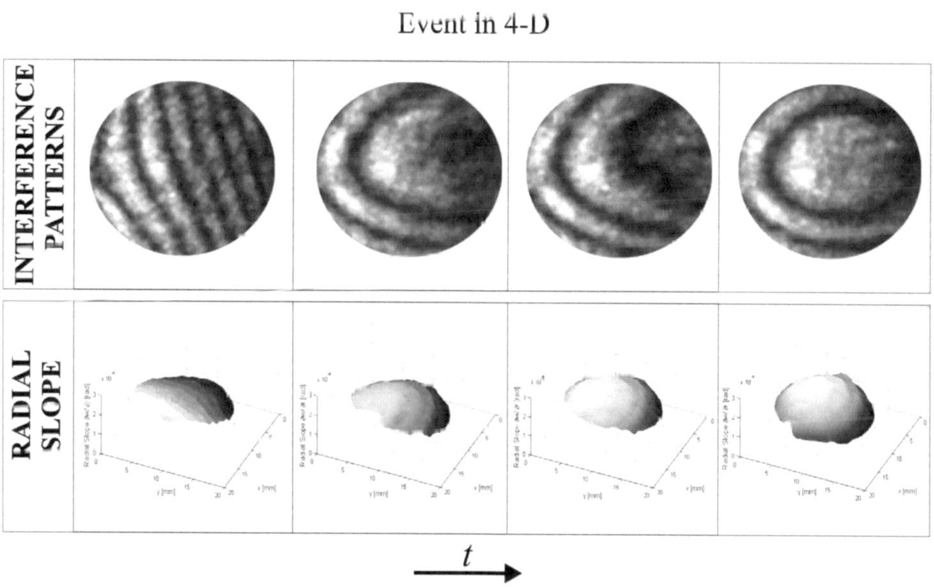

Fig. 11. Dynamic distributions of the radial slope are presented in 4-D. Representative Frames. Upper Row: Interference patters. Lower row: Evolution of the radial slope.

7. Conclusion

With the presented interferometric systems, the slope of phase objects can be obtained in a single shot in real time by using the diffraction properties of high frequency Ronchi gratings. It is suitable to replace the Ronchi gratings used in other systems by phase gratings when it is necessary to implement systems capable of obtaining *N*-interferograms. It should be noted that the elements used are inexpensive and easily accessible; as a result, the system presented can be easily implemented for various applications in single shot polarizing phase shifting interferometry.

For the case of the radial derivative, an adaptation of the MZRI with a phase grid to simultaneously capture several shearograms for slope measurements through the use of phase-shifting techniques has been demonstrated. We present different experimental results obtained by a radial shear interferometer capable of studying events.

The advantage of using these systems is that the lateral or radial slope of the phase object is obtained in real time, allowing us to study transparent samples with this symmetry or dynamic events of transparent samples, which requires obtaining real-time directional derivatives. As future work, we intend to use these systems as wave front sensors, for temperature measurements, or in optical tomography applications.

8. Acknowledgments

Authors thank M. A. Ruiz for his contribution in proofreading the manuscript. Partial support from "Consejo Nacional de Ciencia y Tecnología (CONACYT)" and "Centro de Investigaciones en Óptica A.C. (CIO)" through projects 290597 (CONACYT-CIO) and 124145 (CONACYT-BUAP) is also acknowledged. Author **N. I. Toto-Arellano** expresses sincere appreciation to **Lichita, Miguelon and Damian-El,** for the support provided, and to CONACYT for grant 102137/43055. Author **D. I. Serrano-García** (Grant:227470/31458) is very grateful to CONACyT for the graduate scholarship granted, and expresses sincere appreciation to **Geliztle**.

9. References

Barrientos-García B., Moore A. J., Pérez-López C., Wang L., and Tschudi T.,(1999). Transient deformation measurement with electronic speckle pattern interferometry by use of a holographic optical element for spatial phase stepping, *Appl. Opt.,* Vol. 38, pp. 5944–5947 (1999).

Bhaduri B., Krishna Mohan N., and Kothiyal M. P., (2006). A dual function ESPI system for the measurement of out-of-plane displacement and slope," Opt. Lasers Eng. Vol. 44, pp. 637-644.

Cheng Y.-Y and Wyant J. C., (1984). Two-wavelength phase shifting interferometry, *Appl. Opt.*, Vol. 23, pp. 4539-4543.

Chen L. C., Yeh S. L., Tapilouw a. M. and Chang J. C., (2010). 3-D surface profilometry using simultaneous phase-shifting interferometry, *Opt. Commun.*, Vol. 283, pp. 3376- 3382.

Cordero-Dávila A., Luna-Aguilar E., Vázquez-Montiel S., Zárate-Vázquez S. and Percino-Zacarías M. E. (1998). Ronchi test with a square grid, *Appl. Opt.*, Vol. 37, pp. 672-675.

Geary J. M., R. K. Tyson ed., (2000).Wavefront sensors," C.IV in *Adaptive Optics Engineering Handbook* pp. 123-150.

Goodman, Jooseph W.; J. Wiley & Sons (1988). *Introduction to Fourier optics.*

Hariharan P, Oreb B. F., Wanzhi Z.,(1984). Measurement of aspheric surfaces using a microcomputer controlled digital radial-shear interferometer, *Jour. Mod. Optics,* Vol. 31, pp. 989-999.

Hernandez-Gomez C, Collier J. L. , Hawkes S. J., Danson C. N., Edwards C. B., Pepler D. A., Ross I. N., Winstone T. B. (2000). Wave-front control of a large-aperture laser system by use of a static phase corrector, *Appl. Opt.* Vol. 39, pp. 1954-1961.

Hutchin R. A.,(1985). Combined shearing interferometer and Hartmann wavefront sensor, *Pat. Num.* 4518854.

Kowalik W. W., Garncarz B. E., Kasprzak H. T., (2002). Corneal topography measurement by means of radial shearing interference: part I- theoretical considerations, *Optik* 113, pp. 39-45.

Lago E. L. and Fuente R. ,(2008). Amplitude and phase reconstruction by radial shearing interferometry, *Appl. Opt.*, Vol. 47, pp. 372-376.

Liu D., Yang Y., Wang L., Zhuo Y., (2007). Real-time diagnosis of transient pulse laser with high repetition by radial shearing interferometer, *Appl. Opt.*, Vol. 46, pp. 8305-8314.

Malacara D., (1974). Mathematical interpretation of radial shearing interferometers, *Appl. Opt.*, Vol. 13, pp. 1781-1784.

Malacara D., Servin M., and Malacara Z.; Marcel Dekker (1998). *Interferogram Analysis for Optical Testing.*

Meneses-Fabian C., Rodriguez-Zurita G., and Arrizón V., (2006). Optical tomography of transparent objects with phase-shifting interferometry and stepwise-shifted Ronchi ruling, *J. Opt. Soc. Am. A,* Vol. 23, pp. 298-305

Mercer C. R. and Creath K.,(1996). Liquid-crystal point-diffraction interferometer for wave-front measurements, *Applied Optics*, Vol. 35, pp. 1633.

Neal M.J., Ph.D. Thesis, University of Arizona (2003). *Polarization Phase-Shifting Point-Diffraction Interferometer.*

Ng T. W., (1995). Digital speckle pattern interferometer for combined measurements of out-of-plane displacement and slope, *Opt. Commun.*, Vol. 116, pp. 31–35.

Nomura T., Murata S., Nitanai E., and Numata T, (2006). Phase-shifting digital holography with a phase difference between orthogonal polarizations, *Appl. Opt.*, Vol. 45, pp. 4873–4877.

Novak M., Millerd J., Brock N., North-Morris M., Hayes J. and Wyant J.C., (2005). Analysis of a micropolarizer array-based simultaneous phase-shifting interferometer, *Appl. Opt.*, Vol. 44, pp. 6861-6868.

Rastogi P. K., (1996). Measurement of in-plane strains using electronic speckle and electronic speckle- shearing pattern interferometry, *J. Mod. Opt.*, Vol. 43, pp. 1577–1581.

Ronchi V.,(1964). Forty Years of History of a Grating Interferometer, *Appl. Opt.*, Vol. 3, pp. 437-451.

Rodriguez-Zurita G., Toto-Arellano N. I., Meneses-Fabian C. and Vázquez-Castillo J. F.,(2008).One-shot phase-shifting interferometryfive, seven, and nine interferograms, *Opt Letters,* Vol. 33, pp. 2788-2790.

Rodríguez-Zurita G., Toto-Arellano N. I., Meneses-Fabian C. and Vázquez-Castillo J. F., (2009). Adjustable lateral-shear single-shot phase-shifting interferometry for moving phase distributions, *Meas. Sci. Technol.* Vol. 20, pp. 115902.

Sharma D. K., Sirohi R. S. and Kothiyal M. P., (1984). Simultaneous measurement of slope and curvature with a three-aperture speckle shearing interferometer, *Appl. Opt.*, Vol. 23, pp. 1542–1546.

Serrano-García D.I., Toto-Arellano N.I., Martínez García A., Rayas Álvarez J.A., Téllez-Quiñones A., and Rodríguez-Zurita G.,(2011). Simultaneous phase-shifting cyclic interferometer for generation of lateral and radial shear, Rev. Mex. Fís., Vol. 57, pp. 255–258.

Shirai T. H. Barnes T. G. Haskell (2001),Real-time restoration of a blurred image with a liquid-crystal adaptive optics system based on all-optical feedback interferometry, *Opt. Commun.*, Vol. 118, pp. 275-282.

Steel W. H., (1965). A radial shear interferometer for testing microscope objectives, *J. Sci. Instrum.*, Vol. 42, pp. 102-104.

Toto-Arellano N. I., Rodríguez-Zurita G., Meneses-Fabian C., Vazquez-Castillo J. F., (2008). Phase shifts in the Fourier spectra of phase gratings and phase grids: an application for one shot phase-shifting interferometry, Opt. Express, Vol. 16, pp. 19330- 19341

Toto-Arellano N.I., Rodriguez-Zurita G., Meneses-Fabian C. and Vázquez-Castillo J.F., (2009).A single-shot phase-shifting radial-shearing interferometer, *J. Opt. A: Pure Appl. Opt.*, Vol. 11, pp, 045704.

Toto-Arellano N. I., Martínez-García A., Rodríguez-Zurita G., Rayas-Á lvarez J.A., and Montes-Perez A., (2010). Slope measurement of a phase object using a polarizing phase-shifting high-frequency Ronchi grating interferometer, *Appl. Opt.*, Vol. 49, pp. 6402-6408.

Wyant J.C. (2004). Vibration insensitive interferometric optical testing, in Frontiers in Optics, OSA Technical Digest, OTuB2.

Yamaguchi I. and Zhang T., (1997). Phase-shifting digital holography, *Opt. Lett.*, Vol. 22, pp. 1268–1270.

14

Laser Interferometric Determination of Liposomes Diffusion Through Artificial Membranes

Michał Arabski[1], Sławomir Wąsik[1], Zuzanna Drulis-Kawa[2], Hubert Grześkiewicz[1], Jerzy Gubernator[2] and Wiesław Kaca[1]

[1]Jan Kochanowski University,
[2]University of Wrocław
Poland

1. Introduction

Cellulose membranes have been used in the medical field as wound dressing and artificial skin material for the treatment of skin wounds such as burns, ulcers, and grafts and as an adjuvant in dermal abrasions (Cheng et al., 2009; Fontana et al., 1990). Using liposomes as drug-delivery strategies in connection with a cellulose-based wound dressing might have important impact on increasing the efficacy of wound healing. The concentration of liposomes used in this therapy is crucial. The amount of liposomes might permit increasing the efficacy of their diffusion through the cellulose membrane. In the present study we propose a laser interferometric method as a novel technique to investigate liposome diffusion through membranes and cellulose-based wound dressing to optimize drug delivery.

1.1 Analysis of diffusion by laser interferometry

Laser interferometric methods are widely used for studying diffusion through membranes or in polymeric media (Dworecki et al., 2003; Dworecki et al., 2006; Wąsik et al., 2010). Substance transport through any membrane or release from gel structure are connected with the formation of concentration boundary layers (CBLs). The analysis of thickness and hydrodynamic properties of the CBLs play an important role in determining parameters of substance transport. These parameters might be analyzed by laser interferometry, for example, determination of antibiotics transfer through cellulose biomembrane in the presence of native and O-deacylated (lack of ester-bound fatty acids) forms of lipopolysaccharide (Arabski et al., 2007). Lipopolysaccharides (LPSs), a constituent of the outer membrane of gram-negative bacterial cell walls, are one of the important factors of pathogenicity of bacteria. LPS characterized amphiphilic biopolymeric compounds combining, in a single molecule, hydrophilic (O-specific chains, core oligosaccharide, etc.) and hydrophobic (lipid A) entities. The interferometric technique combined with digital image analysis of the antibiotic CBLs showed that the amount of colistin transported through the cellulose membrane was not influenced by the presence of native LPS, in contrast to presence of O-deacylated form. This effect might be associated with the phenomenon of micelles formed by native LPSs (Arabski et al., 2007).

The modification of laser interferometric technique by immobilising the tested molecules in agarose gel and measuring the amount of released substances allowed investigating interactions of partially insoluble mixtures as lipopolisaccharide with chitosan (Arabski et al., 2009a). The results of studies with laser interferometry indicated that chitosan's binding to LPS and this interaction is weaker than that of colistin. It might be associated with presence of fatty acid residues as colistin components. Chitosan reduced the speed of colistin diffusion from the gel measured by laser interferometry. These results were confirmed with electron microscopy and precipitation assay (Arabski et al., 2009a)

Additionally saponins enhance the interaction of colistin with the S and Re types of *P. mirabilis* LPSs. Analysis by laser interferometry method shows that less colistin diffused from the S1959 and R45 LPSs aggregates pre-incubated with saponins. This effect might be associated with disaggregation of LPSs micelles by saponin and facilitation of colistin binding to the lipid A part of the LPSs. These results were confirmed in a whole bacterial cell experiment (Arabski et al., 2009b).

Because this method was successfully used in our previously studies we decided to apply interferometry to examine the physical properties of PC:Chol:DOTAP (3:4:3) liposomes.

1.2 Liposomes as antibiotic carriers

Liposomes are spherical vesicles consisting of one or more phospholipid bilayers surrounding a water space. The diameter of the liposome varies from 0.02 to 10 μm. Vesicle formulations are usually based on natural and synthetic phospholipids and cholesterol. The structure may also possess lipoproteins (Urlich, 2002). The base property of these compounds is to form bilayer structure in water environment. The physicochemical properties of liposomes can be modified by changing: the types of lipids, the composition and proportions of lipids in the liposomal formulation, the size of the liposome, the charge of the liposomal surface, pH and temperature sensitivity and the fluidity of the liposomal membrane.

Regarding the variety of liposomal formulations, the vesicles are universal carriers for both hydrophilic and hydrophobic compounds. Hydrophilic elements are dissolved in the water space inside the vesicles. The most useful for this are LUVs (*Large Unilamellar Vesicles*) because the volume of encapsulated water is relatively high (Gregoriadis, 1995; Sharma & Sharma, 1997). Hydrophobic compounds are located in the lipid bilayer, and MLVs (*Multilamellar Vesicles*) or SUVs (*Small Unilamellar Vesicles*) may be applied. Charged drugs can be associated to the lipid surface (Gregoriadis, 1995; Sharma & Sharma, 1997). The size of the liposomal vesicles significantly influences drug distribution. Large (>1 μm) MLV formulations are usually not used as drug carriers, but SUVs of ~100 nm exhibited high efficacy in the eradication of bacterial pathogens (Krieger et al., 1999; Drulis & Dorotkiewicz, 2010). Encapsulation of the drugs in lipid vesicles is a good solution for designing the required pharmacokinetic and pharmacodynamic properties (Allen, 1998; Bakker-Woudenberg 1993, 1994, 2002; Swenson et al., 1988). There are many advantages of liposomes as antibiotic carriers: improved pharmacokinetics and biodistribution, decreased toxicity, enhanced activity against intracellular pathogens, target selectivity and enhanced activity against extracellular pathogens, in particular to overcome bacterial drug resistance.

The variety of liposomal formulations allows the design of effective antibiotic forms and subsequent therapeutic success (Abeylath & Turos, 2008; Schiffelers et al., 2001d; Sharma & Sharma 1997; Yimei et al., 2008). There is much evidence of the benefits of liposomes as antibiotic delivery systems. The advantage of liposomal carriers is the possibility of a gradual and sustained release of antibiotics during drug circulation in the body. This allows maintaining the proper drug concentration for a relatively long term. In comparison, administration of the free antibiotic exhibits a quick and short effect and requires several doses per day (Hamidi et al., 2006). Drug encapsulation in liposomal vesicles improves the pharmacokinetics and also protects drug against the hydrolytic activity of enzymes and chemical and immunological deactivation (Allen, 1998; Omri & Ravaoarinoro 1996a,b; Schiffelers et al., 2001). Conventional liposomes applied by intravenous administration are recognized as foreign antigens by the immunological system and are opsonised. This activates nonspecific defence mechanisms and the liposomes are taken up by the mononuclear phagocyte system (MPS), which leads to lower blood circulation time and fast blood clearance. Liposomes accumulate in the liver, spleen, lung, and kidneys (Bakker-Woudenberg, 1994, 2002; Schiffelers et al., 2001). This phenomenon (phagocytosis of liposomes) is desirable for intracellular pathogen eradication, but unfavourable for other kinds of infection (Lasic, 1998; Voinea & Simionescu, 2002). The MPS uptake rate depends on several liposomal properties, such as size, charge, and fluidity. The blood clearance of small vesicles (~100 nm) rises to several hours, in comparison with several minutes for MLV formulations. Rigid and uncharged vesicles circulate longer than fluid and charged ones (Beaulac et al., 1997; Scherphof et al., 1997). The plasma circulation time of antibiotics can be improved by encapsulation in polyethylene glycol-coated ("pegylated") (STEALTH) liposomes. The hydrophilic layer composed of PEG protects the vesicles from the MPS and allows a long liposome circulation in the blood system. Sterically stabilised STEALTH liposomes exhibit sustained release of drug and are able to accumulate selectively at sites of infection (Bakker-Woudenberg et al., 1993; Ceh et al., 1997). Liposomes are currently in common use as universal drug carriers in the cosmetic and pharmaceutical industries. In healthcare there are antitumor anthracyclines such doxorubicin and antifungal amphotericin B liposomal formulations available (Allen & Martin, 2004; Bakker-Woudenberg et al., 1994). Intensive research is focused on antibiotics entrapped in liposomes to enhance their antibacterial activity and pharmacokinetic properties. Lipid vesicles as drug carriers significantly influence drug distribution and reduce toxic side effects during antibiotic therapy (Allen 1998; Bakker-Woudenberg 1993, 1994, 2002; Drulis et al., 2006ab, 2009; Gubernator et al., 2007; Sapra & Allen 2003).

The present study was designed to examine the physical properties of PC:Chol:DOTAP (3:1:3) liposomes, especially the diffusive transport of the liposome through different membranes, as a model of a liposomal drug system for local applications. Interferomentric description of diffusive transport of the liposome through different membranes would be useful, as a model of a liposomal drug system for local applications.

2. Laser interferometric method

The laser interferometry method for the investigations of substance transport was presented previously (Arabski et al., 2007). A sketch of the measuring system is presented in Fig. 1.

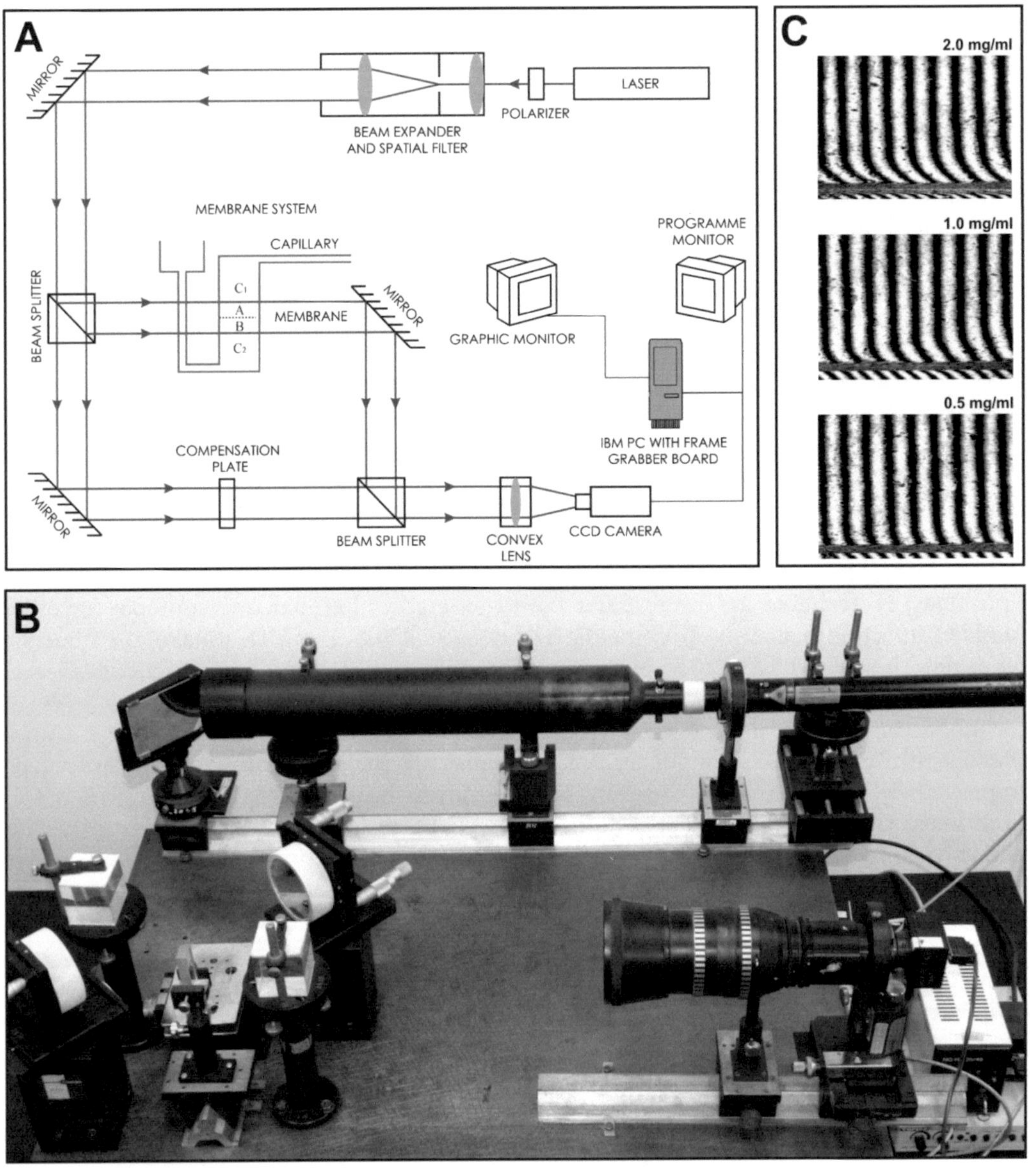

Fig. 1. Experimental set-up (A) of the interferometric investigations of the transport processes in a membrane system and its photo (B). Examples of interferograms (C) obtained for PC:Chol:DOTAP 3:4:3 liposomes at 0.5, 1, and 2 mg/ml initial concentrations transported through BioFill membrane after 10 min. C_1 denotes the initial solution concentration in the upper cuvette (water $C_1=0$), C_2 is the initial solution concentration in the lower cuvette (aqueous liposome solution at 0.5, 1 or 2 mg/ml). B→A- direction of liposomes diffusion (A).

It consist of a Mach-Zehnder interferometer with a He-Ne laser, a membrane system, a TV-CCD camera and a computer with a system for the acquisition and processing of interference images. The membrane system under study consists of two glass cuvettes (internal dimensions: 70 mm high × 10 mm wide, optical path length: 7 mm) separated by the horizontally located membrane. The cuvettes are made of optical glass of high uniformity. In our experiments the upper cuvette held pure water while the lower was filled with aqueous liposome solution (0.5, 1 or 2 mg/ml).

The laser light (Fig.1A,B) is spatially filtered and is transformed by the beam expander into a parallel beam ca. 80 mm wide and then split into two beams. The first beam goes through the investigated membrane system parallel to the membrane surface, while the second goes directly through the compensation plate to the light detection system. As a consequence of the superimposition of these beams, respective interference images are generated (Fig.1C). The images depend on the refraction coefficient of the solute, which in turn depends on the substance concentration. When the solute is uniform the interference fringes are straight, and they bend when a concentration gradient appears. In this system, water and dissolved liposomes diffusing through the membrane will lead to the formation of concentration boundary layers (CBLs). The applied computer program to analyse these images allows, among other things, to find the concentration profiles and the CBL thicknesses. The concentration profile $C(x,t)$ is determined by the deviation of the fringes from their straight line run $d(x,t)$. Since the concentration C and the refraction coefficient are assumed to be linear (Robinson & Reid, 1993), we have:

$$C(x,t) = C_0 + a\frac{\lambda d(x,t)}{hf} \quad (1)$$

where C_0 is the initial liposomes concentration, a the proportionality constant between the concentration and the refraction index, λ the wavelength of the laser light, h the distance between the fringes in the field where they are straight lines, and f the thickness of the solution layer in the measurement cuvette. The CBL thickness δ was defined according the following criterion δ is the distance from the membrane to the point at which the concentration C_δ is a certain part (k) of concentration on the membrane surface C_m:

$$k = \frac{C_\delta}{C_m} \quad (2)$$

In our investigations we took arbitrary k=0.08. The flux J_s of the solute which flows through the membrane is given by:

$$J_S = \frac{N(t)}{St} \quad (3)$$

where $N(t)$ is the amount of liposomes which diffuses after time t through a membrane of area S from one compartment of the membrane system to the other. $N(t)$ at any time t was calculated by integrating the concentration profile according to:

$$N(t) = S\int_0^\delta C_1(x,t)dx \quad (4)$$

and $N(\Delta t)$ in the given time interval Δt was calculated by integrating the concentration profile according to:

$$N(\Delta t) = S\int_0^{\delta'} C_1(x,t+\Delta t)dx - S\int_0^{\delta} C_1(x,t)dx \tag{5}$$

On the basis of (3) and (5) we obtain:

$$J_S = \frac{\int_0^{\delta'} C_1(x,t+\Delta t)dx - \int_0^{\delta} C_1(x,t)dx}{\Delta t} \tag{6}$$

where $C_1(x,t+\Delta t)$, δ' and $C_1(x,t)$, δ denote the concentration profiles and the thickness of CBL for times $t+\Delta t$ and t, respectively.

The membrane permeability coefficient was determined on the basis of Fick's first law. In the case of diffusive transport of binary solutions, this law can be written as:

$$J_S = RT\omega_m\left[C_2(x=0,t) - C_1(x=0,t)\right] \tag{7}$$

where J_s is the solute flux and $C_1(x=0,t)$ and $C_2(x=0,t)$ the concentrations of the solutions on the membrane surfaces at time t. On the basis of Eqs. (6) and (7) we obtain the formula for the membrane permeability coefficient:

$$\omega_m = \frac{\int_0^{\delta'} C_1(x,t+\Delta t)dx - \int_0^{\delta} C_1(x,t)dx}{RT\left[C_2(x=0,t) - C_1(x=0,t)\right]\Delta t} \tag{8}$$

We present method of diffusion coefficient calculations based time evolution of CBLs measured by laser interferometry. The diffusion coefficient D is determined by the CBLs thickness criterion mentioned above and the theoretical concentration profile (Dworecki et al., 2000):

$$C_1(x,t) = C_2\frac{1-\sigma}{2}erfc(\frac{x}{2\sqrt{Dt}}) \tag{9}$$

where σ denotes the membrane selectivity coefficient, C_2 is the initial solution concentration in the lower cuvette, *erfc* is complementary error function defined as:

$$erfc(x) = \frac{2}{\sqrt{\Pi}}\int_x^{\infty} e^{-\eta 2}d\eta \tag{10}$$

On the basis of the Eqs. (2) and (9) we obtain the formula for the diffusion coefficient D(t) after time t for the membrane sytems in whose one compartment held pure solvent.

$$D(t) = \frac{\delta^2(t)}{4t(erfc^{-1}(k))^2} \tag{11}$$

where $erfc^{-1}$ is the inverse function to *erfc*. The formula of diffusion coefficient $D(\Delta t)$ in the time interval Δt is given:

$$D(\Delta t) = \frac{\delta_2^2(t_2) - \delta_1^2(t_1)}{4\Delta t(erfc^{-1}(k))^2} \tag{12}$$

The interferograms were recorded from 120 to 2400 s with a time interval of Δt = 120 s and the profiles for a given initial liposome concentration (0.5, 1 or 2 mg/ml) were reconstructed. Such profiles were used to calculate the time dependencies $N(t)$, $N(\Delta t)$, $J_s(t)$, D(t), D(Δt) and ω_m. All parameters were measured on the basic of interferograms by *KALPROST* application descript by Mr. Grześkiewicz using *VISUAL C++ 2005 EXPRESS EDITION.* All experiments were performed at a temperature of 37°C.

A cationic liposomal formulation (PC:Chol:DOTAP 3:4:3) was prepared using a thin lipid film method. Appropriate amounts of lipids dissolved in chloroform (10 mg/ml) were mixed in a 100-ml round-bottom flask. Then the chloroform was evaporated under vacuum on a rotary evaporator and the resulting thin lipid film was hydrated by agitation with 1 ml of PBS at room temperature. The MLV liposomes were then extruded 10 times through a 100-nm-pore filter using a 10-ml extruder. The size of the liposomes was determined with a ZetaSizer Nano-ZS (Malvern Instruments, UK) and was usually in the range of 105-110 nm. The stability of liposomal formulation was previously determined by carboxyfluorescein release profile. The liposome leakage reached approximately 6, 8, 14, and 24% after 1, 3, 12, and 24 h, respectively. In the experiments of vesicles stability Sephadex columns were used and no lipid exchange was observed (Drulis-Kawa, 2006b). 1,2-dioleoyloxy-3-trimethylammonium-propane (DOTAP) and phosphatidylcholine (PC) were purchased from Northern Lipids Inc. (Vancouver, BC, Canada). Cholesterol (Chol) was obtained from E. Merck (Darmstadt, Germany). HPLC solvents were supplied by J. T. Baker (Deventer, the Netherlands).

The liposome diffusion efficiency was examined using commercial membranes such as Biofill and Nephrophane. The bacterial cellulose membrane (BioFill) is a transduced film with a gram-meter of 9 gm^{-2} to 20 gm^{-2}, pH between 6.0 and 7.0, obtained through biosynthesis from bacteria of the genus *Acetobacter* (Klemm et al. , 2001). Its thickness varies between 20 µm and 52 µm in dehydrated and in hydrated conditions, respectively. Bacterial cellulose has a microfibrous structure and it is a hypoallergic, nontoxic, non-irritant, biodegradable, non-pyrogenic, highly hydrophilic, and biocompatible material (Pitanguy et al., 1988; Kucharzewski et al., 2003). A cellulose acetate membrane Nephrophane is a hydrophilic membrane and its thickness varies from 25 µm in the dehydrated state to 200 µm under hydrated conditions (Kargol, 2001) Nucleopores, a polymeric nuclear track membranes, with different pore diameters (0.092-0.1 µm, 0.2 µm, 0.9 µm, and 1.27-1.35 µm) were purchased from Joint Institute for Nuclear Research in Dubna, Russia.

3. Results

The results of liposome diffusion at three concentrations (0.5, 1, and 2 mg/ml) through nucleopore membranes of different pore diameter, bacterial cellulose (BioFill), and cellulose acetate (Nephrophane) membranes are shown in Table 1.

Membrane	Liposome concentration		
	2 mg/ml	1 mg/ml	0.5 mg/ml
Nucleopore 0.092-0.1 μm	no diffusion observed	no diffusion observed	no diffusion observed
Nucleopore 0.2 μm	0.031 mg of liposomes pass membrane, blocked after 2 min of diffusion	0.030 mg of liposomes pass membrane, blocked after 8 min of diffusion	0.035 mg of liposomes pass membrane, blocked after 12 min of diffusion
Nucleopore 0.9 μm	diffusion of 0.069 mg of liposomes after 40 min. (Fig. 2A•)	diffusion of 0.069 mg of liposomes after 40 min. (Fig. 2A•)	diffusion of 0.036 mg of liposomes after 40 min. Fig. 2A○)
Nucleopore 1.27-1.35 μm	filtration	filtration	filtration
Biofill	diffusion of 0.187 mg of liposomes after 40 min. (Fig. 2B•)	diffusion of 0.082 mg of liposomes after 40 min. (Fig. 2B•)	diffusion of 0.034 mg of liposomes after 40 min. (Fig. 2B○)
Nephrophane	diffusion of 0.144 mg of liposomes after 40 min. (Fig. 2C•)	diffusion of 0.088 mg of liposomes after 40 min. (Fig. 2C•)	diffusion of 0.028 mg of liposomes after 40 min. (Fig. 2C○)

Table 1. Diffusion of liposomes through different membranes.

Filtration of the liposome vesicles was noted for nucleopore membrane of pore diameters 1.27-1.35 μm. In the case of 0.2 μm pores, diffusion was blocked after 2, 8, and 12 min. for the 0.5, 1, and 2 mg/ml concentrations of lipids, respectively, although the vesicle sizes were lower than the pore diameters. No diffusion through 0.092-0.1 μm pores was observed although fluid formulation of liposome was tested and the plastic deformation of vesicles thus the passing across smaller pores would be possible. Mathematical analysis of the time-dependent concentration profiles indicates that the amount of liposomes (0.069 mg) transported through the nucleopore membrane of 0.9 μm pore diameter at the initial concentration of 2 mg/ml was the same as for 1 mg/ml after 40 min in distance of 0.995 mm from membrane (Fig. 2A).

In the first phase of diffusion we observed throw of liposomes through nucleopore membrane of 0.9 μm pore diameter and then (after ~10 min.) diffusion was relatively constant (Fig. 3).

The same kinetics of liposomes transport across nucleopore 0.9 μm after analysis of flux J_s through a membrane area was observed (Fig. 4).

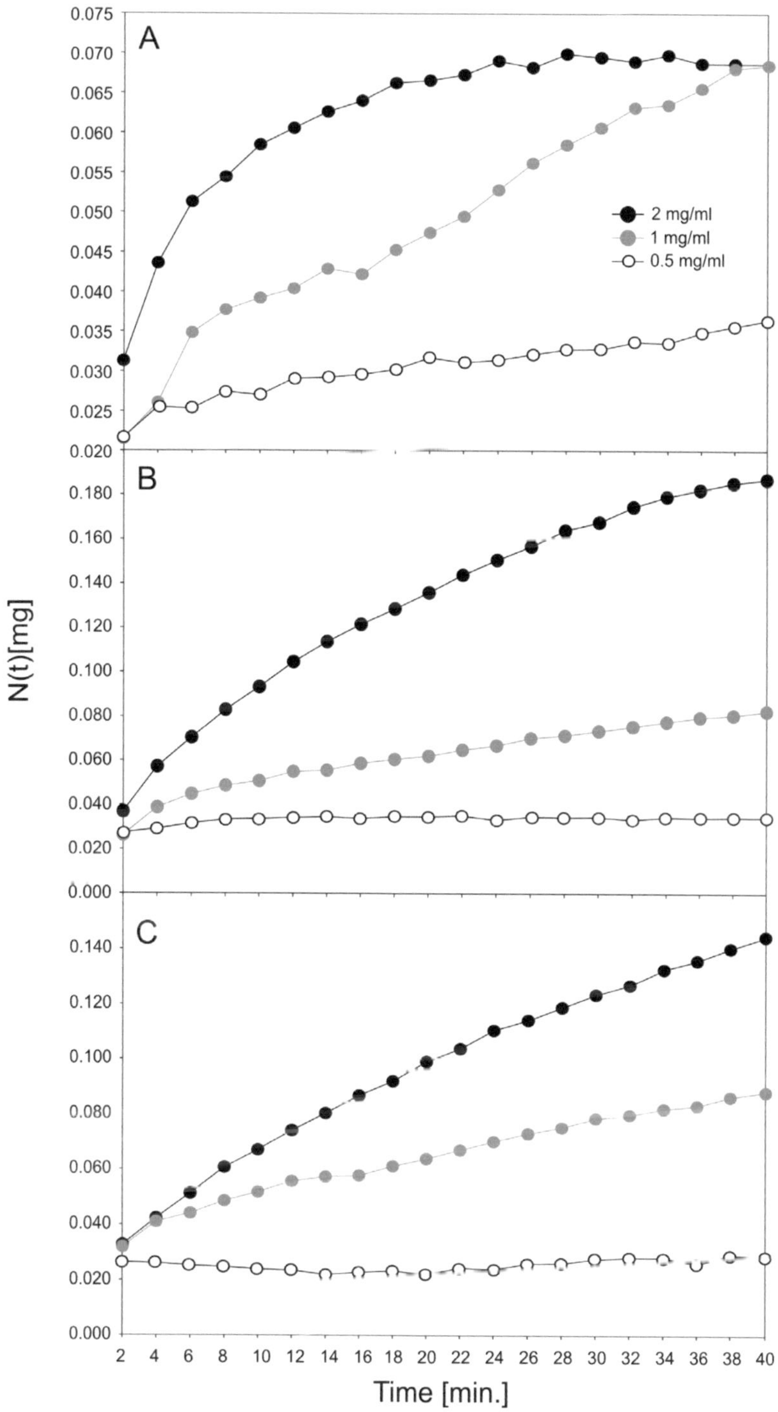

Fig. 2. The amount *N(t)* of PC:Chol:DOTAP 3:4:3 liposomes transported through nucleopore membrane of pore diameter 0.9 μm (A), BioFill membrane (B) and Nephrophane membrane (C) in 40 min at 37°C measured by the laser interferometric system.

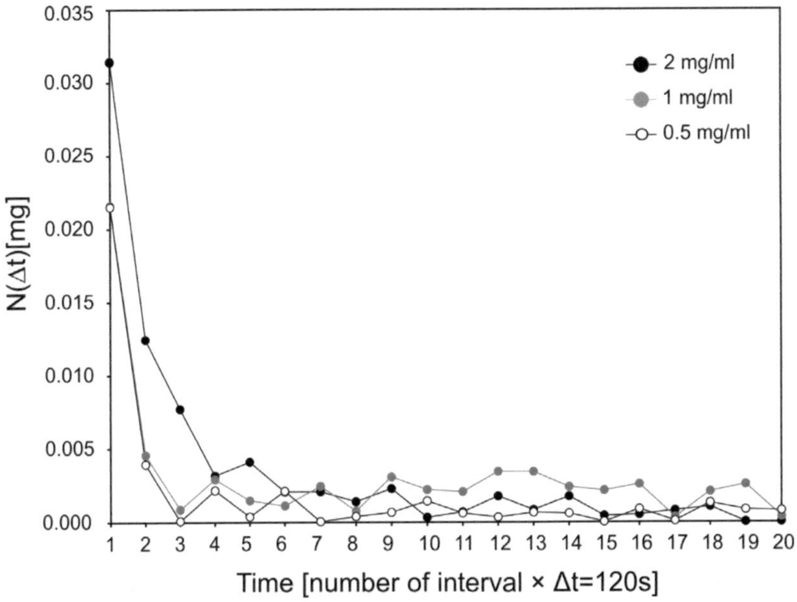

Fig. 3. The amount $N(\Delta t)$ of PC:Chol:DOTAP 3:4:3 liposomes transported through the nucleopore membrane of 0.9 μm pore diameter measured by the laser interferometric system.

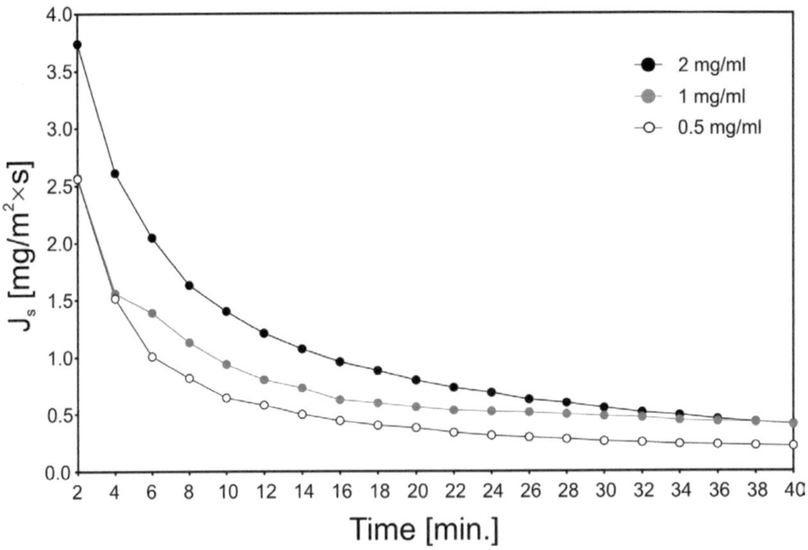

Fig. 4. The flux J_s of the PC:Chol:DOTAP 3:4:3 liposomes which cross the nucleopore membrane of 0.9 μm pore diameter measured by the laser interferometric system.

More efficient diffusion of liposomes was noticed in regards to BioFill (Fig. 2B) and Nephrophane (Fig. 2C) membranes than to 0.9 μm nucleopore (Fig. 2A). For the highest liposome concentration, 0.187 and 0.144 mg of lipids were transported across BioFill and Nephrophane, respectively.

The next step of our mathematical analysis was calculations of the average membrane permeability coefficient for aqueous solution of liposomes. The greater amounts of liposomes transported through BioFill, Nephrophane in contrast to nucleopore membrane was associated with the different values of this coefficients for above membranes (1.21×10^{-10}, 1.06×10^{-10}, and 4.31×10^{-11} mol $N^{-1}s^{-1}$, respectively). We observed different kinetics of liposomes diffusion through both cellulose membranes. The amount of transported substances was positively correlated with the initial concentration of liposomes in dose-dependent manner and increase was regular for 40 min.

Additionally, analysis of the membrane permeability coefficient for different liposomes concentrations shown the lower value of this coefficient for 2 mg/ml than for 1 and 0.5 mg/ml.

The same conclusions are based on analysis of diffusion coefficient D values for liposomes solutions measured both after time t and in the given time interval Δt. Analysis of diffusion coefficient for liposomes solution in the time interval Δt (Fig. 6) was more precise than after time t (Fig. 5). Therefore, we could noted that throw of liposomes from initial concentration 2 mg/ml took place in the first ~10 minutes of diffusion (Fig. 6)

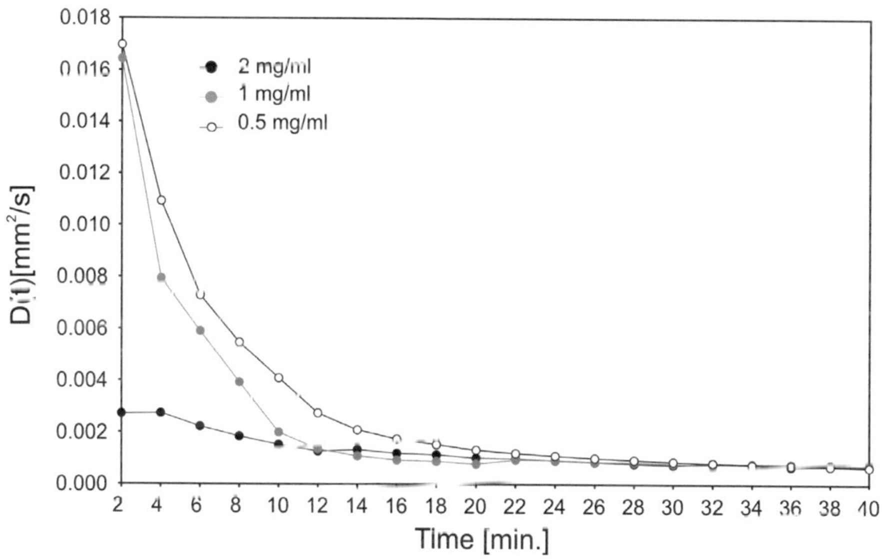

Fig. 5. The diffusion coefficient in water solution of liposomes measured in time *t* by the laser interferometric system.

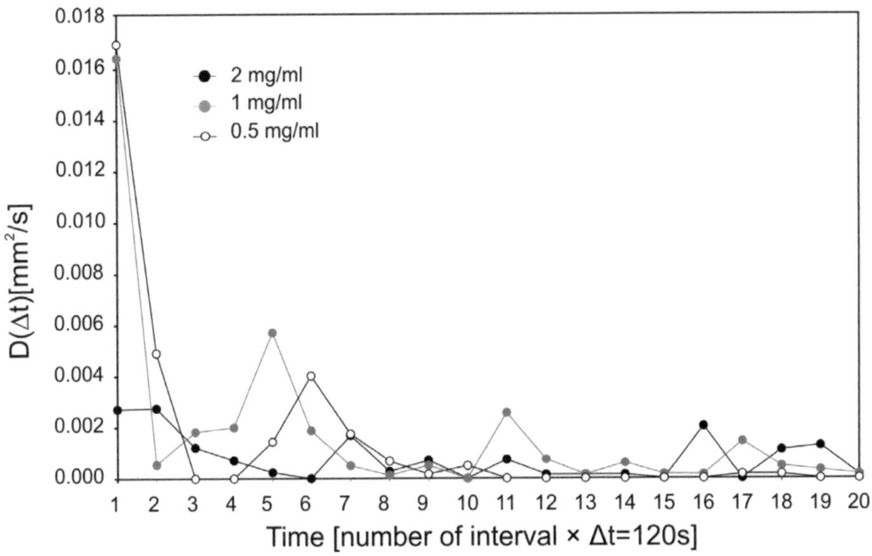

Fig. 6. The diffusion coefficient in water solution of liposomes measured in the given time interval Δt by the laser interferometric system.

4. Conclusions

Liposome diffusion through the membrane separating solutions of lipid concentration leaded to build-up of the concentration boundary layers (CBLs) on both membrane sides measured by laser interferometry. The CBL dependents on the distance from the membrane, the physical properties of the membrane, and the type of solutes. These factors leaded to concentration polarization on both sides of the membrane (Grzegorczyn & Ślęzak, 2007).

In this study the measurements of the concentration profiles and liposome fluxes were done using laser interferometry. The analysis of liposome diffusion was performed for 40 min and in the given time interval Δt. Analysis of diffusion of liposomes through nucleopore and cellulose membranes showed different kinetics of diffusion process which determined amount of transported substance (Fig.7).

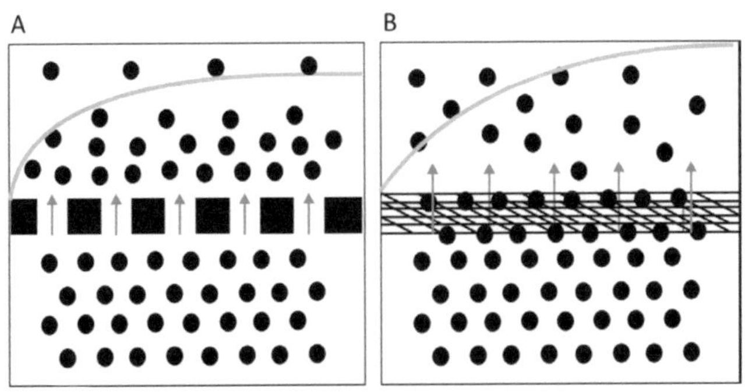

Fig. 7. Schematic presentation of liposomes diffusion through nucleopore (A) and cellulose (B) membranes proposed by the authors. Grey arrow denotes direction of diffusion. Grey curves denote layout of substance concentration.

In the first phase of diffusion across nucleopore membrane of 0.9 μm pore diameter throw of liposomes was observed and then diffusion started to be constant (Fig. 7A and Fig. 2A). This effect might be associated with different polarizations of the liposome concentrations on both sides of the nucleopore membrane. The difference of substance concentrations influenced on physical properties of diffusion process, and amount of transported particles. The lower concentration (1 mg/ml) might promote constant transport of liposomes through the membrane. At the greater difference in liposomes concentration (2 mg/ml) on the two membrane sides, reciprocal inhibition of liposome diffusion was observed. The bacterial cellulose membrane transfered greater amounts of liposomal vesicles in a gradual manner and no inhibition of diffusion was detected (Fig. 7B and Fig. 2BC). In conclusion, in the first step of diffusion process cellulose membranes were saturated by liposomes and thus releasing of substance to aqueous solution in upper side of cuvette was observed. Both commercial membranes exhibited efficient diffusion and no filtration effect.

In conclusion, the determination of dependence between the concentration of the liposome vesicles and membrane diffusion efficiency by laser interferometry method might be important from a clinical point of view. The effect of liposome concentration polarization on diffusion process through a wound dressing and artificial skin materials might be useful in the optimization of therapy using liposome-encapsulated drugs. The wound dressing based on cellulose membrane might promote more effective transport of liposomes as antibiotics carriers in comparison to standard nucleopore membranes. Additionally, laser interferometry is a very precise and quick method for determining of liposomes diffusion.

5. Acknowledgment

This work was supported by a grant from the Jan Kochanowski University, Kielce, Poland (062/R/11 for M. Arabski).

6. References

Abeylath, S.C. & Turos, E. (2008). Drug delivery approaches to overcome bacterial resistance to beta-lactam antibiotics. *Expert Opin Drug Deliv,* Vol.5, No.9, pp.931-949, ISSN 1744-7593

Allen, T.M. (1998). Liposomal drug formulations. Rationale for development and what we can expect for the future. *Drugs* Vol.56, No.5, (November 1998), pp.747-756, ISSN 0012-6667

Allen, T.M. & Martin, F.J. (2004). Advantages of liposomal delivery systems for anthracyclines. *Semin Oncol,* Vol.31, No.6 Suppl 13, (December 2004), pp.5-15, ISSN 1532-8708

Arabski, M., Davydova, V.N., Wąsik, S., Reunov, A.V., Lapshina, L.A., Solov`eva, T.F., Yermak, I.M. & Kaca, W. (2009a). Binding of lipopolysaccharide from *Proteus vulgaris* O25 (48/57) with chitosan. *Carbohydrate Polymers,* Vol.78, No.3, (October 2009), pp.481-487, ISSN 0144-8617

Arabski, M., Wąsik, S., Dworecki, K. & Kaca, W. (2007). Laser interferometric determination of ampicillin and colistin transfer through cellulose biomembrane in the presence of *Proteus vulgaris* O25 lipopolysaccharide. *J Memb Sci,* Vol.299, No.1-2, (August 2007), pp.268-75, ISSN 0376-7388

Arabski, M., Wąsik, S., Dworecki, K. & Kaca, W. (2009b). Laser interferometric and cultivation methods for measurement of colistin/ampicilin and saponin interactions with smooth and rough of *Proteus mirabilis* lipopolysaccharides and cells. *J Microbiol Methods*, Vol.77, No.2, (May 2009), pp.179-83, ISSN 0167-7012

Bakker-Woudenberg, I.A.J.M. (2002). Long-circulating sterically stabilized liposomes as carriers of agents for treatment of infection or for imaging infectious foci. *Int J Antimicrob Agents*, Vol.19, No.4, (April 2002), pp. 299-311, ISSN 1872-7913

Bakker-Woudenberg, I.A.J.M., Lokerse, A.F., ten Kate, M.T., Mouton, J.W., Woodle, M.C. & Storm, G. (1993). Liposomes with prolonged blood circulation and selective localization in *Klebsiella pneumoniae*-infected lung tissue. *J Infect Dis*, Vol.161, No.1, (July 1993), pp.164-171, ISSN 1537-6613

Bakker-Woudenberg, I.A.J.M., Storm, G. & Woodle, M. C. (1994). Liposomes in the treatment of infections. *J Drug Target*, Vol.2, No.5, pp.363-371, ISSN 1029-2330

Beaulac, C., Clement-Major, S., Hawari, J. & Lagace, J. (1997). In vitro kinetics of drug release and pulmonary retention of microencapsulated antibiotic in liposomal formulations in relation to the lipid composition. *J Microencapsul*, Vol.14, No.3, (May/June 1997), pp.335-348, ISSN 1464-5246

Ceh, B., Winterhalter, M., Frederik, P.M., Vallner, J.J. & Lasic, D. D. (1997). Stealth® liposomes: from theory to product. *Adv Drug Deliv Rev*, Vol.24, No.2-3, pp.165-177, ISSN 0169409X

Cheng, K.C., Catchmark, J.M. & Demirci, A. (2009). Enhanced production of bacterial cellulose by using a biofilm reactor and its material property analysis. *J Biol Eng*, Vol.24, (July 2009), pp.3-12, ISSN 1754-1611

Drulis-Kawa, Z. & Dorotkiewicz-Jach, A. (2010). Liposomes as delivery systems for antibiotics. *Int J Pharm*, Vol.387, No.1-2, (March 2010), pp.187-198, ISSN 1873-3476

Drulis-Kawa, Z., Dorotkiewicz-Jach, A., Gubernator, J., Gula, G., Bocer, T. & Doroszkiewicz, W. (2009). The interaction between *Pseudomonas aeruginosa* cells and cationic PC:Chol:DOTAP liposomal vesicles versus outer-membrane structure and envelope properties of bacterial cell. *Int J Pharm*, Vol.367, No.1-2, (February 2009), pp.211-219, ISSN 0378-5173

Drulis-Kawa, Z., Gubernator, J., Dorotkiewicz-Jach, A., Doroszkiewicz, W. & Kozubek, A. (2006a). A comparison of the in vitro antimicrobial activity of liposomes containing meropenem and gentamicin. *Cell Mol Biol Lett*, Vol.11, No.3, (September 2006), pp.360-375, ISSN 1689-1392

Drulis-Kawa, Z., Gubernator, J., Dorotkiewicz-Jach, A., Doroszkiewicz, W. & Kozubek, A. (2006b). *In vitro* antimicrobial activity of liposomal meropenem against *Pseudomonas aeruginosa* strains. *Int J Pharm*, Vol.315, No.1-2, (June 2006), pp.59-66, ISSN 0378-5173

Dworecki, K., Kosztołowicz, S., Mrówczyński, S. & Wąsik, S. (2000). Time evolution of near membrane layers. *Eur J Phys E*, Vol.3, No.4, pp.389-394, ISSN 1292-895X

Dworecki, K., Ślęzak, A., Drabik, M., Ornal-Wąsik, B. & Wąsik, S. (2006). Determination of the membrane permeability coefficient under concentration polarisation conditions. *Desalination*, Vol.168, (August 2006), pp. 326-334, ISSN 0011-9164

Dworecki, K., Wąsik, S. & Ślęzak, A. (2003). Temporal and spatial structure of the concentration boundary layers in a membrane system. *Physica A*, Vol. 326, No.3-4, pp.360-369, ISSN 0378-4371

Fontana, J.D., de Souza, A.M., Fontana, C.K., Torriani, I.L., Moreschi, J.C., Gallotti, B.J., de Souza, S.J., Narcisco, G.P., Bichara, J.A. & Farah, L.F.X. (1990). *Acetobacter* cellulose pellicle as a temporary skin substitute. *Appl Biochem Biotech,* Vol.24/25, No.1, pp.253-263, ISSN 1559-0291

Gregoriadis, G. (1995). Engineering liposomes for drug delivery: progress and problems. *Trends Biotechnol,* Vol.13, No.12, (December 1995), pp.527-537, ISSN 0167-7799

Grzegorczyn, S. & Ślęzak, A. (2007). Kinetics of concentration boundary layers buildup in the system consisted of microbial cellulose biomembrane and electrolyte solutions. *J Memb Sci,* Vol.304, No.1-2, (November 2007), pp.148-155, ISSN 0376-7388

Gubernator, J., Drulis-Kawa, Z., Dorotkiewicz-Jach, A., Doroszkiewicz, W. & Kozubek, A. (2007). In vitro antimicrobial activity of liposomes containing ciprofloxacin, meropenem and gentamicin against gram-negative clinical bacterial strains. *Letters in Drug Design & Discovery,* Vol.4, No.4, (June 2007), pp.297-304, ISSN 1570-1808

Hamidi, M., Azadi, A. & Rafiei, P. (2006). Pharmacokinetic consequences of pegylation. *Drug Deliv,* Vol.13, No.6, (November/December 1995), pp.399-409, ISSN 1521-0464

Kargol, A. (2001). A mechanistic model of transport processes in porous membranes generated by osmotic and hydrostatic pressure. *J Memb Sci,* Vol.191, No.1-2, (September 2001), pp.61-69, ISSN 0376-7388

Klemm, D., Schumann, D., Udhardt, U. & Marsch, S. (2001). Bacterial synthesized cellulose – artificial blood vessels for microsurgery. *Prog Polym Sci,* Vol.26, No.9, (November 2001), pp.1561-1603.

Kucharzewski, M., Ślęzak, A. & Franek, A. (2003). Topical treatment of non-healing venous leg ulcers by cellulose membrane. *Phlebologie,* Vol.32, No.6, pp.147-151, ISSN 0939-978X

Lasic, D.D. (1998). Novel applications of liposomes. *Trends Biotechnol.* Vol.16, No.7, (July 1998), pp.307-21, ISSN 0167-7799

Omri, A. & Ravaoarinoro, M. (1996a) Comparison of the bactericidal action of amikacin, netilmicin and tobramycin in free and liposomal formulation against *Pseudomonas aeruginosa. Chemotherapy,* Vol.42, No.3, (May/June 1996), pp.170-176, ISSN 1421-9794

Omri, A. & Ravaoarinoro, M. (1996b). Preparation, properties and the effects of amikacin, netilmicin and tobramycin in free and liposomal formulations on Gram-negative and Gram-positive bacteria. *Int J Antimicrob Agents,* Vol.7, No.1, (May 1996),pp.9-14, ISSN 1872-7913

Pitanguy, F., Salgado, P. & Maracaja, F. (1988). Utilization of the cellulose pellicle biofill as a biological dressing. *Rev Bras Cir,* Vol.78, No.5, pp. 317-326, ISSN 0034-7124

Robinson, O.W. & Reid, G,T. (1993). *Interferogram analysis, digital fringe pattern measurement techniques,* Institute of Physics Publishing, ISBN 0-7503 0197-X, Bristol, UK

Sapra, P. & Allen, T.M. (2003). Ligand-targeted liposomal anticancer drugs. *Prog Lipid Res,* Vol.42, No.5, (September 2003), pp.439-462, ISSN 1873-2194

Scherphof, G.L., Velinova, M., Kamps, J., Donga, J., van der Want, H., Kuipers, F., Havekes, L. & Daemen, T. (1997). Modulation of pharmacokinetic behavior of liposomes. *Adv Drug Deliv Rev,* Vol.24, No.2-3 (March 1997), pp.179-191, ISSN 0169-409X

Schiffelers, R., Storm, G. & Bakker-Woudenberg, I.A.J.M. (2001). Liposome-encapsulated aminoglycosides in pre-clinical and clinical studies. *J Antimicrob Chemother,* Vol.48, No.3, pp.333-344, ISSN 1460-2091

Sharma, A. & Sharma, U.S. (1997). Liposomes in drug delivery: progress and limitations. *Int J Pharm*, Vol.154, No.2, (August 1997), pp.123-140, ISSN 1873-3476

Swenson, C.E., Popescu, M.C. & Ginsberg, R.S. (1988). Preparation and use of liposomes in the treatment of microbial infections. *Crit Rev Microbiol*, Vol.15, No. Suppl 1, pp.S1-31, ISSN 1549-7828

Ulrich, A.S. (2002). Biophysical aspects of using liposomes as delivery vehicles. *Biosci Rep*, Vol.22, No.2, (April 2002), pp.129-150, ISSN 1573-4935

Wąsik, S., Arabski, M., Dworecki, K., Kaca, W. & Ślęzak, A. (2010). Influence of gravitational field on substance transport in gels. *J Memb Sci*, Vol.365, No.1-2, (December 2010) ,pp.341-346, ISSN 0376-7388

Yimei, J., Joly, H. & Omri, A. (2008). Liposomes as a carrier for gentamicin delivery: development and evaluation of the physicochemical properties. *Int J Pharm*, Vol.359, No.1-2, (July 2008), pp.254-263, ISSN 1873-3476

15

Experimental π Phase-Shifts Observed in the Fourier Spectra of Phase Gratings and Applications in Simultaneous PSI

Noel-Ivan Toto-Arellano[1], Gustavo Rodríguez-Zurita[2], Amalia Martínez-García[1], David-Ignacio Serrano-García[1] and María-Graciela Hernández-Orduña[3]
[1]Centro de Investigaciones en Óptica A.C. León, Gto.,
[2]Benemérita Universidad Autónoma de Puebla,
[3]Instituto Tecnológico Superior de Misanta, Ver.
México

1. Introduction

Through the use of a great variety of interferometric techniques, simultaneous phase shifting interferometry can be implemented to study dynamic events. Most of these techniques use expensive components; therefore, it is convenient to develop simultaneous phase shifting techniques that employ more accessible elements.

Recent studies show that a grating can be used as a beam splitter to attain several interference patterns around each diffraction order. Since each pattern has to show a different phase-shift, a suitable shifting technique must be employed, and phase gratings are attractive to perform the former task due to their higher diffraction efficiencies.

As it is very well known, the Fourier coefficients of only-phase gratings are integer order Bessel functions of the first kind. The values of these real-valued functions oscillate around zero, so they can adopt negative values, thereby introducing phase shifts of π at certain diffraction orders. Because of this, an almost trivial fact seems to have been overlooked in the literature regarding its practical implications.

In this chapter such phase shifts are presented in the description of interference patterns obtained with grating interferometers. These patterns are obtained by means of a two beam interferometer using a Mach Zehnder arrangement. The beams interfere at the object plane of a 4-*f* system with a sinusoidal grating/grid on the Fourier plane. It is shown the corresponding experimental observations of the fringe modulation, as well as the corresponding phase measurements, are all in agreement with the proposed description. A simultaneous phase shifting interferometer is finally proposed taking into account these properties after proper incorporation of polarization modulation.

2. Basic considerations

Phase gratings have been employed as an optical element with more efficiency than absorption gratings to perform a variety of tasks. Among them are beam splitting for interferometry, intensity measuring (Azzam, 1982) and optical shop testing (Ronchi, 1964; Cornejo, 1992). Their performance strongly depends upon their Fourier spectra, and for the case of sinusoidal phase grating has been extensively discussed (Barrekette and Freitag, 1963) It is well known that their Fourier coefficients are Bessel functions of the first kind of integer order q, J_q (Goodman, 1988) ; such functions are real valued, and their values oscillate around zero, so they can eventually introduce π-phase shifts into a given grating Fourier spectrum. These shifts were of little relevance, if any, to applications where power spectrum is the main concern, as is often the case in spectroscopy (Kneubühl, 1969) , but recently, it has been pointed out that a grating interferometer with two windows on the object plane performs a common path interferometer (Arrizón and Sánchez, 2004). Several advantages have been shown about this interferometers, i.e. their mechanical stability (Meneses et al., 2006). Moreover, in conjunction with a suitable modulation of polarization, single-shot phase-shifting interferometric systems can be implemented with phase gratings (Rodriguez et al., 2008). Two-beam phase-grating interferometers (TBPGI) are based on the interference between neighboring diffraction orders. Thus, the fringe modulation of each interference pattern can be affected when π-phase shifts are present (Rodriguez, 2008; Thomas, 1976). Furthermore, these phase shifts have to be taken into account for the overall performance of the system, and their practical advantages can even influence its design. This chapter is first aimed at the phase shifts between diffraction orders that have been observed in the Fourier spectra of a phase grating. An example of phase sinusoidal grating is calculated with a standard FFT routine that is useful for the interpretation of later experimental observations.

Experimental observations in agreement with the previous discussions, are shown by using commercial phase gratings, and finally, as an application of the above, a system able to obtain n-interferograms captured in one shot.

2.1 Experimental setup

The optical system proposed is shown in Fig. 1. Basically, it consists of a combination of a quarter-wave plate Q and a linear polarizing filter P_0 that generates linearly polarized light oriented at a 45° angle entering the Mach Zehnder configuration from a YVO_3 laser operating at 532 nm (see Fig. 1). This configuration generates two symmetrically displaced beams by moving mirrors M and M', enabling one to change the spacing x_0 between beam centers. Two retardation plates (Q_L and Q_R) with mutually orthogonal fast axes are placed in front of the two beams (A, B) to generate left and right nearly-circular polarized light (Novak et al., 2005; Rodrigurez 2008). A phase-grating $G(u/\lambda f, v/\lambda f)$ is placed on the frequency plane (u,v) of the 4-f Fourier optical system that is coupled to the Mach Zehnder configuration, with f being the focal length of each transforming lens. Then, $\mu = u/\lambda f$ and $\upsilon = v/\lambda f$ are the frequency coordinates scaled to the wavelength λ and the focal length. On plane (μ,υ), the period of G is denoted by d, and its spatial frequency by $\sigma = 1/d$. Two neighboring diffraction orders have a distance of $F_0 \equiv \lambda f / d$ at the image plane for a grating; then, $\sigma \cdot u = F_0 \cdot \mu$ and F_0 can be used as a frequency. In the following sections, phase shifts at the image plane of this system due to the grating are discussed.

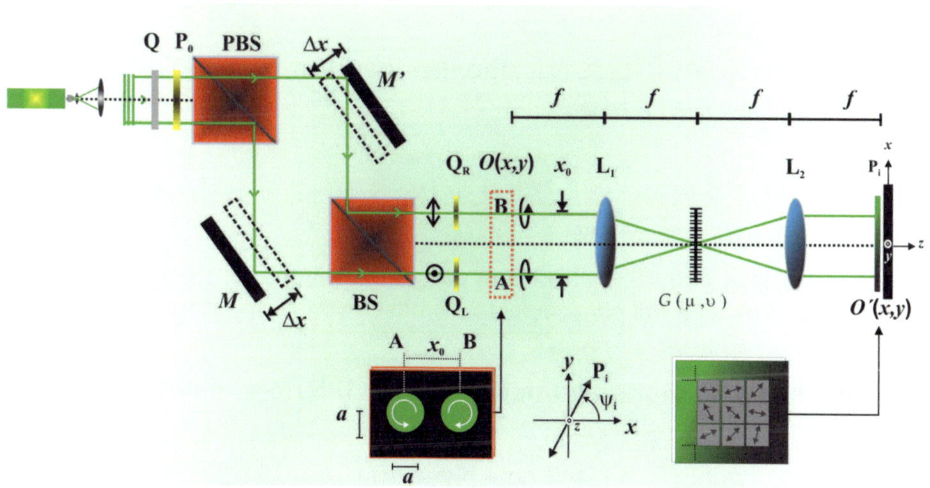

Fig. 1. Experimental setup for the Two Beam Phase Grating Interferometer using a phase grating $G(\mu,\upsilon)$ of period d placed in a 4-f system at the end of a Mach Zehnder configuration.

3. Sinusoidal phase gratings

For simplicity, we will consider a sinusoidal phase grating centered on the Fourier plane of a 4-f system. Its complex amplitude can be expressed as

$$G(\mu,\upsilon)=e^{i2\pi\cdot A_g \sin[2\pi F_0\mu]}=\sum_{q=-\infty}^{\infty} J_q\left(2\pi A_g\right)\cdot e^{i2\pi\cdot qF_0\mu} \tag{1}$$

with $2\pi A_g$ being the grating's phase amplitude and J_q the Bessel function of the first kind of integer order q. The Fourier transform of Eq. 1 is then given by

$$\tilde{G}(x,y)=\sum_{q=-\infty}^{\infty} J_q\left(2\pi A_g\right)\delta(x-qF_0,y) \tag{2}$$

with $\delta(x,y)$ as the two-dimensional Dirac delta function. Thus, the Fourier spectrum of a sinusoidal phase grating comprises point-like diffraction orders of amplitude weighted by Bessel functions. Such spectrum can be detected at the image plane of the 4-f system.

3.1 Phase grids

A sinusoidal phase grid can be generated through the multiplication of two sinusoidal phase gratings whose respective grating vectors are forming a 90-degree angle. Taking the rulings of one grating along the " μ " direction and the rulings of the second grating along the " υ " direction, the resulting centered phase grid can be written as

$$G_2(\mu,\upsilon)=e^{i2\pi A_g\sin[2\pi\cdot X_0\mu]}\,e^{i2\pi A_g\sin[2\pi\cdot Y_0\upsilon]}=\sum_{q=-\infty}^{\infty}J_q\left(2\pi A_g\right)e^{i2\pi\cdot qF_0\mu}\sum_{r=-\infty}^{\infty}J_r\left(2\pi A_g\right)e^{i2\pi\cdot rF_0\upsilon} \quad (3)$$

where the frequencies along each axis direction are taken as $X_0 = Y_0 = F_0$. The Fourier transform of the phase grid becomes

$$\tilde{G}_2(x,y)=\sum_{q=-\infty}^{q=\infty}\sum_{r=-\infty}^{r=\infty}J_q\left(2\pi A_g\right)J_r\left(2\pi A_g\right)\delta\left(x-qF_0,y-rF_0\right), \quad (4)$$

which also consists of point-like diffraction orders distributed on the image plane on the nodes of a lattice with a period given by value F_0.

4. Two-windows phase-grating interferometry: interference-pattern contrasts and modulation

Phase grating interferometry is based on a phase grid placed as the pupil of a 4-*f* Fourier optical system. The use of two windows at the object plane, in conjunction with phase grating interferometry, allows interference between the optical fields associated to each window with higher diffraction efficiency (Arrizón, 2004 and Meneses, 2006); such system performs as a common path interferometer (Fig. 1). A convenient pair of windows for a grating interferometer implies an amplitude transmittance given by

$$O(x,y)=w(x+\frac{x_0}{2},y)+w'(x-\frac{x_0}{2},y) \quad (5)$$

where x_0 is the separation from center to center between two beams. For simplicity, considering a rectangular aperture $w(x,y)$, written as $w(x,y)=rect[x/a]\cdot rect[y/b]$ and $w'(x,y)=w(x,y)\exp\{i\phi(x,y)\}$, with $\phi(x,y)$ as the object phase function. As shown in Fig. 2, a and b represent the side lengths of each beam (A and B). Placing a grating of spatial period $d=\lambda f/F_0$ at the Fourier plane, with a corresponding transmittance given by $G(\mu,\upsilon)$. The image formed by the system consists basically of replications of each window at distances F_0. This image is defined by $O'(x,y)$; that is, the convolution of $O(x,y)$ with the Fourier transform of the phase grating, $\tilde{G}(x,y)$ represented by

$$\begin{aligned}O'(x,y)&=O(x,y)*\tilde{G}(x,y)\\&=w\left(x+\frac{x_0}{2},y\right)*\sum_{q=-\infty}^{\infty}J_q(2\pi A_g)\delta(x-qF_0,y)\\&+\left[w\left(x-\frac{x_0}{2},y\right)e^{i\phi(x-\frac{x_0}{2},y)}\right]*\sum_{q=-\infty}^{\infty}J_q(2\pi A_g)\delta(x-qF_0,y).\end{aligned} \quad (6)$$

where (*) denotes the convolution. By adding the terms of Eq. 6, taking q and q-1 (both located within the same replicated window $w\left(x-qF_0+\frac{x_0}{2},y\right)$), and for the case of matching the beams' positions with the diffraction order's positions ($F_0 = x_0$), the previous equation simplifies to

$$O'(x,y) = \sum_{q=-\infty}^{\infty} \left[J_q(2\pi A_g) + J_{q-1}(2\pi A_g) e^{i\phi\left(x - x_0\left[q-\frac{1}{2}\right], y\right)} \right] w\left(x - x_0\left[q - \frac{1}{2}\right], y\right) \qquad (7)$$

Thus, an interference pattern between fields associated to each window must appear within each replicated window. The fringe modulation m_q of each pattern would be of the form

$$m_q = \frac{2J_q J_{q-1}}{J_q^2 + J_{q-1}^2} \qquad (8)$$

Further phase-shifting techniques must be used in order to introduce proper additional shifts into each pattern. One possible technique to be used is through an appropriate displacement of the grating (Meneses 2006), and an alternative approach is the use of the polarization-induced phase shifting (Rodriguez 2008a) method used and the one depicted in Fig. 1.

4.1 π–shifts of the Fourier spectra of sinusoidal phase gratings

According to Eq. 8, m_q and the contrast of each interference pattern depend on the signs of J_q and J_{q-1}; thus, the signs of J_q are also relevant. For the case of $\phi(x,y) = 0$, this relationship is shown on Table 1 for some cases of q. For $q = 0,-1,-2,-3,-4$, a negative contrast is expected; otherwise, it would be positive.

$q = 0$	$J_0 + J_{-1} = J_0 - J_1$	$q = -1$	$J_{-1} + J_{-2} = -J_1 + J_2$
$q = 1$	$J_1 + J_0$	$q = -2$	$J_{-2} + J_{-3} = J_2 - J_3$
$q = 2$	$J_2 + J_1$	$q = -3$	$J_{-3} + J_{-4} = J_3 + J_4$
$q = 3$	$J_3 + J_2$	$q = -4$	$J_{-4} + J_{-5} = J_4 - J_5$

Table 1. Bessel coeficients. Sign relations of superimposed orders. Case $\phi(x,y) = 0$.

Interference fringe modulation is positive for one half of the diffraction orders if the grating's Fourier coefficients are all positive for $q > 0$, whereas the other half would show alternating fringe modulation due to the odd parity of J_{2q+1}. These results can also be depicted as in Fig. 2 for a hypothetical case, where only the signs of the amplitudes of the diffraction orders from a given grating are separately shown displaced from the origin due to the respective displacement of windows A and B (first two plots from above). The case of $J_4 < 0$ is shown.

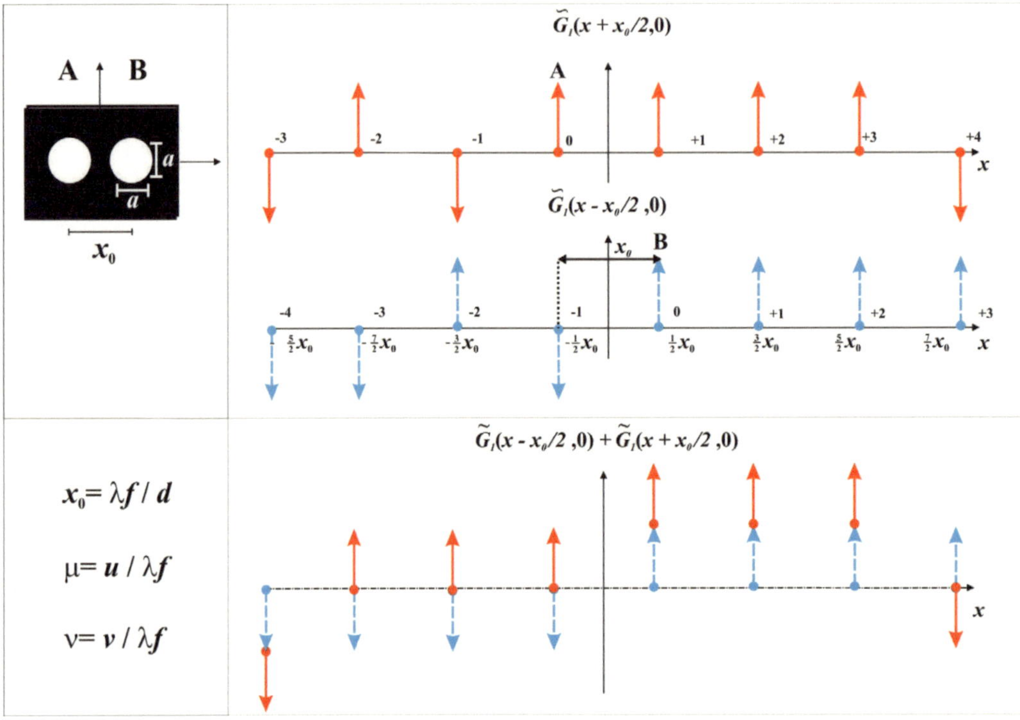

Fig. 2. Amplitude signs of diffraction orders (hypothetical grating) on the image plane of a TBPGI resulting from a window displacement of $\pm x_0 / 2$. Upper left: window configuration.

For simplicity, the replicated windows are not plotted. The third row on the graph exhibits the superposition of the two previous spectra. Fringe modulation changes must appear in both halves of the image plane. In this case, the expected signs of each fringe modulation would be +,-,-,-,+,+,+,- from left to right. The Fig. 2 then shows that odd diffraction orders have odd-order Bessel function parity, and conversely for even diffraction orders and even parity. Figure 3 shows a case of positive J_q values for $q = 0,1,4,5$, and negative for $q = 2, 3, 6$ as an example (left). This situation corresponds to the value of $2\pi A_g = 26.5$, which belongs to a range in the Bessel domain within the 9th lobule of J_0. A phase grating of the type represented by Eq. 1 (with the same amplitude $2\pi A_g$ as before) has a Fourier spectrum whose components show π–shifts accordingly (see the positive orders on the right of Fig. 3). π-shifts such as these can be detected as changes in the signs of fringe modulations in the experimental interferometers discussed later.

4.2 Some examples of calculated Fourier spectra: sinusoidal phase gratings.

In order to show some cases of phase-shifts in grating spectra, several fast Fourier transforms have been calculated for gratings of different phase amplitudes A_g. We use the lobules of J_0 as a reference to indicate the Bessel region within which the amplitude A_g of the corresponding phase grating is to be found. Only seven Bessel functions are shown in Fig. 4. In the first case, within the first lobule of J_0, the associated spectrum results with one half of

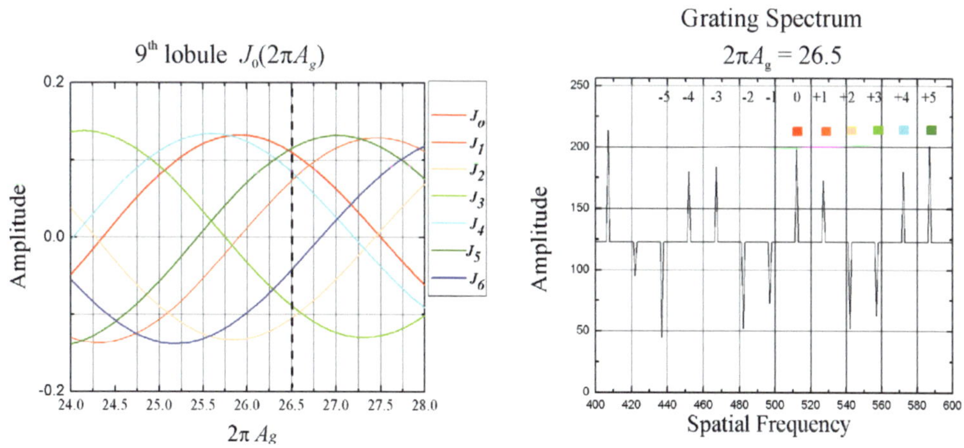

Fig. 3. Bessel domain for the ninth lobule of J_0 (left). Corresponding Fourier spectrum of the phase grating (right).

its Fourier components in phase, while the other half shows phase shifts of π in an alternating fashion. In this region, J_0 is positive and in phase with the Fourier components of the first half above mentioned. For the second case, J_0 is negative. Taking a value of A_g such that the Bessel functions with $q = 1 \ldots 6$ take only positive values, the Fourier spectrum of the resulting grating turns out to be similar to the first case, with J_0 being negative as the only different phase shift. Within the 3th lobule, it is possible to pick up negative values for $q = 1$, 2, 3, with the resulting Fourier spectrum showing π-shifts accordingly. In Fig.4 , a case of positive values for $q = 1,2,3$ and negative for $q = 4,5$ is shown. This situation corresponds to a region within the 9th lobule of J_0.

4.3 π-shifts of the Fourier spectra of sinusoidal phase grids

A rectangular phase grid $G(\mu,\upsilon)$ can be generated with two phase gratings of equal spatial frequency. Fig. 5(a) depicts the signs of the diffraction orders of a grid made up from two crossed gratings having spectra as those of the one in Fig. 2. Positive signs are denoted by hollow circles, whereas negative signs are marked with crosses. The dashed lines form regions enclosing diffraction orders of index pairs $0,n$ or $m,0$. Then, order 0,0 is found at the intersection of these regions. Two possible window configurations can be considered for a TBPGI; these configurations are shown in Fig. 5(b), and they are denoted by $\mathbf{W}_1$ and $\mathbf{W}_2$. The respective displacements of the diffraction patterns are indicated with displaced dashed lines too. For the case of phase grids with windows in configuration $\mathbf{W}_2$, the image can be written as:

$$\begin{aligned} O'(x,y) &= O(x,y) * \tilde{G}_2(x,y) \\ &= \sum_{q=-\infty}^{\infty} w\left(x+\frac{x_0}{2}, y+\frac{x_0}{2}\right) * J_q(2\pi A_g) J_r(2\pi A_g)\delta(x-qF_0, y-rF_0) \\ &+ \sum_{q=-\infty}^{\infty} w'\left(x-\frac{x_0}{2}, y-\frac{x_0}{2}\right) * J_q(2\pi A_g) J_r(2\pi A_g)\delta(x-qF_0, y-rF_0). \end{aligned} \tag{9}$$

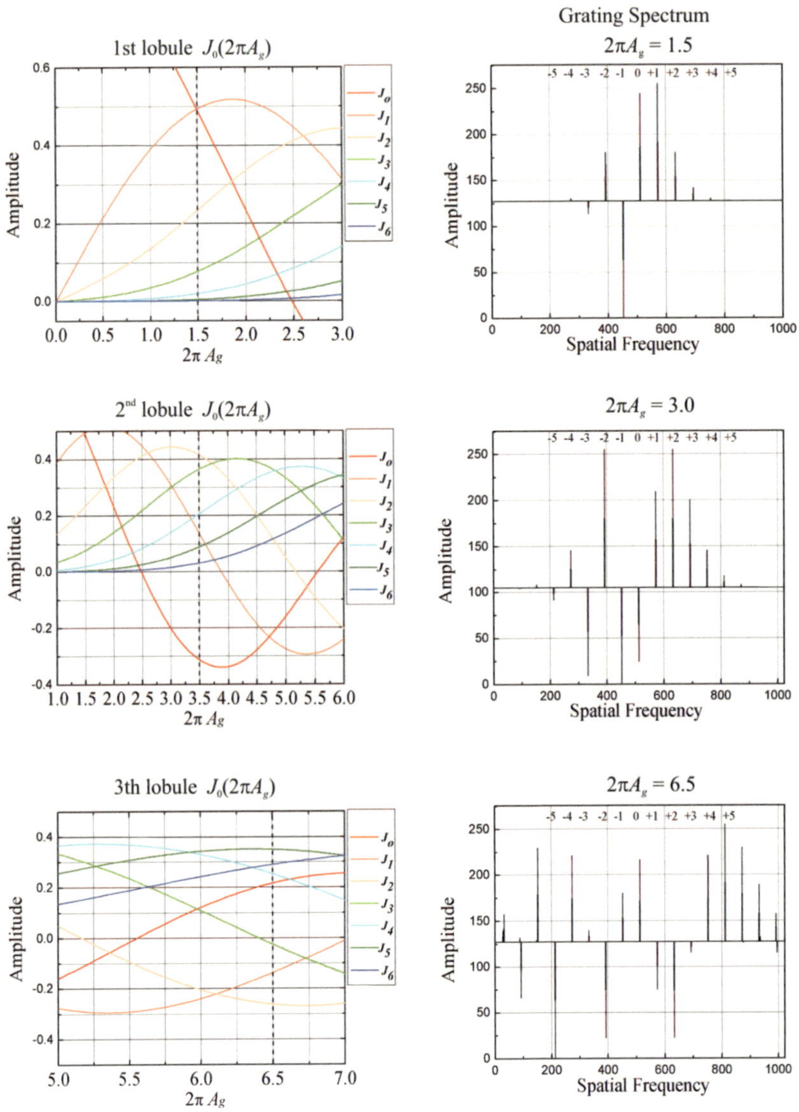

Fig. 4. Fourier spectra of phase gratings: left column, Bessel region (first seven Bessel functions shown); right column, grating spectra.

Again, with $F_0 = x_0$

$$
\begin{aligned}
O'(x,y) = & \sum_{q=-\infty}^{\infty} \sum_{r=-\infty}^{\infty} J_q(2\pi A_g) J_r(2\pi A_g)\, w\left(x + \frac{x_0}{2}[1-2q], y + \frac{x_0}{2}[1-2r] \right) + \\
& \sum_{q=-\infty}^{\infty} \sum_{r=-\infty}^{\infty} J_{q-1}(2\pi A_g) J_{r-1}(2\pi A_g)\, w'\left(x - \frac{x_0}{2}[1-2q], y - \frac{x_0}{2}[1-2r] \right),
\end{aligned}
\qquad (10)
$$

where $O(x,y) = w\left(x+\frac{x_0}{2}, y+\frac{x_0}{2}\right) + w'\left(x-\frac{x_0}{2}, y-\frac{x_0}{2}\right)$. This result simplifies to

$$O'(x,y) = \sum_{q=-\infty}^{\infty}\sum_{r=-\infty}^{\infty}\left[J_q(2\pi A_g)J_r(2\pi A_g) + J_{q-1}(2\pi A_g)J_{r-1}(2\pi A_g)e^{i\phi\left\{\left(x-x_0\left[q-\frac{1}{2}\right]\right),\left(y-x_0\left[r-\frac{1}{2}\right]\right)\right\}}\right] \times w\left(x-x_0\left[q-\frac{1}{2}\right], y-x_0\left[r-\frac{1}{2}\right]\right). \tag{11}$$

As it is for gratings, fringe modulation of the interference pattern within a window centered in $\left(x_0\left[q-\frac{1}{2}\right], x_0\left[r-\frac{1}{2}\right]\right)$ is

$$m_{qr} = \frac{2J_qJ_{q-1}J_rJ_{r-1}}{\left(J_qJ_r\right)^2 + \left(J_{q-1}J_{r-1}\right)^2} \tag{12}$$

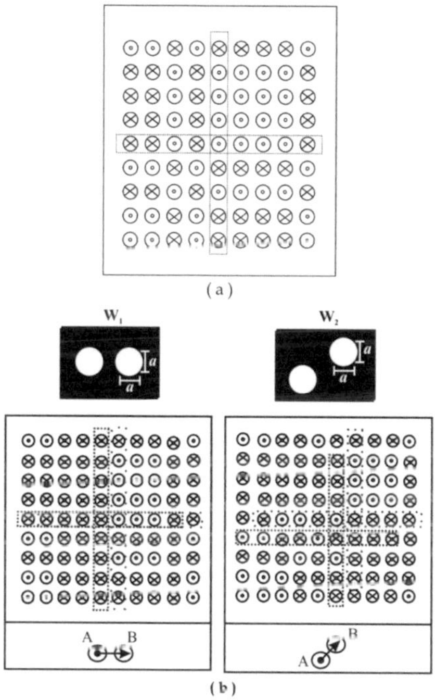

Fig. 5. (a) π-phase distribution of diffraction orders of grids. Dashed lines enclose diffraction orders of indexes $0,n$ or $m,0$. b) TBPGI order superposition: Configuration $\mathbf{W}_1$: interference pattern signs for windows displaced along the horizontal axis. Configuration $\mathbf{W}_2$: interference patterns signs for displaced windows along a line at 45°. Respective displacements of diffraction patterns are highlighted by dashed lines. An explanation of dot and cross patterns is to be found in the text.

Each corresponding fringe modulation depends on the relative phases between J_q, J_{q-1}, J_r and J_{r-1}. These relationships are discussed with the same example of Fig. 5(a). Fig. 5(b) shows the relative phases (and thus, expected signs of m_{qr}) of each diffraction order arising from configuration $\mathbf{W}_1$ or $\mathbf{W}_2$. If positive signs or negative ones coincide in a diffraction order, fringe modulation will be positive, and this is plotted with one symbol (cross or hollow dot). Only when a cross with a hollow dot appears, modulation is expected to be negative. "Vertical" bands whose sign is equal to that of m_{qr} are thus expected in configuration $\mathbf{W}_1$. Regions with a sign that is equal to that of m_{qr} can be seen in configuration $\mathbf{W}_2$.

4.4 Phase shifting interferometry with modulation of polarization

Turning the attention to gratings, in order to introduce additional phase shifts in the interference pattern centered at $\left(x_0\left[q-(1/2)\right],y\right)$ for 4-step phase-shifting (Schwieder et al., 1983), each of the windows is illuminated with different polarizations using retarding plates Q_R and Q_L (Fig. 1). This arrangement introduces Jones polarization vectors $\vec{J}_R$ and $\vec{J}_L$ into the interference terms of Eq. 6 through appropriate modification of Eq. 5. After placing a linear polarizing filter with the transmission axis at an angle ψ, its irradiance results as being proportional to

$$\left\|\vec{J}_T\right\|^2 = A(\xi,\alpha')\left\{\left(J_q^{\,2}+J_{q-1}^{\,2}\right)+2J_qJ_{q-1}\cos[\xi(\psi,\alpha')-\phi(x,y)]\right\} \tag{13}$$

Where ψ is the linear polarizer angle, $\pm\alpha'$ the retardation of each plate, and $\xi(\psi,\alpha')$ denotes the phase shifting term induced by modulation of polarization given by

$$\xi(\psi,\alpha') = ArcTan\left[\frac{\sin[2\psi]\cdot\sin[\alpha']+\sin^2[\psi]\cdot\sin[2\alpha']}{\cos^2[\psi]+\sin^2[\psi]\cdot\cos[2\alpha']+\sin[2\psi]\cdot\cos[\alpha']}\right], \tag{14}$$

and $A(\psi,\alpha')$ is defined as

$$A(\psi,\alpha') = 1+\sin[2\psi]\cos[\alpha'] \tag{15}$$

For phase-shifting interferometry with four patterns, four irradiances can be used, each one taken at a different ψ angle. Denoting each pattern as

$$\left\|\vec{J}_i\right\|^2 = A(\xi,\alpha')\left\{\left(J_q^{\,2}+J_{q-1}^{\,2}\right)+2J_qJ_{q-1}\cos[\xi(\psi_i,\alpha')-\phi(x,y)]\right\} \tag{16}$$

with $i = 1\ldots4$, the relative phase can be calculated as (J. Schwieder 1983)

$$\tan\phi = \frac{\left\|\vec{J}_1\right\|^2-\left\|\vec{J}_3\right\|^2}{\left\|\vec{J}_2\right\|^2-\left\|\vec{J}_4\right\|^2}, \tag{17}$$

where $\|\vec{J}_1\|^2$, $\|\vec{J}_2\|^2$, $\|\vec{J}_3\|^2$ and $\|\vec{J}_4\|^2$ are the intensity measurements with the values of ψ such that $\xi(\psi_1)=0, \xi(\psi_2)=\pi/2, \xi(\psi_3)=\pi, \xi(\psi_4)=3\pi/2$, respectively. Note that $\xi(\psi,\pi/2)=2\psi$ and $A(\psi,\pi/2)=1$, so a good choice for the retarders is quarter-wave retarders, as it is well known. Dependence of ϕ on the coordinates of the centered point has been simplified to *x,y*. The same fringe modulation m_q results as in Eq. 8. Therefore, the discussion about fringe modulation given in previous sections is retained when introducing the modulation of polarization. Such polarization modulation can also be made for grids, resulting in similar conclusions.

5. Experimental testing of the phase-shifts in phase gratings and phase grids

Figure 6 shows the superposition of Fourier amplitude spectra under window configuration W_1 for a phase grating with 110 lines/mm. The contrast of the corresponding experimental interference patterns can be interpreted as if its first four Fourier coefficients had phase relations as the ones sketched in Fig. 2. They were obtained with the system of Fig.1 before placing the retardation plates Q_R, Q_L and the polarizing filters on the image plane. The patterns show the relative phases of the diffraction orders discussed in previous sections. The phase-shifted steps of the experimental fringe patterns can be calculated by applying the algorithm proposed by Kreis, 1986. The resulting mean values are shown in Table 2. It can be seen from the table that they depart by small amounts from π or 0.

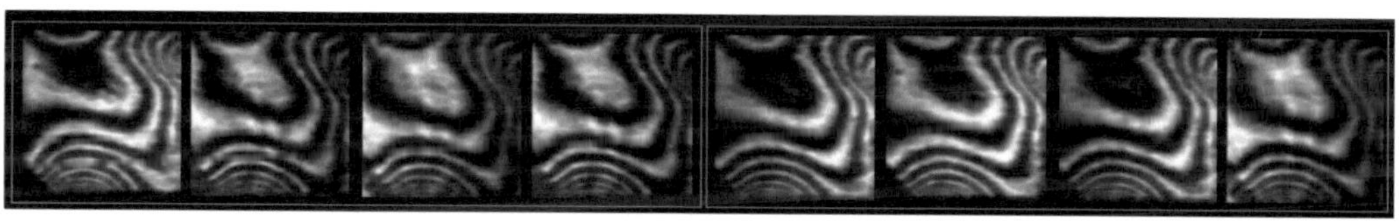

Fig. 6. Experimental results with a phase-grating.

g_1	g_2	g_3	g_4	g_5	g_6	g_7	g_8
0.00	3.143	3.130	3.150	0.010	0.014	0.003	3.145

Table 2. Phase shifts measured from experimental patterns in Fig.6 according to the method from Kreis, 1986

Due to that, the interference patterns are obtained from the interference between the replicas of each beam, centered around each diffraction order. Fig. 7(a) presents the replicas of window A with right circular polarization, and the replicas of window B with left circular polarization; each order is superposed depending on separation x_0. Fig. 7(b) presents the interference pattern generated by the interference of windows AB, due to the cross circular polarization, a linear polarizer is placed to modulate the phase shift on each replicated interference pattern.

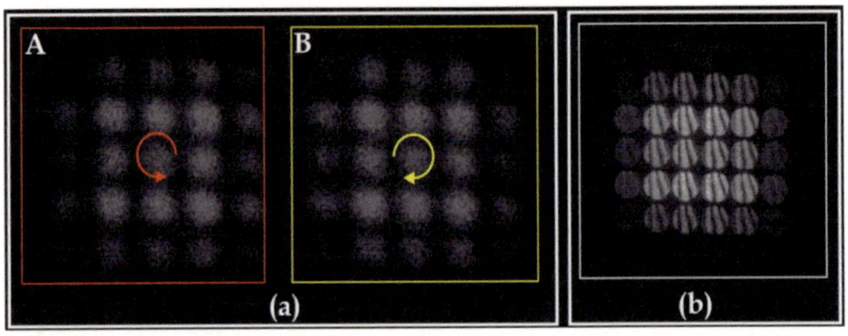

Fig. 7. Replicated windows generated by the Phase grating. (a) Replicas of A and B. (b) Interference pattern obtained through order diffraction superposition. (window configuration $\mathbf{W}_1$).

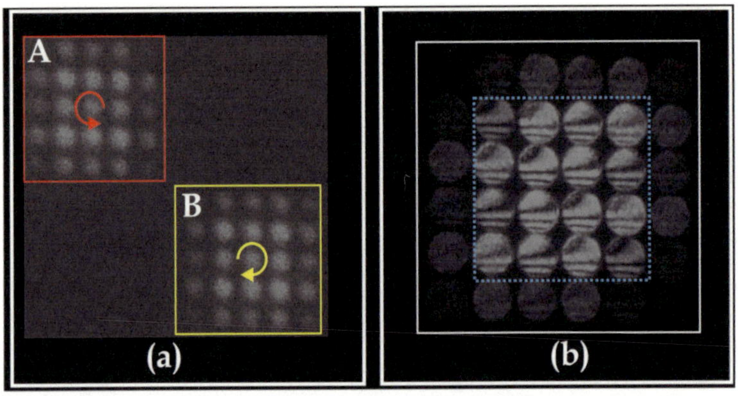

Fig. 8. Replicated windows generated by the Phase grating. (a) Replicas of A and B. (b) Interference pattern obtained through order diffraction superposition. (window configuration $\mathbf{W}_2$).

For the case of the diffraction orders belonging to a phase-grid constructed with two crossed gratings of equal frequency, the corresponding interference patterns are shown in Fig.8 for window configuration $\mathbf{W}_2$. Each grating gives patterns as in Fig. 6 when placed alone in the system of Fig. 1, with neither plate retarders nor polarizers. In each case, contrasts are in agreement with the conclusions derived from Fig. 5(b). The relative phase values of the 16 patterns within the square drawn in the patterns of Fig. 8(b) employing the method from Kreis, 1986 can be seen in Table 3.

g_1	g_2	g_3	g_4
0.010	0.030	3.120	3.152
0.016	0.020	3.140	3.141
3.145	3.150	3.136	0.020
3.139	3.170	3.182	0.043

Table 3. Phase shifts measured by means of the Kreis method for $\mathbf{W}_2$ configuration presented in Fig. 8.

6. Phase-grid interference patterns with modulation of polarization

Incorporating modulation of polarization, a TBPGI can be used for dynamic interferometry measurements. This system is able to obtain n-interferograms phase-shifted with a single shot. Phase evolving in time can be calculated and displayed on the basis of phase-shifting techniques with four or more interferograms. The system performs as previous proposals to attain four interferograms with a simultaneos capture (Barrientos et al., 1999; Novak et al., 2005). The system uses a grid as a beam splitter in a way that resembles the well-known double-frequency shearing interferometer as proposed by Wyant 1986, but our proposal differs from it not only due to its modulation of polarization, the use of a single frequency and the use of two windows, but also in the phase steps that our system introduces. Besides, our system is not a shearing interferometer of any type.

Figure 1 shows the arrangement of a simultaneous phase-shifting grid interferometer including modulation of polarization with retarders for the beams and linear polarizers on the image plane. The system generates several diffraction orders of similar irradiances, but not of equal fringe modulations, as expected (Eq. 12). In order to use Eq. 17 properly, each interferogram image was scaled to the same values of grey levels (from 0 to 255) and normalized. In the previous section, it was shown that a simplification for the polarizing filter array can be attained when using the π–phase shifts obtain values of ξ of 0, π/2, π and 3π/2. For configuration $\mathbf{W_2}$ (Fig. 8), due to the π-shifts, only two linear polarizing filters have to be placed (instead of four filters, as when without the π-shifts). The transmission axes of filter pairs P_1, P_3 and P_2, P_4 can be the same for each as long as they cover two patterns that are 180° phase apart (Fig. 9). Considering the polarizers angles , according with Sec. 4.3, the needed values of ψ can be $\psi_1 = 0°$ and $\psi_2 = 46.577°$ (G Rodriguez 2008a). They are sketched in Fig. 9. The square enclosing the 16 window replicas in the same figure is to be compared with the similar square of Fig. 8(b) (dotted lines).

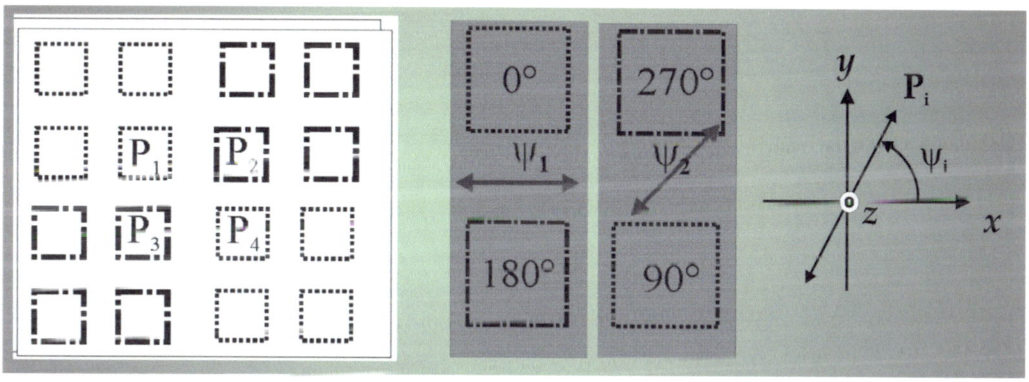

Fig. 9. Polarizing filters array for 90° phase stepping.

7. Simultaneous phase shifting interferometry

In order to extract phase distributions that evolve in time using phase-shifting interferometry, simultaneous capture of several interferograms with a prescribed shift has to be done. The well-known four-step phase shifting algorithm is presented. Due to the capability of TBPGI to capture more than four interferograms in a single shot by placing a phase grid, we have the advantage of analyzing the case of more than four interferograms; the case of $n = (N+1)$ interferograms has been chosen. This method reduces errors in phase calculations when noisy interferograms are involved (Malacara and Servin, 2005). It the next sections of the chapter, we will show experimental results for $n = 9$ interferograms.

7.1 Four interferogram case

A phase dot was placed in the path of beam B, while beam A was used as a reference; the results obtained are shown in Fig. 10. Fig. 10(a) shows the four patterns with $\pi/2$ phase shifts obtained in a single shot, and Fig. 10(b) presents the phase profile of the object, in false color coding. However, more than four interferograms could be used, whether for *N*-steps phase-shifting interferometry (Rodriguez, 2008b) or for averaging images with the same shift.

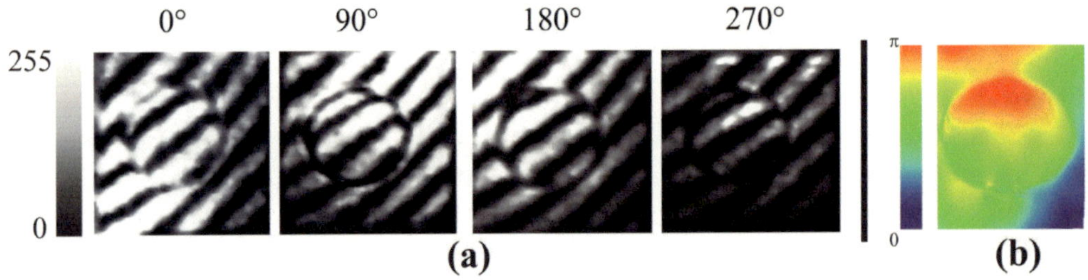

Fig. 10. Static test objects: Phase dot. (a) Four 90° phase-shifted interferograms. (b) Unwrapped phase.

Figure 11 shows the case of moving distributions, corresponding to water flowing on a microscope slide; it shows the phase profile resulting of phase evolution $\phi(x,y,t)$. Fig. 12, by analyzing flow motion by gravity, shows the temporal evolution of oil flow, clearly showing the phase changes induced by the motion of oil; the evolution of the drops as they go by in front of the camera can be shown. Images are presented in 4-D (Wyant 2003).

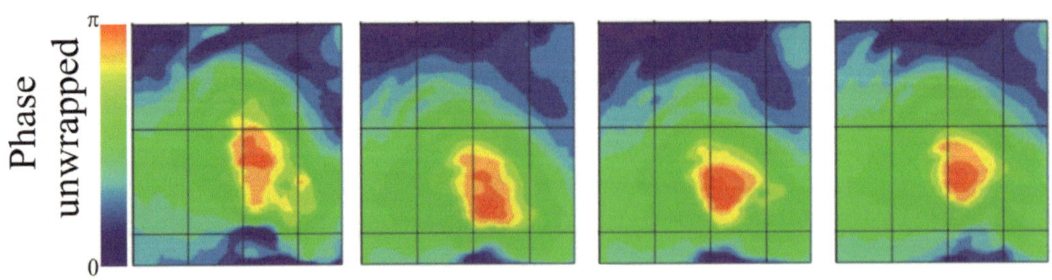

Fig. 11. Moving distributions. Representative Frames. Evolution of the phase, one capture per second.

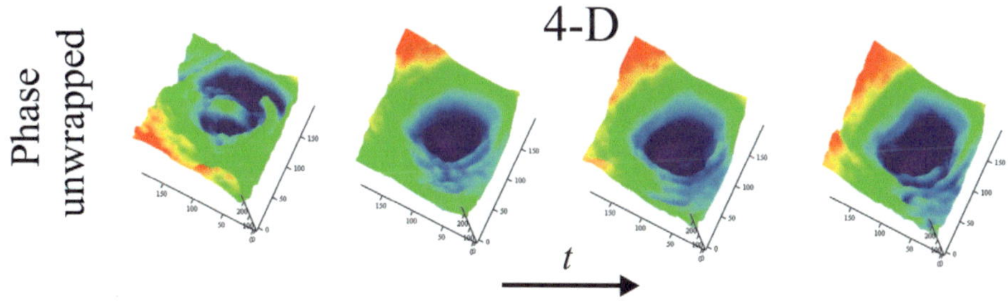

Fig. 12. Moving distributions. Oil drops moving under gravity on a slide. One capture per second.

7.2 Case of nine interferograms

To demonstrate the use of several interferograms, we choose the symmetrical n=N+1 phase steps algorithms, for data processing (case N=8). A constant phase shift value of $2\pi/N$ is employed when using these techniques. The phase shifts of p due to the grid spectra allow the use of a number of polarizing filters that is less than the number of interferograms, simplifying the filter array (See Fig. 13).

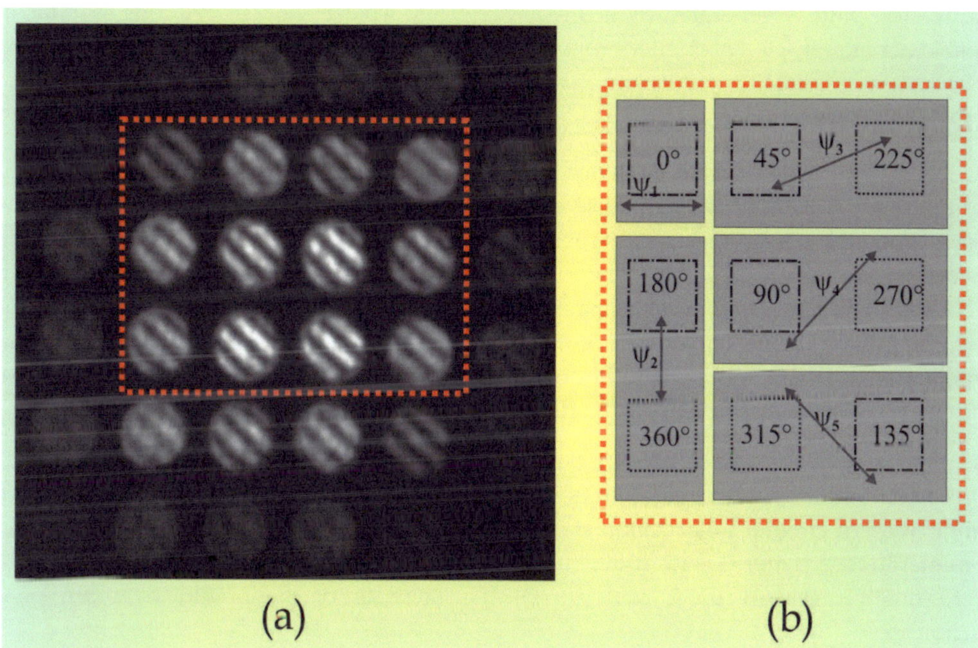

Fig. 13. Interference patterns detected with a polarizing filter at $\psi = 25°$. (a)The interference patterns used are enclosed in the dotted rectangle. (b) Polarizer filter array for N= 8, nine symmetrical interferograms.

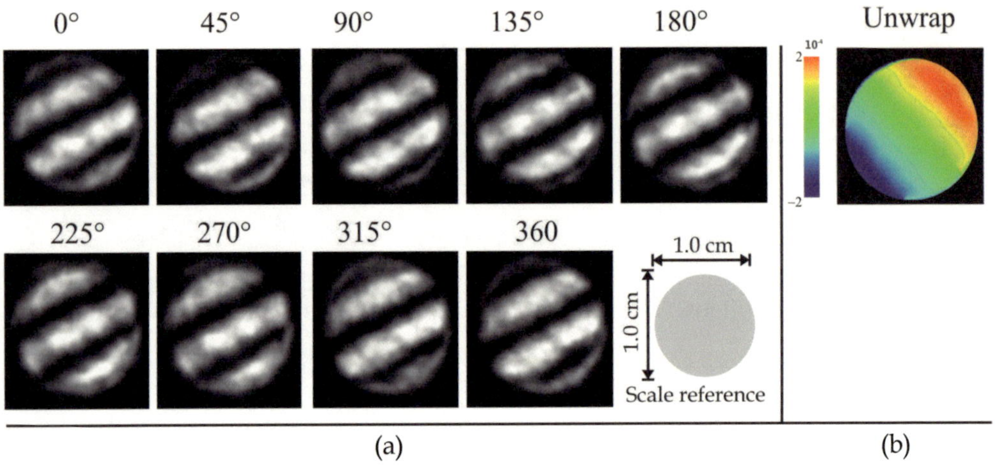

Fig. 14. Tilted wave-front. (a) Nine 45° phase-shifted interferograms (b) Unwrapped phase data map.

Considering the retarders at disposal, according to Rodriguez 2008b, it can be shown that for symmetrical nine, $\psi_1 = 0°$, $\psi_2 = 92.989°$, $\psi_3 = 22.975°$, $\psi_4 = 46.577°$ and $\psi_5 = 157.903°$. In this case, each resulting step ξ corresponds to a 45° phase-shift. The corresponding results and calculated phases are shown in Fig. 14, the experimental results shows the interference pattern generated by a tilted wavefront. That is said before, this systems are capable of obtaining $n = N+1$ interferograms in a single capture (this case, $n \le 12$).

8. Final remark

Theoretical and experimental evidence of π–shifts in the Fourier spectra of phase gratings and phase grids were presented; these shifts are not discussed in the literature as far as we know. When the power spectrum is focused, it tends to hide the effect presented, but it can be of considerable relevance when gratings or grids are used for interferometric applications. For example: in phase shift interferometry, these results are convenient since only one grating displacement is necessary to capture four interferograms, and for the case of polarization phase shifting interferometry, the characteristics of the phase gratings used simplify the placement of polarization filters and allow us to obtain *n*-patterns in one shot.

As a final remark the system presented does not use a physical double window; it generates two beams with variable separation according to the characteristics of the grid used. The combination of circular polarization states and the use of phase-grid enable the generation of *n*-interference patterns with independent phase shifts. This characteristic optimizes the interferometric system used, and allows the analysis of static and dynamic phase distributions.

9. Acknowledgments

Authors thank M. A. Ruiz for his contribution in proofreading the manuscript. Partial support from "Consejo Nacional de Ciencia y Tecnología (CONACYT)" and "Centro de

Investigaciones en Óptica A.C. (CIO)" through projects 290597 (CONACYT-CIO) and 124145 (CONACYT-BUAP) is also acknowledged. Author *NITA* expresses sincere appreciation to *Luisa, Miguel* and *Damian-El* for the support provided, and to CONACYT for grant 102137/43055. Author DISG (Grant:227470/31458) is very grateful to CONACyT for the graduate scholarship granted and expresses sincere appreciation to *Gelizlle.*

10. References

Arrizón V. and Sánchez-De-La-Llave D., (2004). Common-Path Interferometry with One-Dimensional Periodic Filters, *Opt. Lett.* Vol. 29, pp. 141-143.

Azzam R. M. A., (1982). Division-of-Amplitude Photopolarimeter (DOAP) for the Simultaneous Measurement of all Four Stokes Parameters of Light, *Opt. Acta,* Vol. 29, pp. 685-689.

Barrekette E. S., Freitag H., (1963). Diffraction by a Finite Sinusoidal Phase Grating, *IBM Journal*, pp. 345-349.

Barrientos-García B., Moore A. J., Pérez-López C., Wang L., and Tschudi T.,(1999) Spatial Phase-Stepped Interferometry using a Holographic Optical Element, *Opt. Eng.* Vol. 38, pp. 2069-2074.

Cornejo-Rodríguez A. (1992). Ronchi Test, in *Optical Shop Testing.*

Goodman, Jooseph W.; J. Wiley & Sons (1988). *Introduction to Fourier optics.*

Kneubühl F.,(1969). Diffraction Grating Spectroscopy, *Appl. Opt.*, Vol. 8, pp. 505-519.

Kreis T., (1986). Digital Holographic Interference-Phase Measurement Using the Fourier-Transform Method, *J. Opt. Soc. Am. A,* Vol. 3, pp. 847-855.

Malacara D., Servin M., and Malacara Z.; Marcel Dekker (1998). *Interferogram Analysis for Optical Testing.*

Meneses-Fabian C., Rodriguez-Zurita G., and Arrizon V. (2006). Common-Path Phase-Shifting Interferometer with Binary Grating, *Opt. Commun.* Vol. 264, pp. 13-17.

Novak M., Millerd J., Brock N., North-Morris M., Hayes J. and Wyant J.C., (2005). Analysis of a micropolarizer array-based simultaneous phase-shifting interferometer, *Appl. Opt.*, Vol. 44, pp. 6861-6868.

Rodriguez-Zurita G., Meneses-Fabian C., Toto-Arellano N., Vázquez-Castillo J. F. and Robledo-Sánchez C., (2008). One-Shot Phase-Shifting Phase-Grating Interferometry with Modulation of Polarization: case of four interferograms," *Opt. Express,* Vol. 16, 7806-7817.

Rodriguez-Zurita G., Toto-Arellano N. I., Meneses-Fabian C. and Vázquez-Castillo J. F., (2008). One-shot phase-shifting interferometryfive, seven, and nine interferograms, *Opt Letters*, Vol. 33, pp. 2788- 2790.

Ronchi V.,(1964). Forty Years of History of a Grating Interferometer, *Appl. Opt.*, Vol. 3, pp. 437-451.

Schwieder J., Burow R., Elssner K.-E., Grzanna J., Spolaczyk R. and Merkel K., (1983). Digital Wave-Front Measuring Interferometry: some systematic error sources, *Appl. Opt.*, Vol. 22, pp. 3421-3432.

Thomas D. A. and Wyant J. C., (1976). High Efficiency Grating Lateral Shear Interferometer, *Opt. Eng.*, Vol. 15, pp. 477.

Wyant J. C., (1973). Double Frequency Grating Lateral Shear Interferometer, *Appl. Opt.*, Vol. 12, pp. 2057-2060.
Wyant J. C., (2003). Dynamic Interferometry, *Optics & Photonics News*, Vol. 14, pp. 36-41.

16

Length Measurement for Optical Transmission Line Using Interferometry

Takeshi Tsujimura[1], Koichi Yoshida[2] and Kuniaki Tanaka[3]
[1]*Department of Mechanical Engineering, Saga University*
[2]*Department of Information and Systems Engineering, Fukuoka Institute of Technology*
[3]*NTT corporation*
Japan

1. Introduction

More than half a million kilometres of aerial/underground optical telecommunication cable have been installed in Japan. A guaranteed telecommunication service is replacing the best-effort service as real-time streaming delivery increases. Fiber-to-the-home (FTTH) systems are mainly designed by passive optical network (PON) topology (ITU-T Recommendation G.983.1, 1998; Hornung et al., 1990; Sankawa et al., 2006). They are not allowed for even an instantaneous interruption day and night, because numerous customers suffer damage from it as 1 x 10^9 bits of data will be lost every second.

Telecommunication cables are required to be relocated in road construction and work on the water supply. Each optical fiber leading from an optical line terminal (OLT) in a telephone office to a customer's optical network unit (ONU) must be cut and reconnected as shown in Fig. 1. Customers expect real-time transmission for high-quality communications to continue uninterrupted, especially for video transmission services.

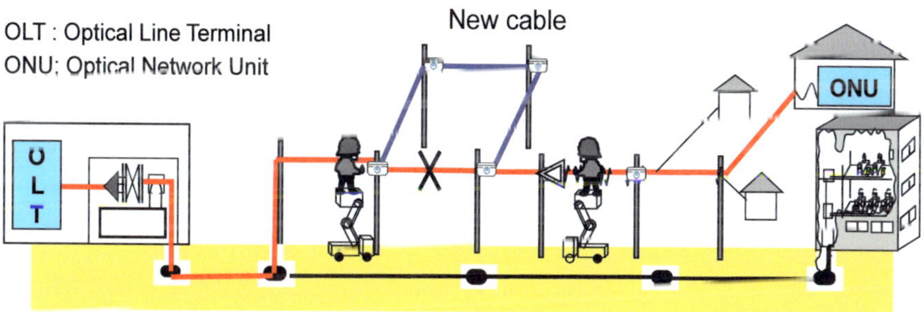

Fig. 1. Optical fiber cable replacement.

Several studies have reported on protection of PON system (Xu & Ho, 2001; Tanaka & Horiuchi, 2008). Electrical transmission apparatus can maintain communication without interruption, even when optical cables are temporarily cut. The system is complicated and any transmission delay during O/E conversion is fatal to real-time communication. Although it is desirable to directly switch the transmission medium itself, it had been thought that some data bits would inevitably be lost during the replacement of optical fibers. An optical fiber cable transfer splicing system has been developed to minimize cable replacing time (Watanabe et al., 1992; Tanaka et al., 2002). It disconnects an optical fiber and exchanges it for another in a flash. It takes 30 ms to switch a transmission line, and more than 2 seconds to restore communications with, for example, a gigabit Ethernet passive optical network (GE-PON) (Azuma et al., 2008). Neither the PON protection technique nor the splicing system avoids communication failures because they cannot reduce the disconnecting time absolutely to zero.

The authors have developed an interruption-free replacement method for in-service telecommunication lines, which can be applied to the current PON system equipped with conventional OLTs and ONUs (Tsujimura et al., 2010; Tanaka et al., 2009; Yoshida & Tsujimura, 2010). Two essential techniques were newly proposed; a measurement method and an adjustment system for the transmission line length. The latter continuously lengthens/shortens the transmission line over very long distances without losing transmitted data based on free space optics (FSO) system (Tsujimura et al., 2009b; Yoshida et al., 2009). The former distinguishes the difference between the duplicated line lengths using interferometry (Tsujimura et al., 2009a). Interferometry is the technique of diagnosing the properties of two or more lasers or waves by studying the pattern of interference created by their superposition. It is an important investigative technique in the fields of astronomy, fiber optics, optical metrology and so on. Studies on optical interferometry are reported to improve tiny optical devices (Saunders & Hardcastle, 1994; Cao & Cartledge, 2002; Torregrosa et al., 2007). We have applied the optical interferometry technique to measure length of several kilometers of transmission lines with a 10 mm resolution in this study.

This paper reveals the problems in replacing optical cables first. It proposes a switching method of optical fiber transmission lines next. We design a shunt system for in-service optical fibers. Then an optical line length measurement method is studied to distinguish the difference of two lines by evaluating interfered optical pulses. An optical line switching system is designed, and a line length adjustment system is prototyped. The proposed system is applied to a 15 km GE-PON optical fiber network while adding a 10 m extension to show the efficiency of this approach when replacing in-service optical cables. Finally, we discuss some proposal for improvement of line length measurement.

2. Optical cable replacement

First of all, fundamental investigations were conducted on the effects of transmission interruption on the telecommunication service.

When an optical fiber line was intensionally disconnected in a short time, transmission broke off and was restored by PON transmission system. Figure 2 and 3 show the experimental results of transmission characteristics when GE-PON optical communication line is switched using high-speed line transfer. Figure 2 expresses the relationship between interruption time and optical network unit (ONU) restoration time for 4 ONUs. Even if an optical fiber line was reconnected only in 100 ns, the transmission system took about 6 seconds to restore the telecommunication service. Up to 8 ONUs can be connected to an OLT. It took more than 10 seconds if eight ONUs were connected to an OLT as shown in Fig. 3, which indicates that the restoration time increases in relation to the number of ONUs.

These results suggest that communication service is inevitably interrupted no matter how quick the conventional method reconnects the transmission line. That is why a new switching method is necessary to maintain communications completely during optical cable replacement and thus provide a highly reliable network.

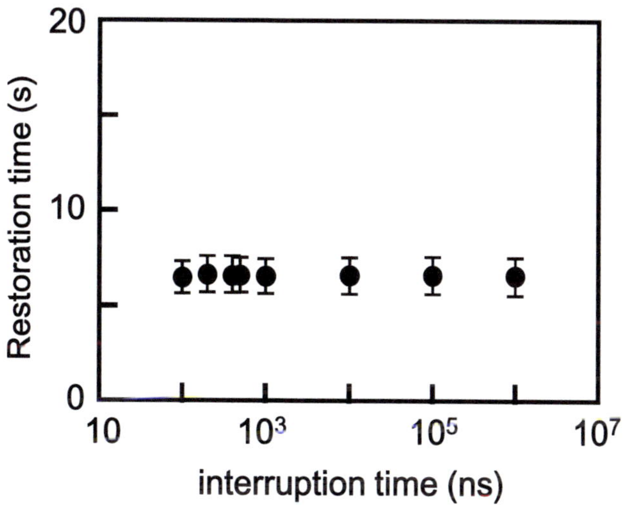

Fig. 2. Relationship between interruption time and restoration time.

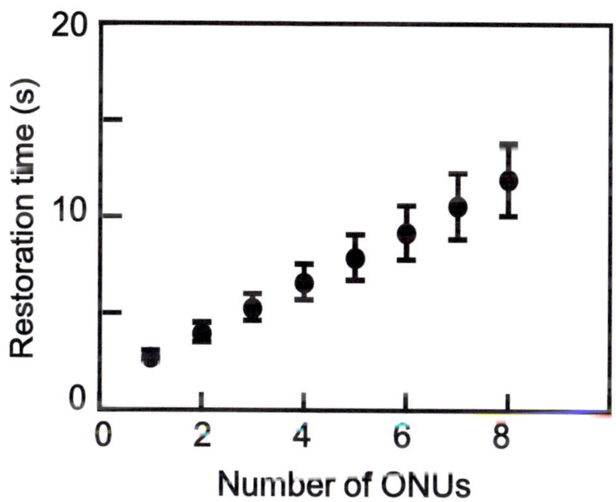

Fig. 3. Relationship between number of ONUs and restoration time.

3. Optical fiber line switching

3.1 Optical line switching technique

We propose an optical line switching technique that uses duplex transmission lines as shown in Fig. 4. Normal transmission system consists of an OLT, ONUs, and an optical fiber labeled 'regular line' in Fig. 4. Signals are bidirectionally transmitted along the line through a wavelength independent optical coupler (WIC) and an optical splitter. Our proposed technique requires a detour optical fiber to temporarily transmit signals. It also uses a test light whose wavelength is different from ones for signal transmission. The test light is created by an optical frequency-chirped pulse light source, and reaches an oscilloscope through the optical splitter, the regular/detour line, and the WIC. An optical line length measuring method detects the optical path difference between the detour and the regular line with these arrangements. An optical line length adjusting method controls the detour line length in virtue of FSO device.

Because either of two lines are managed to keep connection and to transmit signals at any time, the entire transmission system are secured against interruption. One of optical cables can be freely cut or reconnected, while the other cable maintains connected and transmits signals.

In order to avoid suspending transmission, the communication conditions have to be kept in the physical layer, for example, carrier power or signal phase. Thus, the following two conditions must be met while exchanging an in-service line for a spare line,

a. No transmission signals must be lost,
b. Transmission time in a protocol must be maintained.

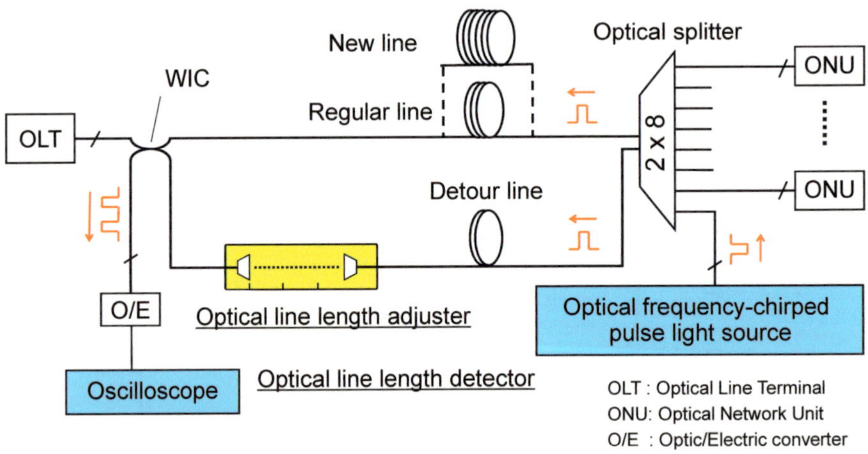

Fig. 4. Duplex optical fiber transmission system.

The conditions are satisfied by equalizing the transmission times of both signals transmitted through the duplicated lines. We accomplish this by tuning the length of the detour line.

The optical line replacement procedure, illustrated in Fig. 5 where the transmission lines are simplified, is as follows:

Step 0. Signals are ordinarily transmitted through an optical fiber (regular line) between an OLT and ONUs through a 2x8 optical splitter.

Step 1. A detour line is established between the WIC and the 2x8 optical splitter. A transmission system in the telephone office commonly equips with a WIC next to an OLT. A 1650 nm LD is connected to one of the branches of the 2x8 optical splitter.

Step 2. The detour line length is measured with a 1650 nm test light using an optical line length measuring method, and is adjusted to the same length as the regular line using an optical line length adjusting method.

Step 3. Once the lengths of the two lines coincide, the transmission signals are also launched into the detour line.

Step 4. The regular line is cut. Transmission keeps connected because the signals travels through the detour line. A long-wavelength pass filter (LWPF) is temporarily installed in the new line.

Step 5. The LWPF passes the test light alone through the new line.

Step 6. The test light measures the lengths of the new line and the detour line. The detour line is adjusted to the new line while communications are maintained. The LWPF keeps preventing the optical pulses for telecommunication from traveling through the new line.

Step 7. The LWPF is then removed and the transmission is duplicated. The detour line is finally cut off.

Step 8. Replacement of the optical fiber line is completed without any interruption.

Step 0: Signals travel through regular line.

Step 1: Set detour line.

Step 2: Adjust detour line to regular line.

Step 3: Multiplex optical signals.

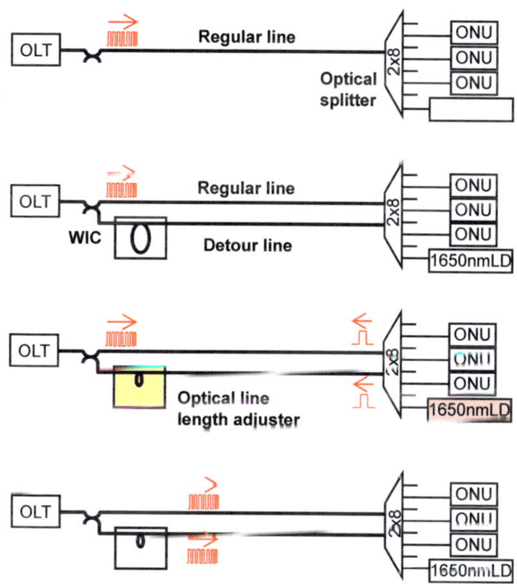

Fig. 5. Part I

Step 4: Cut regular line.

Step 5: Connect new line.

Step 6: Adjust detour line to new line.

Step 7: Cut off detour line.

Step 8: Complete replacement.

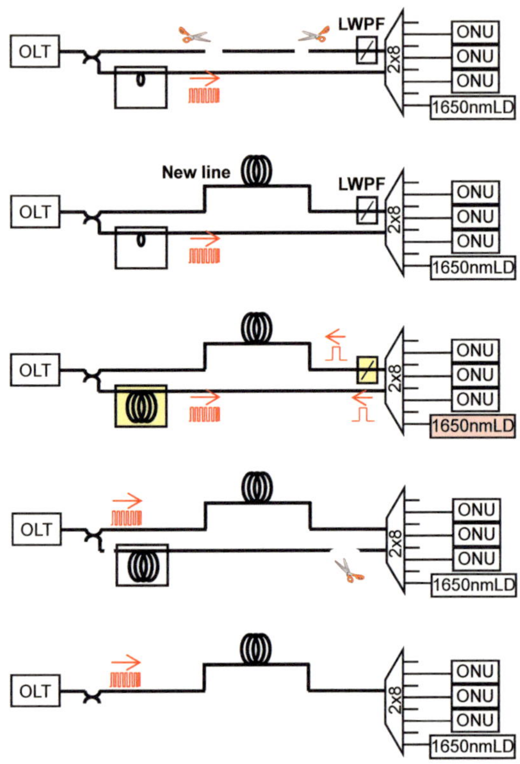

Part II

Fig. 5. Optical line replacement procedure.

3.2 Optical line shunt system

The authors have designed an optical line shunt system, and a switching procedure for three wavelengths, namely 1310 and 1490 nm for GE-PON transmission, and 1650 nm for measurement. The optical line shunt system is installed in a telephone office. It is composed of an optical line length detector and an optical line length adjuster. Figure 6 shows an individual optical fiber line in a GE-PON transmission system. Optical pulse signals at two wavelengths are bidirectionally transmitted through a regular line between customers' ONUs and an OLT in a telephone office via a wavelength independent optical coupler (WIC) and a 2x8 optical splitter (2x8 SP), respectively.

A test light at a wavelength different from those of the transmission signals is sent from one of the optical splitter's ports to the duplicated lines. We use a 1650 distributed feedback laser diode (DFB-LD) whose pulse width is 200 ns as a test light source. An oscilloscope is connected to the optical coupler to detect the test light through a long-wavelength pass filter (LWPF). The optical line length adjuster is an FSO application (Yoshida et al., 2011; Willebrand et al., 1999). Some optical switches (SW) and optical fiber selectors (FS) control the flow of the optical signals managed by a PC. The optical pulses are compensated by 1650 and 1310/1490 nm amplifiers (Fukada et al., 2008).

The proposed method temporarily provides a duplicate transmission line as shown in Fig. 6 to replace optical fiber lines. A detour line is prepared in advance through which to divert signals while the existing line (regular line) is replaced with a new one.

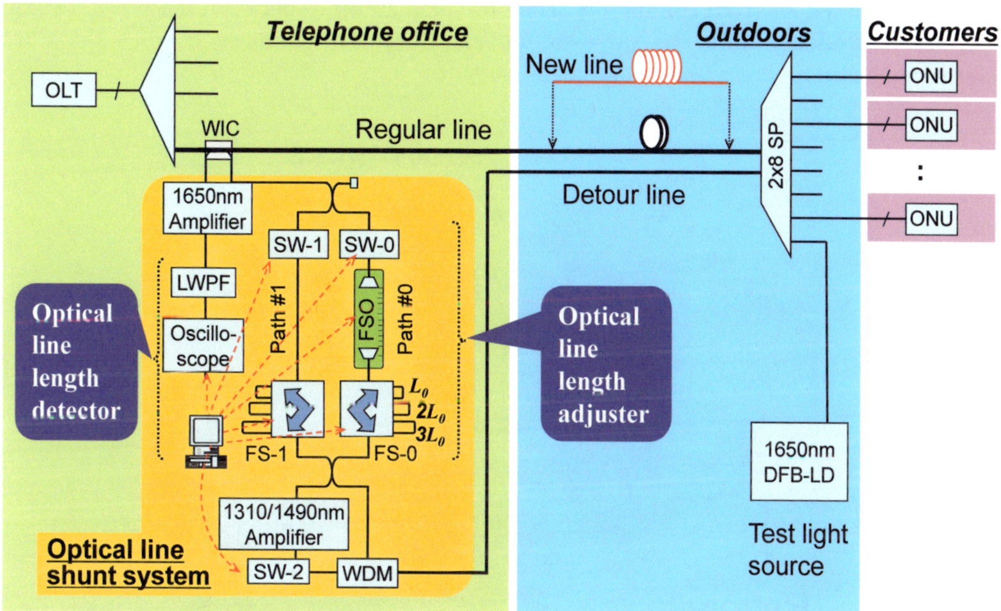

Fig. 6. Shunt system of optical fiber transmission line.

This system transfers signals between the two lines. Signals are duplicated at the moment of changeover to maintain continuous communications. The signals travel separately through the two lines to a receiver. A difference in the line lengths leads to a difference in the signals' arrival times. A communication fault occurs if, as a result of their proximity, the waveforms of the two arriving signals are too blurred for the signals to be identified as discrete. Thus it is important to adjust the lengths of both lines precisely.

We investigated the tolerance of the multiplexed signal synchronicity in advance. The transmission quality is observed by changing the difference between the duplicated line lengths. The results show that the transmission linkage is maintained if the difference is within 80 mm as with GE-PON. A multiplexed signal cannot be perceived as a single bit when the duplicated line lengths have a larger gap for 1 Gbit/s transmission. Because these characteristics depend on the periodic length of a transmission bit, the requirement is assumed to be more severe when the method is applied to higher-speed communication services.

Experiments determined that the tolerance of the difference in line length is 80 mm with regard to the GE-PON transmission system.

Accordingly, the proposed system controls the adjustment procedure so that the difference in length between the detour and regular lines is adjusted within 80 mm.

We next constructed a prototype of the optical line shunt system to apply to a GE-PON optical fiber line replacement according to the procedure described above.

An optical line length adjuster, shown in Fig. 7, was installed along the detour line. The adjuster was equipped with two retroreflectors, which directly faced each other as illustrated in Fig. 8. Optical pulses were transmitted through an optical fiber, divided into three wavelengths by wavelength division multiplexing (WDM) couplers, and discharged separately into the air from collimators. They traveled 10 times between the retroreflectors, and were introduced into the opposite optical fiber. The number of reflections was determined based on the retroreflector arrangement.

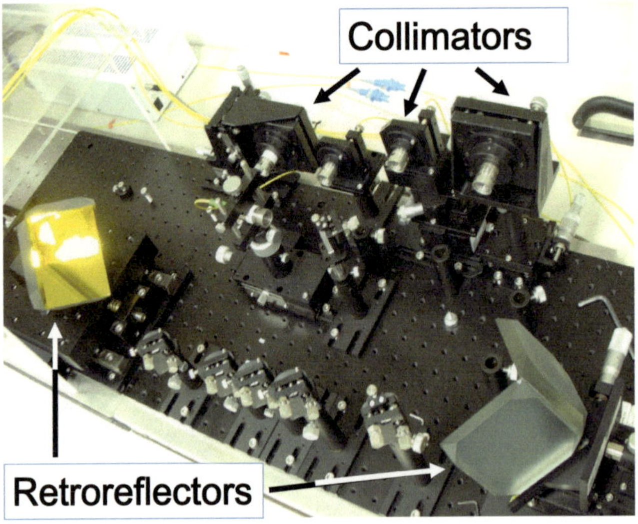

Fig. 7. Photos of optics line length adjuster.

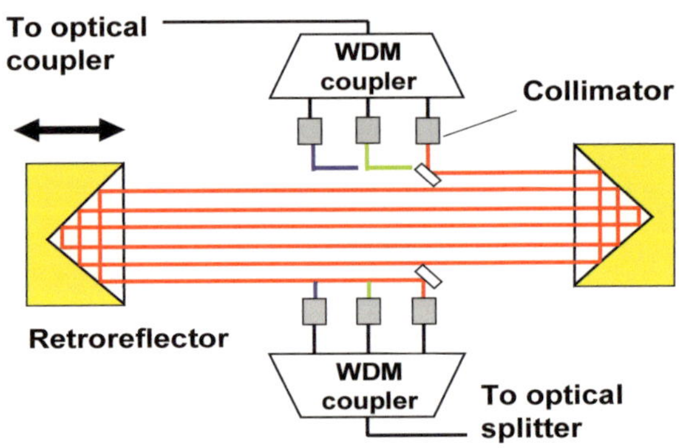

Fig. 8. Free-space optics line length adjuster.

The detour line between the retroreflectors consisted of an FSO system. The detour line length could be easily adjusted by controlling the retroreflector interval with a resolution of 0.14 mm with a motorized sliding stage. Optical pulses travel **n**-times faster in the air than in an optical fiber, where **n** is the refractive index of the optical fiber. Thus the optical line length adjuster lengthens/shortens the corresponding optical fiber length, ***L***

$$L = k\,W / n, \tag{1}$$

where ***k, W, n*** are the number of journeys between the retroreflectors, the retroreflector interval, and the refractive index of optical fiber, respectively.

The FSO lengthens the optical line length up to L_0.

$$L_0 = k\,(W_{max} - W_{min})/\,n, \qquad (2)$$

where W_{max} and W_{min} are the maximum and minimum retroreflector intervals. The retroreflector stroke of our prototype, $W_{max} - W_{min}$, was 0.3 m, the refractive index, n, of the optical fiber was 1.46, the number of journeys, k was 10, and the optical line span, L_0, tuned by the adjuster was 2 m. Our prototype adjusted the detour line length with a resolution of 0.1 mm.

The limit of the adjustable range is a practical problem when this system is applied to several kilometers of access network. Therefore, we designed an optical line length accumulator. The optical line length adjuster contains two optical paths, #0 and #1, as shown in Fig. 6. An optical switch (SW) and an optical fiber selector (FS) are installed in each path. Optical switches control the optical pulse flow. Each optical fiber selector is equipped with various lengths of optical fiber, for example L_0, $2L_0$ and $3L_0$. The path length can be discretely changed by choosing any one of them.

The optical line length adjuster can extend the detour line as much as required using the operation shown in Fig. 9. First, the FSO system lengthens path #0 by L_0 by gradually increasing the retroreflector interval. After the optical fiber selector, FS-1, has selected an optical fiber of length L_0, the active line is switched from path #0 to path #1 at time t_0. The FSO system then returns to the origin, and the optical fiber selector, FS-0, selects an optical fiber of length L_0 instead to keep the length of path #0 at L_0. The FSO system increases the retroreflector interval again at time $2t_0$ to repeat the same operation. In this way the adjuster accumulates spans extended by the FSO system. The scanning time of our prototype, t_0, was 10 seconds, because the retroreflector moved along the motorized sliding stage at 30 mm/s.

The optical line length adjuster enables us to lengthen/shorten the detour line while continuing to transmit optical signals.

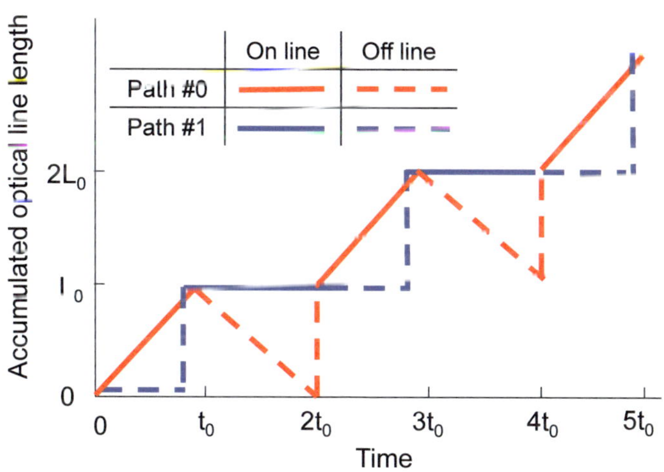

Fig. 9. Accumulation of optics line length.

4. Optical line length measurement

The proposed system uses laser pulses at a wavelength of 1650 mm to measure the optical path length. They are introduced from an optical splitter, duplicated, and transmitted toward the OLT through the active and detour lines.

They are distributed by an optical coupler just in front of the OLT, and observed with an oscilloscope. The conventional measurement method evaluates the arrival time interval between the duplicated signals, and converts it to the difference between the lengths of the regular line and the detour line at a resolution of 1 m.

The difference in line length, ΔL is described as

$$\Delta L = c \cdot \Delta t \;/\; n, \tag{3}$$

where c is the speed of light, Δt is the difference between the signal arrival times for the regular and detour lines, and n is the refractive index of optical fiber.

Figure 10 shows the received pulses observed with an oscilloscope. When the detour line was 100 m shorter than the regular line, pulses traveling through the detour reached the oscilloscope about 500 ns earlier than through the regular line as shown in Fig 10 (a). The proposed system lengthened the detour line using the optical path length adjuster. Consequently, the former pulse approached the latter. Figure 10 (b) is an example when the gap was shorten to 20 m, where the regular line pulse arrived after a delay of 100 ns. This method failed if the difference between the line lengths was less than 1 m, because the two pulses combined as shown in Fig. 10 (c).

That is why the authors have developed an advanced technique for measuring a difference of less than 1 m between optical line lengths. Interferometry enables us to obtain more detailed measurements even when the optical pulses combine. A chirped light source generates interference in the waveform of a unified pulse.

A 1650 nm laser pulse, discharged from the test light source, is devided by the 2x8 optical splitter. Two pulses travels through the detour and regular line separately toward the oscilloscope.

Both pulses along the regular line, $E\,(L_1, t)$, and the detour line, $E\,(L_2, t)$ are expressed as

$$E\,(L_1, t) = A_1 \exp\,[\,-i\,(k \cdot n \cdot L_1 - \omega_1 \cdot t + \phi_0)\,]\,, \tag{4}$$

$$E\,(L_2, t) = A_2 \exp\,[\,-i\,(k \cdot n \cdot L_2 - \omega_2 \cdot t + \phi_0)\,]\,, \tag{5}$$

where A_j, k, n, L_j, ω_j, t, and ϕ_0 denote amplitude, wavenumber in a vacuum, refractive index of optical fiber, line length, frequency, time, and initial phase, respectively. Subscript j represents the regular line, 1, or the detour line, 2.

The intensity of a waveform with interference, I, is calculated by taking the square sum as

$$I = |\,E\,(L_1, t) + E\,(L_2, t)\,|^{\,2}$$

$$= A_1^2 + A_2^2 + 2\,A_1\,A_2 \cos\,(k \cdot n \cdot \Delta L - \Delta\omega \cdot t)\,, \tag{6}$$

where ΔL and $\Delta\omega$ represent the differences between line lengths and frequencies, respectively.

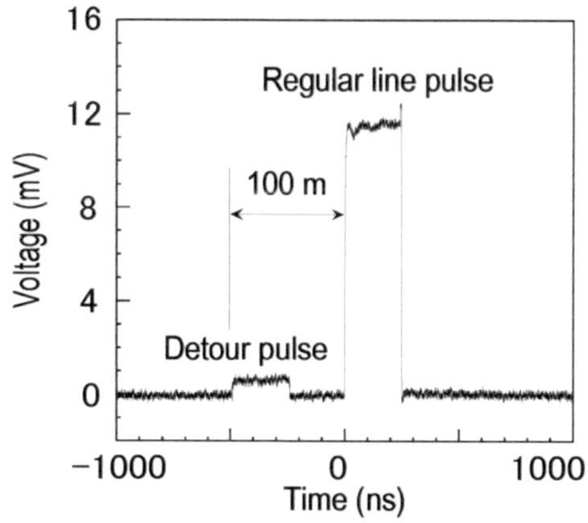

(a) Optical path difference 100m.

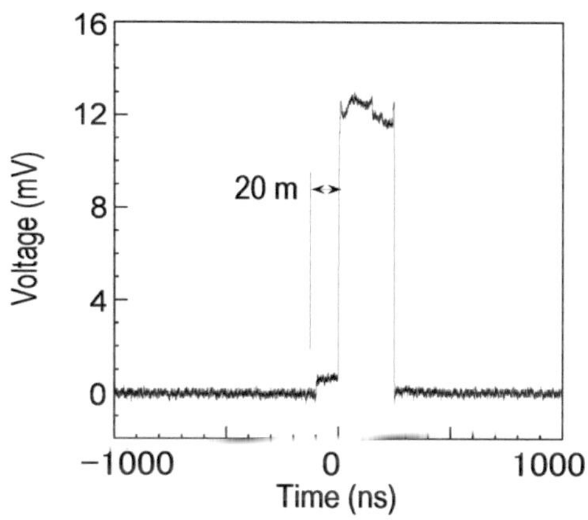

(b) Optical path difference 20m.

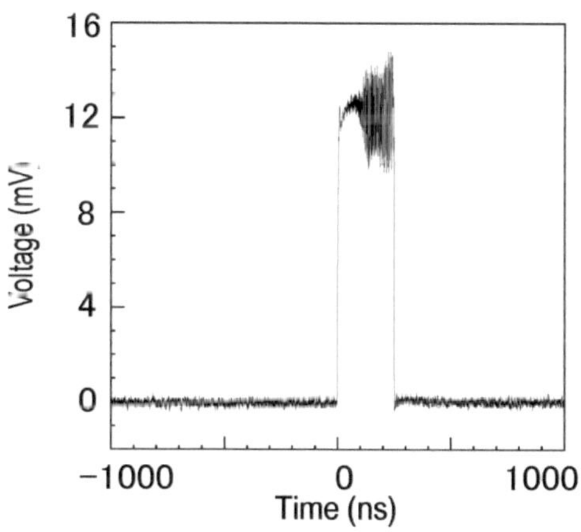

(c) Optical path difference 1m.

Fig. 10. Delay of duplicated received pulse.

The waveform with interference depends on the delay between the pulses' arrival times. Experiments were carried out to obtain time-domain waveforms as shown in Fig. 11. When the optical path difference was 1 m, the waveform contained high-frequency waves as shown in Fig. 11 (a). The less the gap became, the lower-frequency the interfered waveform was composed of. Figures 11 (b) and (c) express that phenomenon. When the lengths of two lines coincided, a quite low-frequency waveform was observed as shown in Fig. 11 (d).

A Fourier-transform spectrum reveals the characteristics. When the optical path difference was 1 m, the waveform with interference was composed of the power spectrum shown in Fig. 12 (a). The peak power represented that the major frequency component was around 700 MHz. Figure 12 (b) and (c) indicate that the peak powers for gaps of 0.5 and 0.1 m were around 400 and 70 MHz, respectively. It became difficult to determine the peak for smaller gaps, because the frequency peak became so low that it was hidden by the near direct-current part of the frequency component. When the lengths of duplicated lines coincided, the power spectrum was obtained as Fig. 12 (d).

An evaluation of the frequency characteristics in the interfered waveforms showed that the peak power frequencies are proportional to the difference between the line lengths from -1 to 1 m as shown in Fig. 13, where the horizontal and vertical axes express the optical path difference and the frequency for the maximum power of Fourier-transformed spectrum, respectively. This result helps us to determine the optimal position for adjustment. The optimal position where the line lengths coincide can be estimated by extrapolating the data.

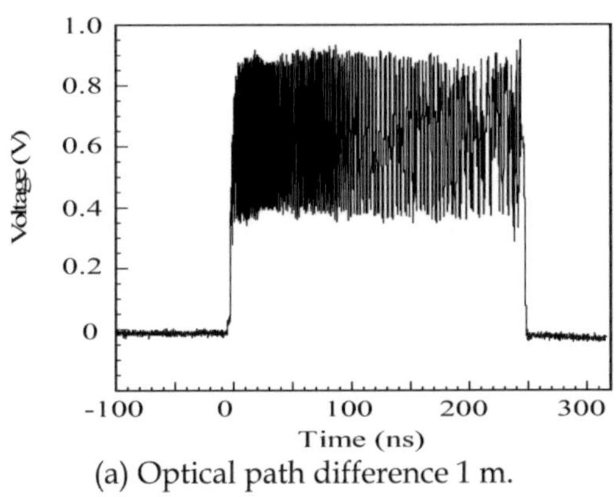

(a) Optical path difference 1 m.

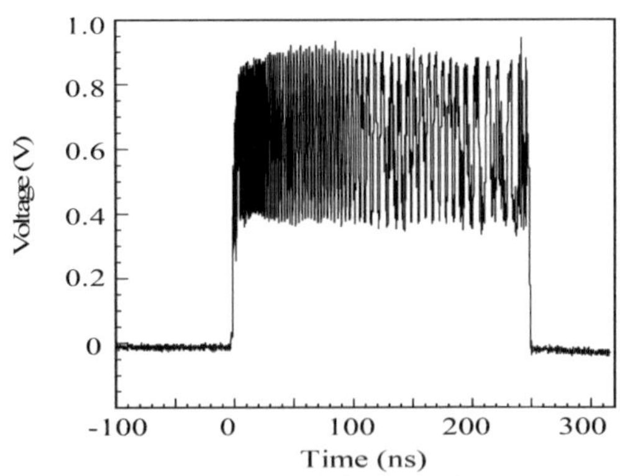

(b) Optical path difference 0.5 m.

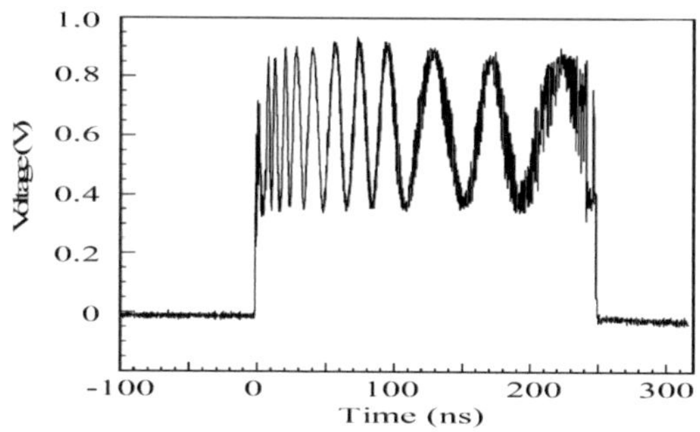

(c) Optical path difference 0.1 m.

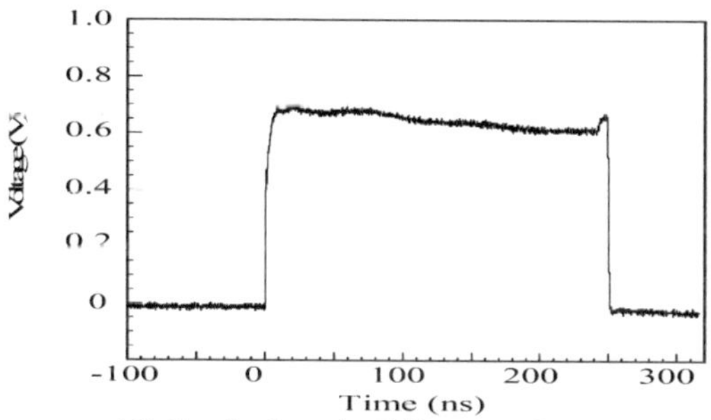

(d) Optical path difference 0 m.

Fig. 11. Time-domain unified optical pulses.

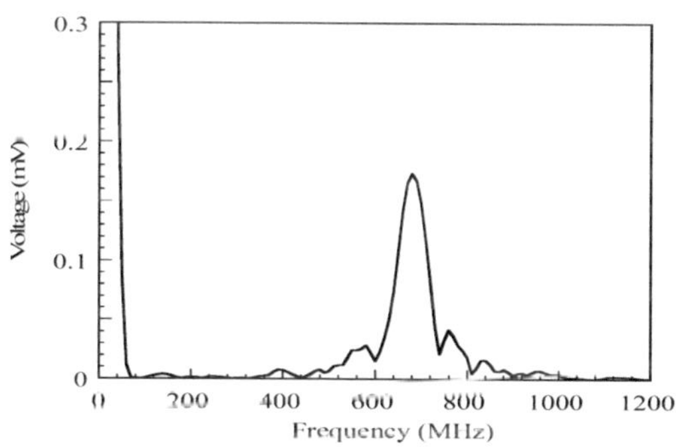

(a) Optical path difference 1 m.

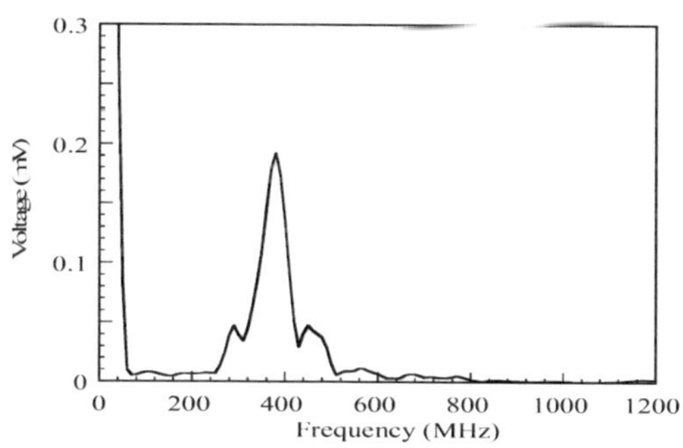

(b) Optical path difference 0.5 m.

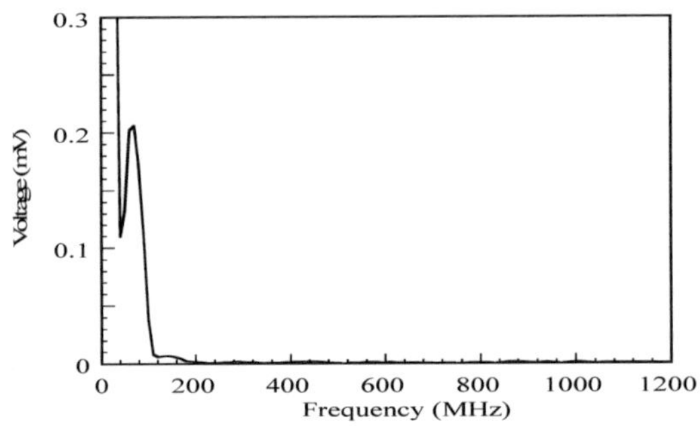

(c) Optical path difference 0.1 m.

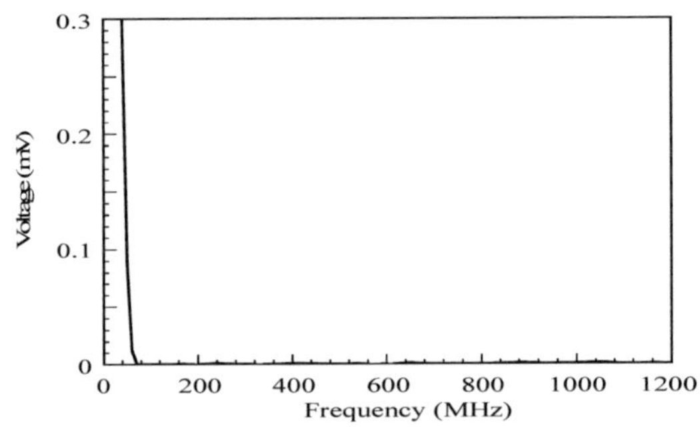

(d) Optical path difference 0 m.

Fig. 12. Frequency-domain unified optical pulses.

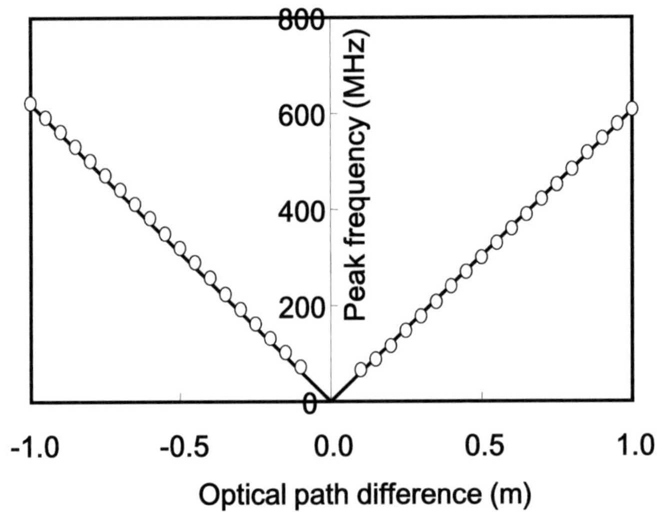

Fig. 13. Estimation of line length coincidence.

Our investigation has established a technique for distinguishing the difference between line lengths to an accuracy of better than 10 mm by analyzing interfering waveforms created by chirped laser pulses. We have realized a complete length measurement for optical transmission lines from 100 m to 10 mm.

We finally applied the prototype of the optical line shunt system to a 15 km GE-PON optical transmission line replacement. Photographs of our experimental setup are shown in Fig. 14. Figure 14 (a), (b), and (c) display a whole system for experimental optical transmission line, the side view, and the perspective view of the optical line length adjuster, respectively.

A 10 m optical fiber extension was added to the transmission line, while optical signals were switched between the duplicated lines during transmission. We have evaluated the frame loss that occurred during optical line replacement by measuring with a SmartBit network performance analyzer (Spilent Communications, 2011). Optical line length difference was purposely set at 0, 50, 80 and 120 mm in order to verify the effects of measurement accuracy. Same optical line replacement operation, illustrated in Fig. 5, was conducted for each case. No frame loss was observed at any stage of the replacement procedure if the difference between the duplicated line lengths was less than 80 mm. If the difference exceeded 80 mm, signal multiplexing caused frame loss at step 3 in Fig. 5.

We also evaluated communication quality of the transmission signals through duplicated lines. A 1488.6 nm optical pulse was actually transmitted along both detour and regular lines. Figure 15 shows the eye diagrams of the received signals with regard to optical path difference, ΔL, 1, 22, 44, 66, 88, and 110 mm. If two line lengths were close, communication quality was satisfactory as the eye opened wide in the diagram. When the difference became larger, the transmitted signals turned out erroneous because binary bit patterns were breaking.

As a result, we confirmed that the optical signals were completely switched between the regular, detour, and new lines on condition that the line length was adjusted with sufficient accuracy. The experimental results proved that our proposed system successfully relocated an in-service broadband network without any service interruption.

5. Improvement of line length measurement

5.1 Sensitivity improvement due to narrow pulse width

We used a 1650 nm DFB-LD as a test light source. Because it was chirped nonlinearly, interference depended on the pulse width. Experiments were conducted to evaluate the effects on the power spectrum of the interfered waveform with regard to several pulse widths 20, 50, 100, 200, and 500 ns.

The results indicated that the test light pulse with narrower width provided higher sensitivity in estimating optical line length coincidence as shown in Fig. 16, where the horizontal and vertical axes express the optical path difference and the frequency for the maximum power of Fourier-transformed spectrum, respectively. Note that in the practical usage of the narrow width pulse with an oscilloscope, measurable range is restricted by the upper limit of the Fourier-transformed frequency. For example, a test light with 20 ns pulse width has a sensitivity of 6 MHz/mm while 1 GHz oscilloscope is necessary to identify an optical path difference of 150 mm.

5.2 Deterministic line length detection using frequency shift method

As discussed in Section 4, it is difficult for the above proposed method to directly identify the point where the lengths of the detour and regular lines coincide because the frequency peak is hidden by the near direct-current part of the frequency component in the power spectrum.

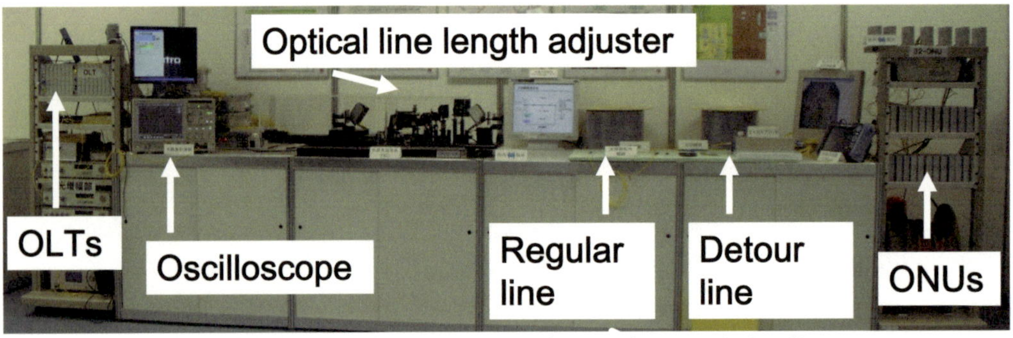

(a) Whole system for experimental optical transmission line.

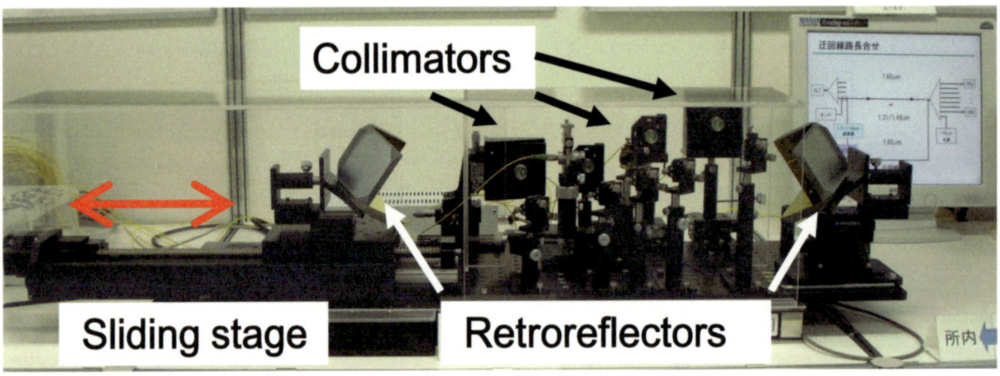

(b) Side view of optical line length adjuster.

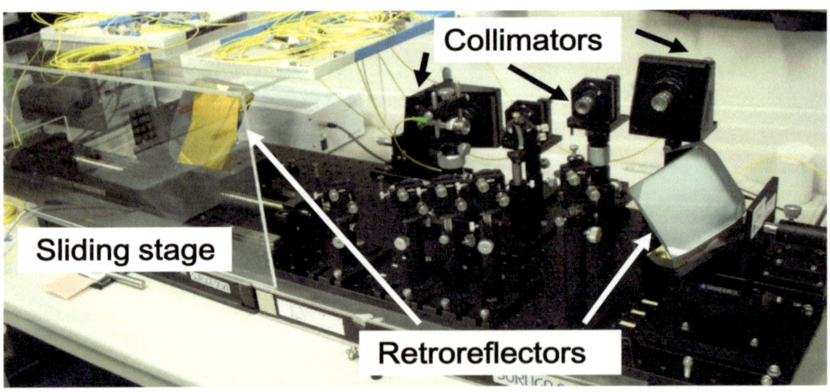

(c) Perspective view of optical line length adjuster.

Fig. 14. Photographs of optical line replacement experiments.

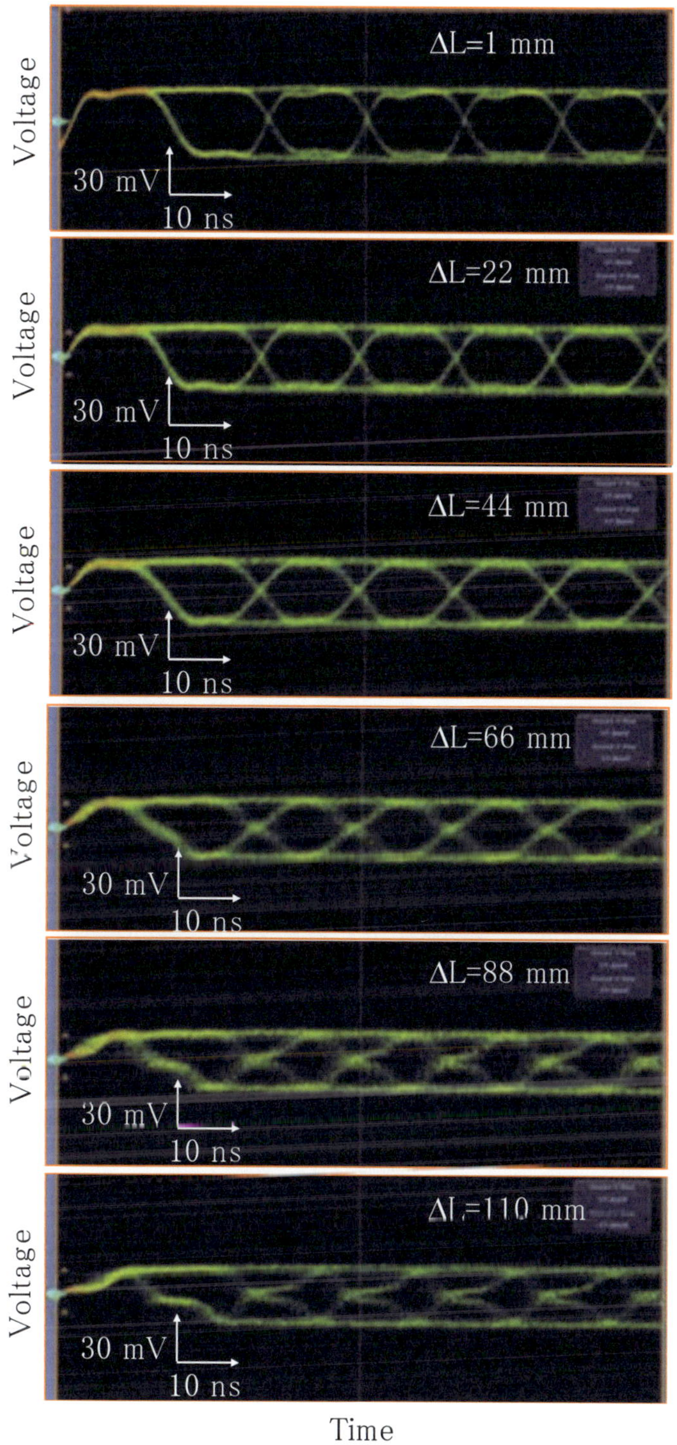

Fig. 15. Eyediagram of duplicated transmission signal.

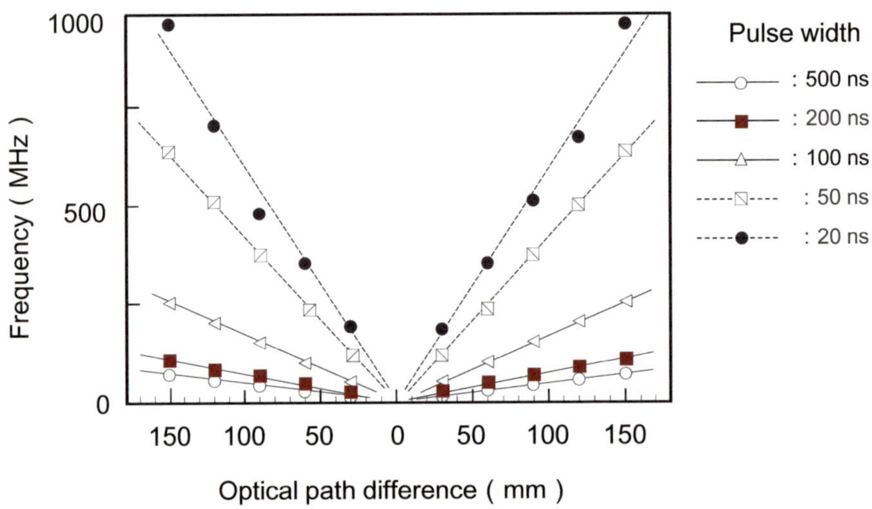

Fig. 16. Estimation of line length coincidence.

We are developing an advanced method to resolve the problem. Our new proposal is to deliberately shift the interference frequency. Figure 17 shows an advanced duplex optical fiber transmission system with frequency shifter. It equips an 80 MHz frequency shifter along the detour line. Other devices are the same as those described in Section 3.2.

We have conducted the same experiments with this advanced transmission system as before. We evaluated interference of the duplicated signals by changing the optical path difference with the optical line length adjuster.

Experimental results are shown in Figs. 18 and 19. Time-domain waveform of unified optical pulses indicates in Fig. 18 that signals along the detour and regular lines have interfered. A Fourier-transformation provided the frequency-domain characteristics of the interference as shown in Fig. 19. It reveals that each interfered waveform has the peak of spectrum power. In case the detour line length agrees with the regular line, the frequency of the peak power is 80 MHz which is as much as we have assigned with the frequency shifter.

An evaluation of the frequency of the peak power in terms of the optical path difference confirmed that the frequencies were proportional to the optical path difference as shown by white circles in Fig. 20, where the horizontal and vertical axes express the optical path difference and the frequency for the maximum power of Fourier-transformed spectrum, respectively.

Black circles in the figure represent the remainders of subtracting 80 MHz from the peak frequency. They also show the linearity as good as we obtained without the frequency filter.

If adopting this method, we can determine the optimal position where the line lengths coincide in high accuracy only by finding the specified value of the frequency shifter instead of extrapolation.

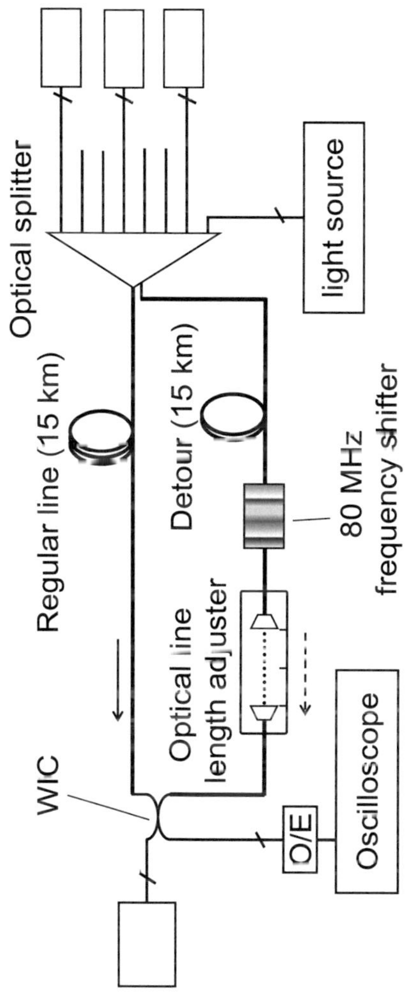

Fig. 17. Advanced duplex optical fiber transmission system with frequency shifter.

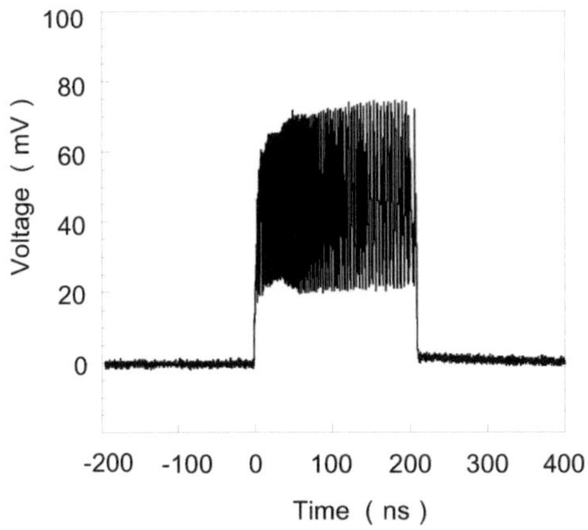

(a) Optical path difference 0.31 m.

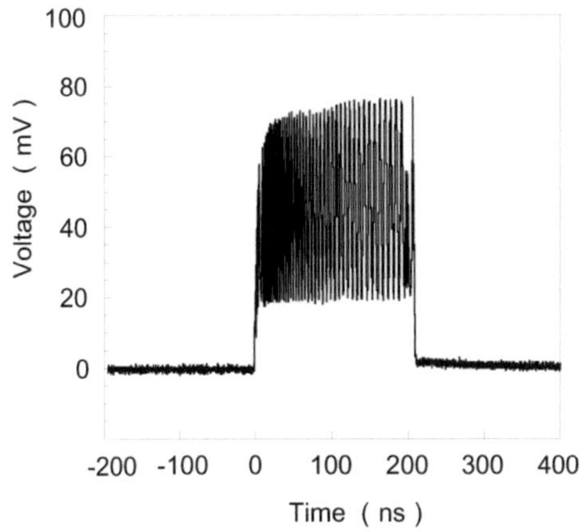

(b) Optical path difference 0.14 m.

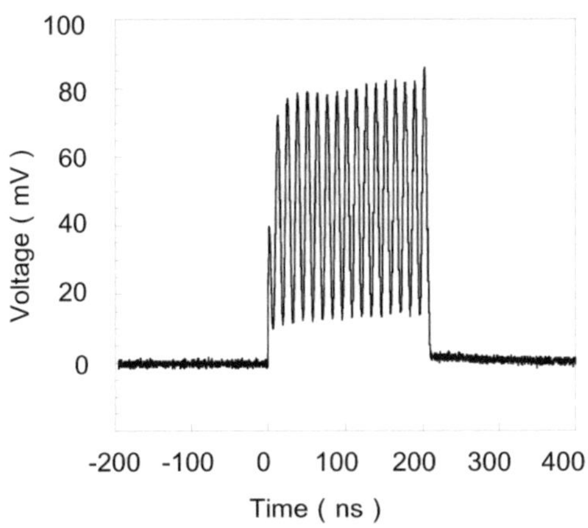

(c) Optical path difference 0 m.

Fig. 18. Time-domain unified optical pulses.

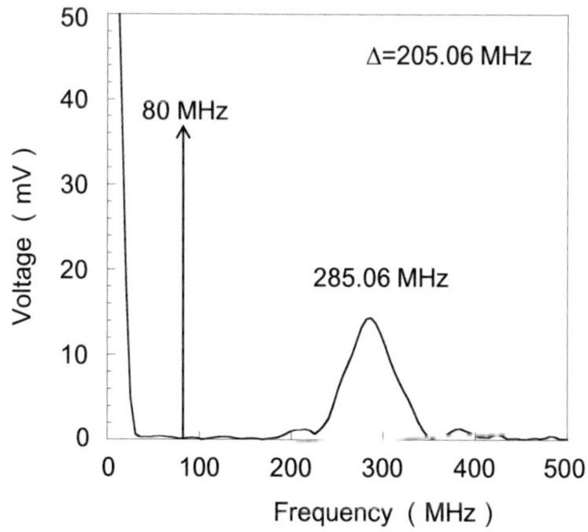

(a) Optical path difference 0.31 m.

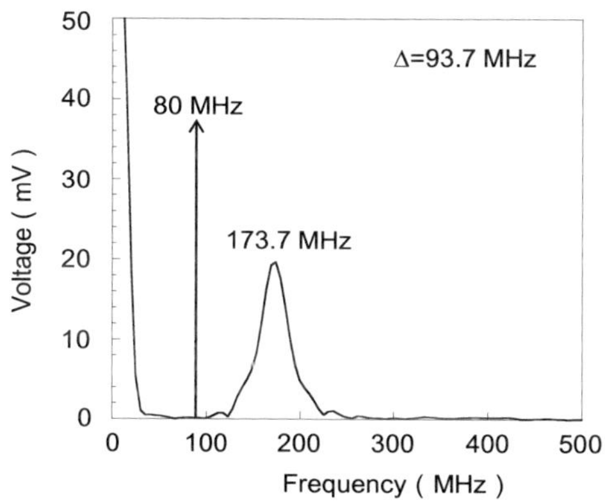

(b) Optical path difference 0.14 m.

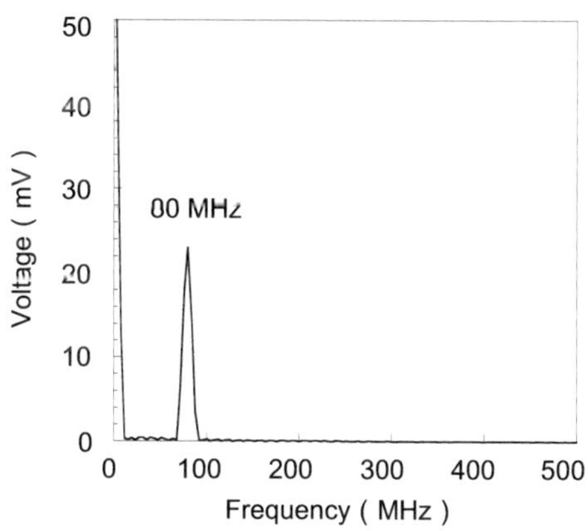

(c) Optical path difference 0 m.

Fig. 19. Frequency-domain unified optical pulses.

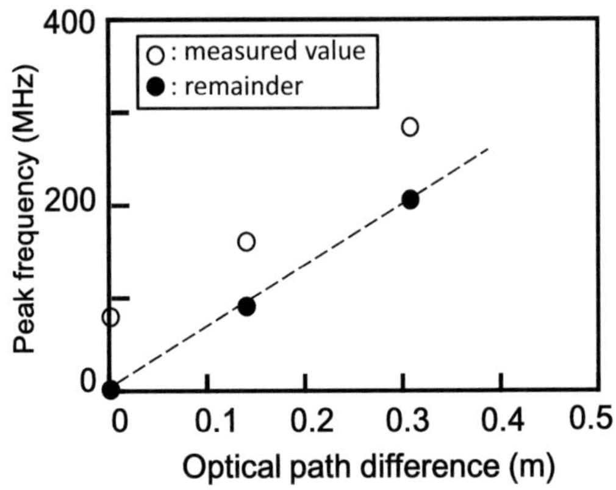

Fig. 20. Deterministic identification of line length coincidence.

6. Conclusion

This paper proposed the high-resolution optical line length measurement technique, which was applied to the switching method for optical transmission lines transferring live optical signals. The method exchanged optical fibers instead of using electric apparatus to control transmission speed.

We have disclosed the requirements for replacement of in-service optical transmission cables after investigating problems of the existing methods.

We next proposed the optical line switching technique and the optical cable replacement procedure that used duplex transmission lines. It was determined that the tolerance of the difference in duplex line length is 80 mm as for the GE-PON transmission system.

The optical line shunt system was designed, and its prototype was actually constructed. An optical line length adjuster, designed based on an FSO system, continuously lengthened the optical line up to 100 m with a resolution of 0.1 mm.

An optical line length measurement technique was investigated in detail by evaluating the duplicated test light pulses. As a result, it successfully identified the difference in length between the duplicated lines from 100 m to 10 mm. An interferometry measurement distinguished the difference between line lengths to an accuracy of better than 10 mm by analyzing interfering waveforms created by chirped laser pulses. This system was applied to a 15 km GE-PON network and succeeded in replacing the communication lines without inducing any frame loss.

We also discussed the improvement of line length measurement. It was suggested that the test light with narrow pulse width could give a higher sensitivity. If using the frequency shifter within the detour transmission line, we could identify the optimal position where the line lengths coincide in high accuracy without extrapolation.

7. References

Azuma, Y., Tanaka, K., Yoshida, K., Katayama, K., Tsujimura, T., & Shimizu, M. (2008). Basic investigation of new transfer method for optical access line, *IEICE Technical Report Japan*, OFT2008-52, (2008), pp. 27-31

Cao, S. C. & Cartledge, J.C. (2002). Measurement-Based Method for Characterizing the Intensity and Phase Modulation Properties of SOA-MZI Wavelength Converters, *IEEE Photonics Technology Letters*, Vol.14, No.11, (2002), pp.1578- 1580

Fukada, Y., Suzuki, K., Nakamura, H., Yoshimoto, N., & Tsubokawa, M. (2008). First demonstration of fast automatic-gain-control (AGC) PDFA for amplifying burst-mode PON upstream signal, *Proceedings of European Conference on Networks and Optical Communications 2008*, paper WE.2.F.4, 2008

Hornung, S., Wood, R., & Keeble, P. (1990). Single-mode optical networks to the home, *Proceedings of IEEE International Conference on Communications.*, vol. 4, pp. 1563-1571, April 1990

ITU-T Recommendation G.983.1 (1998). Broadband optical access systems based on Passive Optical Networks (PON), 1998

Sankawa, I., Yamamoto, F., Okumura, Y., & Ogura, Y. (2006). Cost and quantity analysis of passive double-star optical-access-network facilities for broadband service multiplexing, *Journal of Lightwave Technolgy*, vol. 24, no. 10, (2006), pp. 3625-3634

Saunders, R.A., King, J.P. & Hardcastle, I. (1994). Wideband chirp measurement technique for high bit rate sources, *Electronics Letters*, Vol.30, No.16, (1994), pp. 1336-1338

Spilent Communications. (2011). *SmartBits*, Available from: http://www.spirent.com/Solutions-Directory/Smartbits.aspx

Tanaka, K., & Horiuchi, Y. (2008). 1:N OLT Redundant Protection Architecture in Ethernet PON System, *Proceedings of Optical Fiber Communication Conference/ National Fiber Optic Engineers Conference*, pp. 1-6. San Diego, February 2008

Tanaka, K., Tsujimura, T., Yoshida, K., Katayama, K., & Azuma, Y. (2009). Frame-loss-free Line Switching Method for In-service Optical Access Network using Interferometry Line Length Measurement, *Proceedings of Optical Fiber Communication Conference 2009*, paper PDPD6, 2009

Tanaka, K., Zaima, M., Tachikura, M. & Nakamura, M. (2002). Downsized and enhanced optical fiber cable transfer splicing system, *Proceedings of 51st International Wire and Cable Symposium*, November 2002

Torregrosa, A. J. et al. (2007). Return-to-Zero Pulse Generators Using Overdriven Amplitude Modulators at One Fourth of the Data Rate, *IEEE Photonics Technology Letters*, Vol.19, No.22, (2007), pp.1837- 1839

Tsujimura, T., Tanaka, K., Yoshida, K., Katayama, K., & Azuma, Y. (2009a). Infallible Layer-one Protection Switching Technique for Optical Fiber Network, *Proceedings of 14th European Conference on Networks and Optical Communications/ 4th Conference on Optical Cabling and Infrastructure*, 2009

Tsujimura, T., Yoshida, K., Tanaka, K., & Azuma, Y. (2010). Transmission length measurement for error-free optical fiber line switching system, *Proceedings of International Conference on Networked Sensing Systems*, pp. 149-152, 2010

Tsujimura, T., Yoshida, K., Tanaka, K., Katayama, K., & Azuma, Y. (2009b). Interruption-free Shunt System for Fiber Optics Transmission Line, *SICE Transactions on Industrial Application,* Vol. 8, No. 16, (2009), pp. 125-131

Watanabe, I., Saito, H., Kobayashi, H., & Takashima, S. (1992). Subscriber optical fiber cable transfer splicing using mechanically transferable connectors, *IEEE Journal of Lightwave Technology,* Vol. 10, No. 6, (June, 1992), pp. 720-727

Willebrand, H. et al. (1999), *Free-Space Optics,* Sams Publishing

Xu, D. J., Yen, W., & Ho, E. (2001). Proposal of a New Protection Mechanism for ATM PON Interface, *Proceedings of IEEE International Conference on Communications,* Vol. 7, pp. 2160-2165, 2001

Yoshida, K., & Tsujimura, T. (2010). Seamless Transmission Between Single-mode Optical Fibers Using Free Space Optics System, *SICE Journal of Control Measurement and System Integration,* Vol. 3, No. 2, (2010), pp.94-100

Yoshida, K., Tanaka, K., & Tsujimura, T. (2011). Robotic Waveguide by Free Space Optics, In: *Advances in Mechatronics,* Martínez-Alfaro, H., InTech, ISBN 978-953-307-373-6

Yoshida, K., Tanaka, K., Katayama, K., Tsujimura, T., & Azuma, Y. (2009). Collimator Focus Adjustment for Free Space Optics System Using Single-mode Optical Fiber, *Proceedings of ICROS-SICE International Joint Conference,* pp. 1338-1341, 2009

17

High Order Momentum States by Light Wave Scattering

Xiaoji Zhou and Xuguang Yue
School of Electronics Engineering and Computer Science, Peking University, Beijing China

1. Introduction

Atomic interferometry is very useful in fundamental studies of coherence, decoherence and phase shifts and for practical precision measurements, with the example of gravimeters, gyroscopes, and gradiometers (Cronin et al., 2009). Bose-Einstein Condensate (BEC) based atomic interferometry provides high contrasts, long integration times and the possible use of small devices (Baumgärtner et al., 2010; Sapiro et al., 2009a; Torii et al., 2000). In such atomic interferometry, coherent momentum manipulation is very effective for splitting and recombining the condensate (Deng et al., 1999; Kozuma, Deng et al., 1999; Ovchinnikov et al., 1999), for realizing the interference. In some precision measurements, the accumulated phase is positively correlated to the atomic velocity, so that the larger the atomic momentum is, the more precise the measurements can be if the measuring time is unchanged.

Based on our previous experimental and theoretical work about the interaction between laser pulses and a BEC, in this chapter we mainly present methods of getting high order of momentum states by resonant superradiance (Zhou et al., 2009) and by multi-pulse Kapitza-Dirac scattering (Xiong et al., 2011), corresponding to travelling wave scattering and standing wave scattering, respectively.

In the superradiant Rayleigh scattering (Bar-Gill et al., 2007; Guo et al., 2008; Inouye et al., 1999, Kozuma, Suzuki et al., 1999; Moore & Meystre, 1999; Pu et al., 2003; Sadler et al., 2007; Schneble et al., 2003; Slama et al., 2007; Zobay & Nikolopoulos, 2006), the spatial and time evolutions of superradiant scattering are studied for a weak pump beam with different frequency components traveling along the long axis of an elongated Bose-Einstein condensate. Through the analysis of the mode competition between the different resonant channels and the local depletion of the spatial distribution, we can get a large number of high-order forward modes by resonant frequency components of the pump beam (Zhou et al., 2009).

On the other hand, the atomic diffraction from a standing wave light grating is a primary method of atomic momentum manipulation. Different to usual separation of the Kapitza-Dirac regime, the Bragg regime and the channeling regime (Keller et al., 1999), the scattering process is described by the projection of atomic states between the momentum presentation and Bloch states form by the scattering standing waves. According to this method, we are able to design and realize several specific high order momentum states for the atomic interferometry (Xiong et al., 2011).

2. High-order momentum modes by resonant superradiant scattering

A typical superradiance experiment consists in a far off-resonant laser pulse traveling along the short axis of a cigar-shaped BEC sample (Inouye et al., 1999), the scattered lights, called end-fire modes, propagate along the long axis of the condensate, and the recoiled atoms are refereed to as side modes. A series of experiments (Bar-Gill et al., 2007; Kozuma, Suzuki et al., 1999; Sadler et al., 2007; Schneble et al., 2003; Slama et al., 2007) have sparked related interests in phase-coherent amplification of matter waves (Kozuma, Suzuki et al., 1999; Schneble et al., 2003), quantum information (Bar-Gill et al., 2007), collective scattering instability (Slama et al., 2007), and coherent imaging (Sadler et al., 2007). Several theoretical descriptions of these cooperative scattering in BEC with single-frequency pump have also been presented (Guo et al., 2008; Moore & Meystre, 1999; Pu et al., 2003; Zobay & Nikolopoulos, 2006).

For the long and weak pump beam, we can observe the forward peaks correspond to Bragg diffraction of atoms (Inouye et al., 1999), where the high order scattering is limited by detuning barriers for the end-fire mode radiation (Zobay & Nikolopoulos, 2007). On the other hand, a X-shaped recoiling pattern is demonstrated in a short and strong pulse as Kapitza-Dirac diffraction of atoms (Schneble et al., 2003), where an atom in the condensate absorbs a photon from the pump laser, then emits a photon into an end-fire mode, and recoils forwardly. Meanwhile another atom absorbs a photon from the end-fire modes, emits into the pump beam and finally recoils backwardly. In this case, there is an energy mismatch of four times the one-photon recoil kinetic energy $\hbar\omega_r$ in backward scattering, which then remains very weak unless a short pumping pulse with a broad spectrum is used. Hence, two phase-locked incident lasers with the frequency difference $\Delta\omega$ compensating for the energy mismatch has been used (Bar-Gill et al., 2007; Cola et al., 2009; Yang et al., 2008), which is named resonant superradiance, where a large number of backward recoiling atoms can be produced.

Followed that, it is attractive to extent this idea to achieve a high momentum transfer by overcoming the detuning barriers, by a weak and long pump beams with the resonant frequency. It requires to analysis the competition between the different transition channels and the spatial distribution of different modes. Because the above traditional superradiant-scattering configuration involves many atomic side modes coupled together, to simplify it, we chose another configuration where a pump beam travels along the long axis of the BEC. This scheme is widely studied in photon echo (Piovella et al., 2003), decoherence (Fallani et al., 2005), spatial distribution effects (Li, Zhou et al., 2008) and self-organized formation of dynamic gratings (Hilliard et al., 2008). Since the pulse length is far longer than the initial spontaneous process (Zobay & Nikolopoulos, 2006), we choose the semi-classical theory which can well describe the experimental results (Bar-Gill et al., 2007; Yang et al., 2008; Zobay & Nikolopoulos, 2006).

In this section, we first introduce the semi-classical theory for the superradiance scattering with a several-frequency pump in the weak coupling. Then the spatial and time evolutions of scattered modes are analyzed for two-frequency pump beam, we find the backward first

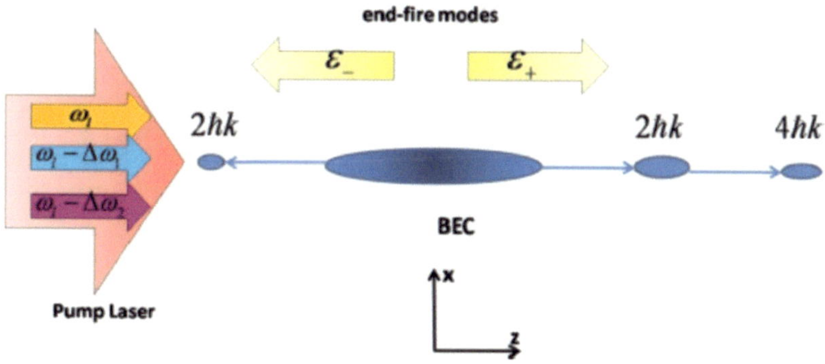

Fig. 1. (Color online) Our experimental scheme. A cigar-shape BEC is illuminated by a far off-resonant laser pulse along its long axis $\hat{\mathbf{z}}$. Collective Rayleigh scattering induces superradiance. Two end-fire modes, which are also along $\hat{\mathbf{z}}$ axis, form in superradiance process and the 1st-order recoiled atoms obtain a momentum of $2\hbar\mathbf{k}$.

order scattering mode is suppressed at the resonant condition $\Delta\omega = 8\omega_r$ and the forward second order mode is enhanced, resulting from the combination of mode competition effects and spatial distribution of the modes. The case of the three-frequency pump beams for a large number of the forward third order scattering modes, and the higher modes for more resonant frequencies are studied, which supplies a new method to get a large number of atoms in higher order forward modes. Finally, some discussion and conclusion are given.

2.1 Model for a multiple-frequency end-pumped beam

We consider the pump laser, with amplitude $\mathcal{E}_l(t)$, polarization $\mathbf{e_y}$, wave vector $\mathbf{k}_l$, frequencies ω_l and $\omega_l \quad \Delta\omega_n$, propagating along the long axis $\hat{\mathbf{z}}$ of an elongated BEC, $\mathbf{E}_l = \mathcal{E}_l(t)\mathbf{e_y}[(1 + \Sigma_n e^{i\Delta\omega_n t})e^{i(k_l z-\omega_l t)} + c.c.]/2$, as shown in Fig. 1. When supperradiant Rayleigh scattering happens, end-fire modes spread along the same axis. The $\mathcal{E}_+$ mode has the same direction as the incident light and mainly interacts with the right part of the condensate, and the $\mathcal{E}_-$ mode overlaps with the left part of the condensate. The atoms are recoiled to some discrete momentum states with momentum $2m\hbar\mathbf{k}$, where m is an integer and the wave vector of end-fire mode light k is approximated as k_l for energy conservation. The total electric field $\mathbf{E}(\mathbf{r},t) = \mathbf{E}^{(+)} + \mathbf{E}^{(-)}$ is given by (Bar-Gill et al., 2007; Li, Zhou et al., 2008; Yang et al., 2008; Zobay & Nikolopoulos, 2006)

$$\mathbf{E}^{(+)}(\mathbf{r},t) = [(1+\sum_n e^{i\Delta\omega_n t})\mathcal{E}_l(t)e^{-\mathrm{i}(\omega_l t-k_l z)}/2 + \mathcal{E}_-(z,t)e^{-\mathrm{i}(\omega t+kz)}]\mathbf{e_y} \quad (1)$$

where $\omega = ck$, $\mathbf{E}^{(-)} = \mathbf{E}^{(+)*}$, and $\mathcal{E}_+$ is ignored because it has the same wave vector as the pump beam but is very small in comparison to $\mathcal{E}_l$. $\Delta\omega_n$ satisfies the condition $\Delta\omega_n \ll \omega_l$ (Bar-Gill et al., 2007) and the initial phases of the different frequency components are assumed to be zero.

Since the BEC is tightly constrained in its short axis ($\hat{\mathbf{x}}$, $\hat{\mathbf{y}}$) in the present superradiance setup and the Fresnel number of the optical field is around 1, one dimensional approximation is usually used (Bar-Gill et al., 2007; Hilliard et al., 2008; Inouye et al., 1999; Li, Zhou et al.,

2008). We expand the wavefunction of the condensate $\psi(\mathbf{r},t)$ in momentum space, $\psi(\mathbf{r},t) = \sum_m \phi_m(z,t) \times e^{-i(\omega_m t - 2mkz)}$, where $\phi_m(z,t) = \psi_m(z,t)/\sqrt{A}$, $\omega_m = 2\hbar m^2 k^2/M$ with M the atomic mass, $m = 0$ corresponds to the residual condensates, $m \neq 0$ denotes the side modes, and A is the average cross area of the condensate perpendicular to $\hat{\mathbf{z}}$. Using the Maxwell-Schrödinger equations, we obtain dynamics equations for $\phi_m(z,t)$,

$$i\frac{\partial \phi_m}{\partial t} = -\frac{\hbar}{2M}\frac{\partial^2 \phi_m}{\partial z^2} - \frac{2im\hbar k}{M}\frac{\partial \phi_m}{\partial z} + \bar{g}\left[\mathcal{E}_-^* \phi_{m-1} e^{-4i(1-2m)\omega_r t} + \mathcal{E}_- \phi_{m+1} e^{-4i(1+2m)\omega_r t}\right], \quad (2)$$

where $\omega_r = \hbar k_l^2/2M$ is the recoil frequency, the coupling between modes is given by

$$\bar{g}(t) = g\left(1 + \sum_n e^{i\Delta\omega_n t}\right), \quad (3)$$

with the coupling factor $g = \sqrt{3\pi c^3 R/(2\omega_l^2 AL)}$, R is the Rayleigh scattering rate of the pump components, and L is the BEC length.

The first term on the right-hand-side of Eq.(2) describes the dispersion of ϕ_m, and the second term gives rise to their translation. The terms in square brackets describe the atom exchange between ϕ_m and ϕ_{m+1} or ϕ_{m-1} through the pump laser and end-fire mode fields. An atom in mode m may absorb a laser photon and emit it into end-fire mode $\mathcal{E}_-$, and the accompanying recoil drives the atom into $m+1$ mode, hence atoms with mode $m+1$ can emerge in forward scattering. On the other hand, in the backward scattering, atoms with mode m absorb one $\mathcal{E}_-$ mode photon, deposit it into the laser mode and go into mode $m-1$. The envelope function of end-fire mode $\mathcal{E}_-$ is given by

$$\mathcal{E}_- = -i\frac{\omega_r \bar{g}}{2c\varepsilon_0}\int_z^{+\infty} dz' \sum_m \phi_m(z',t)\phi_{m+1}^*(z',t) e^{i4(2m+1)\omega_r t}, \quad (4)$$

indicating that the end-fire mode field $\mathcal{E}_-$ is due to the transition between m and $m+1$ mode and the magnitude of $\mathcal{E}_-$ depends on the spatial overlap between the two modes. In addition, there is a frequency difference of $8\omega_r$ between adjacent modes.

2.2 Mode competition for a two-frequency pump beam

In the case of a single-frequency pump in the weak coupling regime, the evolution of the side modes and the end-fire mode indicates that the scattering is a localized process. For this end-pumping configuration, the scattering first starts on the leading edge of the BEC and then moves towards the tailing edge. To investigate the effect of the two-frequency pump beam, the different frequency components of the end-fire mode which indicate the energy change during the scattering are depicted in Fig. 2. The momentum of side mode $m = n$ is $2n\hbar\mathbf{k}$, and its kinetic energy is $4n^2\hbar^2\mathbf{k}^2/2M = 4n^2\hbar\omega_r$. For the pump component with frequency ω_l, atoms from the condensate are pumped to the side mode $m = 1$ and emit end-fire mode photons with frequency $\omega_l - 4\omega_r$ spontaneously. However, in the backward scattering process, an atom in the condensate absorbs the end-fire mode $(\omega_l - 4\omega_r)$ and emits a photon with frequency ω_l back into the pump laser. Since energy is not conserved in backward-scattering, the backward side mode is not populated in weak-pulse regime. Side mode $m = 2$ is also not populated

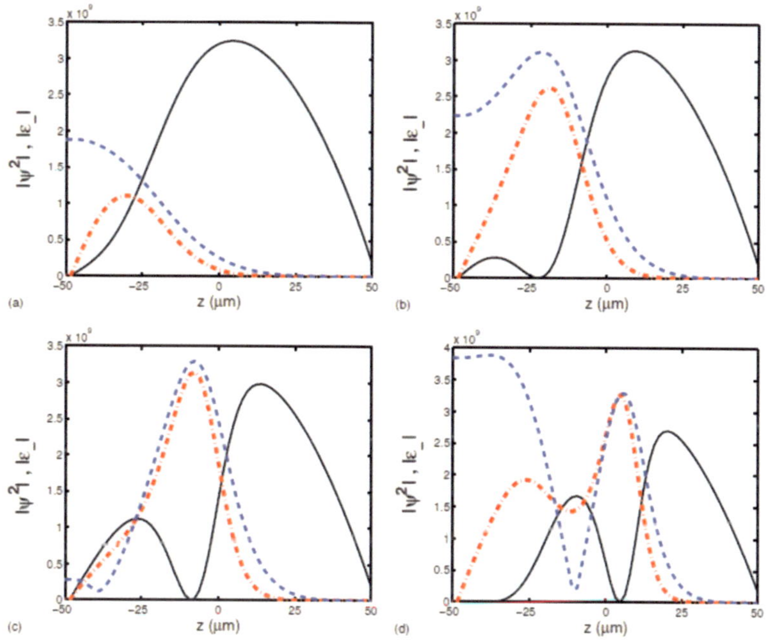

Fig. 2. (Color online) Light-field components of a two-frequency pump laser. The broad arrows are the pump laser and narrow ones are the end-fire mode (scattering optical field). In a spontaneous process, atoms in the condensate absorb photons from the pump laser with frequencies ω_l and $\omega_l - 8\omega_r$, are scattered to side mode $m = 1$ and emit end-fire mode photons with frequency $\omega_l - 4\omega_r$ (dashed arrow) and $\omega_l - 12\omega_r$ (dotted arrow), respectively. Meanwhile, atoms in the condensate can also absorb end-fire mode photons with frequency $\omega_l - 4\omega_r$, be scattered back to side mode $m = -1$ and emit photons with frequency $\omega_l - 8\omega_r$(solid arrow), resonating to one of the pump laser components. The side mode $m = 1$ can absorb pump laser photons with frequency ω_l and be scattered to mode $m = 2$, emitting photons with frequency $\omega_l - 12\omega_r$ resonating to the existing end-fire mode.

due to the energy barrier. However, if we use the two components pump laser with frequency difference $8\omega_r$, i.e. resonant frequency difference, the energy mismatch can be compensated by the pump laser.

Although the resonant condition for the backward mode is satisfied, it should be noticed that two scattering channels exist almost simultaneously. One is atoms scattered from side mode $m = 0$ to $m = -1$ and the other is from $m = 1$ to $m = 2$, resulting in mode competition. The transition from mode $m = 1$ to $m = 2$ requires absorbtion of photons from pump laser, while the backward transition takes photons from the end-fire mode. Because the intensity of the pump laser is far greater than that of the end-fire mode, the transition from $m = 1$ to $m = 2$ has a bigger probability than the transition from $m = 0$ to $m = -1$. Thus the population of the backward mode $m = -1$ is suppressed even at the resonant condition, while the forward mode $m = 2$ is enhanced.

However, the existence of competition between these two channels may not lead to the suppression of the backward mode. If these two channels happen in different spacial parts

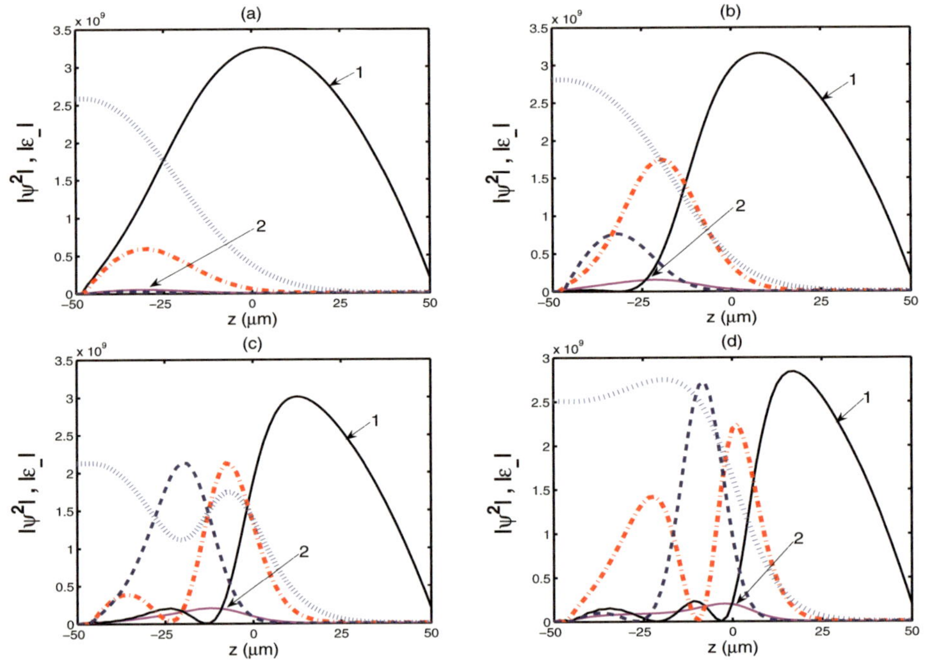

Fig. 3. (Color online) Spatial distribution of the side modes $|\psi^2|$ and the end-fire mode $|\varepsilon_-|$ in the weak coupling regime ($g = 1.25 \times 10^6 s^{-1}$) with the two-frequency pump for different pulse durations: $150\mu s$ (a); $200\mu s$ (b); $250\mu s$ (c); $300\mu s$ (d). Condensate mode $m = 0$ is the solid line-1, backward first-order side mode $m = -1$ is the solid line-2, forward first-order side mode $m = 1$ is the dash-dotted line, forward second-order side mode $m = 2$ is the dashed line, and end-fire mode is the dotted line.

of the condensate, then both of side mode $m = -1$ and $m = 2$ will be enhanced. The suppression of backward mode $m = -1$ and the enhancement of mode $m = 2$ need that these two scattering channels happen in the same area. Therefore, the spatial distribution effect should be considered.

We analyze the spatial effect when second-order forward side mode and backward side mode are populated at the resonant condition $\Delta\omega = 8\omega_r$. The evolution of spatial distribution of side modes and end-fire mode is shown in Fig.3. Superradiance first starts on the leading edge of the BEC, as shown in Fig.3(a). Although the backward first-order side mode $m = -1$ is populated through the overlap between end-fire mode $\mathcal{E}_-$ and side mode $m = 0$, it is very small and emerges at the leading-edge of the BEC. Since the overlap between end-fire mode and side mode $m = 1$ is in the same area, the population of side mode $m = 2$ is obvious on this edge, as shown in Fig.3(b). Side mode $m = 2$ grows more rapidly than side mode $m = -1$, which means more atoms are scattered from side mode $m = 1$ to $m = 2$ than that from $m = 0$ to $m = -1$.Then the first peaks of side modes $m = 1$ and $m = 2$ move to the center of the BEC, as shown in Fig.3(c). Though the movement of the peaks is similar to that in the case of a single-frequency pump laser, one major difference is that the regrowth of side mode $m = 0$ is very little, hence nearly all the atoms on this edge are forwardly scattered. Due to the nearly-complete depletion of the condensate, atoms are mainly transferred between side

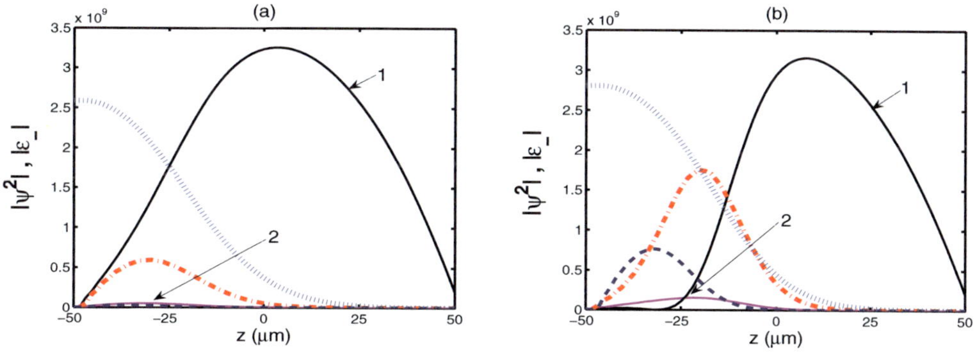

Fig. 4. (Color online) Normalized side mode populations versus time: (a) for a single-frequency pump beam; (b) for a two-frequency resonant pump beam. In both cases the coupling constant is kept $g = 1.55 \times 10^6 s^{-1}$. The side mode are: m=-1 (solid); m=1 (dotted); m=2 (dashed).

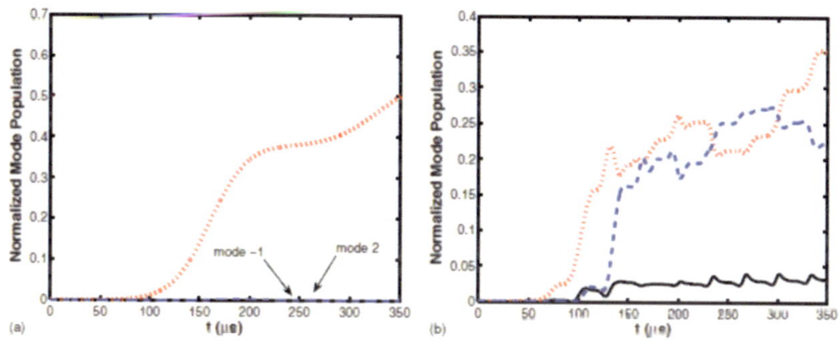

Fig. 5. (Color online) The light-field components of a three-frequency pump laser. The broad arrows are the pump laser and narrow ones are the end-fire mode.

mode $m = 1$ and $m = 2$. The apparent regrowth of side mode $m = 1$ on the leading-edge shown in Fig.3(d) indicates that there are Rabi oscillations between side modes $m = 1$ and $m = 2$ in the depleted area of the condensate.

The time evolution of several side modes populations normalized by the total atom number are depicted by Fig.4. Fig.4 (a) shows that using a single-frequency pump laser cannot produce backward mode $m = -1$ or forward higher mode $m = 2$ in the weak-pulse regime. Using a resonant two-frequency pump beam with the same intensity, modes $m = -1$ and $m = 2$ increased, as shown in Fig.4 (b), however, the forward mode is greatly enhanced while the backward mode remains very small.

2.3 The third order forward modes enhanced with a three-frequency pump beam

The second forward side mode $m = 2$ is greatly enhanced with a resonant two-frequency pump beam, however, the populations of higher forward modes such as $m = 3$ are very small as the channel from the second forward mode to the third forward mode is not resonant with the exiting optical field. To get a large number mode for $m = 3$, Fig.5 depicts the scheme

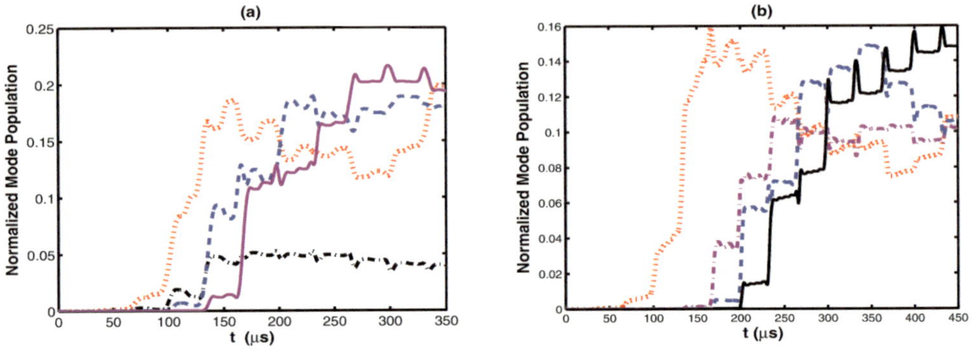

Fig. 6. (Color online) Normalized side mode populations versus time with the coupling constant $g = 1.55 \times 10^6 s^{-1}$: (a) for a three-frequency pump laser: $m = -1$ (dash-dotted), $m = 1$ (dotted), $m = 2$ (dashed), $m = 3$ (solid) ; (b) for a five-frequency pump laser: $m = 1$ (dotted), $m = 3$ (dash-dotted), $m = 4$ (dashed), $m = 5$ (solid) .

of the three-frequency pump beam with the frequencies of the pump laser ω_l, $\omega_l - 8\omega_r$ and $\omega_l - 16\omega_r$. The frequency components ω_l, $\omega_l - 8\omega_r$ and $\omega_l - 16\omega_r$ both have the resonant frequency difference. Hence, there could be two channels to form the backward side mode $m = -1$ but the enhancement of the backward scattering is small because of the formation of higher forward side modes. There are also two channels to form side mode $m = 2$. One thing different from the two-frequency pump beam is that there is also a channel to form side mode $m = 3$ for the reason that atoms in side mode $m = 2$ absorb pump laser photons with frequency ω_l, are then scattered to mode $m = 3$ and eventually emit end-fire mode photons with frequency $\omega_l - 20\omega_r$ which is resonant to an existing end-fire mode. This means that more atoms in side mode $m = 2$ will be pumped to side mode $m = 3$ and less will be transferred back to side mode $m = 1$, a competition between side mode $m = 3$ and $m = 1$ is set up. As a result, side mode $m = 3$ will be enhanced and $m = 1$ will be reduced relatively.

Fig.6(a) is the simulated result of the time evolution of normalized side mode populations for a three-frequency pump beam. We could see that side mode $m = 3$ would be strongly enhanced at long time while side mode $m = 1$ reduced.

2.4 Momentum transfer in the high order forward modes

From the above discussion we know that using multi-resonant frequencies is a promising way to get a large number of higher forward modes. When a pump laser has frequency components $\omega_l, \omega_l - 8\omega_r, \cdots, \omega_l - (n-1) \times 8\omega_r$, satisfying $(n-1) \times 8\omega_r \ll \omega_l$, with the kinetic energy of mode $m = n$ equal to $4n^2\hbar\omega_r$, then after the condensate atoms spontaneously scattered to mode $m = 1$, the end-fire mode will have frequency components $\omega_l - 4\omega_r, \omega_l - 12\omega_r, \cdots, \omega_l - (2n-1) \times 4\omega_r$. For resonance concern, mode $m = 1$ will absorb photons from the pump components $\omega_l, \omega_l - 8\omega_r, \cdots, \omega_l - (n-2) \times 8\omega_r$ and emits end-fire mode photons with frequency $\omega_l - 12\omega_r, \cdots, \omega_l - (2n-1) \times 4\omega_r$ which are resonant with existing end-fire mode, so mode $m = 2$ is produced. Like mode $m = 1$, modes $m = 2, m = 3, \cdots, m = n - 1$ can absorb pump photons and emit photons resonant to the existing end-fire mode. For example, mode $m = n - 1$ will absorb photons with frequency ω_l and emits photons with frequency $\omega_l - (2n-1) \times 4\omega_r$. Therefore atoms could finally be transferred to mode $m = n$. Note that mode $m = n$ cannot emit resonant end-fire mode, so mode $m = n$ will be enhanced. To show it, Fig.6(b) is the simulated result of the time evolution of normalized side mode populations for a five-frequency pump beam. We could see that side mode $m = 5$ would be strongly enhanced.

2.5 Discussion

In the experiment, to get several resonant frequencies, the laser beam from an external cavity diode laser can be split into several parts, and their frequencies are shifted individually by acoustic-optical modulators (AOMs) which are driven by phase-locked radio frequency signals, as demonstrated in the case of two resonant frequency (Bar-Gill et al., 2007; Yang et al., 2008). Therefore, the frequency difference between the beams can be controlled precisely. Furthermore, to avoid the reflection from the glass tube and formation of Bragg scattering in the experiment, the pump beam can actually deviate a few degrees from the long axis, as shown in the experiments (Fallani et al., 2005; Hilliard et al., 2008; Li, Zhou et al., 2008).

Different to the works in the configuration where the pump beam travels along the short axis of the condensate with the resonant frequency (Yang et al., 2008), where a large number of backward scattering is obvious in a two-frequency pump beam, the backward scattering is suppressed and the forward second-order mode is obviously enhanced in our case. This is due to mode competition between the forward second-order mode and the backward mode and local depletion of the superradiant process.

We have not considered the initial quantum process because its time scale is very small, shorter than $1\mu s$. In this quantum process there is also mode competition to form the end-fire modes along the long axis and suppress the emission on the other direction. This is different concept from what has been discussed above, in which case mode competition exists in the different channels satisfying the energy match and spatial condition.

For the pump beam with several resonant frequencies, not only can we obtain the high order momentum transfer which is important in the momentum manipulation for atom interferometry, but also the above analysis is useful to understand the interplay between the matter wave and light in the matter wave amplification (Kozuma, Suzuki et al., 1999; Schneble et al., 2003), atomic cooperative scattering in the optical lattice (Xu et al., 2009), and by the pump with a noisy laser (Robb & Firth, 2007; Zhou, 2009).

3. High order momentum states by scattering of standing wave pulses

In this section, we apply a method for flexible manipulation of the atomic momentum states with the standing wave pulses. The atomic diffraction from standing wave pulses is demonstrated in the experiments and systematically analyzed by the band structure theory of one-dimension optical lattice. With this method, we are able to design and realize several specific momentum states, which may be applied in atomic interferometry. In principle, this method could be used for designing a wide range of possible target states.

3.1 Theoretical model

We consider a non-interacting condensate being diffracted by a sequence of square shaped standing wave pulses with the successive durations $\tau_i(i = 1, 2, ..., s + 1)$, separated by the intervals $\tau_{fi}(i = 1, 2, ..., s)$. The standing wave consists of a pair of laser beams far-detuned enough to suppress the spontaneous emission.

The periodic potential (Denschlag et al., 2002; Morsch & Oberthaler, 2006) introduced by the ac Stark shift can be described as $V(x) = U_0 \cos^2(k_L x)$, with the trap depth U_0 and the laser's wave vector $k_L = 2\pi/\lambda_L$ (λ_L is the wavelength of the laser). The lattice leads to a band structure of the energy spectra, of which the eigenvalues of the energy $E_{n,q}$ and eigenvectors $|n,q\rangle$ (Bloch states) are labeled by the quasi-momentum q and the band index n, and they satisfy the equation:

$$\hat{H}\,|n,q\rangle = E_{n,q}\,|n,q\rangle\,, \tag{5}$$

where the Hamiltonian $\hat{H} = \hat{p}^2/2M + U_0\cos^2(k_L x)$, with the atomic momentum $\hat{p}$ and the atomic mass M. The Bloch states form a quasi-momentum space. In the lattice, the spatial periodicity of the wave function results in separated peaks in momentum space, corresponding to the reciprocal lattice vector $2k_L$.

When a condensate with an initial momentum $p_{m_0} = \hbar(q + 2m_0 k_L)$ ($\hbar$ is the Plank constant, $-k_L \leq q \leq k_L$, $m_0 = ..., -1, 0, 1, ...$) is abruptly loaded into a lattice, the wave packet can be described as a superposition of the Bloch states:

$$|\Psi(t=0)\rangle = \sum_{n=0}^{\infty} |n,q\rangle\langle n,q|p_{m_0}\rangle, \tag{6}$$

where $\langle n,q|p_{m_0}\rangle = c_{n,q}(m_0)$. The nth Bloch state evolves independently as $e^{-iE_{n,q}t/\hbar}$, and the total wave function evolves as

$$|\Psi(t)\rangle = \sum_{n=0}^{\infty} c_{n,q}(m_0)e^{-iE_{n,q}t/\hbar}|n,q\rangle. \tag{7}$$

While the incident light is switched off after the duration τ_1, the wave function is projected back to the momentum space from the quasi-momentum space. The coefficient $b(m_0, m, \tau_1)$ of each $|p_m\rangle$ state ($m = ..., -1, 0, 1, ...$) can be acquainted as:

$$b(m_0, m, \tau_1) = \sum_{n=0}^{\infty} c_{n,q}(m_0)c_{n,q}(m)e^{-iE_{n,q}\tau_1/\hbar}. \tag{8}$$

For a zero initial momentum of the condensate, the subscript q can be omitted for simplification and $m_0 = 0$. For one pulse scattering, the population of the $|p_m\rangle$ state is $P_m^{(1)} = |b(0, m, \tau_1)|^2$. It can be seen that the probabilities of the momentum states after one scattering pulse depend on the lattice depth and the pulse duration. The lattice depth determines the band structure and is reflected in the terms $c_{n,q}$. The pulse duration influences the phase evolution of each Bloch state as $e^{-iE_{n,q}\tau_1/\hbar}$.

The multi-pulse process, which consists of a number of single pulses and intervals can be solved as follows. The wave function of the condensate after the first pulse τ_1 can be derived from Eq. (8) as

$$|\Psi(\tau_1, t)\rangle = \sum_m b(m_0, m, \tau_1)e^{-iE^{(m)}t/\hbar}|2m\hbar k_L\rangle. \tag{9}$$

After the first interval τ_{f1} and the second pulse τ_2, the population of the $|p_m\rangle$ state can be achieved as

$$P_m^{(2)} = \left| \sum_{m_1} b(m_0, m_1, \tau_1) e^{-iE^{(m_1)}\tau_{f1}/\hbar} b(m_1, m, \tau_2) \right|^2 . \tag{10}$$

As shown in Eq. (10),the population is affected by the two pulses, the first one corresponding to $b(m_0, m_1, \tau_1)$, and the second one corresponding to $b(m_1, m, \tau_2)$. During the interval τ_{f1}, the phase of the $|p_{m_1}\rangle$ state evolves along the time as $e^{-iE^{(m_1)}t/\hbar}$, where $E^{(m_1)} = (2m_1\hbar k_L)^2/2M = 4m_1^2 E_R$ is the kinetic energy, and $E_R = (\hbar k_L)^2/2M$ is the single photon recoil energy. The interval τ_{f1} produces a phase shift $e^{-iE^{(m)}\tau_{f1}/\hbar}$ and contributes to the momentum distribution.

In the same way, the population of the $|p_m\rangle$ state after $(s+1)$ pulses can be achieved as:

$$P_m^{(s+1)} = \left| \sum_{m_1, m_2, \cdots, m_s} \prod_{i=1}^{s+1} b(m_{i-1}, m_i, \tau_i) \prod_{i=1}^{s} e^{-iE^{(m_i)}\tau_{f_i}/\hbar} \right|^2 , \tag{11}$$

with $m_0 = 0$, and $m_{s+1} = m$.

From the analysis above, the momentum distribution after a sequence of pulses' scattering is influenced by not only the lattice pulses with the term $\prod_{i=1}^{s+1} b(m_{i-1}, m_i, \tau_i)$, but also the intervals among the pulses as reflected in the term $\prod_{i=1}^{s} e^{-iE^{(m_i)}\tau_{f_i}/\hbar}$. Although the populations of the momentum states do not change during the intervals, the phase-evolution rates of the momentum states with different kinetic energies are not identical. The phase deviations between the states oscillate from 0 to 2π with the interval, and the heterogeneously accumulated phases change the distribution of the condensate in the quasi-momentum space.

3.2 Experiments of standing wave pulse sequences

We performed the experiments of a condensate in a magnetic trap (MT) (see Fig. 7(a)) being scattered by a sequence of standing wave pulses (see Fig. 7(b)). As shown in Fig. 7(c), after pre-cooling, a cigar shaped ^{87}Rb condensate of 2×10^5 atoms in $5^2S_{1/2}\,|F = 2, M_F = 2\rangle$ state was achieved by the radio frequency (RF) cooling in the magnetic trap, of which the axial frequency is 20 Hz and the radial frequency is 220 Hz (Yang et al., 2008; Zhou et al., 2010). A pair of counter-propagating laser beams, of which the durations were controlled by an acousto-optical modulator, and the amplitudes were adjusted by the injection current of a tapered-amplifier, were applied to the condensate along the axial direction. The linear polarized incident light at the wavelength $\lambda_L = 852$ nm was focused with a waist of 110 μm to cover the condensate. The trap depth, which was calibrated by Kapitza-Dirac scattering experimentally, reached $120E_R$, corresponding to the light power of 320 mW. The incident light and the magnetic trap were simultaneously shut after the BEC-light interaction. After 30 *ms* free falling and ballistic expansion, the atomic gas was pictured by absorption imaging. Since the minimum gap between different momentum states is $2\hbar k$, which is much larger than the momentum width of a single momentum state, the components with different momenta will be separated in the TOF (time of flight) signal (see Fig. 7(d)), and the atomic number of each momentum state is possible to be read separately as N_m (the momentum

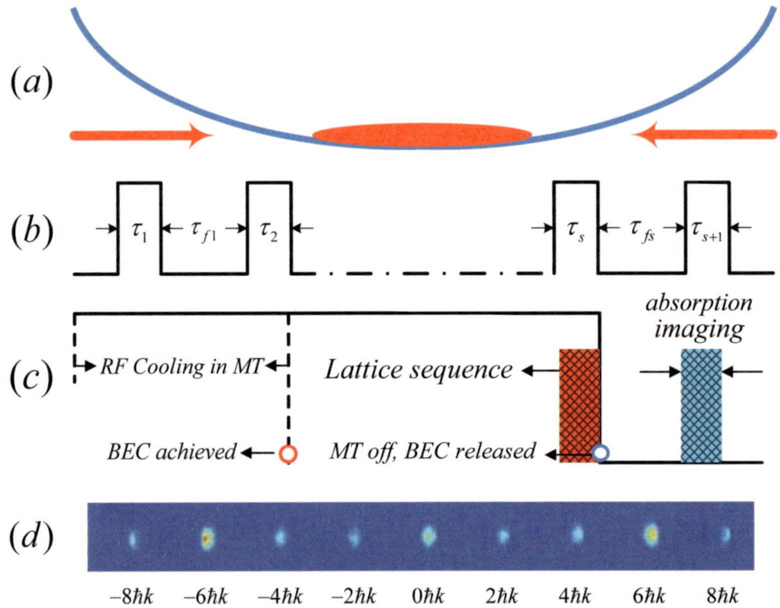

Fig. 7. (Color online) (a) A pair of counter-propagating light beams are applied to a condensate in magnetic trap. (b) The scattering process consists of a sequence of standing wave pulses, which contained s intervals with widths τ_{fi} (i=1,2,...s) and $s+1$ light pulses with durations τ_i (i=1,2,...$s+1$). The incident light's wavelength is 852 nm and its maximum intensity can reach $120E_R$. (c) The procedure for the experiments is shown. The condensate is exposed to a sequence of standing wave pulses and then released from the magnetic trap. The absorption images of the condensate can be observed after the free falling. (d) A TOF signal obtained in our experiment.

order $m = ..., -2, -1, 0, 1, 2$). The relative population of the momentum state $|2m\hbar k\rangle$ can be evaluated as $N_m / \sum N_m$.

The lattice in our experiments is quite deep, so we concentrate on the short-pulse diffractions to avoid the de-coherence and heating effects of long pulses relevant for Bragg scattering. However, for more flexible momentum manipulation, our pulses are not so short as the Raman-Nath pulses (Huckans et al., 2009) used in previous works.

A brief introduction to the Raman-Nath regime is given in the following for comparison. In the scattering process, the evolution during the free evolution intervals is analyzed as in previous section, while the effect of lattice with adequately short duration τ can be analytically solved by the *Schrödinger* equation $i\hbar\partial\,|\Psi(t)\rangle\,/\partial t = \hat{H}\,|\Psi(t)\rangle$, after omitting the atomic kinetic energy term $\hat{p}^2/2M$ in the Hamiltonian. This approximation can be made while the displacement of the scattered atoms during the interaction time is much smaller than the spatial period of the standing wave. Equivalently, the standing wave duration τ and the single photon recoil frequency $\omega_r = \hbar k_L^2/2M$ have to fulfill $\tau \ll 1/\omega_r$. The pulse is able to split a stationary condensate into components with symmetrical momenta $p_n = 2n\hbar k_L(n = 0, \pm1, \pm2, ...)$, with corresponding populations $P_n = J_n^2(U_0\tau/2\hbar)$, where $J_n(z)$ are Bessel functions of the first kind.

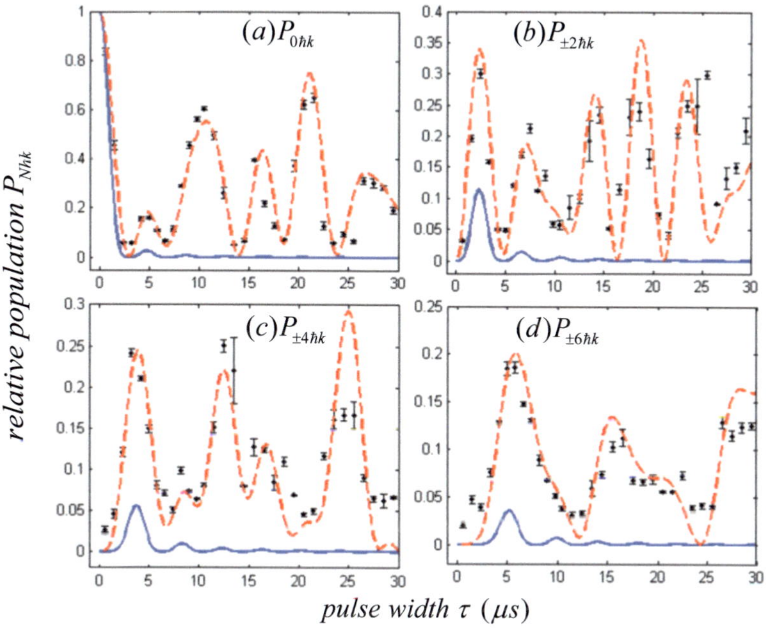

Fig. 8. (Color online) Single pulse scattering of condensate: The black dots represent the experiment results. The blue solid line is the theoretical analysis with Raman-Nath approximation. The red dashed line is the numerical simulation with band structure theory. Figure (a), (b),(c) and (d) correspond to the relative populations of the condensates with the momenta $0\hbar k$, $\pm 2\hbar k$, $\pm 4\hbar k$ and $\pm 6\hbar k$ respectively.

First we demonstrate a one-pulse scattering experiment. A condensate is exposed to a standing wave pulse with depth $100E_R$ and duration varying from 0 to $30\mu s$. The relative populations of the condensates with the momenta $0\hbar k$, $\pm 2\hbar k$, $\pm 4\hbar k$ and $\pm 6\hbar k$, corresponding to Fig. 8(a), (b), (c) and (d) respectively, are measured and theoretically analyzed. In addition, the theoretical analysis with the Raman-Nath approximation, is also shown in the figure for comparison. It can be seen that within $3\mu s$ the theoretical analysis with the Raman-Nath approximation (blue solid line) is close to the experimental results (black dots), and so is the theoretical analysis with band structure theory (red dashed line). When the pulse duration exceeds $3\mu s$, the analysis with the Raman-Nath approximation gradually goes far away from the experimental results, while the numerical simulation with band structure theory still agrees with the experimental results along the entire time scale. As shown in Fig. 8, the probability of each momentum state oscillates with the pulse duration as described by the band structure theory. It is clear that, in the single pulse scattering process, the band structure theory works well not only for the short pulse but also for the longer pulse, because the atomic motion has been taken into account. So the atomic diffraction by a single standing wave pulse can be predicted in a wider range of pulse duration with the band structure theory.

Then we increase the number of pulses in the experiments to explore the extra factors influencing the momentum distributions. In every sequence, all the pulses are the same and all the intervals are identical to make the experiments more convenient to carry out. For further comparison between the band structure theory and the analysis in the Raman-Nath regime, every single pulse is made short enough for the Raman-Nath approximation.

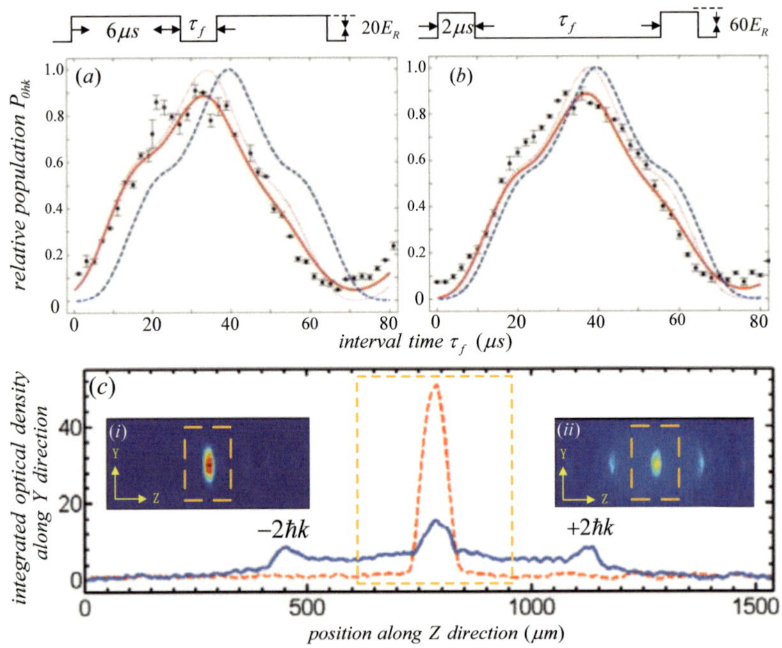

Fig. 9. (Color online) Two-pulse scattering of the condensate: The relative populations of the stationary condensate $P_{0\hbar k}$ versus the varied intervals τ_f is shown. The parameters of the experiments are described above each figure. The black dots are the experimental results. The blue dashed line is the analysis with Raman-Nath approximation. The magenta dotted line is the numerical simulation with band structure theory. The red solid line is a numerical simulation taking into account the momentum dispersion. (c) A demonstration of momentum expansion introduced by *s*-wave scattering. The TOF signal (i) corresponds to the dashed curve and pictures the momentum distribution before the former lattice pulse. The signal (ii) corresponds to the solid curve and shows the momentum distribution after the former pulse. The momentum width along Z direction (the lattice direction) is evaluated based on the parts in the dashed boxes which include the momenta from $-\hbar k$ to $\hbar k$.

Two experiments of two-pulse scattering are demonstrated in Fig. 9, in which the relative populations of the stationary condensate $P_{0\hbar k}$ versus the varied intervals τ is shown. The parameters of the scattering pulses used in different sequences are chosen to be of the same products of the lattice depth and the pulse duration, so that each pulse affects the condensate equivalently. As shown in the figure, the intervals actually affect the final momentum distribution, and the theoretical analysis with the band structure theory and Raman-Nath approximation both picture well the evolution of the atomic distributions versus the interval between the two pulses. The results of two-pulse scattering can be explained as the fact that since the phase shift accumulated during the interval varies harmonically from 0 to 2π, the probability of the stationary condensate oscillates between the minimum and the maximum. When the phase shift is 2π with the interval $\pi\hbar/2E_R$ (around 80μs), the wave function is little affected by the interval and the two pulses diffract the condensate as one combined pulse to make the probability $P_{0\hbar k}$ the minimum. While the phase shift is π with the interval $\pi\hbar/4E_R$ (about $40\mu s$), the second pulse produces an effect opposite of the first one and diffract the non-stationary components of the condensate back to the stationary one and make the probability $P_{0\hbar k}$ the maximum.

It also can be acquainted from Fig. 9 that the numerical simulation with band structure theory is much closer to the experimental results than the analytical solution with Raman-Nath approximation. It is conjectured that the phase evolution during the scattering process makes the difference. The phase evolution in the scattering process is neglected in the Raman-Nath approximation, but not in the numerical simulation with band structure theory. Although the duration of the scattering process is short, the phase shifts in the scattering process still increase. The phase shift in the scattering process needs to be taken into account and influences the final momentum distribution. As a result, the longer the scattering pulse is, the larger the difference is. Although the maximum of the probability $P_{0\hbar k}$ corresponds to the interval $\pi\hbar/4E_R$, the two-pulse experiments in Fig. 9 can clearly show that the longer pulse leads to the larger difference. In Fig. 9 (a), the pulse duration is $6\mu s$, the probability $P_{0\hbar k}$ reaches the maximum with the interval $34\mu s$. In Fig. 9 (b), the pulse duration is $2\mu s$, the probability $P_{0\hbar k}$ gets to the top with the interval $38\mu s$.

Nevertheless, there is still some obvious deviation between the simulation and the experimental results. It is observed that the momentum width has been expanded after the former pulse (see Fig. 9(c)), because of the s-wave scattering between the different momentum states. Consequently, this dispersion process is approximated to an initial momentum width of $\sim 0.1\hbar k_L$ on average to optimize the numerical simulation. Unlike the analysis without momentum width, phase evolution is different for different initial momenta and results in a phase dispersion. The quasi modes obtained at the end of the diffraction process result from the linear superposition of final states obtained after time evolutions of the different momenta populated the initial BEC. It can be seen from Fig. 9 that the approximation is effective.

As discussed in (Li, Deng et al., 2008), the maximum of the probability $P_{0\hbar k}$ will never reach 1 thanks to the imperfect optical lattice. In our case, the momentum expansion is an explanation of the similar situation as shown in Fig. 9. Since the momentum width is considered, the phase shift is populated around π with a width, instead of a definite π, with the interval $\pi\hbar/4E_R$. In other words, there is no interval that accumulates a phase shift exactly equal to π, so with any interval, the second standing wave pulse is not able to diffract all the condensates back to the stationary part.

3.3 Manipulate the momentum states as design

The experiments and the numerical simulations above have shown the possibility and feasibility of the manipulation of a condensate's momentum states. We manage to design several two-pulse sequences to achieve high contrast momentum states such as $|\pm 2\hbar k\rangle$, $|\pm 4\hbar k\rangle$ and $|\pm 6\hbar k\rangle$, which may be useful in atomic interferometry (Beattie et al., 2009; Rohwedder, 2001). For each state, we apply two totally different two-pulse sequences to show the flexibility of the method. The general method to achieve the target states is to find out the condition of the minimum of the square deviation $\Delta^2 = \sum_{m=-\infty}^{+\infty} \left(P_m^g - P_m\right)^2$, where P_m^g is the probability of $|2m\hbar k_L\rangle$ in the goal state, and P_m is that generated by the sequence. A second method, as the target is to obtain the highest population of some certain momentum state, consists in scanning the set of initial conditions and choose the one corresponding to the maximum value of the desired population. We apply the two methods above separately and obtain the same pulse sequences. As shown in Fig. 10, the experimental results (the black

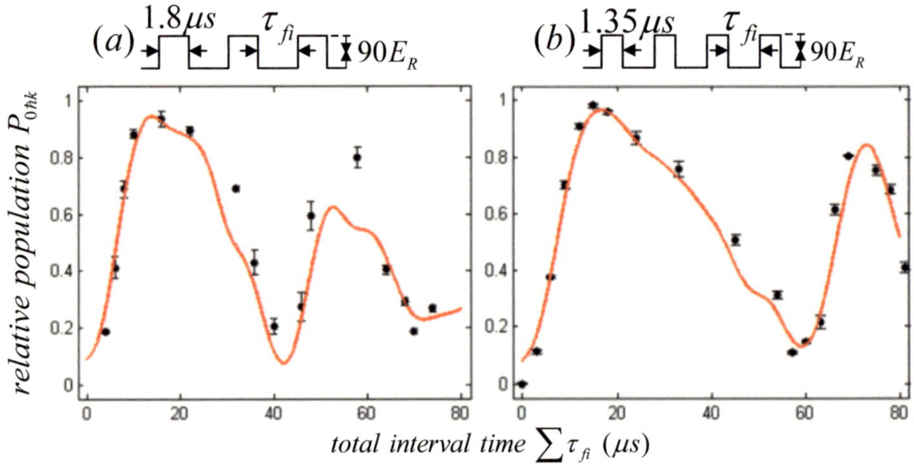

Fig. 10. Experimental realization of designed momentum states. The expected momentum state is $\pm 2\hbar k_L$((a) and (b)), $\pm 4\hbar k_L$((c) and (d)) and $\pm 6\hbar k_L$((e) and (f)). The pulse sequences are shown above each figure. The black round dots are experimental results. The blue diamond dots are the expectations based on the design. The red square dots are the modified design with momentum width, which agree with the experiments better.

round dots) agree well with the expectations of the designs (the blue diamond dots), whether the pulses are in the Raman-Nath regime (see Fig. 10(b)) or not (see others in Fig. 10). When the momentum dispersion is being considered, the expected momentum distributions (the red square dots) get closer to the experiments, where the figures only display the relative populations of the target states and omit the others for the figures being more clear.

It can be seen from Fig. 10 that the momentum width correction can improve the precision of the prediction with our method. The average relative deviation between the experimental results and the expected values without including the momentum width is 25.03%, while the deviation is decreased to 13.15% with the correction.

An asymmetry of the momenta can be observed in Fig. 10, and it may ascribe to the following factors. Besides the measurement error, there is an imperfection of the standing wave, brought forth by the unbalanced intensity of the laser beams. External field (such as the magnetic trap) fluctuations during the scattering process may also affect the momentum distribution.

3.4 Discussion

The band structure theory is a global method to deal with the standing wave scattering a condensate, while Bragg and Raman-Nath scattering are two special situations which can be analytically solved with their respective approximations. In the Bragg regime, the potential height introduced by the standing wave is restrained below $4E_R$ and that leads to the difficulty of generating higher order momentum states. In the Raman-Nath regime, the intensity of the standing wave is not limited so that higher order momentum states can be generated symmetrically (Gadway et al., 2009; Sapiro et al., 2009b). However the pulse duration has to be short enough to neglect the atomic motion, so the momentum states can not be predicted in this regime if the pulse duration is slightly longer. In our work, the scattering can be well explained and numerically analyzed within a much wider range of pulse intensity and duration. So it is natural that more momentum states can be generated.

Some valuable states, such as $|\pm 2\hbar k\rangle$, $|\pm 4\hbar k\rangle$ and $|\pm 6\hbar k\rangle$ states with high contrast, can not be realized by the single pulse scattering, while they can be realized by a sequence of standing wave pulses. A sequence of lattice pulses is a more effective and flexible tool for momentum manipulation. It can generate many useful momentum states, in addition to the ones demonstrated in our work.

Although the numerical simulation is corrected to take into account the momentum dispersion, some deviations between the experiments and the simulation still exist. The inaccuracy of the lattice-depth calibration, which is 5% at least, is one of the reasons. The phase shift introduced by the magnetic trap is another one, while the influence is estimated to be within 0.03%, which is below the experimental uncertainty. The heating and momenta exchange during the s-wave scattering may also lead to some differences.

4. Conclusion

We have shown in this chapter two methods of getting high order momentum states by resonant superradiant scattering and by a sequences of pulsed standing waves, corresponding to traveling wave scattering and standing wave scattering, respectively. Different to Bragg scattering, we can get a large number of high-order forward modes by resonant frequency components of the pump beam by the former method through the mode competition between the different resonant channels and the local depletion of the spatial distribution. While with the latter method a symmetric momentum distribution is generated, with one pair of momentum states designed to be mostly populated. Both methods are beneficial to a lager momentum transfer in atom manipulation for the atom interferometry and atomic optics.

5. References

Bar-Gill, N., Rowen, E. E. & Davidson, N. (2007). Spectroscopy of strong-pulse superradiance in a Bose-Einstein condensate, *Phys. Rev. A* 76: 043603.

Baumgärtner, F., Sewell, R. J., Eriksson, S., Llorente-Garcia, I., Dingjan, J., Cotter, J. P. & Hinds, E. A. (2010). Measuring energy differences by BEC interferometry on a chip, *Phys. Rev. Lett.* 105: 243003.

Beattie, S., Barrett, B., Chan, I., Mok, C., Yavin, I. & Kumarakrishnan, A. (2009). Atom-interferometric studies of light scattering, *Phys. Rev. A* 80: 013618.

Cola, M. M., Volpe, L. & Piovella, N. (2009). Accelerated superradiance and collective atomic recoil lasing with a two-frequency pump, *Phys. Rev. A* 79: 013613.

Cronin, A. D., Schmiedmayer, J. & Pritchard, D. E. (2009). Optics and interferometry with atoms and molecules, *Rev. Mod. Phys.* 81: 1051.

Deng, L., Hagley, E. W., Denschlag, J., Simsarian, J. E., Edwards, M., Clark, C. W., Helmerson, K., Rolston, S. L. & Phillips, W. D. (1999). Temporal matter-mave-dispersion Talbot effect, *Phys. Rev. Lett.* 83: 5407.

Denschlag, J. H., Simsarian, J. E., Häffner, H., McKenzie, C., Browaeys, A., Cho, D., Helmerson, K., Rolston, S. L. & Phillips, W. D. (2002). A Bose-Einstein condensate in an optical lattice, *Journal of Physics B: Atomic, Molecular and Optical Physics* 35: 3095.

Edwards, M., Benton, B., Heward, J. & Clark, C. W. (2010). Momentum-space engineering of gaseous Bose-Einstein condensates, *Phys. Rev. A* 82: 063613.

Fallani, L., Fort, C., Piovella, N., Cola, M., Cataliotti, F. S., Inguscio, M. & Bonifacio, R. (2005). Collective atomic recoil in a moving Bose-Einstein condensate: From superradiance to Bragg scattering, *Phys. Rev. A* 71: 033612.

Gadway, B., Pertot, D., Reimann, R., Cohen, M. G. & Schneble, D. (2009). Analysis of Kapitza-Dirac diffraction patterns beyond the Raman-Nath regime, *Opt. Express* 17: 19173.

Guo, R., Zhou, X. & Chen, X. (2008). Enhancement of motional entanglement of cold atoms by pairwise scattering of photons, *Phys. Rev. A* 78: 052107.

Gupta, S., Leanhardt, A. E., Cronin, A. D. & Pritchard, D. E. (2001). Coherent manipulation of atoms with standing light waves, *Comptes Rendus de l'Acadl̨emie des Sciences - Series IV - Physics* 2: 479.

Hilliard, A., Kaminski, F., le Targat, R., Olausson, C., Polzik, E. S. & Müller, J. H. (2008). Rayleigh superradiance and dynamic Bragg gratings in an end-pumped Bose-Einstein condensate, *Phys. Rev. A* 78: 051403.

Huckans, J. H., Spielman, I. B., Tolra, B. L., Phillips, W. D. & Porto, J. V. (2009). Quantum and classical dynamics of a Bose-Einstein condensate in a large-period optical lattice, *Phys. Rev. A* 80: 043609.

Inouye, S., Chikkatur, A. P., Stamper-Kurn, D. M., Stenger, J., Pritchard, D. E. & Ketterle, W. (1999). Superradiant Rayleigh scattering from a Bose-Einstein condensate, *Science* 285: 571.

Keller, C., Schmiedmayer, J., Zeilinger, A., Nonn, T., Dĺźrr, S. & Rempe, G. (1999). Adiabatic following in standing-wave diffraction of atoms, *Applied Physics B: Lasers and Optics* 69: 303.

Kozuma, M., Deng, L., Hagley, E. W., Wen, J., Lutwak, R., Helmerson, K., Rolston, S. L. & Phillips, W. D. (1999). Coherent splitting of Bose-Einstein condensed atoms with optically induced Bragg diffraction, *Phys. Rev. Lett.* 82: 871.

Kozuma, M., Suzuki, Y., Torii, Y., Sugiura, T., Kuga, b.-g. T., Hagley, E. W. & Deng, L. (1999). Phase-coherent amplification of matter waves, *Science* 286: 2309.

Li, J., Zhou, X., Yang, F. & Chen, X. (2008). Superradiant Rayleigh scattering from a Bose-Einstein condensate with the incident laser along the long axis, *Phy. Lett. A* 372: 4750.

Li, K., Deng, L., Hagley, E. W., Payne, M. G. & Zhan, M. S. (2008). Matter-wave self-imaging by atomic center-of-mass motion induced interference, *Phys. Rev. Lett.* 101: 250401.

Meystre, P. (2001). *Atom optics*, Springer series on atomic, optical, and plasma physics, AIP Press/Springer.

Moore, M. G. & Meystre, P. (1999). Theory of superradiant scattering of laser light from Bose-Einstein condensates, *Phys. Rev. Lett.* 83: 5202.

Morsch, O. & Oberthaler, M. (2006). Dynamics of Bose-Einstein condensates in optical lattices, *Rev. Mod. Phys.* 78: 179.

Ovchinnikov, Y. B., Müller, J. H., Doery, M. R., Vredenbregt, E. J. D., Helmerson, K., Rolston, S. L. & Phillips, W. D. (1999). Diffraction of a released Bose-Einstein condensate by a pulsed standing light wave, *Phys. Rev. Lett.* 83: 284.

Piovella, N., Beretta, V., Robb, G. R. M. & Bonifacio, R. (2003). Photon echo in the superradiant light scattering from a Bose-Einstein condensate, *Phys. Rev. A* 68: 021801.

Pritchard, D. E., Cronin, A. D., Gupta, S. & Kokorowski, D. A. (2001). Atom optics: old ideas, current technology, and new results, *Ann. Phys.* 10: 35.

Pu, H., Zhang, W. & Meystre, P. (2003). Wave mixing of optical pulses and Bose-Einstein condensates, *Phys. Rev. Lett.* 91: 150407.

Robb, G. R. M. & Firth, W. J. (2007). Collective atomic recoil lasing with a partially coherent pump, *Phys. Rev. Lett.* 99: 253601.

Rohwedder, B. (2001). Multiple atomic wave interferometry with standing-waves of light, *The European Physical Journal D - Atomic, Molecular, Optical and Plasma Physics* 13: 255.

Sadler, L. E., Higbie, J. M., Leslie, S. R., Vengalattore, M. & Stamper-Kurn, D. M. (2007). Coherence-enhanced imaging of a degenerate Bose-Einstein gas, *Phys. Rev. Lett.* 98: 110401.

Sapiro, R. E., Zhang, R. & Raithel, G. (2009a). Atom interferometry using Kapitza-Dirac scattering in a magnetic trap, *Phys. Rev. A* 79: 043630.

Sapiro, R. E., Zhang, R. & Raithel, G. (2009b). Reversible loss of superfluidity of a Bose-Einstein condensate in a 1D optical lattice, *New Journal Of Physics* 11: 013013.

Schneble, D., Torii, Y., Boyd, M., Streed, E. W., Pritchard, D. E. & Ketterle, W. (2003). The onset of matter-wave amplification in a superradiant Bose-Einstein condensate, *Science* 300: 475.

Slama, S., Krenz, G., Bux, S., Zimmermann, C. & Courteille, P. W. (2007). Cavity-enhanced superradiant Rayleigh scattering with ultracold and Bose-Einstein condensed atoms, *Phys. Rev. A* 75: 063620.

Torii, Y., Suzuki, Y., Kozuma, M., Sugiura, T., Kuga, T., Deng, L. & Hagley, E. W. (2000). Mach-Zehnder Bragg interferometer for a Bose-Einstein condensate, *Phys. Rev. A* 61: 041602.

Wu, S., Wang, Y.-J., Diot, Q. & Prentiss, M. (2005). Splitting matter waves using an optimized standing-wave light-pulse sequence, *Phys. Rev. A* 71: 043602.

Xiong, W., Yue, X., Wang, Z., Zhou, X. & Chen, X. (2011). Manipulating the momentum state of a condensate by sequences of standing-wave pulses, *Phys. Rev. A* 84: 043616.

Xu, X., Zhou, X. & Chen, X. (2009). Spectroscopy of superradiant scattering from an array of Bose-Einstein condensates, *Phys. Rev. A* 79: 033605.

Yang, F., Zhou, X., Li, J., Chen, Y., Xia, L. & Chen, X. (2008). Resonant sequential scattering in two-frequency-pumping superradiance from a Bose-Einstein condensate, *Phys. Rev. A* 78: 043611.

Zhou, X. (2009). Cooperative atomic scattering of light from a laser with a colored noise spectrum, *Phys. Rev. A* 80: 023818.

Zhou, X., Fu, J. & Chen, X. (2009). High-order momentum modes by resonant superradiant scattering, *Phys. Rev. A* 80: 063608.

Zhou, X., Yang, F., Yue, X., Vogt, T. & Chen, X. (2010). Imprinting light phase on matter-wave gratings in superradiance scattering, *Phys. Rev. A* 81: 0136151.

Zobay, O. & Nikolopoulos, G. (2007). Sequential superradiant scattering from atomic Bose-Einstein condensates, *Laser Physics* 17: 180.

Zobay, O. & Nikolopoulos, G. M. (2006). Spatial effects in superradiant Rayleigh scattering from Bose-Einstein condensates, *Phys. Rev. A* 73: 013620.

18

Thickness Measurement of Photoresist Thin Films Using Interferometry

Fotis Kossivas, Charalabos Doumanidis and Andreas Kyprianou
University of Cyprus, Mechanical and Manufacturing Engineering Department, Nicosia, Cyprus

1. Introduction

Photoresists are polymers sensitive to light, usually in the ultraviolet (UV) range of the electromagnetic radiation spectrum. They are classified into positive and negative. In positive types, the exposed to light region of the photoresist becomes solvable to a developer, while the unexposed one remains unsolvable. In negative types the opposite happens. Photoresists are widely used in micro and nano fabrication such as photo and e-beam lithography. In these techniques the substrate, usually a silicon (Si), wafer, is spin - coated with a thin film of photoresist and then is selectively exposed to the proper radiation. The substrate is then developed with a suitable chemical solution which etches the exposed (positive) or unexposed (negative) regions, resulting in transferring the desired pattern on it. For these, knowledge of the photoresist exact thickness is crucial for the calculation of the necessary radiation energy dose.

Traditionally in photolithography practice, the thickness of photoresist films is deduced indirectly by the volume of the resin dispensed during spin coating, divided by the wafer surface. However, this method is subject to errors due to uncertainty of the solvent evaporation and thickness distribution nonuniformities, e.g. near the wafer edges due to wetting effects etc. Recently, the measurement of liquid crystal (LC) films thickness was achieved using a relative simple processing of common interferometer images (optical profilometer), considering the phase shift of light as propagated through LCs. The technique was tested and verified with atomic force microscopy (AFM) in micron and sub micron film thickness, and was found to be in good agreement to the AFM measurements [1].

This work proposes an extension of this technique to thin films made from photoresists. Thin films of various thicknesses have been fabricated, measured with the proposed method and verified with AFM technique. The results further confirmed the validity of the proposed technique, as the difference of the two methods was within the statistical error bounds. Further testing has been also performed in standard photolithography processes to investigate if the proposed technique affects the performance of photoresists. It was found that the influence in photoresist performance is negligible as they are sensitive only to the i-line of the spectrum from a mercury vapour lamp used for photolithography. The interferometer used here was operated with a low intensity green (550 nm wavelength) laser source.

2. Optical profilometry theory and thickness measurement methodology

This section starts with an overview of how light interference is implemented into a single-wavelength optical profilometer and it is used for the extraction of three dimensional (3D) topography images of substrates. Then it analytically describes the methodology of using the phenomenon of phase shift that appears in light propagated through transparent films, together with optical profilometry to extract the thickness of the film. Finally an overview of fringe order theory is given, which is useful in the determination of the maximum possible height the instrument is capable to measure.

2.1 Existing knowledge

In general, optical profilometry exploits the interference between a reference light beam and a measuring beam reflected from the sample surface in order to extract a 3D topographical image. If the travelled distances of the two beams differ by an even number of half-wavelengths then a bright fringe (constructive interference) is created, whilst a difference equal to an odd number of half-wavelengths creates a dark one (destructive interference). In intermediate cases, gray lines are created. The height of a particular point on the sample surface is determined by the type of interference, which in the resulting image of the surface is captured by the colour of the corresponding pixel. In this mode, the number of fringes defines the maximum possible measured height. It is generally recommended to use less than three fringes (which correspond to a height difference equal to three times the source wavelength), in order to avoid problems with the coherence length of the laser source. However, with some care slightly larger numbers of fringes can also be used.

Measurements of thicknesses and refractive indices of transparent thick polymer films based on light interference phenomena are well known and developed. They are based on the fact that fringes from the upper and lower boundaries of the film are well separated, and hence the film thickness or the refractive index can be calculated from the distance between the envelopes of the two fringe patterns. However, as the film thickness decreases, the envelopes of the two fringes come closer to each other, and for thicknesses less than about 1μm they overlap. This complicates significantly the calculation of the distance between fringes, and thus the film thickness. Recently, the helical conjugate field function (HCF) has been utilized in coherence correlation interferometry to circumvent this limitation [8, 9,10].

In many areas of engineering and physics, the above technique has been incorporated in methodologies of measuring physical properties and their response to the presence of external stimuli. In the field of mechanical engineering, this constituted the underlying principle in holographic interferometry and Moire inteferometry. The former concerns measuring the properties of transparent specimens [3], whereas the latter is used to measure small deformations of non-transparent solids due to external forces and thermal expansions, as for example is described in [4, 5]. Recently, a methodology which uses two sample probes for simultaneously measuring the refractive index and thickness of transparent specimens has been developed and demonstrated in [2]. Interferometry has also been used to measure parameters of transparent films made from nematic liquid crystals, such as birefringence, dielectric anisotropy, refractive index change in response to applied DC voltage [6] and thickness [1]. In this article, the methodology of [1] is extended on other transparent materials that are mainly used as photoresists in micro-fabrication.

2.2 Initial imaging of the borders of a transparent film

Generally, in optical profilometry (or interference profilometry) the measurement starts with the operator focusing to the point where the fringe pattern occurs. Then this pattern is subjected to appropriate processing by the software of the instrument in order to extract the 3D topographical image of the sample. In the case of a transparent film with thicknesses bigger than about one micron, two fringe patterns are generated (figure 1), from exactly the same area of the sample. The first is taken by focusing on the upper surface of the transparent film and the second by focusing on the substrate. Furthermore, for thicknesses less than a micron there is only one fringe pattern which may be attributed to the substrate, as its reflectivity is significantly higher than that of the upper boundary of the film.

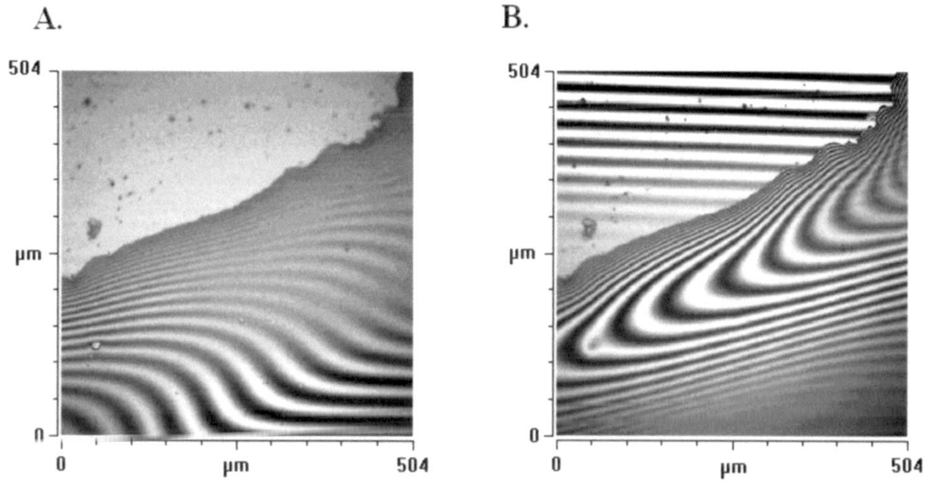

Fig. 1. A) First fringe pattern generated from the upper surface of the transparent film.
B) fringe pattern generated from the substrate. The right image is distorted in the region of film because of the phase shift that light undergoes while it travels through the film.

In the case where the first fringe pattern is used, the topology of the upper boundary of the film is taken. However in the case where the second pattern is used, an image which corresponds to the substrate is obtained. Here the region that is coated by the transparent film appears to be at a lower height than the non-coated substrate. Such an image is shown in figure 8 in section 6. Definitely, this does not reflect the reality. This happens because to the instrument the existence of the transparent film is unaccounted for, and hence the phase difference registered by the instrument is taken to be due to height differences and not due to the delay of light propagation in the film. A method that considers the optical path changes in the film in order to correct the initially measured image follows.

2.3 Understanding the initial image

Here, the principle of operation of the instrument and the optical path of its light source are considered for the case of measuring the boundary of a transparent film. Figure 2 shows a schematic presentation of an optical profilometer, operated both on a flat substrate and a transparent film. In both cases, monochromatic light from the light source impinges on the beam splitter where the beam splits into two halves. One half of the beam travels to the objective lens and then to the second beam splitter. At this point, the light beam splits into two identical beams. The one beam travels upwards to the reference mirror and reflected back towards the second beam splitter (optical path A'B'C' in figure 2). The other travels downwards and reflected from the sample surface also back to the second beam splitter (optical path A'D'C'). The two beams interfere at point C' and the interference pattern is captured by the camera. Further processing of the fringes pattern image from the instrument software, defines the type of interference in each pixel according to its intensity, and therefore its height. Now for the typical case of the non coated sample, the optical path difference of the two beams is given by A'B'C' - A'D'C'. Since the distance A'B'C' and the type of interferences are known, the distance A'C'D' and therefore the height of point D' can be extracted as follows:

$$Z_D' = 0.5(A'B'C' - A'D'C') \tag{1}$$

In the second case, where the substrate is coated by a transparent film, of refractive index n, the optical path difference of the two beams becomes ABC-(AD +nDEF +FC). In this situation the problem with the instrument is that it takes the delay of light travelling in the transparent film as a larger distance travelled in the air. However, the delay of light is due to the presence of a transparent film with a refractive index higher than that of air. As a result, in the final topographical image the region of the substrate that is coated with the thin transparent film, erroneously appears to be at a lower position than the non coated substrate. The height of point E in the initial image is calculated as:

$$Z_E = 0.5(ABC - (AD + nDEF + FC)) \tag{2}$$

Subtracting the two heights and taking that A'B'C' = ABC and for the case of a non tilted substrate A'D'C' = AEC, then:

$$Z_E - Z_{D'} = (n - 1)DE \tag{3}$$

2.4 Correcting the initial image

In subsection 2.3, the optical paths of instrument light during the measurement of both substrate and transparent film were explained and computed. In what follows, these optical paths are considered in a methodology that corrects the images presented in section 2.2 and thereby extracts the thickness of transparent films. This methodology can be used for instruments that use single wavelength optical (interference) profilometry.

The goal of this methodology is the correction of the z height values of pixels that correspond to the substrate area that is coated by a transparent film. Figure 3 depicts schematically the correction step. Figure 3 (A) represents the situation after the measurement but before the correction where the mean height of the film erroneously appears lower than the average height of non coated substrate, $Z_{\text{substrate average}}$. The correction equation 4 is obtained from the application of equation 3 to the general case of figure 3:

$$Z_{\text{film corrected}} = (n - 1)^{-1}(Z_{\text{substrate average}} - Z_{\text{film}}) \tag{4}$$

where $Z_{film\ corrected}$ is the actual height, z, value of the transparent film. This correction is implemented at the pixels corresponding to the film region. The correction step is depicted in figure 3 (B).

The benefit of this technique is that it does not need any instrumentation modification, but only a relatively simple image processing. Furthermore, only the lower fringe pattern is required. This overcomes the common interferometer issue where for very thin films only one fringe pattern appears, which is taken to be that of the lower boundary (substrate). The intensity of the fringe pattern due to the upper boundary (transparent film) is much weaker.

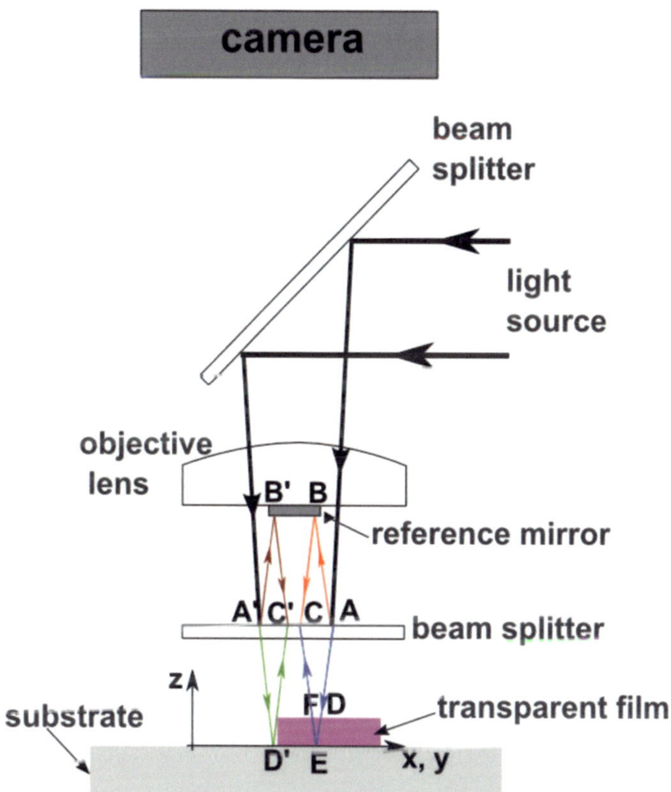

Fig. 2. Schematic representation of the principle of operation of the optical profilometry for both non coated substrate and coated with thin transparent film substrate.

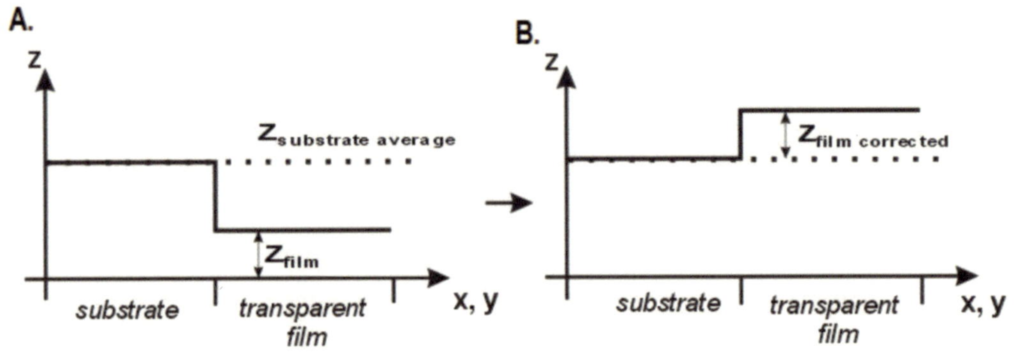

Fig. 3. 2D schematic presentation of (A) the initial image and of (B) the corrected image.

2.5 Fringe order

Let's consider figure 2, where for example the distances between the beam splitter and the reference mirror are taken as equal to the distance between the beam splitter and a sample point, while the profilometer is in focus range. In such a case, the optical path difference is equal to zero and constructive interference occurs, and therefore it is the zeroth order bright fringe. The zero order bright fringe appears with maximum intensity, compared to other fringes. Now consider an increase in the optical path equal to the wavelength, λ, due to sample moving a corresponding distance away from the beam splitter. Now, the first order bright fringe appears with lower brightness than the zeroth order fringe. The same will happen with the second, third and higher order fringes until moving out of the coherence length of the microscope where the intensity will drop to zero. A schematic curve that shows this variation in light intensity, I, with respect to the distance between the sample and lens, dx, is shown in figure 4. The optimum condition occurs when the instrument is focussed as near as possible to the zeroth order bright fringe. However as it is discussed in section 5, the confusion in fringe order for non smooth samples may lead to wrong measurements.

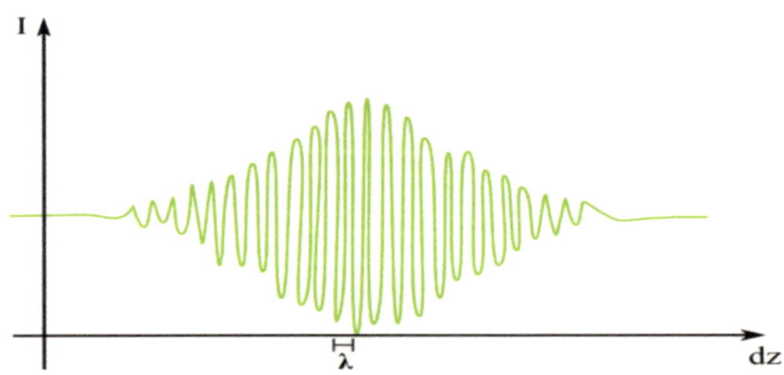

Fig. 4. Schematic representative of light intensity (I) with respect the distance between the sample and lens.

3. Experimental preparation

The optical profilometer used in thickness measurements was an Ambios Xi-100 with monochromatic light source. The light source was a green laser with wavelength equal to *550nm*. The atomic force microscope (AFM) used for the verification of film thickness measurements was a Quesant Qscope 350, operated in contact mode, whose the cantilever stiffness was *50N/m* and *10nm* tip diameter. The influence of the thickness measurement to photolithography performance was tested using a Suss MicroTech MA6 mask aligner. The spin coating machine (has been also supplied from Suss MicroTech). The negative AZ® nLOF 2070 supplied by MicroChemicals was used as photoresist. The value for the photoresist refractive index was taken from the supplier and was $n=1,64$. The photoresists were spin coated at 4000 rpms for 30sec. For the formation of thinner films the photoresist was thinned using AZ® EBR solvent. AZ® 826 MIF was finally used as a developer. The exact photolithography process started with soft bake at 100°C in a hot plate for 5min, exposure 15sec, post-exposure bake 115°C for 2sec and develop for 45sec.

4. Height difference limit

This section, based on the transparent film boundary and geometry, clarifies the rationale for selecting the correct operation mode and identifies the limitations associated with height measurements. This is very important step for correct measurement of thicknesses in general. Consider the maximum height difference that an optical profilometer may accurately measure in the *z* range. In the case where only one interference pattern is used for the extraction of the sample topography, the maximum *z* range that may be measured is different for the various cases described below:

1. Surfaces with high aspect ratio: Consider surfaces where the height of features increases (or decreases) abruptly, i.e. regions that are separated with only very few microns along any horizontal direction and their height difference is higher than half the wavelength. In such a case, the instrument confuses the fringe order and determines erroneous values for heights. The example in figure 5, represents a graded by strips silicon wafer where the edges of the strips are very sharp. The strips were fabricated on the surface by etching. The height of the graded steps is 1.2 microns and the sample was measured with an optical monochromatic profilometer of a light source generating light of 550 nm wavelength (figure 6). Therefore, by neglecting the sample tilt, the height difference at the edges of the strips is equal to about 4.18 times $\lambda/2$. The problem here is that the instrument can not determine if the optical path difference is 0.18 or 4.18 times $\lambda/2$, i.e. it is not possible to determine the fringe order with this mode. The instrument is programmed to consider smooth surfaces where all fringes are of the same order and therefore the step is taken as about 0.18 times $\lambda/2$. That is why the gradation in topographical images appears with wrong heights. As a result this mode is not suitable for surfaces with high aspect ratio.
2. Smooth surfaces: In this case, the features on the surface are separated by relative small inclinations. Therefore, the fringe order is always the same and the height of all features can be accurately measured. Here the *z* range may extend from few nanometers to several tens of micrometers without losing any precision. In figure 7, an example of measuring height differences of several microns on a smooth spherical sample is shown. The apex and base of a spherical surface are separated by a relatively large distance; however, the order of all fringes remains the same and this is reflected on the correct value of their intermediary distance.

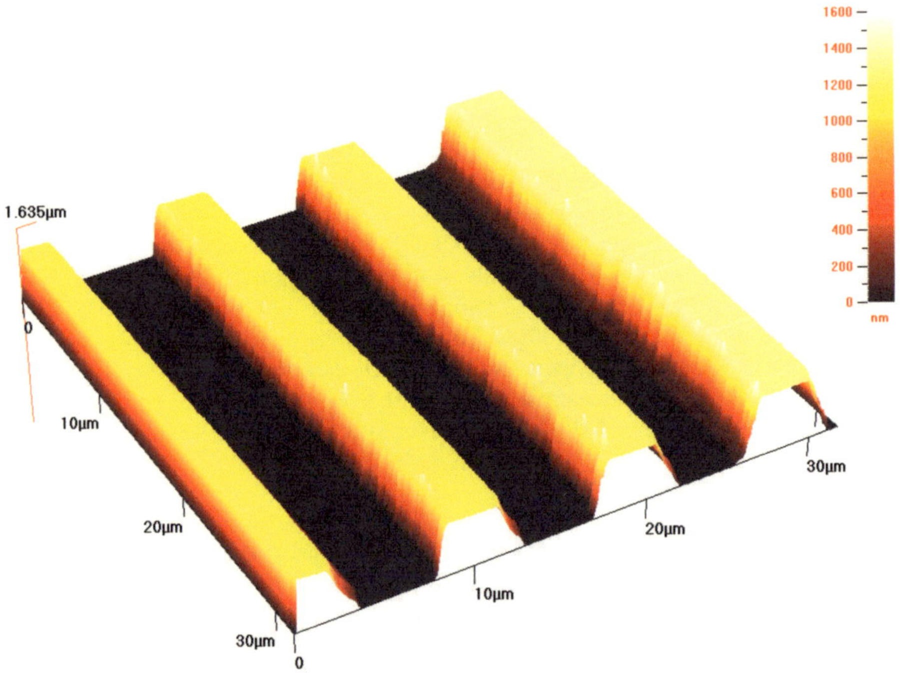

Fig. 5. 3D topography image of the sample with graded strips taken by atomic force microscopy. The steps are formed by sharp edges of 1.2µm height

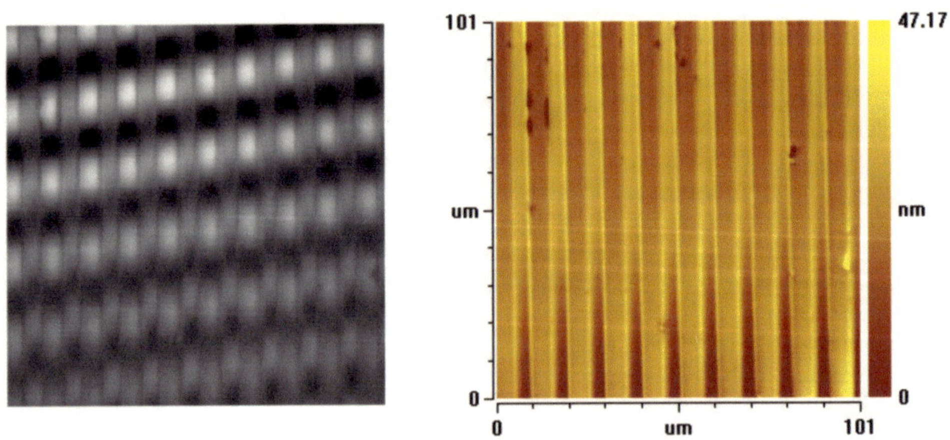

Fig. 6. Optical profilometer fringe pattern and topography images of the sample of figure 5. It is clear that the height dimensions are wrong because of the confusion of fringe order.

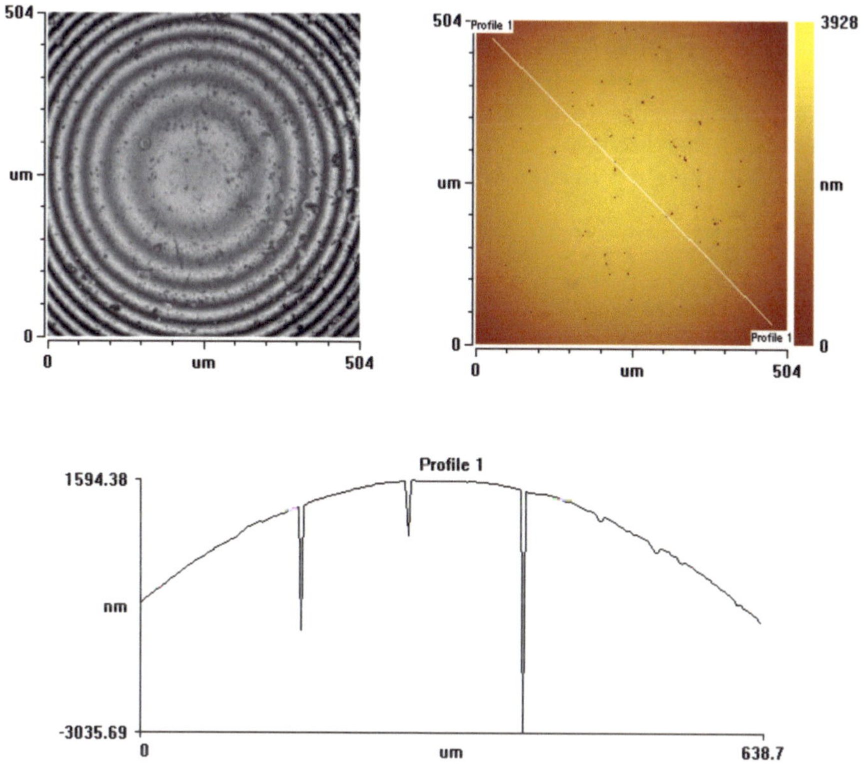

Fig. 7. The fringe pattern and the extracted topography image and profile of a smooth sphere with optical profilometry (diameter D= 10mm).

Finally, it is worth mentioning that further to the mode where only one fringe pattern is used, there is another mode usually called scanning mode. In this mode, a piezoelectric transducer accurately moves the beam splitter while the camera monitors the change in each pixel colour with respect to the travelled distance of the beam splitter. In this way, the instrument identifies the position of the zeroth order fringe (maximum intensity) for each pixel. The height of each pixel is defined by the position of the zeroth order fringe. The scanning mode *z* range is much higher and may be used up to several hundreds of microns.

The general conclusion of this section is that the technique for transparent thin film thickness measurement presented is applicable only in the smooth mode. Therefore, the technique can be applied only to films with thickness below $\lambda/2$, or to films of which the boundaries are smooth enough to avoid jumps on fringe order.

5. Method implementation to photoresist films

This section implements the technique described in section 2 by measuring the thickness of a photoresist thin film. It follows a detailed description of the steps required to implement this technique.

It starts by imaging a region at the boundaries of the transparent film of which the thickness has to be measured. In this step it is important to choose a region which: (a) is smooth enough to overcome the height limitation mentioned in section 5 and (b) includes both the substrate and the flat (thickness height equilibrium) area of the film. In this step, an image such as the one given in figure 8 is obtained. As it can be seen, the film appears to be lower than the substrate, as discussed in section 2.2. In order to circumvent this, the implementation of the height correction technique described in section 2.3 is applied.

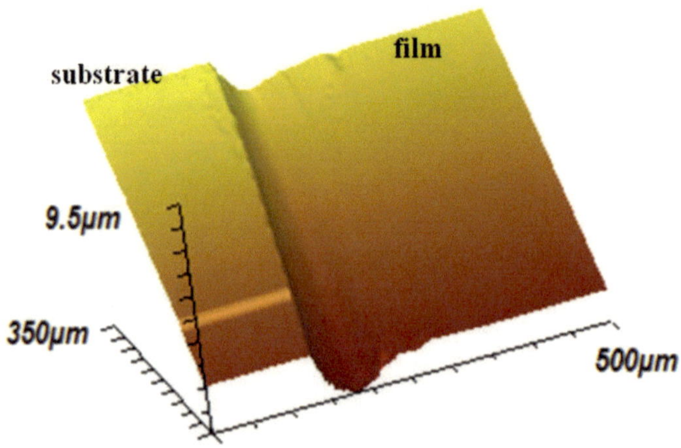

Fig. 8. Initial image taken by the optical profilometer at the boundaries of a photoresist thin film. Both the film and the substrate are also represented.

The second step includes the extraction of the profilometer image to a suitable format (.txt) for Matlab® processing. This removes the tilt that always appears in such images, due to the relative small (but significant for nano-scale measurements) natural tilt of the instrument base, or sometimes the sample apex with respect to the base surface. Tilt removal is performed with a code that fits image lines that are transverse to film boundaries with straight lines, but only to the region of the substrate. This fitting is demonstrated for a single profile in figure 9. The line extracted from this fitting is then subtracted from the corresponding image profile.

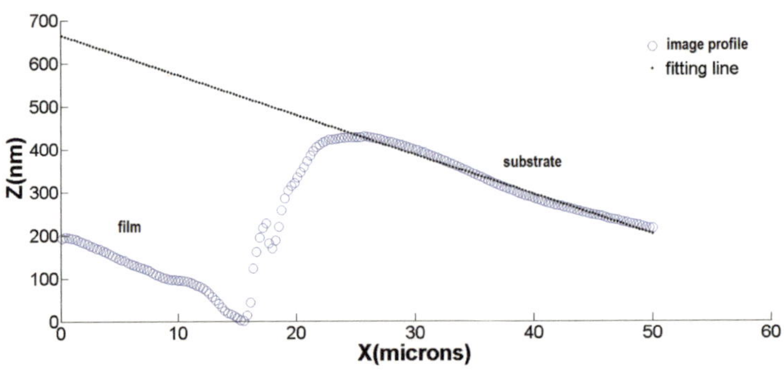

Fig. 9. Schematic representative of image fitting for tilt removal.

In the third step the boundaries of the transparent film are identified in the profilometer image using a code which scans across image lines that transverse the film boundaries and identifies where the tilt between three pixels of the image is higher than a threshold value. This value varies according to each image, and usually several values have to be tested to extract the optimal one. The image of figure 8 after tilt removal and film boundary definition are presented in figure 10.

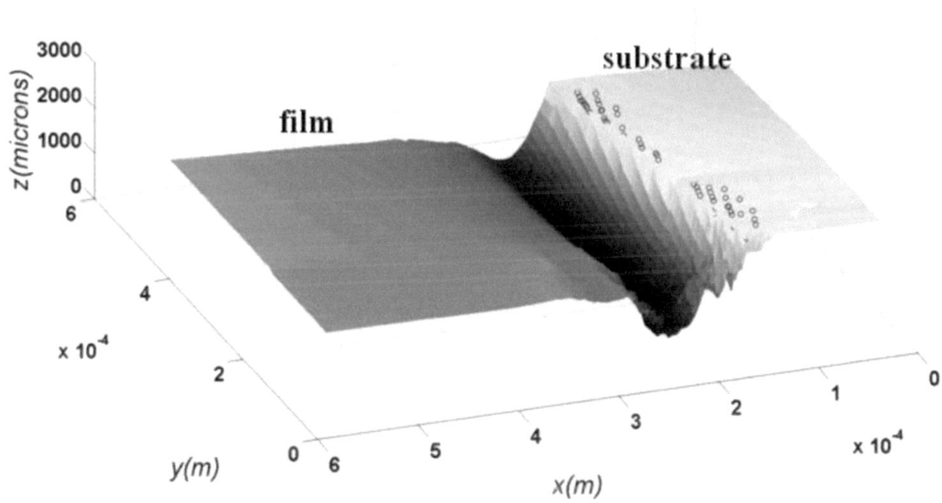

Fig. 10. The image of figure 8 with no tilt and film boundaries identified, with the curve consisting of small cycles.

Finally, the height of all pixels that belong to the transparent film region are subjected to the correction of equation (4) and demonstrated in figure 3. The result of this height correction code after implementation on the image in figure 8 is shown in figure 11. Furthermore the film thickness is calculated by subtracting the average height of an indicative film region from that of the substrate. Indicative examples of these regions are presented in figure 11 with red and blue colours respectively.

The image processing is not always necessary. In the general case where the exact topography of the film boundaries is not needed and only a thickness value is sought, the image processing may be avoided. In such a case the height difference of the substrate and film may be measured in the uncorrected optical profilometer image and then entered into equation (4) for the thickness measurement determination. This speeds up the process significantly as the image processing usually requires several hours while the alternative method needs only a few seconds. For thickness measurement both methods are equally accurate and therefore in the following section the image processing is avoided.

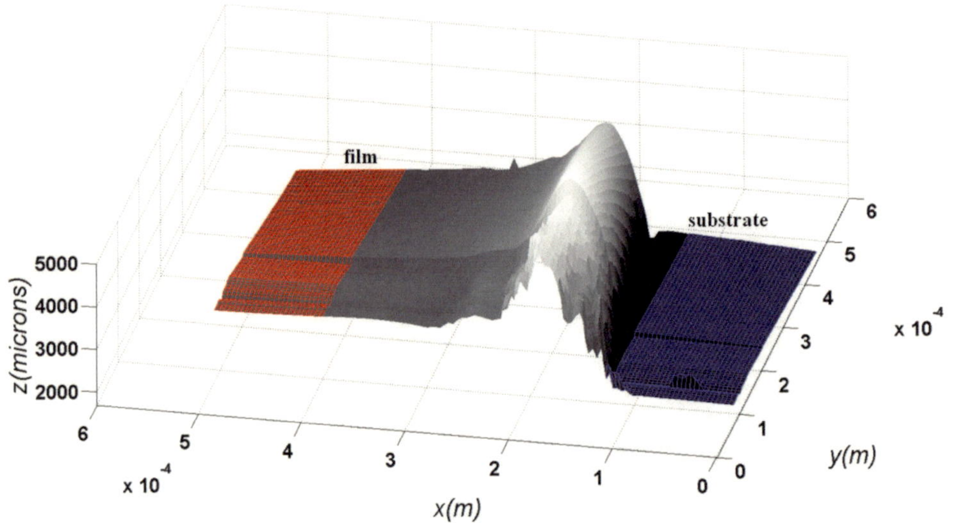

Fig. 11. Corrected image depicted by processing the image of figure 8. The regions used for thickness measurement are also highlighted.

6. Validation using atomic force microscopy

The technique for measuring transparent thin films such as photoresists, already described in section 6, has been validated here using atomic force microscopy (AFM). The atomic force microscope belongs to the scanning probe microscope (SPM) family of instruments. It mainly consists of a cantilever with a sharp tip (radius ~ 10nm), a scanner element and a laser with photo-diodes. The scanner consists of three piezoelectric elements and moves the tip in the x-y-z directions with sub nanometer precision. While the cantilever scans the sample surface, the laser light is reflected its light from the cantilever into a split photo-diode and thus the cantilever deflections at every sample point can be monitored. The whole procedure is controlled and monitored by dedicated software in a computer which reconstructs the three dimensional topographical image of the sample surface.

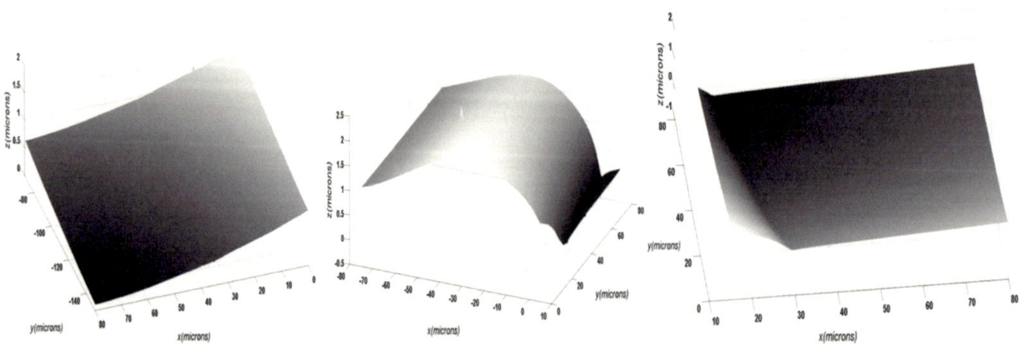

Fig. 12. Successive atomic force microscopy images of the boundaries of the photoresist film at the same region as in figure 11.

As an alternative method to measure the thickness of the photoresist film, AFM was used to scan the region in the boundary of the film with the substrate. A typical set of AFM images taken with this method is presented in figure 12. In this figures, it is obvious that due to the relatively small scanning area of the AFM image it is not possible to measure an area that completely contains both uncoated substrate and a film region with smooth and flat surface. It should be considered in this case that the film forms a hill at its borders as a result of the spin coating process. The width of this hill is typically more than 100μm, while the AFM maximum scanning area is 80μmx80μm. This limitation is overcome by scanning three adjacent areas of the sample using the motorized x-y translation stage of the AFM instrument. The three images were then subjected to the same tilt removal and tiled together using an image code developed in Matlab®. The image resulted from tiling of the three images of figure 12 is shown in figure 13. The thickness was calculated by subtraction of the substrate average height from the film average height using the same image. It was important to exclude the upward projected (hill) region of the film from the average film height calculation.

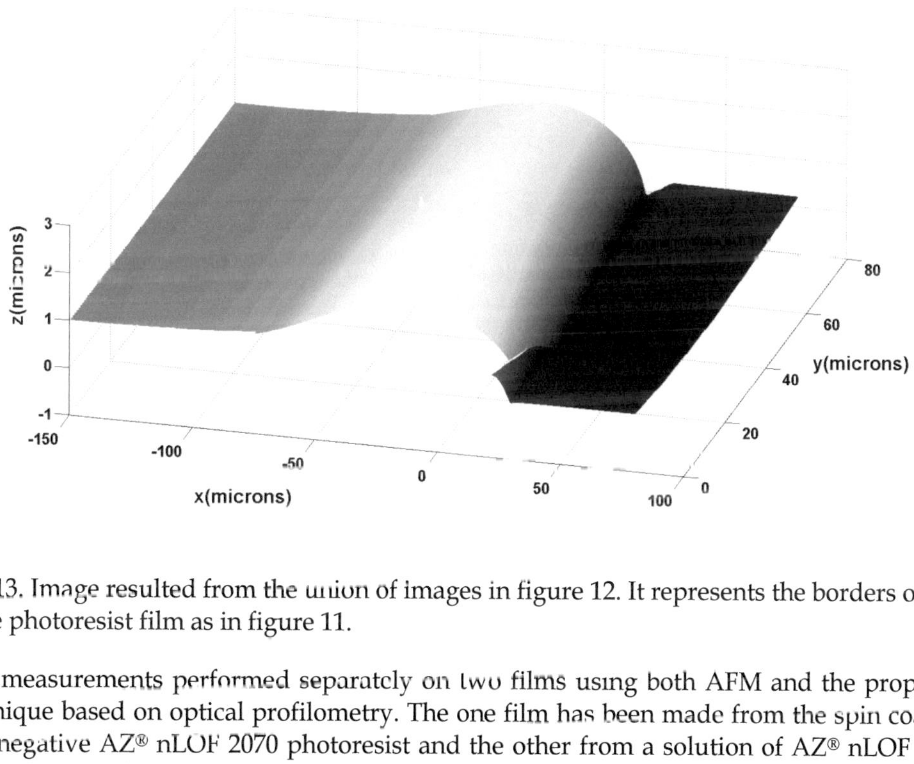

Fig. 13. Image resulted from the union of images in figure 12. It represents the borders of the same photoresist film as in figure 11.

Five measurements performed separately on two films using both AFM and the proposed technique based on optical profilometry. The one film has been made from the spin coating of a negative AZ® nLOF 2070 photoresist and the other from a solution of AZ® nLOF 2070 with AZ® EBR (volume concentration 1:1). As shown in table 1, the results were in good agreement and inside the statistical error. Further to the good agreement of the two techniques in the photoresist film thickness measurement, it is important to note further agreement of the profile and geometry of films boundaries such as the hill shape and its dimensions.

	Thickness obtained with optical Profilometer (nm)	Thickness obtained with AFM (nm)
AZ® nLOF 2070	2059±97	1966±90
Solution of AZ® nLOF 2070 and AZ® EBR (1:1)	531±87	551±39

Table 1. Average values of film thickness measured on the same film (n=5) with both AFM and optical profilometry.

The advantages and disadvantages of optical profilometry in comparison with AFM in film thickness measurement and in general are the following:

Pros:

- Optical profilometry has a much higher imaging area than AFM. With the appropriate lens of small magnification it can image areas with several millimetre width.
- It is a non destructive method as further shown in section 8. AFM is also non destructive for this application. However in other soft or wet transparent films like liquid crystals and elastomer polymers, the interaction and adhesion forces of AFM cantilever with the film may cause changes to the sample surface or even make imaging impossible.
- Optical profilometry is easier, simpler and faster.

Cons:

- The imaging of films with thickness of several hundreds of nanometers, according to section 5, is not possible by optical profilometry if the film edges are not sufficiently smooth.
- Optical profilometry in general requires surfaces with good reflectivity. Also if the sample surface consists of more than one material with different reflectivity this may result in poor imaging.
- The resolution of two techniques in film thickness measurement is practically the same. However in optical profilometry the usage of lenses with small magnification to increase the image area significantly decreases resolution.

7. Effect to photolithography performance

In the previous sections, the applicability and validity of a technique for measuring transparent films thickness has been proven based on optical profilometry on unexposed photoresists. As discussed in the introduction, photoresists are photo-sensitive polymers usually used in micro and nano patterning methods. Therefore it is possible that the exposure of photoresist films to the optical profilometer laser source for several minutes that measurements take, may influence the photolithography process. This section studies the influence of the thickness measurement technique in photolithography performance.

To investigate this, a plate substrate made from silicon wafer was spin coated with negative AZ® nLOF 2070 resist. The sample was subsequently exposed to the light of the optical profilometer. The exposure duration was 60sec in each point while the source scanned the sample at the successive points in lines to cover the area where the mask pattern was going to be transferred, figure 14. Finally, the sample has been exposed to i-line (wavelength equal to 365nm) using a mask with a pattern consisted of black lines with 20μm width separated by 40μm spacing. The sample was then developed and imaged with optical microscopy, figure 14.

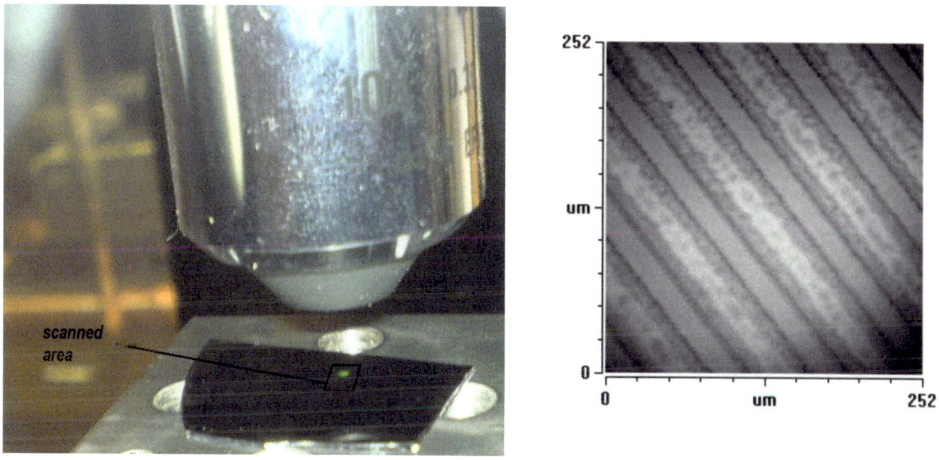

Fig. 14. Left image represents the light spot from the instrument and the sample area that was exposed to it. Right image shows an optical microscopy image of the transferred to the sample pattern after the exposure to instrument light.

As depicted from figure 14, the influence of optical profilometry was negligible. This is reasonable as in this case single wavelength optical profilometry was used with wavelength equals to 550nm while the photoresist used is sensitive to 365nm.

8. Conclusions

This work has further developed and tested on photoresist thin films a technique of measuring the thickness of LCs films [1]. It was shown that the thcikness of thin films of transparent films, like photoresists, can be measured accurately and the limitation of the the technique was also pointed out and analysed. This provides a cheap and quick non-destructive methdology of measuring photoresist thickness that could easily be adopted in the MEMS fabrication industry. In particular, the optical profilometry-based non-contact method introduced here for measurement of photoresist and other thin films is well suited not only to single and batch wafer fabrication, but also to continuous roll-to-roll (R2R) manufacturing on flexible substrates. Such R2R manufacturing, based on new deposition processes other than spin coating (e.g. doctor blading, transfer and jet printing etc). is becoming prevalent for flexible electronics and organic/hybrid photovoltaics production. The uncorrected (without image processing) version of the profilometric thickness measurement technique is fast enough for in-process monitoring of R2R production and quality control of the thin films. Such a real-time sensing tool therefore shows promise and could be valuable for feedback control of the photoresist deposition process, with considerable improvements in throughput and cost.

9. References

[1] Kossivas, F. & Kyprianou, A. (2010). Measurement of liquid crystal film thickness using interferometry, *Meausurement Science and Technology*, 105707

[2] Park, J. S. ; Park, S. K. ; Kim, H. Y & Lee, B. H. (2011). Simultaneous measurements of refractive index and thickness by spectral-domain low coherence interferometry having dual sample probes, *IEEE Photonics Technology Letters*, Vol. 23, No. 15, pp. 1076-1078

[3] Xia, T. H. ; Guo, X. R. ; Fan B. Z.; Cheng, M. H. & Yang C. B. (2011). Non-invasive mechanical measurement for transparent objects by digital holographic interferometry based on iterative least squares phase unwrapping, *Experimental Mechanics*

[4] Czarnek, R. ; Skrzat, A. & Lin, Y. S. (2011). Application of Moire interferometry to reconstruction of residual stress in cut railroad car wheels, *Measurement* Vol. 44 pp. 569-579

[5] Chen, B. & Basaran, C (2011). Measuring Joule heating and strain induced by elctrical current with Moire interferometry, *Journal of Applied Physics* Vol. 109 074908

[6] Inam, M. ; Singh, G. ; Srivastava, V. ; Prakash, J ; Joshi, T. ; Biradar, B. I. & Mehta, S. D. (2001) Two dimensional cell parameter measurement of nematic liquid crystal using optical interferometry and Fourier transform fringe analysis technique, *Optics Communications*, Optics-16416

[7] Ambios Technology Corporation (2009). Xi-100 plus operator's manual, V1.00

[8] H. M. Shabana (2004). Determination of film thickness and refractive index by interferometer, *Polymer Testing* Vol. 23 pp. 695-702

[9] M. Conroy (2009). Advances in thick and thin film analysis using interferometry, *Wear* Vol. 266 pp. 502-506

[10] D. I. Mansfield (2006). Thin film extraction from scanning white light interferometry, *Proceedings of the 21st Annual ASPE Meeting*

19

Real-Time Heterodyne Interferometry with Correlation Image Sensor

Akira Kimachi
Department of Engineering Informatics, Osaka Electro-Communication University
Japan

1. Introduction

Optical heterodyne interfereometry (HI) (R. Dändliker, 1973; N.A. Massie, 1979) is an interferometric technique that employs two interfering beams of slightly different frequencies. Interference of these beams yields a time-varying sinusoidal intensity signal whose frequency equals the difference between the frequencies of the two beams. The signal and frequency are called beat signal and beat frequency, respectively. One of the most significant properties of HI is that the phase difference between the two beams, which is usually the most important quantity in interferometry, equals the phase of the beat signal. In conventional HI, the phase of the beat signal is measured by a single photodetector and a demodulation circuit or instrument. It is said that accuracy of phase measurement in HI reaches $2\pi/1000$ rad. Because of the nature of point-wise measurement, however, this method requires two-dimensional (2-D) scanning of the measured point to realize 2-D measurement, resulting in a trade-off between temporal and spatial resolution.

The objective of this article is to introduce a real-time 2-D HI technique. The key idea is the use of the time-domain correlation image sensor (CIS) (S. Ando, 2003) as a 2-D array of phase-demodulating pixels for incident heterodyne beams (A. Kimachi, 2007; 2010a). The CIS produces temporal correlations between the intensity signal of incident light and global reference signals over a frame period and outputs them as images at an ordinary frame rate. This property enables the CIS to simultaneously demodulate the amplitude and phase of the wavefront of incident heterodyne beams at each pixel in every frame using sinusoidal reference signals of the beat frequency. Consequently, as the most significant advantage, this real-time 2-D HI can ensure both maximal temporal and spatial resolution of the image sensor and high accuracy of phase measurement. Note that other than real-time 2-D HI, the CIS has opened a very wide range of novel applications such as image feature extraction (A. Kimachi, 1998), three-dimensional measurement (A. Kimachi, 2001; T. Kurihara, 2003; S. Ando, 2008a; A. Kimachi, 2009), optical flow measurement (S. Ando, 2008b), spectral matching imaging (A. Kimachi, 2002; 2004; 2010b), magneto-optic leakage flux imaging (S. Ando, 2007), and so forth.

The remainder of this article is organized as follows. Section 2 reviews previous studies on real-time 2-D HI and phase-shifting interferometry (PSI). Section 3 describes the principle of the CIS and the real-time 2-D HI. Section 4 shows experimental results of the real-time 2-D HI for a polarizing Michelson interferometer and speckle pattern interferometers for in-plane and out-of-plane deformation measurement. Section 5 briefly describes ongoing research on a networked sensing system for health monitoring of large-scale structures based on multi-zero beams and the real-time 2-D HI. Finally, Section 6 concludes this article.

2. Review of real-time 2-D heterodyne and phase-shifting interferometry

A great number of studies have been devoted to the problem with HI mentioned in Section 1. Realization of real-time 2-D HI can be divided into two categories: 1) development of real-time PSI methods using image sensors, and 2) integration of phase-demodulating sensors into 2-D arrays. The following subsections briefly review studies belonging to each category.

2.1 Real-time PSI using image sensors

PSI, a widely-used interferometry technique realized with a commercial image sensor, is easier to deal with than the development of 2-D arrays of phase-demodulating sensors. PSI captures a sequence of fringe images formed by two interfering beams of the same frequency, while the phase difference between the beams is shifted stepwise with a piezoelectric actuator or a variable retarder in each frame (K.J. Gåsvik, 2002; J.H. Bruning, 1974). The original phase difference is estimated by fitting the intensity of the image sequence to a sine wave individually at each pixel (C.J. Morgan, 1982). PSI is used not only in ordinary interferometry for specularly reflective objects or transparent objects, but also in speckle pattern interferometry (SPI) (K. Creath, 1985; J. Kato, 1993; M. Takeda, 1994), optical coherence tomography (M. Akiba, 2003), or holography (I. Yamaguchi, 1997) for objects with rough surfaces or made of turbid media.

PSI requires three images at the minimum, and frequently employs four images with a phase shift increment by 90°. Higher accuracy is achieved for phase demodulation with a larger number of images; however, it results in a longer time for image capture. In PSI, temporal resolution and measurement accuracy are thus in a trade-off.

It seems that this problem with PSI could be solved by capturing fringe images with different phase shifts at the same time. North-Morris *et al.* (M. North-Morris, 2005) developed a phase mask consisting of an array of 2 × 2-pixel retarder matrices with $\pi/2$ steps of retardance for phase-shifting SPI. Four-step PSI is thus realized from a single-frame image by attaching this mask in front of an image sensor. A similar phase mask was proposed for phase-shifting holography by Awatsuji *et al.* (Y. Awatsuji, 2004). Kiire *et al.* (T. Kiire, 2008) proposed a method for simultaneously producing four separate images of speckle interference for different phase shifts on an image sensor using a polarizing prism, a diffraction grating, and two lasers of different wavelengths. The price for these single-frame PSI methods is, of course, loss of spatial resolution down to half of the image sensor.

Akiba *et al.* (M. Akiba, 2003) proposed a system of optical coherence tomography using two pairs of a liquid-crystal shutter and a CCD (charge coupled device) camera. The beat signal of low coherence interference exerted by a translating mirror is sampled in synchrony by the two shutters and integrated over a frame period of the CCDs at each pixel. This system captures a pair of phase-shifted fringe images at a video frame rate, though it may be categorized into 2-D discrete sampling of a continous beat signal as discussed in the next subsection rather than phase shifting. The phase of the beat signal can be obtained from the pair of integrated images and a dc component image captured beforehand, though phase demodulation was not performed in the paper. The accuracy of this method, however, can be degraded because of temporal sampling with a rectangular window, which has a certain degree of sensitivity to higher-order harmonic components in the beat signal. Moreover, the two shutter-CCD pairs make the system rather bulky and demand cumbersome optical alignment.

2.2 Integration of 2-D phase-demodulating pixel arrays

Other studies focus on integrating real-time phase-demodulating pixels for HI into 2-D arrays. It is a straightforward idea to employ high-speed cameras as 2-D phase-demodulating arrays for sampling continuous beat signals at discrete time intervals (P. Haible, 2000; M.V. Aguanno, 2007). The higher the frame rate, the higher the accuracy of phase demodulation because of a larger number of samples, while the spatial resolution of the image sensor is maintained. The use of a high-speed camera, however, must be accompanied by a high-power laser source; otherwise it degrades the accuracy of phase demodulation because of shorter exposure time, lower light intensity, and hence lower signal-to-noise ratio (SNR) in the interference images. Moreover, high-speed cameras require high-speed frame grabber interface with large memory, which gives rise to a problem in terms of commercial availability.

Another group of studies employ 2-D arrays of phase-demodulating sensors developed on their own. A type of phase-demodulating image sensor called lock-in CCD, originally proposed by Spirig *et al.* (T. Spirig, 1995), outputs a set of discrete samples of the incident beat signal from each pixel, instead of a sequence of fringe images. The multiple samples are transferred to separate bins within a pixel as charge packets and accumulated over a frame period. The phase of the beat signal is demodulated from the set of samples at each pixel on a computer by the same algorithm as used in PSI. Dändliker *et al.* (R.D. Dändliker, 1998) used this sensor for distance interferometry. Earlier than these reports Povel *et al.* (H. Povel, 1990) devised a similar sensor for imaging polarimetry, though it was not named lock-in CCD.

It follows that in lock-in CCDs, the accuracy of phase demodulation, which increases for a larger number of samples or a higher sampling rate, is in a trade-off with spatial resolution because storing a large number of samples requires more space within the pixel on the sensor chip. Note also that the nature of discrete temporal sampling in lock-in CCDs as well as in high-speed cameras, represented by a rectangular temporal window function, again brings about the problem of sensitivity to higher-order harmonic components in the incident beat signal. This degradation becomes severer for a smaller number of samples or a lower sampling rate.

Another type of 2-D phase-demodulating pixel array was developed by Bourquin *et al.* for optical coherence tomography (S. Bourquin, 2001a). The pixel consists of a photodetector followed by a feedback circuit with a low-pass filter (LPF) and a rectifier (S. Bourquin, 2001b). For an incident beat signal of a frequency above the LPF cutoff frequency, the pixel can demodulate the signal amplitude as a result of open-loop operation of the feedback circuit. This sensor, however, cannot perform phase demodulation because it does not have the circuit for it.

Now the real-time 2-D HI presented in this article is compared to the techniques discussed so far. The CIS used in the real-time 2-D HI is exactly an ideal type of 2-D phase-demodulating pixel array. As a significant advantage over other techniques employing 2-D phase-demodulating pixel arrays and high-speed cameras, the CIS demodulates the phase of a continuous beat signal by continuous operation on analog circuitry under an ordinary frame rate, instead of discrete sampling and off-line computation. This property makes the real-time 2-D HI immune to the problems of rectangular windowing, light SNR, and frame

Type	Method/tool	Resolution		Accuracy	System issue
		Temporal	Spatial		
2-D heterodyne	Photodetector, 2-D scan	*	*	Good	
Phase shifting	Image sensor, phase shifter	*	Good	*	
Phase shifting	Two pairs of image sensor and electronic shutter	Good	Good	Fair	Alignment, bulkiness
Phase shifting	Image sensor with phase mask	Good	Poor	Fair	
Phase shifting	Phase-shifted images on one image sensor	Good	Poor	Fair	
2-D heterodyne (discrete)	High-speed camera	Good	Good	Fair	Light S/N, camera I/F
2-D heterodyne (discrete)	Lock-in CCD	Good	*	*	
2-D heterodyne	Correlation image sensor	Good	Good	Good	

Table 1. Comparison of 2-D HI and PSI techniques. The boxes with an asterisk in each row indicate a trade-off relation.

grabber interface. Moreover, spatial resolution can be maximally utilized as far as the pixel size permits, without any loss by sharing among multiple phase-shifted images, as is the case with the real-time PSI methods. The real-time 2-D HI and PSI techniques discussed above are summarized in Table 1.

3. Principle

This section first describes the principle of the CIS, and then explains how the CIS demodulates amplitude and phase images of heterodyne beams in real time.

3.1 Correlation image sensor

Figure 1(a) depicts the architecture of the CIS, which consists of a pixel array and three external inputs of reference signals supplied to all of the pixels. The pixel circuit for passive readout is illustrated in Figure 1(b) (S. Ando, 2003). The photodiode (PD) converts the incident light into a photocurrent $I(t)$ in proportion to its intensity $f_{ij}(t)$ at pixel (i,j). $I(t)$ is divided into three currents $I_k(t)$ ($k = 1,2,3$, $\sum_{k=1}^{3} I_k(t) = I(t)$) that flow through three NMOS transistors Q_k, and accumulated in three capacitors connected to the drains of Q_k over a frame period T. The balance among $I_k(t)$ depends on the gate voltages $V_k(t)$ as

$$I_k(t) = \rho g_k(t) I(t) + \frac{1}{3} I(t) \quad (k = 1,2,3), \tag{1}$$

where $g_k(t) = V_k(t) - \overline{V}$ denotes the differential component of $V_k(t)$ from the average $\overline{V} = \frac{1}{3}\sum_{k=1}^{3} V_k(t)$, and ρ is a constant. The signals $g_k(t)$, called reference signals, are supplied to the gates of Q_k in each pixel under a constraint $\sum_{k=1}^{3} g_k(t) = 0$. In Eq. (1), integrating $I_k(t)$

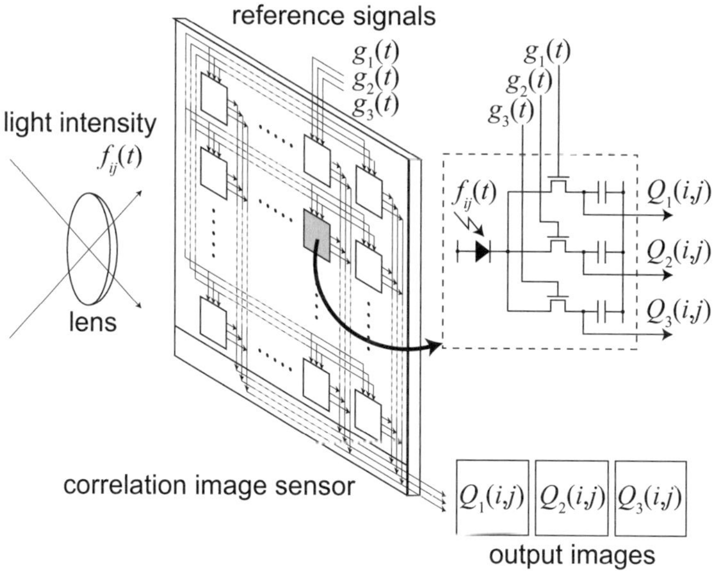

(a)

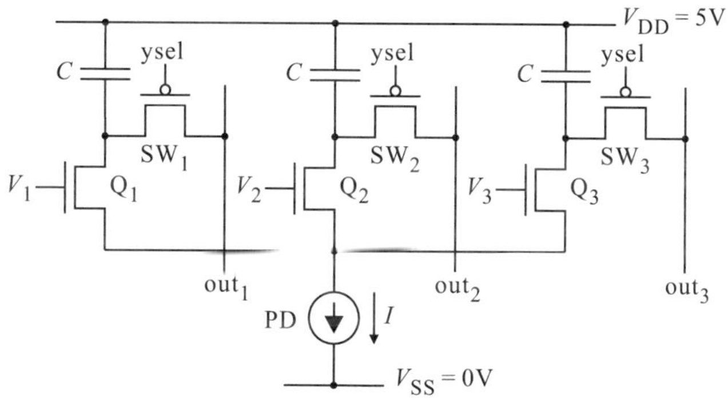

(b)

Fig. 1. Schematic of the CIS. (a) Overall architecture. (b) Pixel circuit for passive readout.

over T while redefining $\rho g_k(t)$ as $g_k(t)$ and replacing $I(t)$ with $f_{ij}(t)$ yields an expression for the accumulated charges $Q_k(i,j)$ as

$$\begin{bmatrix} Q_1(i,j) \\ Q_2(i,j) \\ Q_3(i,j) \end{bmatrix} = \begin{bmatrix} \langle I_1(t) \rangle \\ \langle I_2(t) \rangle \\ \langle I_3(t) \rangle \end{bmatrix} = \begin{bmatrix} \Delta Q_1(i,j) \\ \Delta Q_2(i,j) \\ \Delta Q_3(i,j) \end{bmatrix} + \begin{bmatrix} \frac{1}{3}\langle f_{ij}(t) \rangle \\ \frac{1}{3}\langle f_{ij}(t) \rangle \\ \frac{1}{3}\langle f_{ij}(t) \rangle \end{bmatrix}, \quad (2)$$

where $\langle \cdot \rangle$ denotes an integral over T. The term $\langle f_{ij}(t) \rangle$, obtained by

$$\langle f_{ij}(t) \rangle = \sum_{k=1}^{3} Q_k(i,j), \quad (3)$$

(a) (b)

Fig. 2. Photographs of CIS cameras with CMOS chips of (a) 64 × 64-pixel and (b) 200 × 200-pixel sensors.

is proportional to the average light intensity, as in conventional image sensors. The most significant difference from them is the term

$$\begin{aligned}\Delta Q_k(i,j) &= \langle f_{ij}(t) g_k(t) \rangle \\ &= Q_k(i,j) - \frac{1}{3}\langle f_{ij}(t)\rangle \quad (k = 1,2,3),\end{aligned} \tag{4}$$

which represents the temporal correlation between the intensity of the incident light $f_{ij}(t)$ and each of the reference signals $g_k(t)$. The three temporal correlations $\Delta Q_k(i,j)$ have two degrees of freedom under a constraint $\sum_{k=1}^{3} \Delta Q_k(i,j) = 0$ caused by $\sum_{k=1}^{3} g_k(t) = 0$. The charges $Q_k(i,j)$ are read out either passively through PMOS switches as shown in Figure 1(b) (S. Ando, 2003), or actively with voltage source followers (S. Han, 2010). Figure 2 shows photographs of CIS cameras with 64 × 64-pixel (S. Ando, 2003) and 200 × 200-pixel (S. Ando, 2007) CMOS chips of passive type.

3.2 Real-time amplitude and phase demodulation of heterodyne beams

Consider two heterodyne coherent beams with complex wavefronts u_{ij}, v_{ij} and a beat frequency $\Delta\nu$ to interfere on the focal plane of the CIS. The intensity $f_{ij}(t)$ of the interfering beams is given by

$$\begin{aligned}f_{ij}(t) &= |u_{ij} + v_{ij} \exp[\mathrm{i}2\pi\Delta\nu t]|^2 + n_{ij}(t) \\ &= A_{ij}\cos(\omega t + \theta_{ij}) + B_{ij} + n_{ij}(t), \\ A_{ij} = 2|u_{ij}||v_{ij}|, \quad \theta_{ij} &= \angle v_{ij} - \angle u_{ij}, \quad B_{ij} = |u_{ij}|^2 + |v_{ij}|^2, \quad \omega = 2\pi\Delta\nu,\end{aligned} \tag{5}$$

where i denotes the imaginary unit ($\mathrm{i}^2 = -1$) and background illumination $n_{ij}(t)$ is taken into account. The beat frequency $\Delta\nu$ is chosen to be a multiple of the frame rate of the CIS, $\Delta\nu = mT^{-1}$ (m: integer), or much higher than the frame rate, $\Delta\nu \gg T^{-1}$. The interference amplitude A_{ij} and phase θ_{ij} are simultaneously demodulated from single-frame temporal correlation images $\Delta Q_k(i,j)$, by supplying the CIS with three-phase reference signals $g_k(t)$ of the same angular frequency ω expressed by

$$\begin{bmatrix} g_1(t) \\ g_2(t) \\ g_3(t) \end{bmatrix} = \begin{bmatrix} \cos\omega t \\ \cos\left(\omega t + \frac{2}{3}\pi\right) \\ \cos\left(\omega t + \frac{4}{3}\pi\right) \end{bmatrix}. \quad (6)$$

Noting that the average intensity B_{ij} and the background illumination $n_{ij}(t)$ are both uncorrelated with the sinusoudal reference signals $g_k(t)$, i.e. $\langle g_k(t)B_{ij}\rangle = \langle g_k(t)n_{ij}(t)\rangle = 0$, substitution of Eqs. (5) and (6) into Eq. (4) under the condition $\Delta\nu = mT^{-1}$ or $\Delta\nu \gg T^{-1}$ yields the single-frame temporal correlation images $\Delta Q_k(i,j)$ as

$$\begin{bmatrix} \Delta Q_1(i,j) \\ \Delta Q_2(i,j) \\ \Delta Q_3(i,j) \end{bmatrix} = \frac{TA_{ij}}{2} \begin{bmatrix} \cos\theta_{ij} \\ \cos\left(\theta_{ij} - \frac{2}{3}\pi\right) \\ \cos\left(\theta_{ij} - \frac{4}{3}\pi\right) \end{bmatrix} + \begin{bmatrix} \xi_1(i,j) \\ \xi_2(i,j) \\ \xi_3(i,j) \end{bmatrix}, \quad (7)$$

where noise contributions $\xi_k(i,j)$ are taken into account. The interference amplitude A_{ij} and phase θ_{ij} are obtained from $\Delta Q_k(i,j)$ by least-squares estimation as

$$\theta_{ij} = \tan^{-1} \frac{\sqrt{3}(\Delta Q_2 - \Delta Q_3)}{2\Delta Q_1 - \Delta Q_2 - \Delta Q_3}, \quad (8)$$

$$A_{ij} = \frac{2\sqrt{2}}{3T}[(\Delta Q_1 - \Delta Q_2)^2 + (\Delta Q_2 - \Delta Q_3)^2 + (\Delta Q_3 - \Delta Q_1)^2]^{1/2}, \quad (9)$$

where (i,j) is dropped for notational symplicity.

4. Experimental results

The theory on real-time 2-D HI using the CIS described in Section 3 are verified on two types of interferometer—a Michelson interferometer and speckle pattern interferometers for in-plane and out-of-plane deformation measurement.

4.1 Real-time heterodyne Michelson interferometry

To study the feasibility of the proposed real-time 2-D HI, a heterodyne polarizing Michelson interferometer was constructed with the CIS as depicted in Figure 3(a) (A. Kimachi, 2007). A pair of orthogonally-polarized heterodyne beams coaxially enter the interferometer through a beam expander L1. The beams are split by the polarizing beam splitter PBS1 and reflected on the mirrors M1 and M2 while passing through the quarter wave plates QWP1 and QWP2 twice, which rotate the polarization of each reflected beam by 90°. The reflected beams are merged by PBS1 and interfere on the CIS through an analyzer POL1 and an imaging lens L2. A reference beat signal is generated by taking out a small portion of the heterodyne beams with a beam splitter BS and detecting the interference intensity with a photodiode PD. This signal is converted into three-phase sinusoidal reference signals for the CIS. The heterodyne beams are emitted from a commercial two-frequency He-Ne laser source shown in Figure 3(b) (wavelength λ = 632.8 nm), which consists of two acousto-optic modulators driven at slightly different frequencies to yield a beat frequency of $\Delta\nu$ = 25 kHz. The mirror M1 is longitudinally moved with a piezoelectric actuator under open-loop control to change the interference phase θ_{ij} between the two beams linearly to the displacement δz.

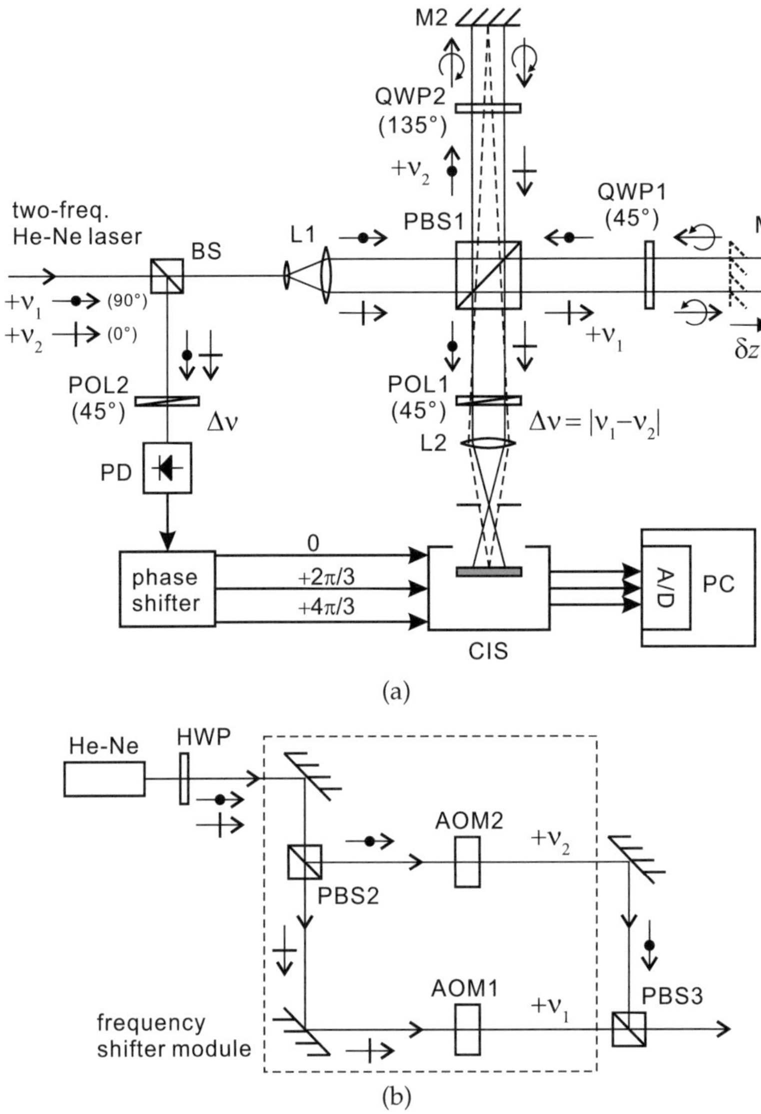

Fig. 3. Schematic of the real-time heterodyne polarizing Michelson interferometer with the CIS (A. Kimachi, 2007). (a) Interferometer optics. BS: beam splitter; PBS: polarizing beam splitter; QWP: quarter-wave plate; POL: polarizer/analyzer; PD: photodiode. (b) Two-frequency He-Ne laser source. AOM: acousto-optic modulator; HWP: half-wave plate.

The 64×64-pixel CIS in Figure 2(a) was used in the interferometer in Figure 3(a), operating at a frame rate of $T^{-1} = 30$ fps ($\ll \Delta\nu$). Figure 4 shows an example of average intensity, demodulated amplitude, and demodulated phase images computed from the raw output images $Q_k(i,j)$ with Eqs. (3), (4), (8) and (9) (A. Kimachi, 2007). The demodulated phase θ_{ij} in Figure 4(c) is coded in gray scale as depicted in Figure 4(d). The abrupt edge from black to white in Figure 4(c) corresponds to a 2π jump.

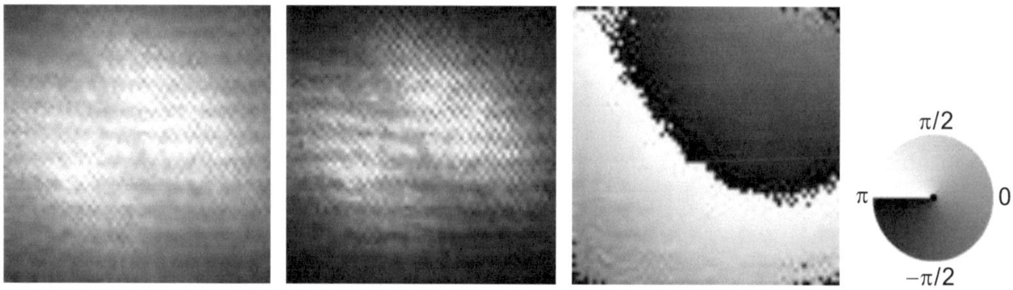

Fig. 4. An example of output images from the 64 × 64-pixel CIS in the real-time heterodyne Michelson interfereometer (A. Kimachi, 2007). (a) Average intensity $\langle f_{ij}(t) \rangle$. (b) Demodulated amplitude A_{ij}. (c) Demodulated phase θ_{ij}. (d) Gray-scale representation of phase.

Figure 5 shows the results when the mirror M1 was moved by the deformation of the piezoelectric actuator for changes in voltage (A. Kimachi, 2007). Figure 5(a) shows a sequence of demodulated phase images for voltages from 30.0 V to 36.5 V in 0.5-V steps. The difference of each phase image from that for 30.0 V is shown in Figure 5(b). The phase difference for the pixel at the image center is unwrapped and plotted in Figure 5(c) up to 50.0 V, along with a linear fit. The plot in Figure 5(c) exhibits a good linear response to the applied voltage, which is ideally linear to the displacement of M1, despite the open-loop control of the piezoelectric actuator. The slope of the linear fit in Figure 5(c) amounts to 65 nm/V in terms of optical path length, which is close to the characteristic of the piezoelectric actuator (61 nm/V).

The real-time heterodyne Michelson interfereometer in Figure 3 was applied to monitoring dynamic change in the interference phase of the heterodyne beams, which was caused by silicone oil inserted in front of the mirror M1 as depicted in Figures 6(a) and (b). Figure 6(c) shows a real-time sequence of phase difference images in every 0.2 s, i.e. every six frames, with respect to the initial state, while the silicone oil was heated by the heating cable (A. Kimachi, 2007). It is observed that gradation is buidling up near the surface of the silicone oil as time passes, suggesting changes in refractive index due to increase in temperature as well as convection that brings heated silicone oil up toward the surface.

4.2 Real-time heterodyne speckle pattern interferometry

The real-time 2-D HI was also applied to SPI. Figure 7(a) depicts a real-time heterodyne SPI system for in-plane deformation measurement (A. Kimachi, 2010a). The heterodyne beams are generated by a He-Ne laser and an electro-optic modulator (EOM), which shifts the phase between a pair of orthogonally-polarized coaxial beams from 0 to 2π linearly to time at a frequency of $\Delta\nu = 100$ Hz by a sawtooth-shaped driving voltage signal. The heterodyne beams are split by the polarizing beam splitter PBS and projected onto the object from symmetrically opposite directions with an inclination angle $\alpha = 45°$. The interference speckle pattern generated on the object is imaged through an analyzer POL onto the CIS, which is supplied with three-phase reference signals of the same frequency $\Delta\nu$ as that of the EOM driving signal. The CIS demodulates the phase of the interference speckle pattern from single-frame output images, yielding a random pattern. The difference of the demodulated phase from that of the object at an initial state changes linearly to the in-plane deformation δx

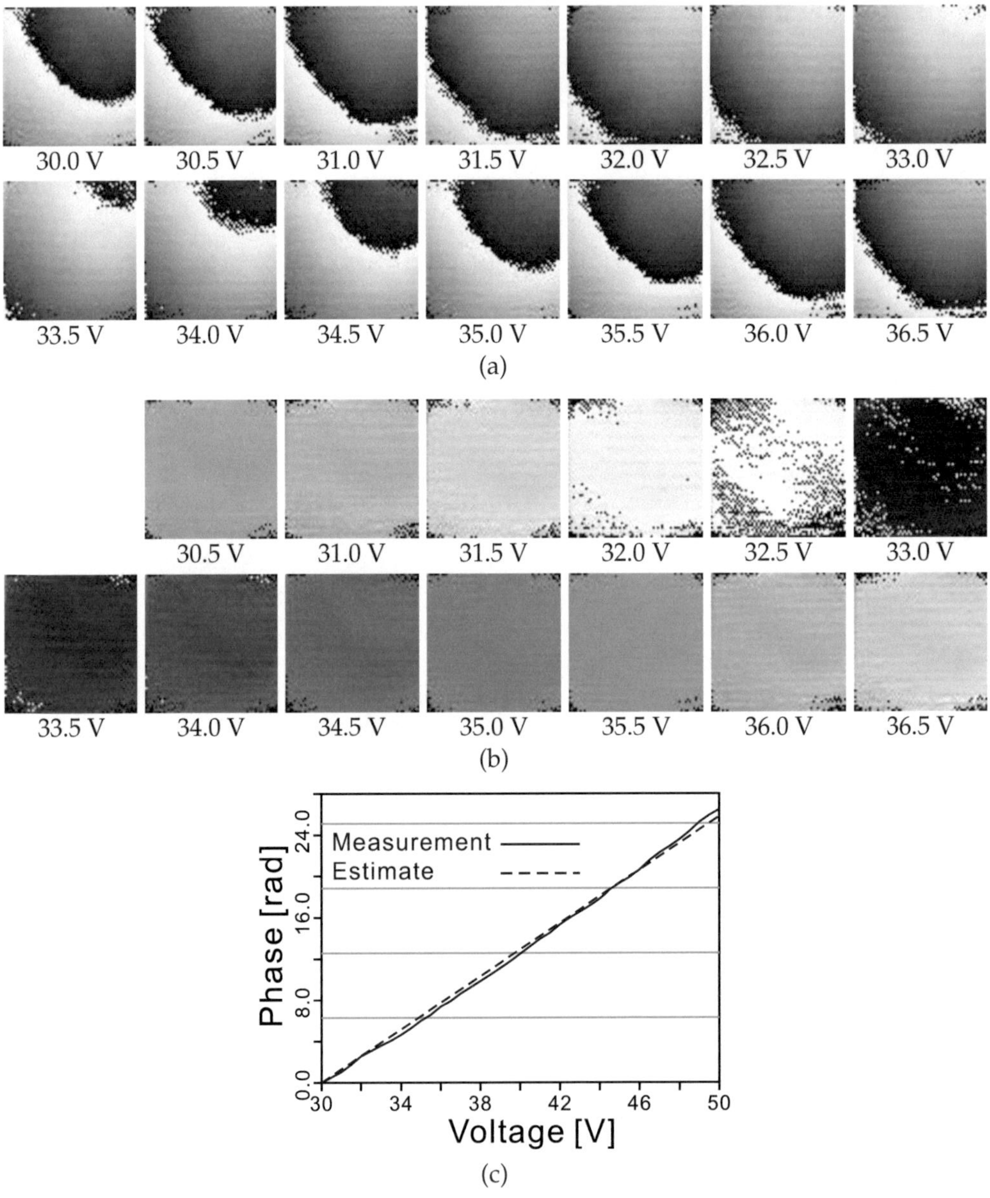

Fig. 5. Results of the real-time heterodyne Michelson interferometer for changes in the piezoelectric actuator's voltage (A. Kimachi, 2007). (a) Demodulated phase images. (b) Images of phase difference with respect to 30.0 V. (c) Plot of the unwrapped phase difference at a pixel at the image center.

at each pixel, as long as the deformation does not exceed the average speckle size to escape speckle decorrelation.

The 200 × 200-pixel CIS camera in Figure 2(b) was used in the SPI system in Figure 7(a), operating at a frame rate of $T^{-1} = 8.33$ fps ($= \Delta\nu/12$). Figure 8 shows the results for a

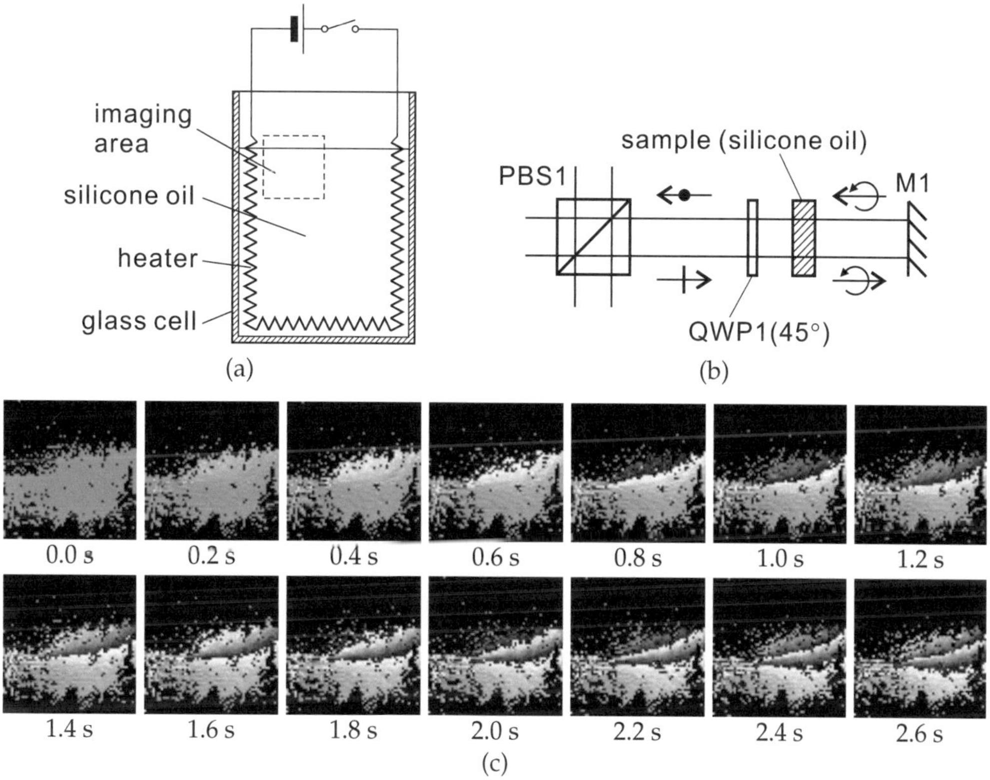

Fig. 6. Results on the real-time heterodyne Michelson interferometer for an object under dynamic behavior (A. Kimachi, 2007) (a) Illustration of the object. (b) Optical arrangement of the object. (c) Real time sequence of phase difference images with respect to 0.0 s while the object was heated.

flat object with a rough surface while it was moved in an in-plane direction in 40-nm steps by a nano-scale linear stage (A. Kimachi, 2010a). Figure 8(a) shows a sequence of phase difference images with respect to the initial position of the object, smoothed with a 5 $\times$ 5-pixel Gaussian filter. The value of phase difference is coded again as in Figure 4(d). The images exhibit gradual changes in phase difference in accordance with the in-plane displacement. The average and standard deviation of unwrapped phase difference over the central 50 $\times$ 50-pixel region are plotted in Figure 8(b) along with a theoretical response, up to the displacement of 2 μm. The plot of the average phase difference agrees well with the theoretical response. The standard deviation was 0.13 rad, or 9 nm in terms of displacement, in average over the plotted range.

A similar experiment was carried out as well for the real-time heterodyne SPI system for out-of-plane deformation measurement depicted in Figure 7(b), which is based on a Michelson interferometer with the reflection mirrors replaced by the object and a reference plate with a rough surface (A. Kimachi, 2010a). The same 200 $\times$ 200-pixel CIS camera, heterodyne laser source, and flat object were used as in the system in Figure 7(a). Figure 9(a) shows a sequence of phase difference images with respect to the initial position of the object, while it underwent out-of-plane displacement along the optical axis in 40-nm steps. The images exhibit gradual

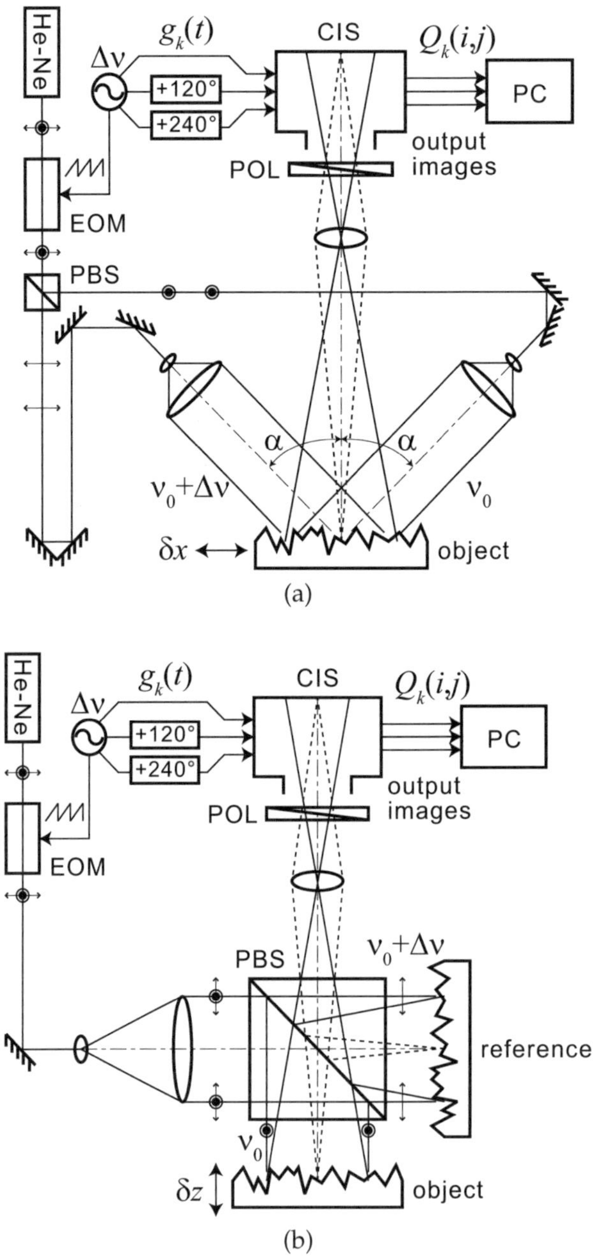

Fig. 7. Schematics of the real-time heterodyne SPI systems with the CIS for (a) in-plane, and (b) out-of-plane deformation measurement (A. Kimachi, 2010a).

changes in phase difference, as observed in Figure 8(a), though they look a little noisier. The average and standard deviation of unwrapped phase difference over the central 50 × 50-pixel region are plotted in Figure 9(b), along with a theoretical response. The standard deviation was 0.33 rad, or 17 nm in terms of displacement, in average over the plotted range, which is larger than the plot in Figure 8(b). Moreover, the agreement between the experimental and

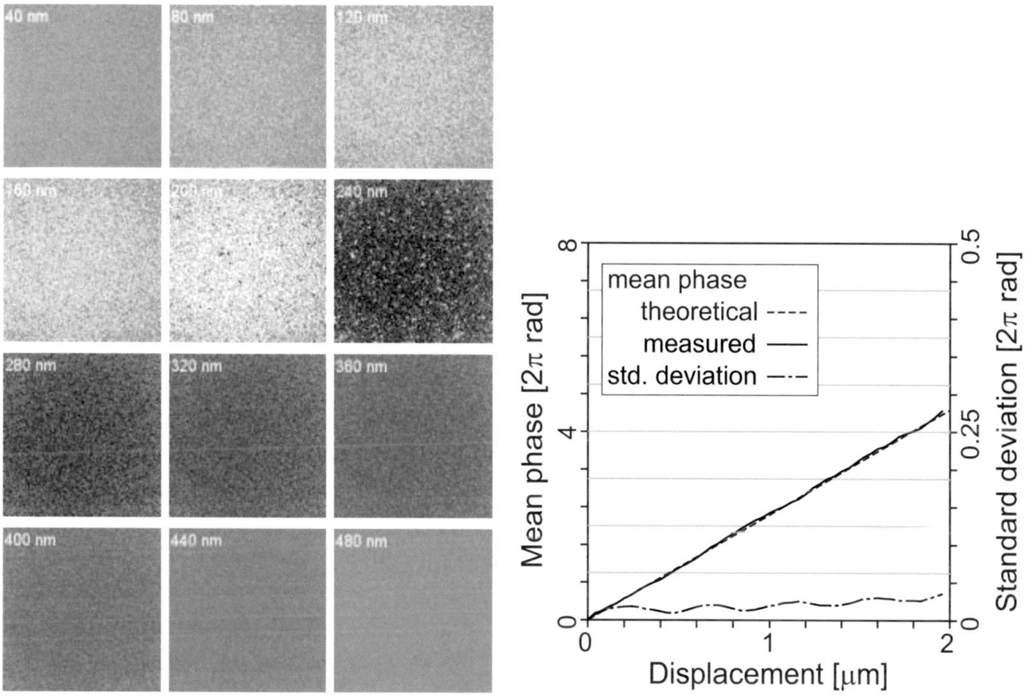

(a) (b)

Fig. 8. Results of the real-time heterodyne SPI system for in-plane displacement of a flat object with a rough surface (A. Kimachi, 2010a). (a) Sequence of phase difference images with respect to the initial position of the object, for displacement increasing in 40-nm steps. (b) Average and standard deviation of the images of unwrapped phase difference within a central 50 × 50-pixel region, along with a theoretical response.

theoretical plots is not as good as that for the in-plane displacement in Figure 8(b), exhibiting a drift of the experimental plot from the theoretical one. These observations can be attributed to the nature of the Michelson interferometer the SPI system in Figure 7(b) is based on, which involves two separate reflective objects and is thus more sensitive to disturbance than the SPI system for in-plane deformation in Figure 7(a).

The real-time heterodyne SPI system for in-plane deformation in Figure 7(a) was employed for monitoring an object under dynamic deformation. The object is a piezoelectric actuator placed horizontally and attached to a fixed frame, as depicted in Figure 10(a). Figure 10(b) shows a real-time sequence of phase difference images in every 0.12 s frame with respect to the initial state, featuring the junction between the object and fixed frame. The profile of the images along the horizontal white line is plotted in Figure 10(c). The left-hand side of the plot, corresponding to the actuator, exhibits a continuous increase in the slope as well as a gradual upward shift. The increasing slope indicates that the actuator is expanding toward the left, whereas the upward shift implies that the actuator itself also slides to the left. The right-hand side of the plot, corresponding to the fixed frame, exhibits a gradual downward shift of much smaller steps than that for the actuator. This suggests the existence of a strong force by the actuator that slightly pushes the fixed frame in the opposite direction to its expanding direction.

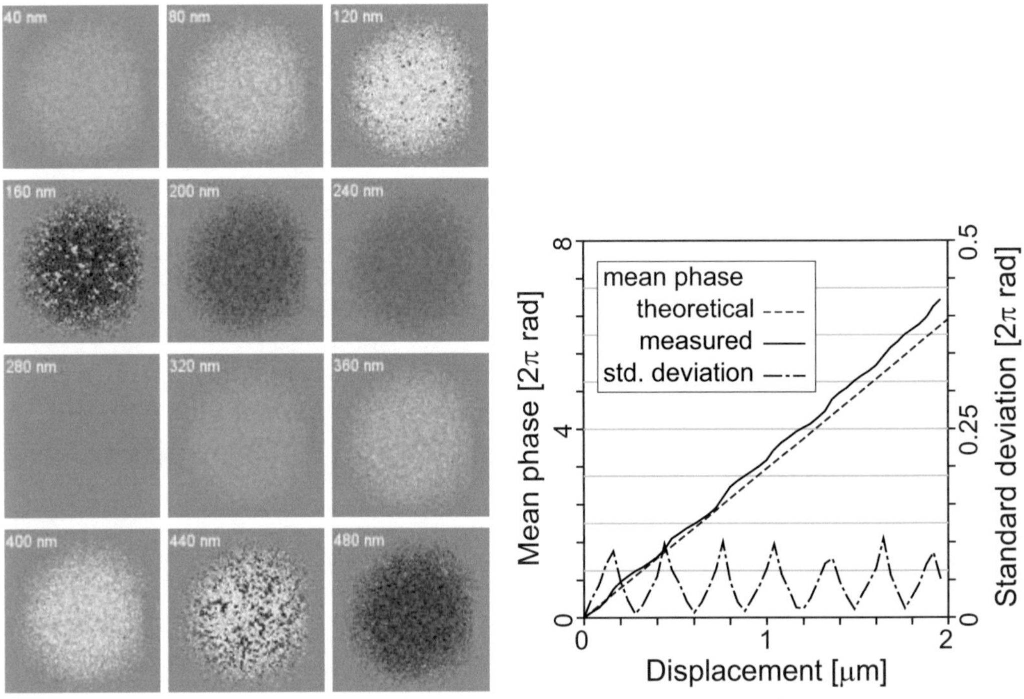

(a) (b)

Fig. 9. Results on the real-time heterodyne SPI system for out-of-plane displacement of a flat object with a rough surface (A. Kimachi, 2010a). (a) Sequence of phase difference images with respect to the initial position of the object, for displacement increasing in 40-nm steps. (b) Average and standard deviation of the images of unwrapped phase difference within a central 50 × 50-pixel region, along with a theoretical response.

5. Application to a health monitoring system for large-scale structures

The real-time 2-D HI described in the previous sections is particularly effective in applications that require not only high accuracy but also real-time 2-D phase measurement. As one of such applications, a research project on health monitoring of large-scale structures has been promoted by Ando *et al.* (S. Sato, 2008a;b;c; S. Ando, 2009). This project aims at developing a networked sensing system that can measure relative deformation of large-scale structures such as bridges, tunnels, industrial plants and tall buildings in real time and at high accuracy over three axes of displacement and three axes of rotation. This kind of sensing system is drawing increasing attention because large-scale structures can cause catastrophic disasters when they are broken down from aging, fatigue, or defects. Health monitoring of large-scale structures, however, is not easy because they are located outdoors in tough enviroment and their size makes it very laborious to inspect the whole entity for defects.

As a solution to this problem, a networked sensing system that employs the real-time 2-D HI has been proposed (S. Sato, 2008a;b;c). The sensor network is comprised of a train of sensing units placed over a large-scale structure. The sensing unit consists of a pair of a CIS chip and a heterodyne laser source, and transmits coaxial heterodyne laser beams to one neighboring unit while receiving coaxial heterodyne laser beams from another neighboring unit. The

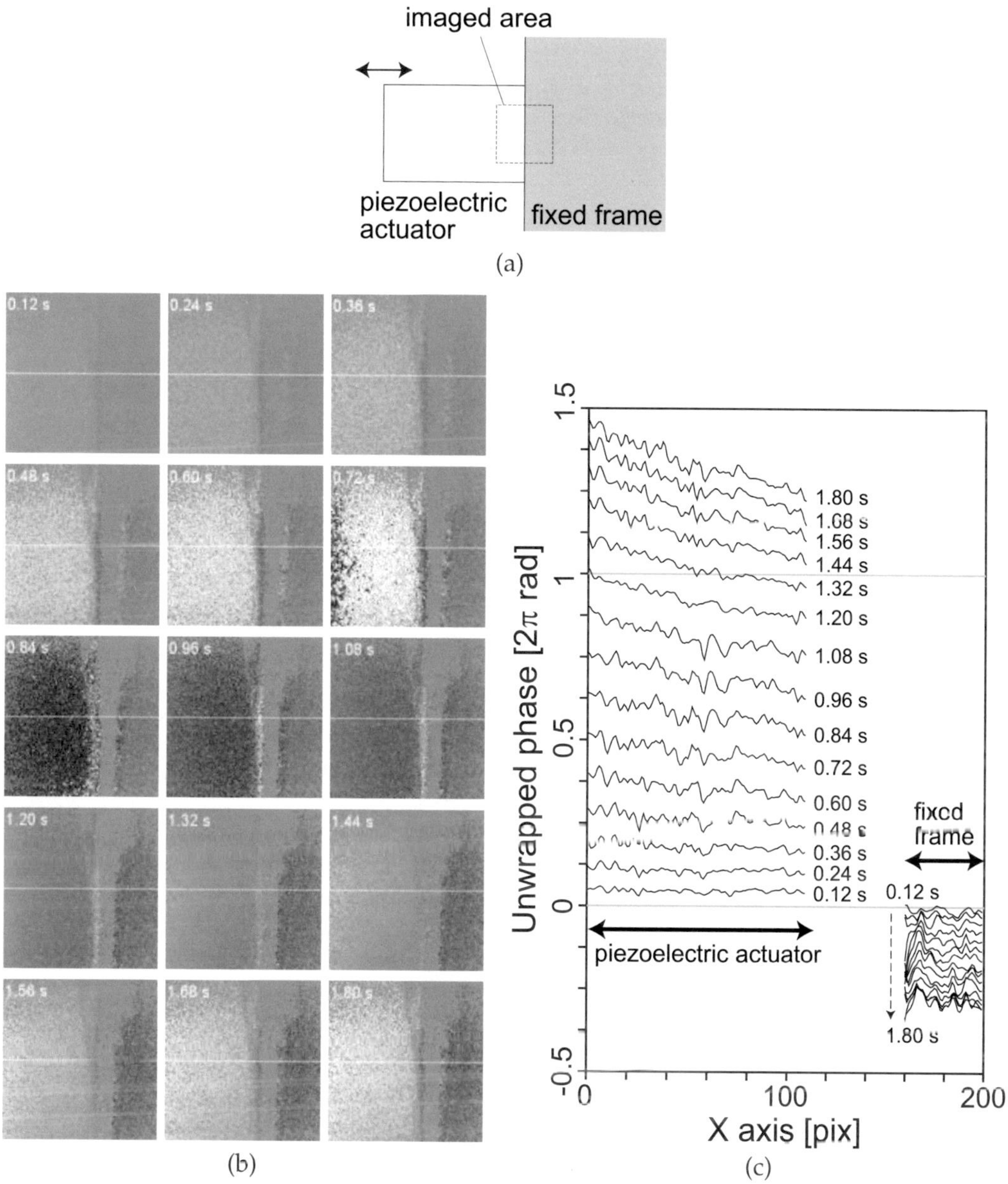

Fig. 10. Results of real-time heterodyne SPI for an object under dynamic in-plane deformation (A. Kimachi, 2010a). (a) Illustration of the object. (b) Real-time sequence of phase difference images with respect to 0.0 s while the object was horizontally expanded. (c) Profiles of unwrapped phase difference along the horizontal white lines in the images.

entire deformation of a large-scale structure can be obtained by integrating the relative six-axis deformation between each transmitter-receiver pair of sensing units over the network.

The key idea for measuring the relative deformation between a pair of sensing units lies in the use of multi-zero beams (S. Sato, 2008a;b;c), which have a wavefront with multiple zeros distributed over the beam area. The relative deformation can be known from a change in the amplitude and phase pattern of a multi-zero wavefront, which is easily detected on the CIS chip by the real-time 2-D HI at high accuracy. The multi-zero beams, which are generated by multiplexing Laguerre-Gaussian (LG) beams of different orders, have been found from extensive studies to possess many characteristics suitable for measurement and sensing (S. Sato, 2008a;b;c).

Based on these ideas, several experimental systems have been so far developed and tested in experiments on real-time tracking of multi-zeros (S. Ando, 2009). This project is currently at the stage of the development of a prototype system and the experimental evaluation on actual large-scale structures.

6. Conclusion

The real-time 2-D HI technique presented in this article is based on the use of the CIS as a two-dimensional array of pixels, each of which simultaneously demodulate the amplitude and phase of incident heterodyne beams at an ordinary frame rate. As the most significant advantage over other real-time 2-D HI or PSI techniques, the CIS-based technique ensures both maximal temporal and spatial resolution and high accuracy of phase measurement. The experimental results obtained on a polarizing Michelson interferometer and speckle pattern interferometers for in-plane and out-of-plane deformation measurement confirm the feasibility and usefulness of the real-time 2-D HI. The real-time 2-D HI is expected to find many application areas such as the health monitoring system for large-scale structures based on multi-zero beams.

7. References

R. Dändliker, B. Ineichen and F.M. Mottier, "High-Resolution Hologram Interferometry by Electronic Phase Measurement," *Opt. Commun.*, vol. 9, pp. 412-416, 1973.

N.A. Massie, R.D. Nelson and S. Holly, "High-Performance Real-Time Heterodyne Interferometry," *Appl. Opt.*, vol. 18, pp. 1797-1803, 1979.

S. Ando and A. Kimachi, "Correlation Image Sensor: Two-Dimensional Matched Detection of Amplitude-Modulated Light," *IEEE Trans. Electron Devices*, vol. 50, pp. 2059-2066, 2003.

A. Kimachi, "Real-Time Heterodyne Imaging Interferometry: Focal-Plane Amplitude and Phase Demodulation Using a Three-Phase Correlation Image Sensor," *Appl. Opt.*, vol. 46, pp. 87-94, 2007.

A. Kimachi, "Real-Time Heterodyne Speckle Pattern Interferometry Using the Correlation Image Sensor," *Appl. Opt.*, vol. 49, pp. 6808-6815, 2010.

A. Kimachi, R. Imaizumi and S. Ando, "Intelligent Image Sensor with a Vibratory Mirror Mimicking Involuntary Eye Movement," in *Tech. Digest of 16th Sensor Symp.*, pp. 171-176, Kawasaki, Japan, 1998.

A. Kimachi, T. Kurihara, M. Takamoto and S. Ando, "A Novel Range Finding System Using Correlation Image Sensor," *Trans. IEE Japan*, vol. 121-E, pp. 367-375, 2001.

T. Kurihara, N. Ono and S. Ando, "Surface Orientation Imager Using Three-Phase Amplitude-Modulated Illumination and Correlation Image Sensor," in *Proc. SPIE*, vol. 5013, pp. 95-102, 2003.

S. Ando and N. Ono, "Spatio-Temporal Phase-Encoding Profilometry Using Correlation Image Sensor," in *Proc. IEEE Int. Symp. Industrial Electronics*, pp. 786-791, Cambridge, U.K., 2008.

A. Kimachi and S. Ando, "Real-Time Phase-Stamp Range Finder Using Correlation Image Sensor," *IEEE Sensors J.*, vol. 9, pp. 1784-1792, 2009.

S. Ando, T. Kurihara and D. Wei, "Optical Flow Sensor with Maximal Spatio-Temporal Resolution," in *Proc. 25th Sensor Symp.*, pp. 683-688, Ginowan, Japan, 2008.

A. Kimachi, T. Imaizumi, A. Kato and S. Ando, "Spectral Matching Imager Using Correlation Image Sensor," *Trans. IEE Japan*, vol. 122-E, pp. 200-206, 2002.

A. Kimachi, H. Ikuta, Y. Fujiwara, M. Masumoto and H. Matsuyama, "Spectral Matching Imager Using Amplitude-Modulation-Coded Multispectral Light-Emitting Diode Illumination," *Opt. Eng.*, vol. 43, pp. 975-985, 2004.

A. Kimachi, S. Ando, M. Doi and S. Nishi, "Spectral Matching Imager with Three-Phase Quadrature Detection," in *Proc. SPIE*, vol. 7528, pp. 75280E-1-9, 2010.

S. Ando, T. Nara, N. Ono and T. Kurihara, "Real-Time Orientation-Sensitive Magnetooptic Imager for Leakage Flux Inspection," *IEEE Trans. Magnetics*, vol. 43, pp. 1044-1051, 2007.

K.J. Gåsvik, *Optical Metrology*, 3rd ed. Wiley: Chichester, 2002.

J.H. Bruning, D.R. Herriott, J.E. Gallagher, D.P. Rosenfeld, A.D. White and D.J. Brangaccio, "Digital Wavefront Measuring Interferometer for Testing Optical Surfaces and Lenses," *Appl. Opt.*, vol. 13, pp. 2693-2703, 1974.

C.J. Morgan, "Least-Squares Estimation in Phase-Measurement Interferometry," *Opt. Lett.*, vol. 7, pp. 368-370, 1982.

K. Creath, "Phase-Shifting Speckle Interferometry," *Appl. Opt*, vol. 24, pp. 3053-3058, 1985.

J. Kato, I. Yamaguchi and Q. Ping, "Automatic Deformation Analysis by a TV Speckle Interferometer Using a Laser Diode," *Appl. Opt*, vol. 32, pp. 77-83, 1993.

M. Takeda and H. Yamamoto, "Fourier-Transform Speckle Profilometry: Three-Dimensional Shape Measurements of Diffuse Objects with Large Height Steps and/or Spatially Isolated Surfaces," *Appl. Opt.*, vol. 33, pp. 7829-7837, 1994.

M. Akiba, K.P. Chan, and N. Tanno, "Full-Field Optical Coherence Tomography by Two-Dimensional Heterodyne Detection with a Pair of CCD Cameras," *Opt. Lett.*, vol. 28, pp. 816-818, 2003.

I. Yamaguchi and T. Zhang, "Phase-Shifting Digital Holography," *Opt. Lett.*, vol. 22, pp 1268 1270, 1997.

M. North-Morris, J. Millerd, N. Brock, J. Hayes and B. Saif, "Dynamic Phase Shifting Electronic Speckle Pattern Interferometer," in *Proc. SPIE*, vol. 5869, 58691B, 2005.

Y. Awatsuji, M. Sasada and T. Kubota, "Parallel Quasi-Phase-Shifting Digital Holography," *Appl. Phys. Lett.*, vol. 85, pp. 1069-1071, 2004.

T. Kiire, S. Nakadate and M. Shibuya, "Simultaneous Formation of Four Fringes by Using a Polarization Quadrature Phase-Shifting Interferometer with Wave Plates and a Diffraction Grating," *Appl. Opt.*, vol. 47, pp. 4787-4792, 2008.

P. Haible, M.P. Kothiyal and H.J. Tiziani, "Heterodyne Temporal Speckle-Pattern Interferometry," *Appl. Opt.*, vol. 39, pp. 114-117, 2000.

M.V. Aguanno, F. Lakestani, M.P. Whelan and M.J. Connelly, "Heterodyne Speckle Interferometer for Full-Field Velocity Profile Measurements of a Vibrating Menbrane by Electronic Scanning," *Opt. Lasers Eng.*, vol. 45, pp. 677-683, 2007.

T. Spirig, P. Seitz, O. Vietze and F. Heitger, "The Lock-in CCD—Two-Dimensional Synchronous Detection of Light," *IEEE J. Quantum Electronics*, vol. 31, pp. 1705-1708, 1995.

R. Dändliker, Y. Salvadé and E. Zimmermann, "Distance-Measurement by Multiple-Wavelength Interferometry," *J. Opt.*, vol. 29, pp. 105-114, 1998.

H. Povel, H. Aebersold and J.O. Stenflo, "Charge-Coupled Device Image Sensor as a Demodulator in a 2-D Polarimeter with a Piezoelastic Modulator," *Appl. Opt.*, vol. 29, pp. 1186-1190, 1990.

S. Bourquin, P. Seitz, and R.P. Salathé, "Optical Coherence Topography Based on a Two-Dimensional Smart Detector Array," *Opt. Lett.*, vol. 26, pp. 512-514, 2001.

S. Bourquin, P. Seitz, and R.P. Salathé, "Two-Dimensional Smart Detector Array for Interferometric Applications," *Electron. Lett.*, vol. 37, pp. 975-976, 2001.

S. Han, T. Sawada, T. Iwahori, S. Kawahito and S. Ando, "Three-Phase Time-Correlation Image Sensor Using Pinned Photodiode Active Pixels," in *Proc. SPIE*, vol. 7536, 2010.

S. Sato, I. Fujimoto, T. Kurihara and S. Ando, "Remote Six-Axis Deformation Sensing with Optical Vortex Beams," in *Proc. SPIE*, vol. 6877, 2008.

S. Sato, I. Fujimoto, T. Kurihara and S. Ando, "Optical Vortex and Correlation Image Sensor for Networked Deformation Sensing of Infrastructures," in *Proc. 5th Int. Conf. Networked Sensing Systems*, pp. 39-42, Kanazawa, Japan, 2008.

S. Sato, I. Fujimoto, T. Kurihara and S. Ando, "Remote Full-Axis Deformation Sensing with Optical Vortex Beam for Health Monitoring of Infrastructures," in *Proc. 3rd Int. Conf. Sensing Technology*, pp. 452-456, Tainan, Taiwan, 2008.

S. Ando, S. Sato and T. Kurihara, "Real-Time Tracking Experiment of Higher-Order Laguerre-Gaussian Beam for Remote Six-Axis Deformation Sensing," in *Proc. Sixth Int. Conf. Networked Sensing Systems*, pp. 1-4, Pittsburgh, PA, 2009.

20

Application of Interferometry to Analysis of Polymer-Polymer and Polymer-Solvent Interactions

Veronica Makarova and Valery Kulichikhin
A.V. Topchiev Institute of Petrochemical Synthesis, Russian Academy of Sciences, Russia

1. Introduction

For prediction of polymer systems behavior on processing and exploiting stages knowledge of phase diagrams is very useful and in some cases - necessary. They characterize equilibrium phase states and for particular systems - phase structure. Among other methods to estimate compatibility of two substances including polymeric one, and consequently to obtain reasonable information about phase state and structure of the final heterophase system, the optical interferometry method in a version of optical wedge (Borovskii et al, 1973; Malkin et al., 1983; Malkin&Chalykh, 1979; Avdeev, 1990; Chalykh & Gerasimov, 2004; Chalykh et al., 1998; Makarova, 2007) engages specific space. Each of components forms the intrinsic interference pattern with a pitch reciprocally proportional to its refraction index. Two components form two interference patterns separated by interface. As a result of interdiffusion, interference fringes close to the interface are bending. This process indicates on formation of solutions of various concentrations with their own refraction indices. Evolution of fringes shape in time gives the information about intensity of interpenetration of components allowing us to calculate the interdiffusion coefficients. Plotting concentration profiles at different temperatures and accepting that concentrations at the interface are equilibrium ones, it is possible to construct phase diagrams of binary systems.

In addition, kinetics of interference patterns evolution reflects constants of mass-transfer, i.e. kinetic stability of systems. This means that in one experiment we can obtain information on interaction kinetics and, finally, on composition of equilibrium phases. Such combination of kinetic and thermodynamic information is exclusively important for correct organization of blending stage and processing of polymer solutions, blends and alloys.

Simultaneously with high informativity, this method has some additional advantages compared with other methods for investigation of interdiffusion processes: simplicity, compactness of equipment, high enough preciseness of measurements, small amount of testing material. The main condition of this method consists in requirement that at least one of interacting components has to be transparent and difference of refraction indices should be not less than 10^{-3}.

Method of optical interferometry in application to interaction of two polymeric components is based on multipath interference from two surfaces of the film (or films) under investigation (Fig.1). Films of testing polymers are placing in side-by-side (edge-to-edge) manner between two glass plates with a small wedge angle (~2°) (Fig.2). Preliminary, semi-transparent metallic layer (Ag, Au, Ni-Cr alloy) is applied on inner surfaces of glass plates for reinforcing their reflection capability. At passing bunch of monochromatic beams through films of alternating thickness, the interference pattern appears as a result of path-length difference.

In the beginning of experiment two systems of interference fringes (bands) with different pitches corresponding to refraction indices of each component are separated by interface. As a result of mutual dissolution, concentration gradients appear which lead to refraction index gradients and, consequently, to fringes bending. Evolution of fringes shape in time gives information about redistribution of components concentration in the diffusion zone as well as about the direction and the rate of the diffusion stream.

The optical interferometer used in this research was designed on a base of UV microscope (Avdeev, 1990; Makarova, 2007). The interferometer consists of light source, optical system, diffusion cell, oven and microscope equipped with video-camera. The diffusion cell consists of frame with optical channel, heating chamber, and the cell as it. Surface of glass plates with a thickness of 3-5 mm and plain dimensions 20x25 mm is treated with the accuracy rating 14. The necessary wedge angle and gap thickness was done by means of metallic (foil) ribbons with a thickness from 80 to 110 μ.

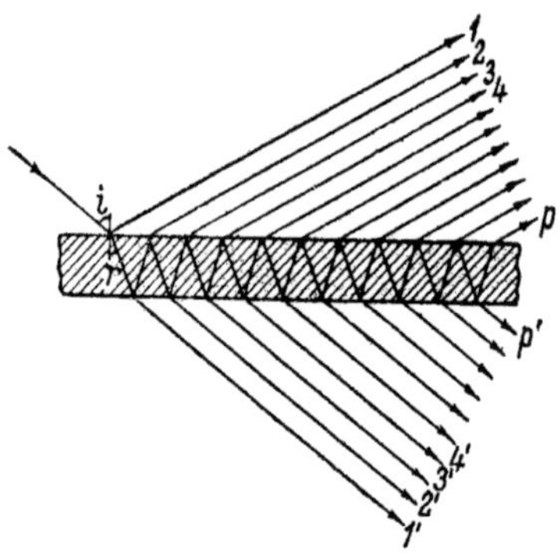

Fig. 1. Interference at multipass light refelection.

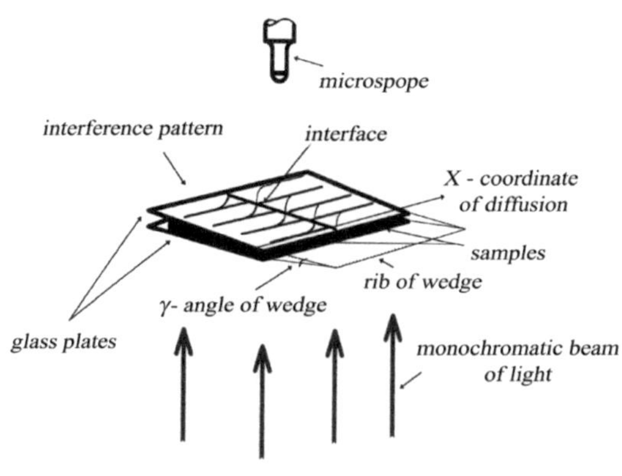

Fig. 2. Schematic image of the experimental procedure.

2. Results and discussion

Binary polymer systems are thermodynamically reversible and submit to phase rule as low-molecular weight formulations. The following kinds of the phase equilibrium are typical for these systems (the rank with increase of phase ordering) (Chalykh&Gerasimov, 2004; Chalykh et al., 1998; Papkov, 1981; Tager, 1978; Lipatov, 1984; Klenin, 1995):

- amorphous (liquid-liquid) equilibrium realizing in the domain of bounded solubility of components where the system decomposes on two phases, and each of them is the saturated solution of one component into another;
- liquid-crystalline (mesomorphic) equilibrium characterizing with transition of isotropic solution to liquid crystal phase via biphasic zone where disordered and ordered phases are in equilibrium;
- crystalline equilibrium in that the saturated solution of one component into another coexists with crystalline or crystal solvate phase of this component.

In many cases the different kinds of the phase equilibrium can superimpose leading to complex phase equilibrium. Let consider in short each individual kind of the phase equilibrium and typical cases of their superposition.

3. Amorphous equilibrium in polymer systems

At mixing of two partially compatible liquids in the definite temperature-concentration domain a system decomposes on two liquid coexisting phases - saturated solutions of one component to another. Change of temperature results in change of components solubility and causes variation of coexisting phase compositions. At definite temperature, known as critical, boundaries between phases disappear and a system passes from heterophase to homophase. Return to previous conditions causes phase decomposition again. Such transitions can be performing unrestricted times without any shift of critical temperature.

Phase equilibria in binary systems are expressed by phase diagrams. Usually they are plotted in coordinates "temperature-composition" at constant pressure. Schematic images of basic kinds of phase diagrams of binary polymer systems characterizing with amorphous equilibrium are presented in Fig. 3. The equilibrium values of coexisting phases compositions form unique curve - binodal. Convex upwards binodal (Fig. 3a) corresponds to the case of increasing compatibility with increase of temperature. Its cupola reflects the limiting temperature higher that a system becomes single-phase at any concentration of components. This temperature is known as the Upper Critical Solubility Temperature (UCST). The phase composition corresponding to UCST is also named as critical.

If compatibility of components increases with decrease of temperature, the binodal has a shape of convex downwards curve (Fig. 3b) and critical temperature is named as Low Critical Solubility Temperature (LCST). Some systems can have both critical temperatures and their amorphous phase equilibriums are expressed by circle-like or closed (UCST>LCST) diagram (Fig. 3c), non-closed diagram (UCST<LCST) (Fig. 3d) or the diagram of sand-glass shape (UCMT~LCMT) (Fig. 3f).

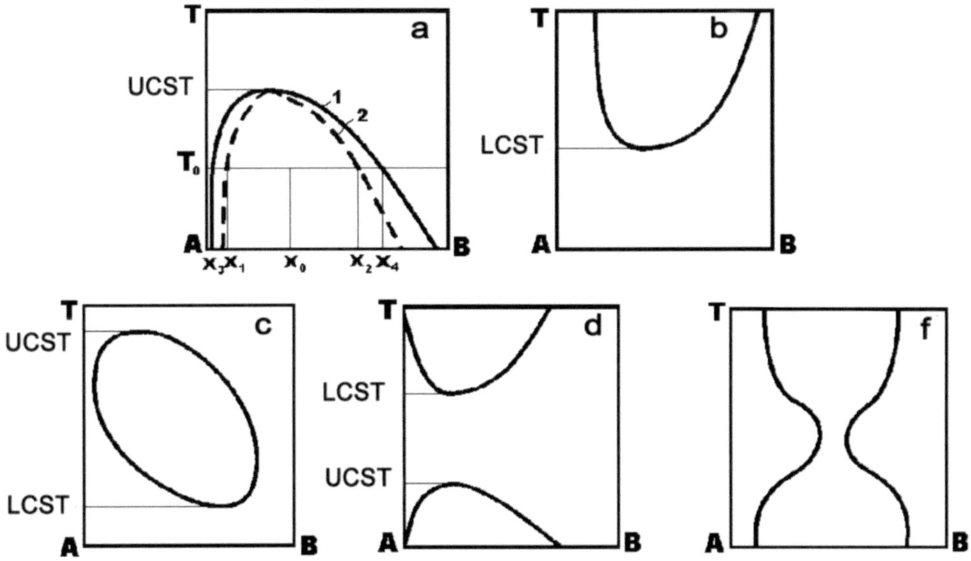

Fig. 3. Basic types of the phase diagrams of polymer system with amorphous equilibrium according to (Chalykh et al., 1998). See explanations in the text.

Under binodal a system decomposes on two phases. Analysis of kinetics of new phases formation indicates on unusual behavior of microheterophase system in vicinity of binodal and necessity to draw additional line - spinodal (Fig. 3a, curve 2). This line separates unstable (under binodal) and meta-stable (between binodal and spinodal) compositions. The critical point is the same as for binodal, as for spinodal. In meta-stable domain formation of novel phases proceeds very quickly at small fluctuations of composition.

In some cases also useful to show on phase diagrams boundaries of relaxation states of binary system (glass and brittleness points, flow temperature, limits of thermal stability of components, etc) (Papkov, 1971; Chalykh et al, 1990). Such diagrams are named "generalized phase diagrams" or "diagrams of phase and relaxation states".

Equilibrium curves shown in Fig. 3 are idealized. Actually, the total binodal curve cannot be registered ever because of temperature limitations. For example, in whole temperature diapason either the single-phase domain (complete compatibility of components) can exist or its coexistence with biphasic domain but without achievement of critical temperatures (partial compatibility of components). The most typical are phase diagrams with UCST because for majority systems increase of temperature leads to increase of kinetic mobility of molecules and macromolecules improving their mutual solubility. Diagrams with LCST meet for aqueous solutions of various polysaccharides

In the case of amorphous equilibrium at T>UCST (or T<LCST) the continuous change of composition at transition from one component to another is realized in diffusion zone. At temperatures of phase transitions in diffusion zone the border between two solutions appears. We will consider further some examples of amorphous phase equilibrium registered by interference method. As a rule, optical measurements were added by

rheological data that allows us to obtain more informative picture of mass-transfer in vicinity of interfaces.

Amorphous phase equilibrium will be considered on examples of polycarbonate (PC) - polymethylmethacrylate (PMMA) pair. PC and PMMA are transparent polymers (the coefficient of light transmission is ~90%) with similar exploiting properties. That is why fields of their application in many cases coincides (light-technical products, engineering plastics, materials for medicine, etc.). However, mechanical characteristics of PC, partially impact strength, are much better compared with PMMA. Keeping in mind that PC is more expensive than PMMA, development of blending materials based on these polymers can be interesting and important, but blends should be transparent and this means that the main task is to find temperature-composition regions of their compatibility.

Literature data concerning compatibility of these components are rather contradictive (Chiou et al, 1987; Kim&Paul, 1992; Butzbach&Wendorff, 1991, Landry & Henrichs, 1989; Debier, 1994; Kyu, 1993; Nishimoto, 1991; Rabeony, 1992; Woo&So, 1996): from conclusion about their immiscibility to partial or even total compatibility. Basic methods used in cited papers were light transmission, light scattering, DSC, and specimens for investigation were prepared from mutual for both polymers solutions. The nature of solvent, thermal prehistory and long relaxation times could influence significantly on formed heterophase structure (Woo&So, 1996). Such uncertainty was the driving force for this investigation using interferometric method allowing us to observe direct penetration of melted components (Avdeev et al., 2001; Eastmond&Kotomin, 1994).

Samples of PC (Mw=$2.5 \cdot 10^4$; T_g=150°C, T_m=267°C) and series of PMMA (M_w=$1.5 \cdot 10^4$; $7.4 \cdot 10^4$; $30 \cdot 10^4$; T_g=114°C; T_{flow}=180°C) were used as objects under investigation. Experiments were carried out in temperature diapason from 150°C (glass point of PC) to 280°C (decomposition point of PMMA). Films with a thickness of 100-120 μ were prepared by hot pressing at 150°C (PMMA) and 240°C (PMMA) with subsequent cooling with a rate of 5-10 °C/min.

Viscosities of the neat and blended samples PC-PMMA (M_w=$1.5 \cdot 10^4$) were measured by method of squeezing flow allowing us to work with micro-amount of materials. Interval of shear rates and stresses realized in this method at 240°C corresponds to Newtonian flow. Blends for rheological measurements were obtained via mutual solvent - chloroform that was removed in vacuum at 60°C during 8 h.

Interferograms obtained present two interference patterns with a different pitches corresponding to the refraction indices of each component. In vicinity of interface "bands of equal thickness" bend that indicates on formation of solutions of various compositions PC in PMMA, on the one hand (Fig. 4, left) and PMMA in PC, on the other hand (right). Based on connection of the refraction index with composition, the profiles of concentration distribution in diffusion zone were plotted for different temperatures and observation times. Components concentrations reached on both sides of interface were accepted for solubility one into another.

Various curvature of bands near the interface for PMMA of different molecular weight reflects strong dependence of compatibility on M_w PMMA. For interaction of PC with PMMA (M_w=$30 \cdot 10^4$) the bands bending is practically absent and this means that the mutual dissolution of components is not higher than 0.5%. In the case of PMMA with M_w=$7.4 \cdot 10^4$ the curvature is more significant but calculation shows that maximum solubility does not exceed

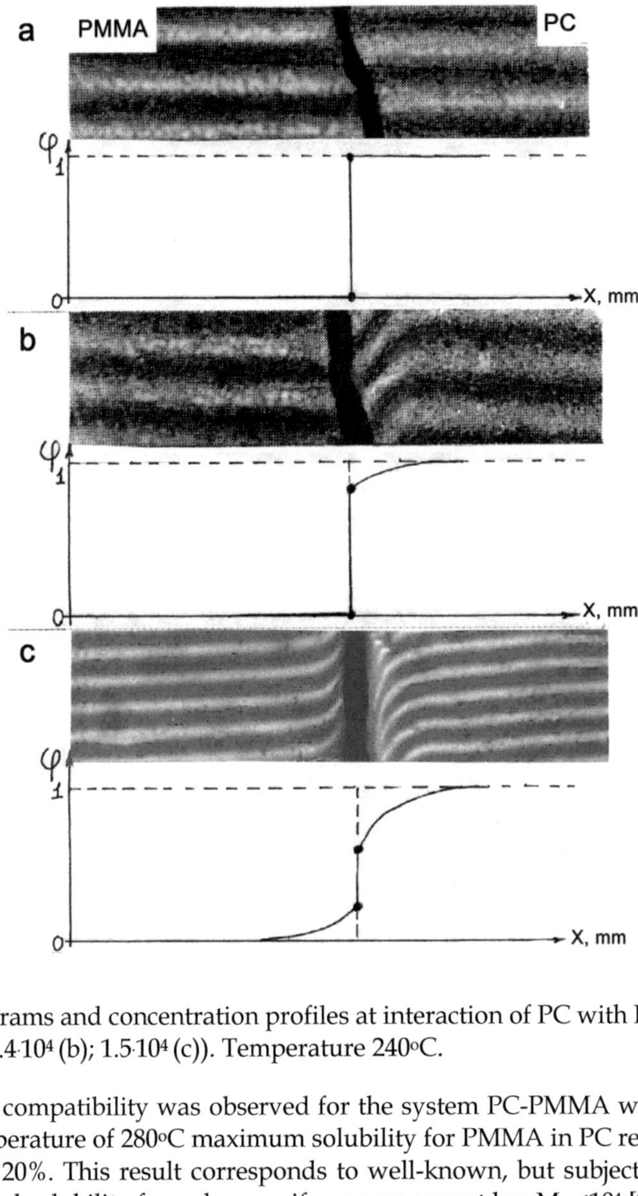

Fig. 4. Interferograms and concentration profiles at interaction of PC with PMMA ($M_w=30\cdot10^4$ (a); $7.4\cdot10^4$ (b); $1.5\cdot10^4$ (c)). Temperature 240°C.

3%. The highest compatibility was observed for the system PC-PMMA with $M_w=1.5\cdot10^4$. At permissible temperature of 280°C maximum solubility for PMMA in PC reaches 40% and for PC in PMMA - 20%. This result corresponds to well-known, but subjective opinion about increase of mutual solubility for polymers, if one component has $M_w<10^4$ [3].

Using data on solubility of components at different temperatures the phase diagrams were plotted (Fig. 5). The temperature diapason was limited by glass point of PC and decomposition point of PMMA. One can see that there exists domains of mutual solubility of components and with growth of temperature the compatibility increases. Consequently, these diagrams belong to amorphous equilibrium with UCMT though the critical point was not achieved experimentally for any system because of decomposition of PMMA. Since the solubility of PMMA in PC is higher than vice versa, the binodal is asymmetric and shifted in direction of higher content of PMMA.

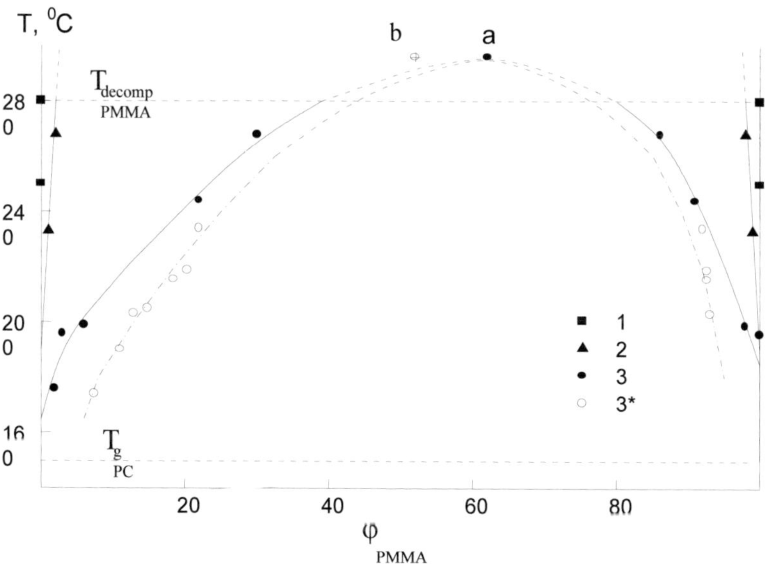

Fig. 5. Phase diagrams of PC-PMMA ($M_w \cdot 10^{-4}$=30 (1), 7.4 (2) and 1.5 (3 - binodal, 3* - spinodal) systems. *a* and *b* – calculated values of UCST with accepting of concentration dependence of χ, and ignoring it, respectively.

For the system with lowest M_w of PMMA the spinodal was determined as well. For this aim the blend was heated to 270°C for significant compatibility of components (Fig. 6a, b), and then it was cooled step-by-step keeping at each temperature several hours (Fig. 6c, d).

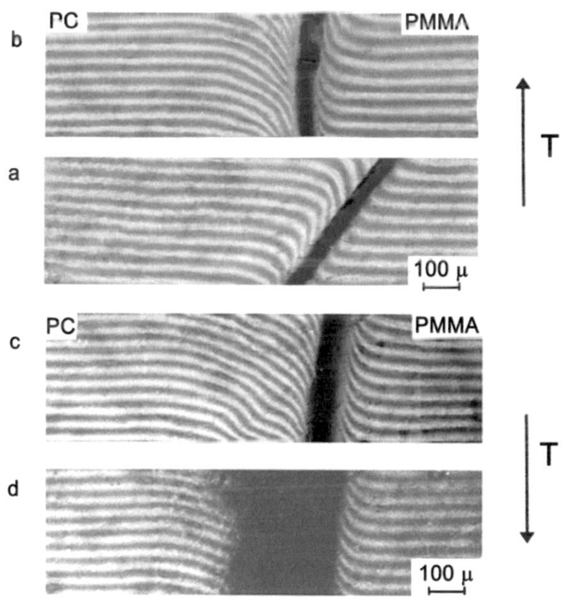

Fig. 6. Interferograms of PC-PMMA (Mw=1.5·10^4) system obtained at increase of temperature from 200 (a) to 260°C (b) and at decrease of temperature from 260 to 220 (c) and 160°C (d).

At cooling the compatibility decreases and in the solutions domain close to the interface the concentration of one component becomes higher than thermodynamically equilibrium (super-saturation) that causes appearance of the disperse phase enriched by this component. Due to high dispersity degree this two-phase region scatters light intensively and it looks more dark than single-phase regions. With time on boundaries between homophase and heterophase domains the concentrations corresponding to spinodal branches are realized (Fig 5, curve 3*) and bending of interference bands decreases (Fig. 6 d). The concentration width of the meta-stable region located between binodal and spinodal monotonously decreases at moving from low to higher temperatures: from 8% at 190°C to 2% at 270°C from the side of the higher content of PC of and from 5 to 2% at the same temperatures from the side of PMMA.

Since the compatibility of PC and PMMA of higher molecular weights is very low, the further calculations and experiments on kinetics of dissolution and measuring the viscous properties were carried out for the system PC-PMMA with M_w=1.5·10^4.

For the quantitative interpretation of experimental data on interdiffusion the theory of Flory-Huggins-Scott (FHS) (Scott, 1949; Tompa, 1956) was applied. Partially, from classical expression for the Gibbs free energy the interaction parameter χ was calculated. Calculation of interaction parameter was performed assuming its linear dependence on concentration (Koningsweld& Kleintjens, 1977; Palatnik&Landau, 1961):

$$\chi = \chi_0 + \alpha\varphi_1 , \tag{1}$$

where χ_o - the interaction parameter at φ_1=0, α - the constant, reflecting influence of blend composition on intermolecular interaction.

Based on concentrations corresponding to binodal branches at definite temperatures, the values of χ_0 and α were calculated using FHS equation and the expression (1). The family of linear dependences $\chi(\varphi_1)$ for different temperatures is shown in Fig. 7. The calculated values of the interaction parameter χ lie in limits of 0.017-0.030, that is inherent for good compatibility of components. At moving from PMMA to PC, the χ value decreases that confirms higher solubility of PMMA in PC than vice versa.

At relatively low temperatures the concentration dependence of the interaction parameter χ is strong enough (the value α in Eq. (1) at 212°C is equal to 0.004), but it decreases with increase of temperature reaching the constant level 0.002 (Fig. 8). Using this fact as well as zero-meaning of the second and the third derivatives in critical point, the expression for φ and χ at UCMT was obtained (the dependence $\varphi(\chi)$ was also taken into account):

$$\varphi_{2кр} = \frac{\sqrt{\frac{x_1}{x_2} - 6\alpha x_1}}{1 + \sqrt{\frac{x_1}{x_2} - 6\alpha x_1}} \tag{2}$$

$$\chi_{12} = \frac{1}{2}\left(\frac{1}{\sqrt{x_2}} + \frac{1}{\sqrt{x_1 - 6\alpha x_1}}\right)^2 + \alpha\left(5\varphi_{2кр} - 1\right) \tag{3}$$

These equations allowed us to calculate critical values: φ_{cr}=0,61, χ_{cr}=0,017 and by extrapolation of χ *at* $\varphi=\varphi_{cr}$ to χ_{cr}, determine the critical temperature equal to 296°C. The calculated value of UCMT allowed us to plot the complete phase diagram of this system.

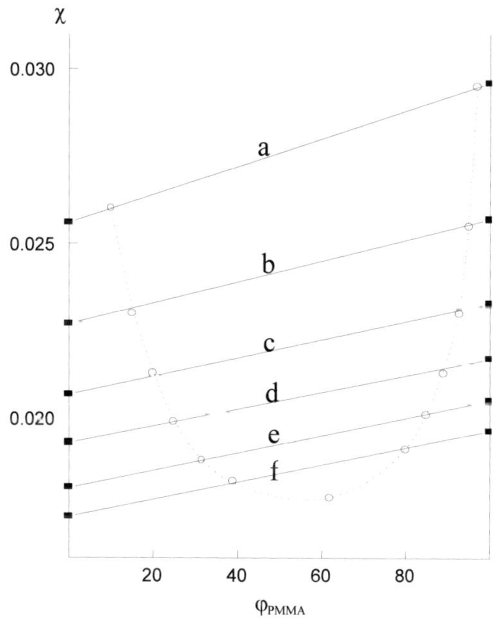

Fig. 7. Concentration dependence of the interaction parameter for the system PC-PMMA (M_w=1.5·10[4]). T=212 (a), 226 (b), 240 (c), 254 (d), 268(e) и 282°C (f). The dotted line corresponds to the χ values on the binodal.

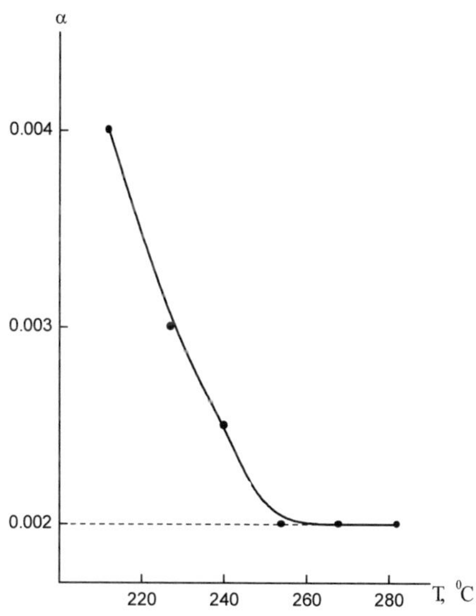

Fig. 8. Temperature dependence of the α constant in Eq. (1) for the same system.

The φ_{cr} value (Fig. 5, point *a*) calculated with taking into account the concentration dependence of the interaction parameter differs almost on 10% on φ_{cr} value estimated if this dependence is ignored (Fig. 5, point *b*) (Koningsweld& Kleintjens, 1977; Palatnik&Landau, 1961). This means that the critical concentration depends not only on dimensions of macromolecules, but on their nature as well.

For description of kinetics of mixing the interdiffusion coefficients D_v, characterizing the rate of interpenetration of components, were calculated. The calculation was done by graphic method from the curves of concentration distribution by means of Matano-Boltzman method (Borovskii, 1973) at three temperatures and concentration domains of compatibility (Fig. 10). Interdiffusion coefficients are rather high for polymer pairs and change from $1{,}5\cdot10^{-12}$ (220°C) to $4\cdot10^{-11}$ m^2/c (260°C). D_v values in the region of predominant PMMA content are a little lower than for the region of higher PC content.

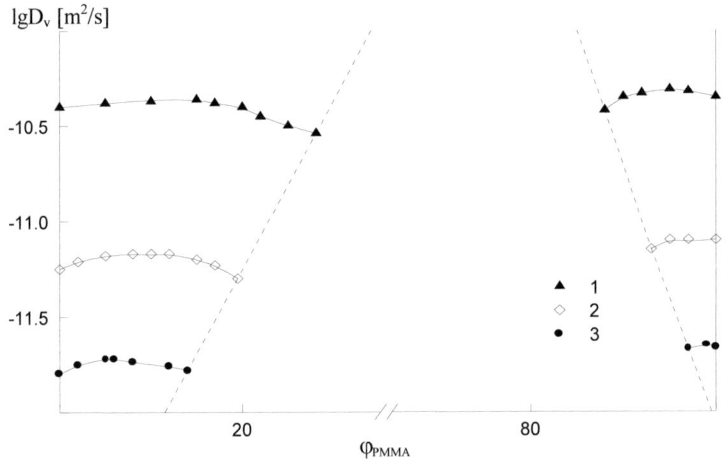

Fig. 9. Interdiffusion coefficients vs. composition for PC-PMMA (M_w=$1.5\cdot10^4$) system at 260 (1), 240 (2) and 220°C (3). The domain restricted with dotted line corresponds to the biphasic region of the phase diagram.

The concentration dependence of the interdiffusion coefficient far from binodal is rather weak. With increase of the one component content in another D_v values a little increase but at approaching to binodal decrease again. Deceleration of diffusion process in vicinity of saturation state for both solutions is natural, since D_v is proportional to derivative of chemical potential on concentration that becomes equal to zero on the spinodal.

With increase of temperature interdiffusion coefficients become higher for all compositions (Fig. 9). Activation energy of diffusion is high enough for polymer systems and equal to 170 kJ/mole. Extrapolation of the temperature dependence of D_v to 160°C gives a value of $8.1\cdot10^{-15}$ m^2/c. Probably, the high difference in diffusion mobility of components at high ($4\cdot10^{-11}$ m^2/c at 260°C) and low temperatures close to the glass point could be the explanation of false LCST's [14-19]. This situation can be realized in attempts to create the homophase mixed solutions of PC and PMMA in a common solvent. After removing a solvent the thermodynamically unstable system at heating suffers very slow noticible decomposition at $T^*>T_g$ only, because of low D_v values. The T* value could be accepted as LCST.

One of the practical application of obtained data on interdiffusion coefficients and mutual solubilities of partially compatible components consists in estimation of time needed for formation of the interphase layer of the definite thickness that should influence on properties of the heterophase system including rheological and mechanical ones. A volume of the interphase layer was accepted as the tenth part of disperse particle volume. Calculation was done using the model of dissolution of a particle of spherical shape (Crank, 1956):

$$\frac{\varphi}{\varphi_0} = \frac{a}{r} erfc \frac{r-a}{2\sqrt{D_v \tau}}, \tag{4}$$

where φ and φ_0 – concentrations of the component on the inner and the outer boundaries of the interphase layer, a – a radius of a particle, r – a distance from particle center to the border of the layer, $erfc$ – empirical probability function (taken from the reference book). The PMMA was chosen as the disperse phase and PC as the disperse medium. The concentration of PC onto interface was accepted as in e times less than the initial concentration. Results of the calculation for 240°C are presented in Table 1. One can see that time of spontaneous formation of such a layer depends strongly on the particle dimensions and can differ from a few minutes to several years. This difference underlines again the determining role of dispersity at dissolution of polymers.

Thickness of the diffusion layer, μ	Diameter of a particle, μ	Time of formation of the diffusion layer of the definite thickness
0,3	10	5 min
3	100	8,3 h
15	500	8,7 days
30	1000	34,7 days
60	2000	4,6 months
150	5000	2,4 year

Table 1. Time of the diffusion layer formation at mixing of PC with PMMA (M_w=1.5·10^4).

Diameter of a particle, μ	T=240°C	T=160°C
1	10 s	30 h
10	20 min	4 months
100	30 h	30 years

Table 2. Time of complete dissolution of the PMMA particle in PC.

The time of total dissolution of PMMA particle in PC was estimated as well. For comparison, calculations were performed for two temperatures: 160 and 240°C. Results shown in Table 2, illustrate significant delay of interpenetration process at approaching to the glass point. Simultaneously, these data allow us to make choice for the mixing regime to obtain either homogeneous solution of PMMA in PC or three-phase system consisting of PMMA, PC and solution of PMMA in PC (as the interphase layer of the definite thickness).

Data on kinetics (interdiffusion coefficients) and thermodynamics (phase equiblibria) of polymers under investigation have been compared with rheological results obtained by squeeze flow method (Kotomin&Kulichikhin, 1996) - the concentration dependence of the Newtonian viscosity, presented in Fig. 10. The matter is that in both processes (diffusion and flow) a segmental motion takes place. In the case of diffusion it is stipulated by difference of chemical potentials, and in the case of rheology - applying the external force. That is why segments participating in the diffusion and the rheology can be different, but mechanisms are similar (Malkin&Chalykh, 1979), and joint consideration of data of both methods are undoubtedly valuable.

The concentration dependence of viscosity of PC-PMMA (M_w=1.5·10^4) system at 240°C at majority compositions excluding specific domain is located higher of the logarithmic additivity line that indicates on active interaction of components (Kuleznev, 1980). At small content of PMMA the characteristic minimum was observed. At comparing with the phase diagram it is evident that its position is nearst binodal. The viscosity decrease can be explained by appearance of microemulsion with high fraction of interfacial layer with lower density compared with neat polymer melts (Kuleznev, 1980; Kuleznev&Kandyrin, 2000; Kuleznev, 1987). Presumably, this minimum is connected with meta-stable state of the system where a great number of heterophase fluctuations exists.

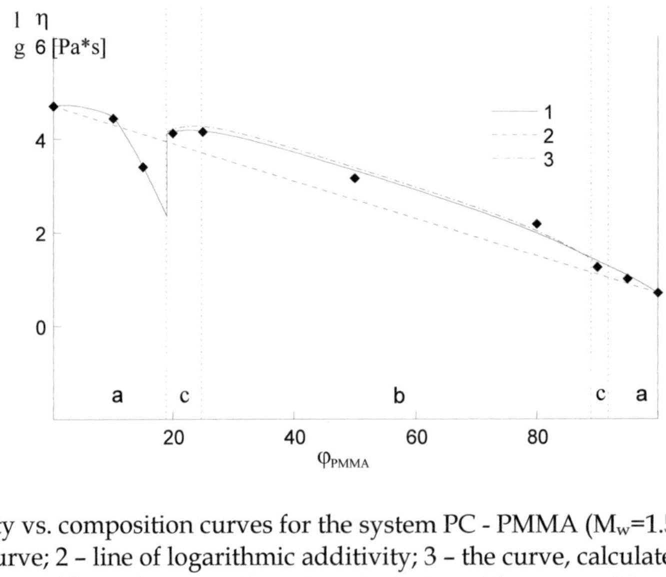

Fig. 10. Viscosity vs. composition curves for the system PC - PMMA (M_w=1.5·10^4): 1 – experimental curve; 2 – line of logarithmic additivity; 3 – the curve, calculated using formulae (5). Vertical lines designate domains of solutions (a), meta-stable state (b) and biphasic (c).

The concentration dependence of viscosity in biphasic region was analyzed based on knowledge of the phase diagram. For this aim the empirical formulae was used (Kuleznev&Kandyrin, 2000):

$$\lg\eta = \psi_2^2 \lg\eta_2 + \left(1-\psi_2^2\right)\lg\eta_1, \quad (5)$$

where η_1 and η_2 – viscosities of disperse phase and medium, respectively, ψ_2 - the volume fraction of dispersion medium that for compatible polymers coincides with relative content of the component forming the medium. But if at least partial mutual dissolution proceeds, this coincidence is absent. It is why the ψ_2 value was found from the phase diagram using the meanings of concentrations on binodal and "the lever rule". The calculated curve 3 (Fig. 11) coincides in limits of equipment error with the experimental one.

So, the couple PC – low molecular weight PMMA is a good example of polymer systems with amorphous phase equilibrium. Method of interferometry is very useful and fruitful for such kind of systems due to informativity and combination of joint approaches: kinetics of mass-transfer and thermodynamics of phase states. Increase of the mutual solubility with temperature indicates on UCST that could not be measured experimentally but was calculated using evolution of the interaction parameter with concentration.

Comparison of the phase diagram with concentration dependence of viscosity allowed us to observe local minimum of viscosity near the binodal (sooner in meta-stable region between binodal and spinodal where microemulsion appears as the first step of the phase decomposition) and to apply the empiric equation for calculation of the blends viscosity by means of accurate determination of the composition from binodal branches. Joint consideration of diffusion and rheological properties seems to be very fruitful for judging about structure, kinetic and thermodynamic behaviors of complex, multiphase polymer systems. This approach will be used further for analysis of systems with LC and crystalline equilibria.

Calculations of the interphase layers thickness depending on dimensions of drops and time, based on the knowledge of interdiffusion coefficients can be very useful for correct organization of technological process of mixing and, consequently, development of new materials with three-phase structure and new properties.

4. Crystalline and complex phase equilibria in polymer systems

Crystalline equilibrium is inherent for polymer systems if at least one component is capable to crystallize. At cooling from the region of the total compatibility (isotropic solution) at definite temperature the system decomposes on two equilibrium phases: deposition of the crystalline component and its saturated solution into the other component.

The curve describing sum of transition (melting or crystallization) points corresponding to various compositions is named "liquidus". Higher of the liquidus line the single-phase region (liquid solution) exists, and lower the liquidus line – biphasic system (crystalline phase distributed in solution). Basic kinds of phase diagrams with crystalline equilibrium are shown in Fig. 11.

Cases a) and b) are typical for components with a weal interaction. If one component of such a system is amorphous and the other – crystalline, the melting point of a system as whole decreases monotonously as a result of dilution of the crystalline component by amorphous one (depression of melting points) as is seen in the graph a). In systems containing two components capable to form crystalline phases two liquidus line cross in the eutectic point

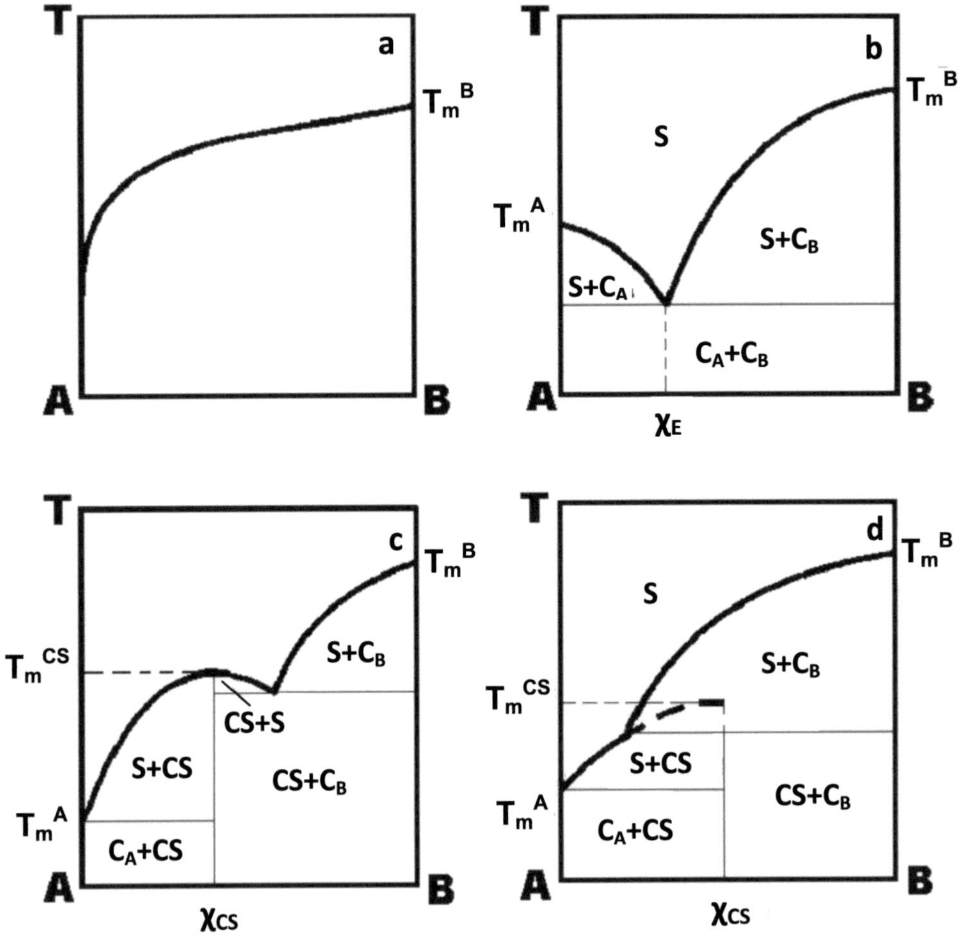

Fig. 11. Basic types of phase diagrams with crystalline equilibrium [5,6,9].

located in position lower than melting points of neat components (type b). Under the left liquidus curve up to x_e point the equilibrium of the crystalline phase A and its saturated solution exists, and after x_e – the equilibrium of the crystalline phase B with its saturated solution. In the temperature region lower of the eutectic point the biphasic domain consists of crystals A and B only.

Phase equilibria of types c) and d) are inherent for systems with polar components and their interaction leads to formation of crystalline additive compounds - crystal-solvates (CS). At congruent melting of CS (without its decomposition on components) the maximum melting point T_m^{CS} exists (type c). But if the energy of the additive compound formation is low, (incongruent melting) its decomposition takes place until reaching T_m^{CS} (type d). Under liquidus curve in region of compositions less than x_{CS} there exist in equilibrium phases of CS of x_{CS} composition and saturated solution (S) B in A ($T>T_m^A$) or crystals (C) A ($T<T_m^A$). Passing through x_{CS} in the temperature region lower the eutectic point, the CS phase coexists with saturated solution of A in B.

The maximum number of phase diagrams for polymer systems with crystalline equilibrium is devoted to polymer-solvent systems (Papkov, 1971; Tager, 1978; Papkov, 1981). One of the classical examples is the phase diagram of poly-p-phenyleneterephthalamide (PPTA) - H_2SO_4 schematically shown in Fig. 12. (Iovleva&Papkov, 1982; Papkov et al., 1986; Kulichikhin, 1989). But in the particular case of stiff-chain polymers solutions the LC phase appears at definite concentration of polymer separated from isotropic solutions by narrow biphasic "corridor". The situation with LC equilibrium will be considered later, but we should indicate here on the right part of the diagram where formation of the CS phase with congruent melting is realized. In this case the knowledge of the phase diagram was used for practical aim of choosing the most appropriate concentration of dopes and the temperature regime of spinning the super-strong fibers of Kevlar (Terlon) type (Papkov&Kulichikhin, 1977).

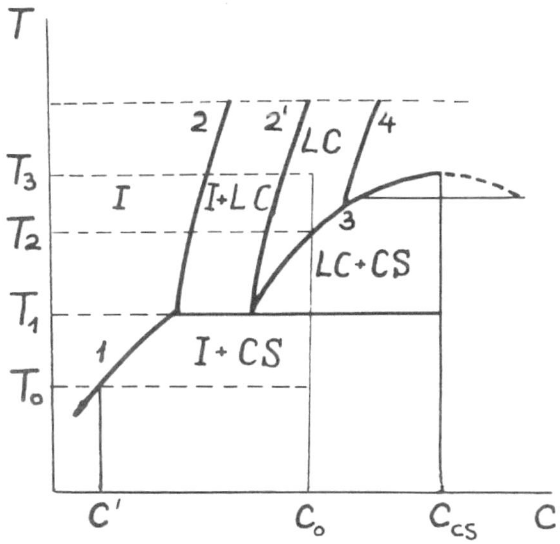

Fig. 12. Schematic image of the phase diagram for PPTA-H_2SO_4 system.

Coming back to polymer-polymer interaction let us consider now several binary systems based on polyesters. These polymers are of interest from the viewpoint of mixing just to reach the new complex of properties hoping at least on partial physical compatibility. Simultaneously, we should keep in mind a possibility of chemical interaction of polyesters via mechanisms of acidolysis or alkoholysis which involve ester-ester exchange. So, in this case interdiffusion can be stipulated as chemical similarity as chemical interaction leading to formation of copolyesters. Let us accept that 280-290°C is the border temperature that separates the low-temperature region where physical interaction (diffusion) prevails, and high-temperature region where chemical interaction becomes significant.

4.1 Interaction between LC- and isotropic polyesters melts

At low-temperature region two couples of polyesters have been studied. In the first one two components were in interaction: polydecamethyleneterephthaloil-bis-4-hydroxybenzoate LC polyester and polybutyleneterephthalate (PBT) (Kulichikhin et al., 1994). PBT is semi-crystalline polymer. This pair has some specific features from the viewpoint of application of method of the optical interferometry. In the case of LC polyester-PBT couple the melt of the first polymer has LC ordering up to clearing (isotropization) point. This means that LC melt is not transparent that is why the evolution of interference bands during interaction of two above indicated melts was controlled from one side - transparent PBT melt only.

At room temperature both polymers are non-transparent. At 230°C PBT melts and intrinsic for it interference bands appear while LC polyester even in melt state remains dark. Bending of PBT interference bands in vicinity of the interface indicates on dissolution of LC polyester in PBT, and in concentration profile gently sloping section is seen. At T~290°C LC polyester becomes isotropic and the interference picture indicates on complete compatibility of components in these conditions. The inereference data allowed us to construct the phase diagram (Fig. 13).

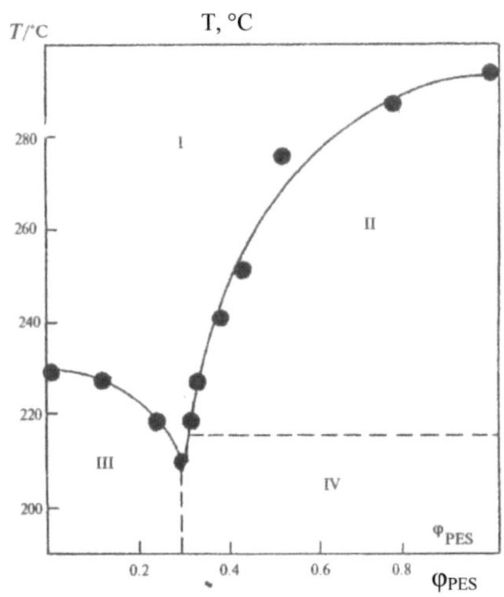

Fig. 13. Phase diagram for the system LC polyester - PBT. I - isotropic solutions, II - LC state of LC polyester solutions in PBT, III crystalline state of PBT saturated solutions, IV - crystalline state of LC polyester.

There are two liquidus lines with eutectics at volume fraction of PBT=0.7. The presence of eutectics confirms by preservation of transparent zone for this composition up to 170°C, i.e. much lower of crystallization points of neat components and solutions of other compositions (Fig. 14). In heating cycle the eutectic zone is clearing the first and then zones of other compositions. This situation is very similar to the classical case of crystalline equilibrium expressed by pattern b) in Fig. 11.

The second system was a pair of polyehyleneterephthalate (PET) and LC copolyester of PET with p-hydroxybenzoic acid (PET/HBA) one half of the interference picture is dark, but sloping the interference bands of transparent PET near interface indicates on partial dissolution of PET/HBA anisotropic melt in PET at 290°C forming interphase layer with a thickness increasing in time. Unfortunately, the clearing point of PET/HBA copolyester is higher than 350°C and higher than decomposition point of this polymer. That is why we cannot say definitely about degree of compatibility, but an increase of bending is observed

a
b
c
d
e

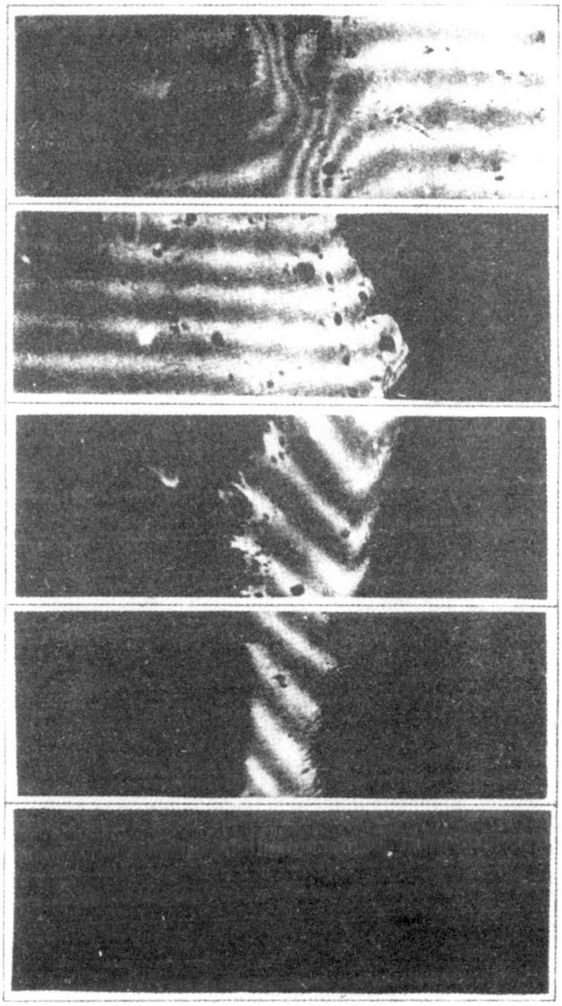

Fig. 14. Evolution of the interference patterns in the diffusion zone at cooling the system LC polyester - PBT. Temperature: 310 (a), 270 (b, LC polyester crystallizes), 220 (c, PBT crystallizes), 210 (d, the eutectic zone is lighted up), and 20°C (e).

that indicates on continuation of diffusion interaction in this time interval (Tereshin at al., 2000). So, in both cases the method of observation on one half of the interference pattern related with transparent polymer melt was developed. This approach could be very useful for analysis of diffusion interaction between two polymers, of one of them is turbid.

So, in previous cases the experiments on optical interferometry were carried out up to temperature of 290°C. Presumably, this is lower than it is possible to expect active chemical interaction between two polyesters, especially if one of them is LCP. The matter is that due to orientation of LC domains onto interface and formation more dense boundary layer, both processes - diffusion and chemical interaction can be slowed. For isotropic polyesters the situation with physical and chemical interactions should be more clear.

4.2 Pair of PET with polyethylenenaphtalate (PEN)

In the case of PET-PEN pair a special attention was devoted to superposition of two processes: interdiffusion and typical for polyesters interchain chemical reaction of the ester-ester exchange. These components are miscible completely, finally forming copolyester, and the main problem consisted in comparison of diffusion and chemical reaction rates. In our opinion, the first stage of interaction of molten polyesters is interdiffusion, and the second one - chemical trans-esterification. Transition from one stage to another is reflected on kinetics of concentration profile suffering significant change that time.

PET is a large-capacity thermoplastic used for the production of fibers, films, engineering plastics, and packaging articles. PET articles demonstrate excellent mechanical properties, with their manufacture technique being relatively simple. Nevertheless, some properties of PET do not suit consumers. Specifically, this refers to the phase state of PET, which belongs to the family of semi-crystalline polymers and possesses a high crystallinity degree (40–45%). Crystallites resulting from the rapid cooling of thin-walled articles are too small in size to affect the optical properties of PET articles. However, in the case of bulky thick-walled products, crystalline structures forming in the course of cooling lead to the loss of transparency. As a consequence, the scope of potential applications of PET appears to be limited. Moreover, a relatively low heat resistance (the glass transition temperature of PET is ~80°C) provides no way of using PET articles at high temperatures.

It is advisable to use as a partner for PET poly(ethylenenaphthalate) (PEN) both for physical (interdiffusion at relatively low melt temperatures) and chemical interaction (at high temperatures). The most universally adopted mechanism of trans-esterification relies on the decisive role of hydroxyl end terminal groups that participate in alcoholysis and acidolysis reactions and enter to neighboring chains of polyesters (Fakirov, 1999). The block copolymers thus produced improve the compatibility of polyesters with different chemical structures. Subsequently, neighboring polyester chains undergo scission and recombination reactions (direct ester–ester exchange), making copolymers random in structure (Stewart et al., 1993).

Over many years, it has been agreed that this reaction is responsible for compatibility of blend components (Litmanovich et al., 2002; Ihm et al., 1996). However, Guo&Brittain, 1998, revealed that trans-esterification is not prerequisite to compatibility. According to solid-state NMR spectroscopy, a mixture of melts became compatible during the first 1.5 min of injection molding and only after that could signs of the trans-esterification reaction be detected. The authors of the cited work proposed that, for the chemical interaction to occur, chains should be situated close to each other (at least within the action of dispersion forces). It is natural that for the incompatible blends of molten polyesters, the existence of interfaces limits coming together of potentially reactive macromolecules.

Thus, polymers making up the blend should form single-phase regions so that the effective chemical interaction can take place. Okamato&Kotaka, 1997, have demonstrated that the rapid cooling (quenching) of a PET/PEN blend being extruded at 280°C leads to the heterophase melt morphology (probably, because of spinodal phase decomposition). Further melting and annealing initiate the occurrence of trans-esterification at the boundaries of domains and their gradual disappearance as a result of compatibility. The use of a random PET–PEN copolymer as a third component accelerates compatibility.

Analogous conclusions concerning the incompatibility of the physical blend and the decisive role of trans-esterification in the compatibility of PET and PEN were reported in (Ihm et al., 1996). If the degree of interchain exchange is greater than 50%, blends do not crystallize and a single glass transition point is observed only. The rate of trans-esterification proves to be temperature- and time-dependent; however, it is unaffected by the blend composition.

Thus, the above seemingly simple and evident approach to modifying the properties of PET, which involves its blending with a certain amount of PEN, is complicated by the chemical interaction of these two polyesters. However, kinetics of physical and chemical processes and its determining factors, allowing us to estimate all stages of interaction are still unknown completely. The compatibility of molten PET and PEN calls for further investigation, and in this Chapter we focus our attention just on this problem (Makarova, 2005).

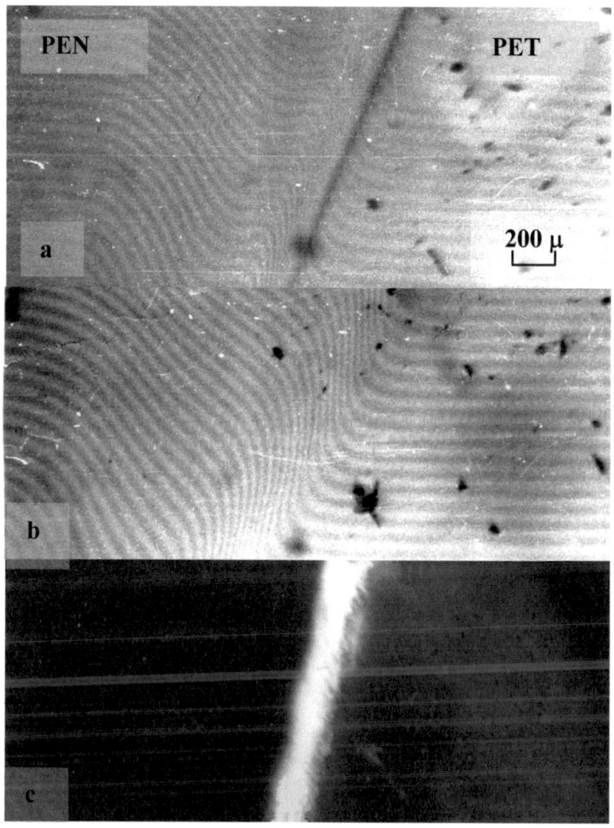

Fig. 15. Interferograms of PET - PEN system at 290°C in 5 (a) and 10 min (b) after contact. The pattern (b) corresponds to the combined specimen cooled to ambient temperature from 290°C.

At contacting PET and PEN after the melting of PET (~260°C) the weak bending of the interference bands occur that is caused by penetration to PET of macromolecules of still crystalline PEN (T_m=266°C). A solubility of PEN in PET in such conditions is not more 5-10%. After PEN melting a lot of sloped bands appear from both sides of the interface, and number of them increases with time (Fig. 15a). In 10 min (this time depends on temperature) the interface disappears completely (Fig. 15b). After cooling the cell to ambient temperature the region in vicinity of the former interface remains light (Fig. 15c). Presumably, either amorphous solution of components is formed in this region (eutectic point) or amorphous transparent copolyester.

Presence of the definite time lag (or "time barrier") as well continuous increasing a number of interference bands in isothermal conditions is not typical for the case of the physical compatibility, since at full compatibility the interface is absent already at first contacting of components. So, such behavior can be caused by simultaneous passage of interdiffusion and chemical interaction, promoting compatibility of components in time.

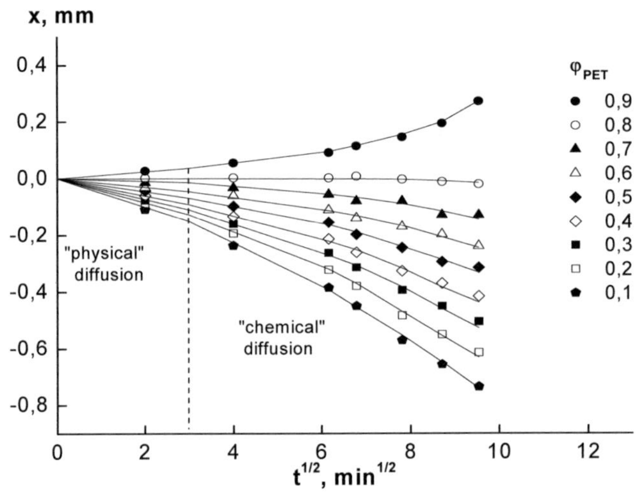

Fig. 16. Movement of iso-concentration fronts for the system PET - PEN at 300°C.

Plotted on the base of interferograms, concentration profiles in the transient zone are asymmetric from the PEN side that witnesses about prevail penetration of PET into PEN. Movement of the iso-concentration plains $x(\varphi)$ calculated from concentration profiles, follows to diffusion laws ($x \sim t^{0.5}$) on the initial sections only (Fig. 16) which length does not exceed 10 min. This time is comparable with disappearance of the interface. The further non-linearity can be connected with chemical interaction. In the figure this stage is noted as "chemical diffusion" because traditional diffusion rules do not work for chemically different elementary units.

For analysis of the difference in behavior of copolymers and complementary blends the interdiffusion was investigated on specimens consisting of three components: PET, PEN and copolyester or blend of the same composition. Copolyesters were synthesized via chemical reaction of PET with 2,6-hydroxynaphthoic acid (HNA) (Makarova et al, 2005). For the combined specimen "PET-PEN-copolyester" the interferograms are shown in Fig. 17. After PEN melting the interface is presented only from the PET side, while from copolyester side the continuous concentration profile is observed. Probably, the above discussed "time lag" is stipulated by necessity of copolyesters formation accelerating the further compatibility.

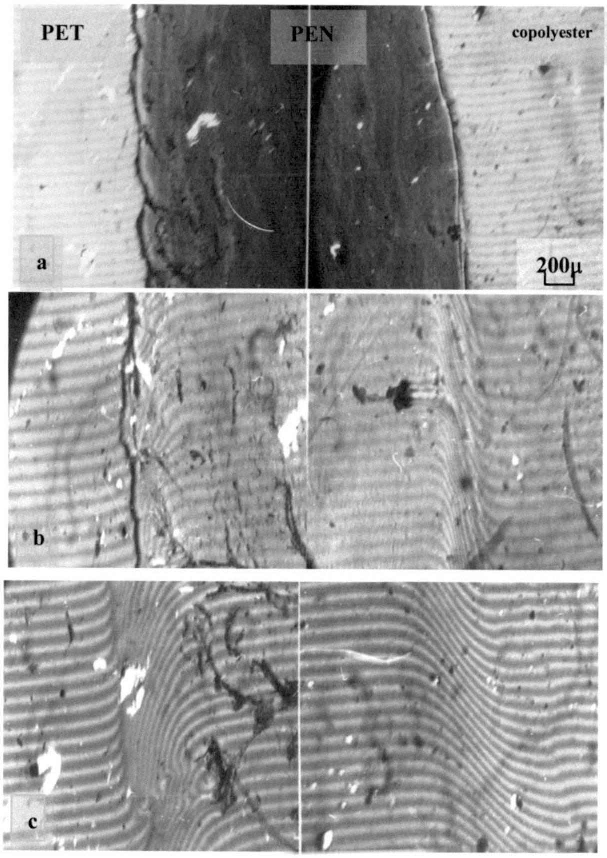

Fig. 17. Interferograms of PET - PEN - copolyester PET/HBA=80/20 at 268 (a) and 272°C in 10 (b) and 30 min (c).

Combination of PEN with mechanical blend PET/PEN=90/10 and copolyester containing 10% of HNA at 280°C (Fig. 18) has shown an absence of interfaces from both sides. However at duration of observation 20 min the bending of interference bands preserves from a side of the blend. Disappearance of such bending from a side of the copolyester indicates on last redistribution of concentrations in this diffusion zone.

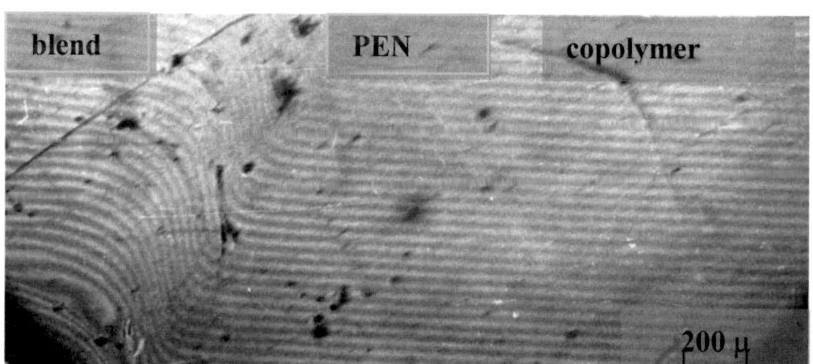

Fig. 18. Interferogram of the combined system: mechanical blend PET/PEN=90/10 - PEN - copolymer PET/HNA=90/10 at 280°C in 20 min after contact.

In some cases, at slow cooling as PET itself as the combined specimen PET - PEN the formation of tree-like morphology was observed similar to fragments of dendrite spherolites. "Branches" of PET were more monolithic and dense compared with graceful branches formed in PET containing 5% of PEN (Fig. 19). This fact can be explained by decrease of crystallinity degree of PET at adding minor amount of PEN and formation of amorphous copolyester. In this circumstances PET crystallites can grow and form spherolites.

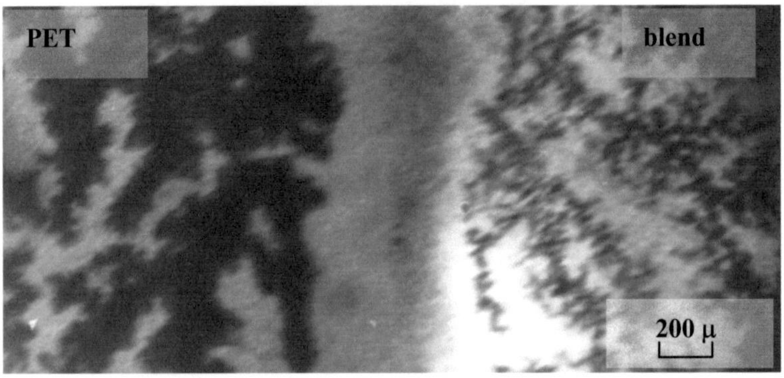

Fig. 19. Micrograph of the diffusion zone at contact of PET with mechanical blend PET/5% of PEN.

Using concentration profiles in the diffusion zone PET-PEN, interdiffusion coefficients were calculated at different temperatures. The D_v values in this temperature range lie in diapason $1.4 \cdot 10^{-12} \div 5.4 \cdot 10^{-11}$ m^2/s. The apparent activation energy of the diffusion process is closed to 410 kJ/mole. It is likely that so high value of the temperature coefficient can be explained by additional input of trans-esterification into the process of mutual penetration of components. Since dependence of D_v on concentration is rather weak, further calculations for various compositions at one temperature were carried out assuming independence of D_v on composition. The algorithm of calculations was the same as for PC-PMMA system using the probability diagram (Crank, 1956; Avdeev, 1990).

Calculated values of D_v initial polyesters and their combination with mechanical blends and copolyesters of complementary compositions are presented in Table 3. These data reflect more fast interpenetration of homo- and copolyesters compared with blends. The corresponding D_v values differ in 2-10 times.

It is rather strange that at contact of homopolyesters with mechanical blends the diffusion penetration is less intensive even than for homopolymers. Since blends were obtained at temperatures higher 280°C during 10 min and more, it is reasonable to expect the definite fraction of copolyesters formation. Nevertheless their interaction with homopolymers is not too intensive as for pure copolyesters. Probably, this can be explained by chaotic morphology of blends that leads to formation of heterogeneous interfaces and to interaction with homopolymer melts different species. So, presence in the mixture of copolyesters not ever accelerates the interpenetration. They should be just at interface as in the case of interaction of homo- and copolyesters.

The hypothetic mechanism of the general interaction in polyesters should be the following. First of all, the interdiffusion between components proceeds resulting in interpenetration of

Systems	Dv ·10¹², м²/c
PET - PEN	2,0
PET - blend PET/PEN=80/20	0,2
PET - copolyester PET/HNA=80/20	3,0
PEN - blend PET/PEN=90/10	3,5
PEN - copolyester PET/HBA=90/10	7,0
PEN - blend PET/PEN=80/20	0,5
PEN - copolyester PET/HBA=80/20	4,5

Table 3. Interdiffusion coefficients for different interacting components at 280°C.

macromolecules of different sorts. At their closing on a distance permitted for covalent exchange, the chemical process starts leading to formation of copolyesters in vicinity of interface. Further, copolyesters promote as diffusion interaction, as propagation of chemical reaction front in depths of both reacting components.

We can estimate the dissolution time of a particle of PEN or PET in an unrestricted volume of another polymer. Calculations were done in the same way as in the case of PC-PMMA pair (Avdeev, 2001; Makarova, 2007). The corresponding data are presented in Table 4.

d, μ	T=280°C
1	0,1 s
10	10 s
100	20 min
1000	30 h

Table 4. Time of complete dissolution of PEN drop of various size in PET.

These data show that in any industrial mixing method only droplets with a size of <10 μm can be dissolved completely. For larger droplets the interphase layer of the definite thickness will be formed consisting of mutual solutions of homopolymers and copolymers. At cooling the system will contain 3 or 4 phases, influencing its exploring properties.

5. LC phase equilibrium in polymer systems

Liquid-crystalline state is stable, thermodynamically equilibrium and engages a gap between amorphous and crystalline states (Papkov et al., 1974, Papkov&Kulichikhin, 1977; Collyer, 1996; Kulichikhin, 2000]. In LC state macromolecules form uni-dimensional or two-dimensional order that distinguishes it from full disorder in amorphous state and three-dimensional order in crystalline state. Such state is named often as anisotropic since presence of the order in macromolecular arrangement leads to appearance of anisotropy of many physical properties, partially optical, rheological (majority of these systems are capable to flow) and many others.

LC phase in polymers can be realized either in melt state, if 3D crystalline order disappears not completely, but step-by-step destruction at melting (thermotropic LC polymers) or in solutions of stiff-chain polymers (lyotropic LC state). In the last case LC equilibrium takes place at full compatibility of components. Formation of LC solutions is stipulated predominantly by geometric asymmetry of rigid macromolecules which can engage any position in solution volume up to definite concentration only. After critical concentration adding new macromolecule is possible at condition of partial (in the beginning) or full ordering of other macromolecules. Using a term "partial ordering" means that the transition from isotropic to LC state along concentration axis passes via biphasic region where isotropic and LC state coexists.

The general view of the state diagram for solution of stiff-chain polymer with rod-like macromolecules, based on theoretical paper of Flory (Flory, 1956) and Papkov (Papkov et al., 1974; Papkov, 1992) is shown in Fig. 20. The diagram contains a combination of narrow and wide biphasic regions separating isotropic and anisotropic solutions. According to Flory, the critical concentration of LC phase appearance is related with molecular stiffness by equation:

$$\varphi_2^* \approx \frac{8}{x}\left(1-\frac{2}{x}\right) \tag{6}$$

where x is asymmetry degree of macromolecules or, in common case, the ratio of the Kuhn segment to diameter $x = \frac{A}{d}$.

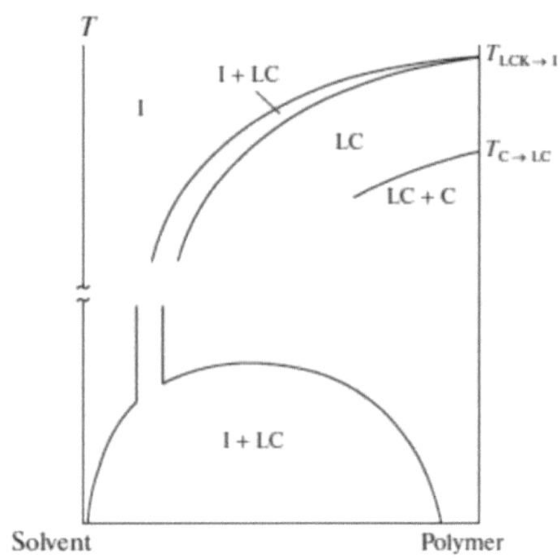

Fig. 20. Theoretical phase diagram for solutions of stiff-chain polymers according to (Flory, 1956) and (Papkov, 1992)

For rod-like macromolecules the location of the narrow biphasic region practically does not depend on temperature. But decrease of macromolecular stiffness leads to increase of the critical concentration φ_2^* and to appearance of significant slope of the equilibrium line T*(φ_2^*). Papkov, 1992, hypothetically extended this line in direction of 100% polymer (dotted line in Fig. 20). According to his opinion, the width of the biphasic region has to be narrower at closing to 100% polymer phase, and in a limiting case crosses the temperature axis inherent for polymer. The crossing point is the temperature of thermotropic transition, i.e.

T_{LC-I}. In addition appearance of crystalline phase is possible in this region and equilibrium (LC+C) - LC should be considered.

Roughly speaking, the phase diagram for polymer systems with LC equilibrium is combination of LC-liquidus line and binodal. With increase of macromolecular flexibility the critical concentration growths and the width of biphasic domain extends. In the limiting case this phase diagram transforms to state diagram typical for amorphous equilibrium (biphasic part disappears completely).

For several polymer systems the superposition is possible of the different types of phase equilibria in various combinations: amorphous and crystalline, amorphous and LC, LC and crystalline. The principle of the mutual independence of at least two versions of phase equilibria is fulfilled (Papkov&Kulichikhin, 1977; Papkov, 1981). In a system non-equlibrium in relation to one kind of phase equilibrium, the other kind can be realized. The sequence of phase states exchange depends on an ordering degree of forming phase. The higher ordering level, the slower a process of the new phase nuclei formation proceeds. The fastest case - formation of amorphous phases since for this case just fluctuating collision of molecules is enough without their any ordering. LC phase is more ordered that is why its nuclei form slower. Much more inhibition is observed for creation of a crystalline phase.

Do not discuss in details different partial versions of superposition of different kinds of phase equilibria there is sense to show one of the most complex to present time phase diagrams (Fig. 21). It combines LC and crystalline (CS) equilibria, as well as two binodals with UCST and LCST (Iovleva et al., 1989).

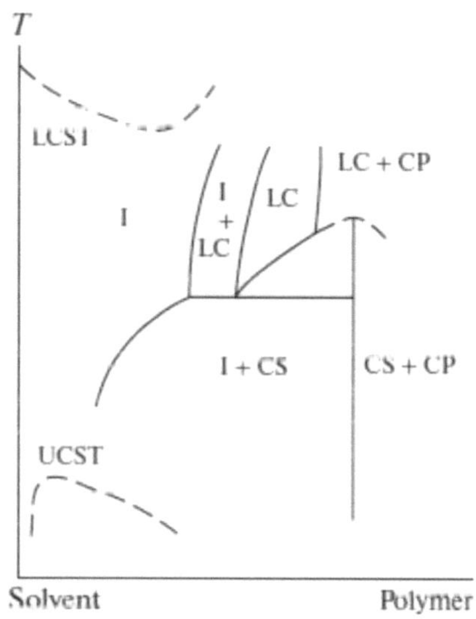

Fig. 21. Superposition of different kinds of phase equilibria for a system polymer-solvent.

In principle, an existence in one polymer a capability to form both thermotropic and lyotropic phases meets very rarely because of high values of T_{LC-I}, lying higher decomposition points of polymers or difficulty to obtain very concentrated solutions for semi-stiff-chain polymers. Hydroxypropylcellulose (HPC) is one of such polymers.

5.1 LC and complex phase equilibria in hydroxypropylcellulose solutions

So, HPC is among few polymers capable of forming both thermotropic and lyotropic LC phases. This is why HPC, as none other natural polymer, is accessible for analysis of phase equilibrium both in melt and solutions in a wide concentration range and in the region of easily achievable temperatures. This situation is primarily explained by the chemical structure of HPC, which contains primary and secondary hydroxypropyl and residual hydroxyl group moieties.

Therefore, a system of H- bonds inherent for cellulose is destabilized and the level of intermolecular interaction is reduced. As a result, the polymer shows good solubility in solvents of various nature. The above features of the chemical structure are responsible for a rather high skeletal rigidity of HPC macromolecules (the Kuhn segment length is ~13 nm in dimethylaceteamide (Kulichikhin&Golova, 1985), which ensures transition to the LC melt at rather low temperatures (~120°C) with an isotropization (clearing) temperature of ~205°C (Makarova, 2007; Makarova et al., 2007).

HPC of the Klucel EF brand (Hercules-Aqualon, Unites States) with $M_w = 8{\cdot}10^4$ was used. Solvents were: two oligomeric PEG samples (Spectrum, Unites States) with Mw=400 (PEG 400) and 1500 (PEG 1500), distilled water, propylene glycol (PG) (Labtex, Russia), triethyl citrate (TEC) (Morflex, United States), and DMSO (Gaylord Chemical Corp., United States).

Solvent	Dipole moment, D	Dielectric constant, F/m	Surface tension, mN/m	Solubility parameter $(cal/cm^3)^{0.5}$ [28]
Water	1.85	80	72.9	21.2
DMSO	3.96	47	44.0	12.8
PG	3.63	32	36.5	12.6
TEC	4.52	11	41.5	8.2
PEG	4.20	14	43.5	8.4

Table 5. Physical-chemical characteristics of solvents used.

Table 5 lists some characteristics of the used solvents (Woolley, 1974; Kolotyrkin, 1974; Allan, 1990; Geller et al., 1996; Uusi-Penttila, 1997; Kulvinder, 2000; Gallyamov, 2009; Mali, 2007) which may be useful for analysis of phase equilibria in HPC--solvent systems. Solutions of HPC with a HPC concentration of 30, 35, 40, 45, 50, 55, 60, 70, 75, and 80 wt % were prepared in various solvents in accordance with (Fischer et al., 1995).

The phase equilibrium in solutions was studied by the microinterference method at 18-210°C which renders it possible to analyze concentration profiles in the diffusion zone. On the basis of interferograms and relationships between refractive index and composition in the transition zone, the profiles of component concentration distributions were constructed and the lines of phase equilibria were derived from the boundary concentrations of components at various temperatures. To confirm the equilibrium state and reversibility of the boundary concentrations, measurements were performed at both ways: increasing and decreasing temperatures. The LC phase in solutions was identified by polarization microscopy.

In parallel, the viscosity of solutions of different concentrations was measured. This information can be very useful for understanding not only compositions of various phases, but their properties as well.

For all systems under consideration, an increase in the concentration of HPC is accompanied by formation of the LC phase; in some cases, this process is complicated by amorphous separation of the system. In addition, for a number of systems, the CS phase can form in the high-concentration range. It is reasonable to examine various systems separately and then to generalize the data on phase equilibrium and rheological properties of HPC-based solutions.

Glycols are among the solvents used in this study. Let us first all analyze the phase equilibrium in low-molecular-weight liquids and oligomers containing terminal hydroxyl groups.

5.1.1 HPC-PG system

The interferograms measured at 75°C (Fig. 22) exhibit two phase boundaries (dotted lines): the first boundary separates isotropic and two-phase solutions, while the second boundary separates fully LC solutions from the crystalline (or the CS) phase. The presence of the LC phase is confirmed by the polarization microscopy data. Thus, under crossed polarizers the region of the LC phase of 60% solution exhibits a characteristic rainbow luminescence typical for cholesteric liquid crystals. Cellulose derivatives containing the asymmetric carbon atom (the chiral center) form just the cholesteric mesophase. As temperature is increased, the fully LC solution transforms into the two-phase region, in which LC and isotropic phases coexist. Given this, the solution gradually darkens and, after full isotropization, the field of vision becomes dark.

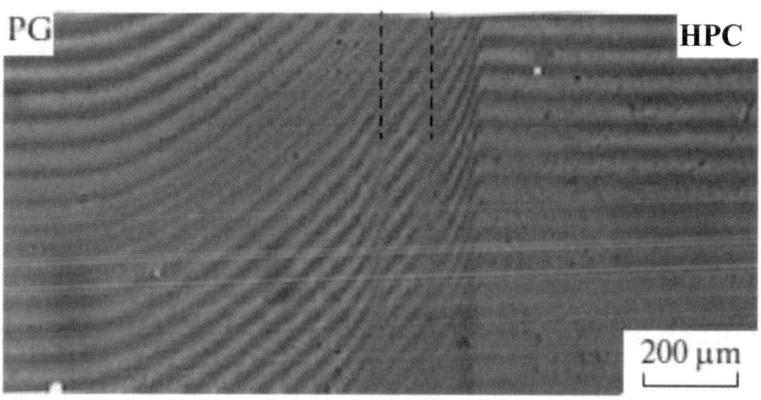

Fig. 22. Interferogram of the transition zone for the HPC-PG system at 75°C. The phase boundaries are marked with dotted line.

The resulting phase diagram is the superposition of LC and crystalline equilibria (Fig. 23). The results of polarization microscopy measurements made it possible to refine position of the LC liquidus and to supplement the diagram with the second boundary of the two-phase corridor separating the two-phase region and the region of fully LC solutions.

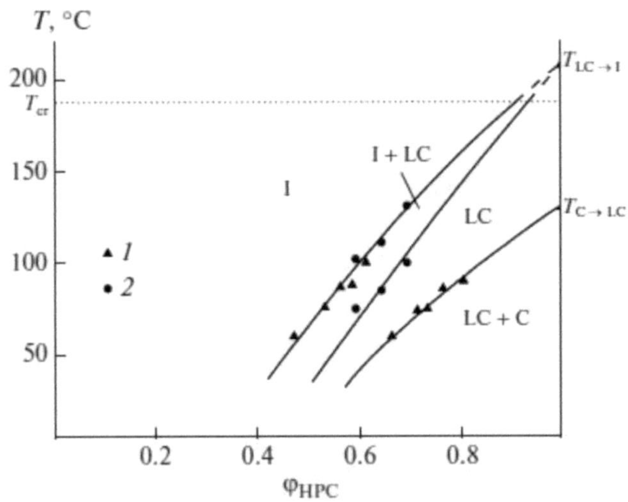

Fig. 23. Phase diagram of HPC-PG system, determined by interferometry (1) and polarizing microscopy (2). I - isotropic, LC - liquid-crystalline, C - crystalline state.

As opposed to extremely rigid-chain polymers with rod-like macromolecules, for which the I-LC transition is almost independent of temperature and concentration (the line of phase equilibrium is almost parallel to the ordinate axis), in the case of HPC solutions in PG, the position of boundary lines strongly depends on temperature and concentration of HPC. This finding is indicative of the semi-rigid nature of HPC in this solvent,

At high concentrations of the polymer, the LC phase occurs in equilibrium with the crystalline phase of HPC (C) or the CS phase. Note that in the neat HPC the degree of crystallinity does not exceed 20% (Shimamura, 1981). On the whole, the pattern of the phase diagram corresponds to the upper part of the hypothetical diagram shown in Fig. 20.

5.1.2 HPC-PEG 400 system

The phase diagram for HPC-PEG 400 solutions was described in (Makarova, 2007; Makarova et al., 2007). Throughout the studied temperature range, this system is characterized by LC and crystalline transitions (Fig. 24). On the left of the boundary line, the solutions are isotropic, while on the right to the boundary line, they are mesomorphic. As in the case of HPC solutions in PG, the position of the boundary line is markedly temperature-dependent.

Given almost full qualitative analogy, there are marked quantitative differences between HPC-PEG 400 and HPC-PG systems. This primarily concerns smaller critical concentration corresponding to formation of the LC phase: ~25% for HPC solutions in PEG 400 compared to ~40% solutions of HPC in PG.

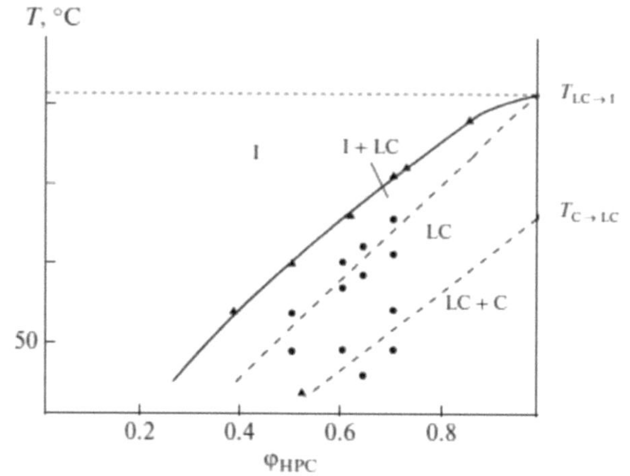

Fig. 24. Phase diagram of HPC-PEG 400 pair.

5.1.3 HPC-PEG 1500 system

The phase equilibrium changes substantially with an increase in the molecular weight of PEG. The main feature of HPC solutions in PEG 1500 is a superposition of LC and amorphous phase equilibria (Fig. 25) (Makarova, 2007; Kulichikhin et al., 2010; Tolstykh et al., 2010). Separation on amorphous phases is observed in the region of small and moderate concentrations with a UCMT of ~185°C, while the liquid-crystalline equilibrium occurs at higher concentrations of HPC. Within the experimental error, the position of the liquidus for a given system coincides with that for the HPC-PEG 400 system.

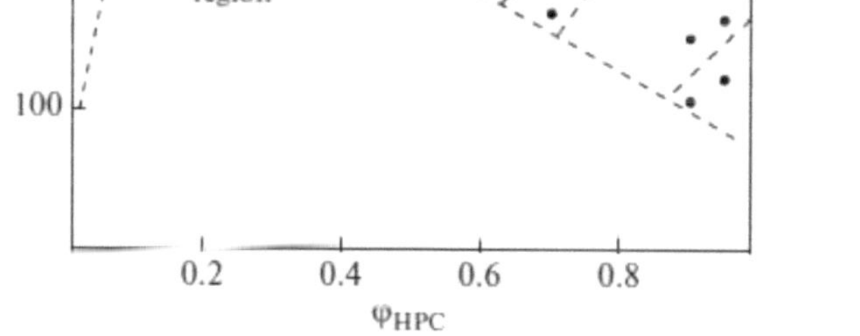

Fig. 25. Phase diagram of HPC-PEG 1500 system.

Nevertheless, PEG 1500 is a less strong solvent for HPC than PEG 400. This is evidenced by the absence of solubility in the concentration and temperature region below the binodal. Such behavior can be explained by the intramolecular association of PEG molecules, as confirmed by the IR data (Makarova et al., 2007). As molecular weight is increased, this effect becomes more pronounced and the probability of interaction between terminal hydroxyls of PEG and the functional groups of HPC tends to decrease.

Thus, solvents containing end hydroxyl groups dissolve well HPC and give rise to isotropic and anisotropic solutions. Owing to the presence of oxygen atoms in oligomeric PEG 400 chains and their interaction with ether or residual hydroxyl groups of HPC, the rigidity of a HPC macromolecule is somewhat higher; as a result, the transition to the LC state occurs at a lower concentration than that in the case of the low-molecular-weight PG. However, an increase in the molecular weight of PEG to 1500 causes a significant reduction in the dissolving capacity apparently due to cyclization of sufficiently long oligomeric molecules which worsens the efficiency of interaction between functional groups of the solvent and the polymer. It is not inconceivable that cyclization likewise depends on the concentration of solution since at high concentrations of HPC the LC equilibrium predominates and gives rise to LC solutions.

The hydroxyl group is also contained in a molecule of TEC, citric ester. Moreover, PEG and TEC have the highest dipole moments and the lowest dielectric permittivities among the used solvents (Table 5). Therefore, in what follows, it is advisable to examine the phase equilibrium in the HPC-TEC system.

5.1.4 HPC-TEC system

Fig. 26 shows the interferograms of the transition zone obtained for this system. At temperatures below 70°C, two phase boundaries appear in the diffusion zone.

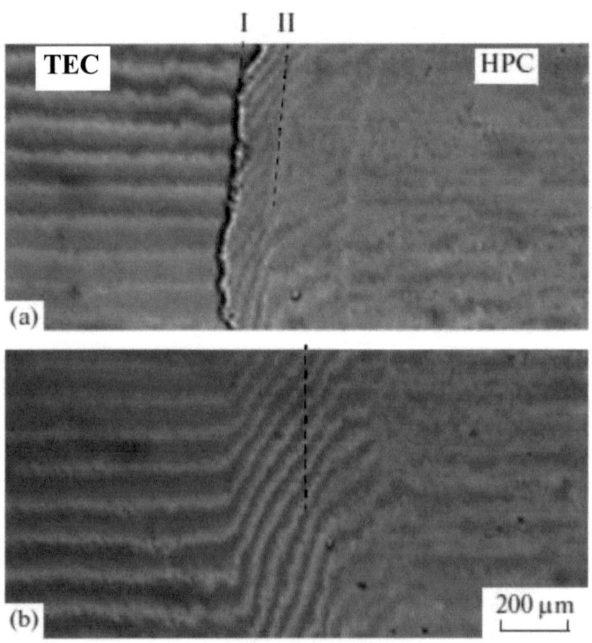

Fig. 26. Interferograms of transition zone for the HPC-TEC system registered at 55 (a) and 77°C (b). Phase boundaries are marked with dotted lines.

One of them (I) is related to the amorphous separation of the components, while the other boundary separates isotropic and mesophase regions. Above 70°C, phase boundary I disappears, whereas boundary II is preserved, thus indicating full dissolution of HPC in TEC at elevated temperatures with the concomitant formation of LC solutions.

On the basis of the interferograms with the use of the boundary concentrations of components estimated at various temperatures, the phase diagram was constructed (Fig. 27). As the HPC-PEG 1500 system, the HPC-TEC pair is characterized by the superposition of amorphous and LC equilibria. This situation is reflected on the diagram as the binodal with the HCST and the LC liquidus, respectively. Separation on amorphous phases is observed in the region of small and moderate concentrations, while the liquid-crystalline equilibrium is seen at high concentrations of HPC.

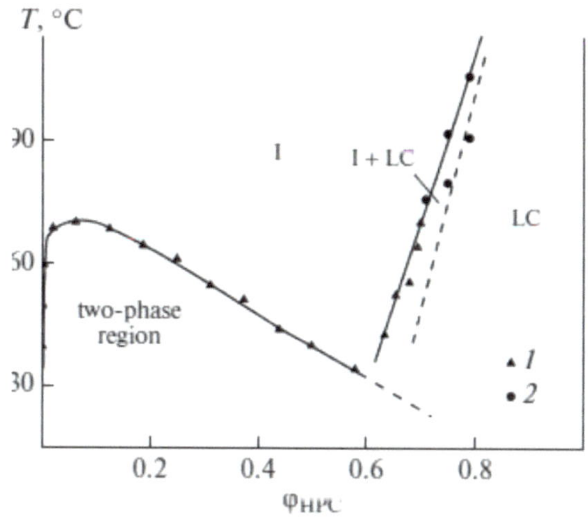

Fig. 27. Phase diagram of the HPC-TEC system constructed from data of microinterference (1) and polarizing microscopy (2)

An analysis of the phase equilibrium in the HPC-TEC system shows that this system behaves as HPC solutions in PEG 1500. It appears that TEC molecules are characterized by self-association via formation of H-bonds between hydroxyl and ester groups. Furthermore, the steric factor may play an important role in this case. These structure features of TEC molecules affect the dissolving capacity of this solvent, which depends on solution concentration, as demonstrated by amorphous separation in the dilute-solution region and formation of the LC phase at increased concentrations of HPC.

Water is the next solvent containing hydroxyl groups, and solubility in water is a indubitable advantage of HPC. However, water possesses the lowest dipole moment and the highest dielectric constant; therefore, a certain specifics of phase equilibrium in aqueous solutions of HPC should be expected.

5.1.5 HPC-water system

The phase behavior of HPC-water solutions has more complex than those described above. Several versions of diagrams for this system were obtained by DSC, X-ray diffraction, NMR, nefelometry, and rheology. A set of some versions of diagrams is shown in Fig. 28. At temperatures below 40°C, with an increase in the concentration of HPC, solutions transform from the isotropic region (I) to the two-phase region containing isotropic and LC phases (II) and then to fully anisotropic region (III). At high concentrations of HPC, CS (IV) forms.

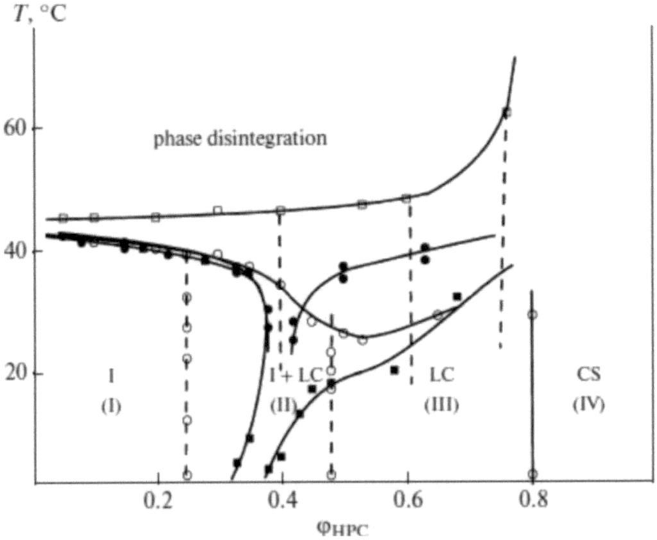

Fig. 28. A set of phase diagrams for the HPC-water system according to (Fischer et al., 1995; Guido, 1995; Larez et al., 1995; Vshivkov et al., 2007)

Although the diagrams follow the general pattern, the positions of phase equilibrium lines differ appreciably; moreover, it is unclear which is the character of phase separation at temperatures above 40-45°C. It is known that elevated temperatures solutions experience reversible gel-formation; however, the mechanism controlling formation of the physical network remains vague up to now. Two mechanisms of this process are discussed in the literature: phase decomposition of the liquid-liquid type and formation of the $HPC{\cdot}6H_2O$ crystal solvate. Unfortunately, structural methods do not assist in understanding this mechanism. This circumstance forced us to refine the lines of phase equilibrium for this system and to gain insight into the reasons of gelation at elevated temperatures.

At temperatures below 37°C, the interferograms show all four above-mentioned regions in the transition zone (Fig. 29). As temperature is increased, the first two regions disappear and the phase boundary responsible for amorphous separation appears. The phase diagram constructed from concentration profiles in the diffusion region (Fig. 30) is the superposition of three types of phase equilibria, namely, amorphous, LC, and crystalline.

From the refined phase diagram, the mechanism of gelation of aqueous solutions at elevated temperatures becomes understandable, namely, amorphous separation and, as a result, "incomplete separation into phases" that leads to formation of the structural network. The fact that in the case of TEC and PEG 1500, solubility increases with temperature (the binodal with the HCST) and, in the case of water, decreases (the binodal with the LCST) suggest that temperature dependences of the interaction parameter are different. However, it should be mentioned that the Flory-Huggins theory cannot predict the existence of systems with the

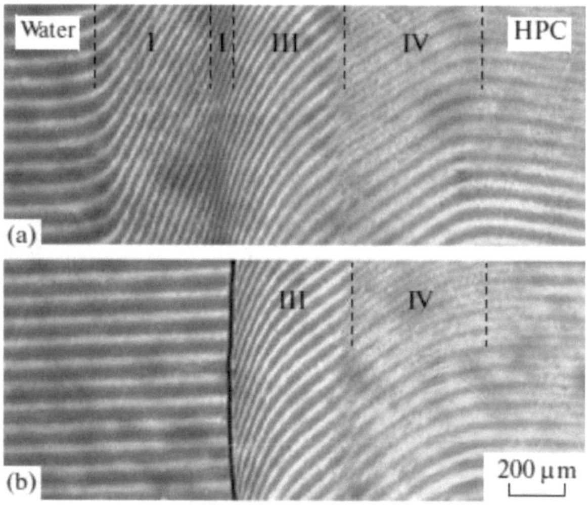

Fig. 29. Interferograms of transition zones for the HPC-water system at 29 (a) and 43°C (b). Figures denote regions corresponding to different phase states (see explanations in text).

LCST. As a rule, such phase diagrams are observed for polymer-solvent components with strong intermolecular interaction and corresponds to the upper part of the theoretical phase diagram shown in Fig. 21. In this sense, glycols and esters are more active solvents in relation to formation of H-bonds with the functional groups of HPC than water.

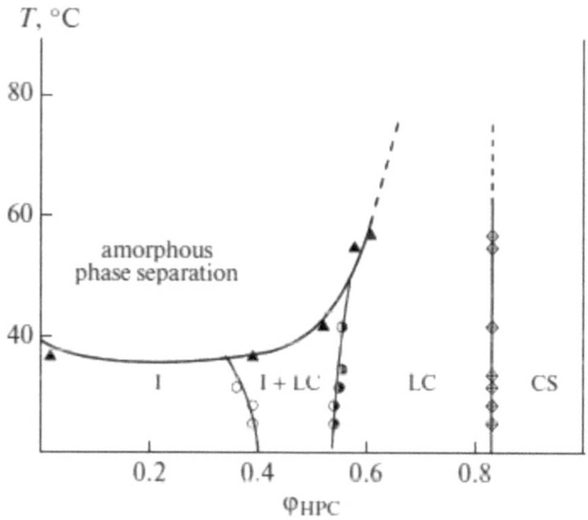

Fig. 30. The corrected phase diagram of the HPC-H_2O system .

The next solvent from the selected series is aprotonic dipolar DMSO, which stands somewhat apart from other solvents since it contains no hydroxyl groups. The DMSO molecule has unshared electron pairs on oxygen and sulfur atoms, and this fact explains high dissolving capacity of this donor solvent. In terms of dipole moment and dielectric constant, DMSO is rather close to PG.

5.1.6 HPC-DMSO system

The typical interferogram of the region of interdiffusion of HPC and DMSO is shown in Fig. 31. Throughout the studied temperature range, the phase boundary (marked by the dotted line) is seen which separates the regions of isotropic and mesomorphic states.

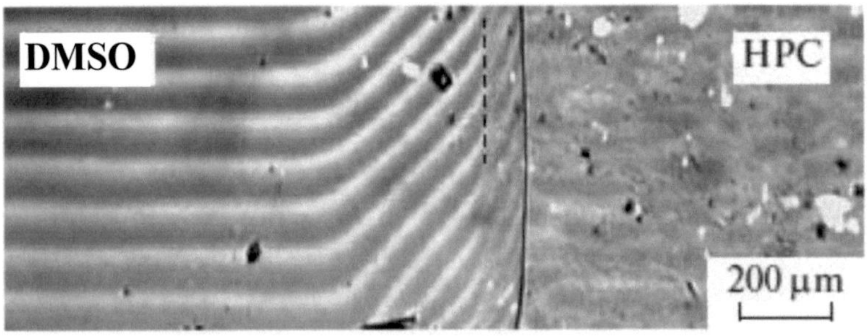

Fig. 31. Interferogram of the transition zone for the HPC-DMSO system measured at 55°C. The phase boundary is denoted with a dotted line.

The phase diagram of the HPC-DMSO system constructed from the data of interference rheology and polarizing microscopy is characterized by the LC equilibrium (Fig. 32). Throughout the temperature range under study, the two-phase corridor manifests itself which separates the regions of isotropic and LC solutions.

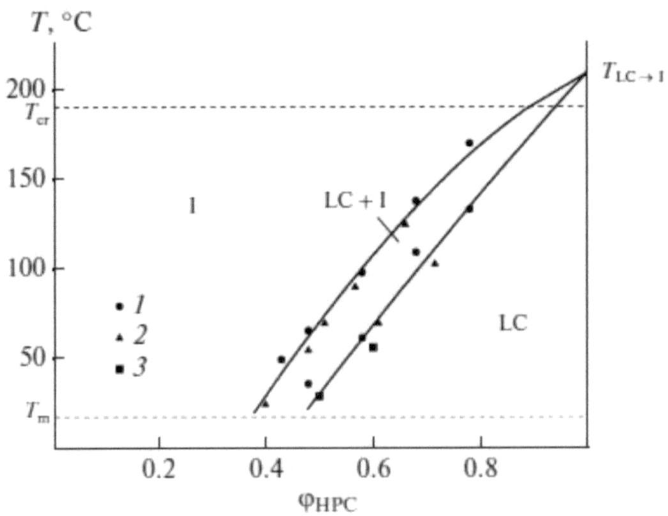

Fig. 32. Phase diagram for the HPC-DMSO system constructed from the data of microinterference (1), microscopy (2) and rheology (3).

Tables 6 and 7 list some parameters of the phase diagrams obtained at 25 and 140°C, respectively. Let compare these data with the characteristics of the tested solvents (Table 5). Strong intermolecular interactions are responsible for the highest solubility parameter δ of water, as calculated in accordance Hildebrand and Small [62], δ = 21.2 (cal/cm^3)$^{0.5}$. This value is much higher than the δ of HPC (approximately10.72 (cal/cm^3)$^{0.5}$ (Allan, 1990).

Solvent	C^*, vol fractions	C^{**}, vol fractions	C^{**}/C^*	C_c, vol fractions	Critical mixing temperature, °C
PG	0.38	0.47	1.24	0.55	–
PEG 400	0.25	0.37	1.48	0.54	–
PEG 1500	–	–	–	–	185 (UCST)
PEC	–	–	–	–	71 (UCST)
Water	0.39	0.54	1.38	0.83	37 (LCST)
DMSO	0.38	0.48	1.26	–	–

Table 6. Critical concentrations of phase transitions for considered systems at 25°C.

It is possible that this fact explains why HPC is not dissolved in water at temperatures above 36.5°C (the LCST) but it cannot explain why HPC shows excellent solubility at reduced temperatures. The latter phenomenon may be attributed to redistribution of H-bonds formed by water molecules to those formed between water molecules and functional groups of HPC.

As regards other solvents, their solubility parameters range from 8.2 to 12.8; that is, they are rather close to the δ of HPC. Note that for TEC and PEG 1500, solubility parameters are somewhat smaller than the δ of HPC (8.2 and 8.4), while for PG and DMSO, they are somewhat higher (12.6 and 12.8). It is pertinent to note that for the first pair of solvents, the superposition of amorphous (with the HCST) and LC equilibria is observed, while for the second pair, only the LC equilibrium is implemented. PEG 400 stands somewhat apart from the rest solvents since in this case no amorphous separation is detected. These differences may be due to different polarizabilities of molecules: for TEC and PEG, high dipole moments coexist with low dielectric constants, while for PG and DMSO, the reverse situation is observed (disregarding water).

If we examine the critical concentrations corresponding to formation of the LC phase (C*) and the achievement of its 100% content in the system (C**), then the highest rigidity of HPC macromolecules is realized in PEG, as evidenced by the lowest value of C* (0.25 vol. fractions at 25°C). In the case of water, PG, and DMSO, these parameters are in the range 0.38-0.39. At a temperature of 140°C, the values of C* are 0.64, 0.70, 0.73, and 0.85 vol. fractions for PEC, DMSO, PG, and TEC, respectively.

Solvent	C^*, vol fractions	C^{**}, vol fractions	C^{**}/C^*
PG	0.73	0.79	1.08
PEG 400	0.64	0.74	1.16
PEG 1500	0.65	0.74	1.14
TEC	0.85	0.87	1.02
DMSO	0.70	0.79	1.13

Table 7. Critical concentrations of phase transitions for considered systems at 140°C.

For all systems, the width of the two-phase corridor, which was estimated from the ratio of critical concentrations C**/C*, is much higher at 25°C (1.24-1.48) than that at 140°C (1.02-1.16). This narrowing of the corridor with temperature, first, is indicative of the semi-rigid nature of HPC (the rigidity of chains is strongly temperature-dependent) and second, is consistent with Papkov's predictions (Fig. 20) that lines C*(C_{HPC}) and C**(C_{HPC}) come closer and in the limit merge on approach to the 100% polymer.

For such solvents as PG, PEG 400, PEG 1500, and water, the crystalline phase (CS or the crystalline HPC) is detected at a concentration of C_c. It appears that the rigidity of HPC macromolecules implemented in various solvents may affect the parameters of amorphous separation of systems. At least, this is true for HPC-PEG 1500 and HPC-TEC systems. In PEG 1500, chain rigidity, as estimated from C*, is higher and its HCST (185°C) is situated at a concentration of 0.18 vol. fractions. For HPC solutions in TEC, C* is lower and the HCST is 71°C at a concentration of 0.09 vol. fractions.

The above analysis does not pretend to be full and does not reveal all features of phase equilibrium in HPC-solvent systems. Nevertheless, it demonstrates the peculiar phase behavior of these systems in relation to the structure and properties of solvents.

5.1.7 Rheological properties of HPC solutions

The rheological properties of LC systems are characterized by some distinctive features (Papkov et al., 1974; Kulichikhin et al., 1982; Kulichikhin, 1989). As opposed to isotropic polymers, LC polymers manifest the abnormal viscosity throughout the studied shear rate range. This effect is associated with the polydomain structure of LC systems and the presence of a system of disclinations at rest. With an increase in the shear rate as a result of orientation processes, the system gradually transforms into the monodomain structure. Furthermore, in the two-phase region, interphase boundaries between drops of isotropic phase and LC disperse medium, or LC drops and isotropic matrix make a substantial contribution to the rheological response.

The main result at analysis of rheological properties of solutions under consideration consists in conclusion that viscosity is a parameter sensitive to LC transitions. The flow curves reflect the existence of the polydomain structure in the initial LC system and its transformation into the monodomain structure with an increase in the intensity of shear deformation. Presence of polydomain structure and such defects as disclinations is stipulated the viscoplastic behavior with a yield stress. However, the most vivid dependence that makes it possible to estimate the critical concentrations of transitions is the concentration dependence of viscosity. In this sense rheological data are good addition to results of interferometric measurements to prove the location of boundary lines between liquid phases.

Fig. 33 displays dependences of this type for several systems under study in the region of the Newtonian (isotropic solutions) or quasi-Newtonian (anisotropic solutions) behavior. As is seen, these dependences contain extreme points. Despite of different scales of maxima and minima, extremes are observed for all systems. As is seen from comparison with phase diagrams, the maximum of viscosity corresponds to the critical concentration, at which the anisotropic phase appears in solutions. When the system contains isotropic and LC phases, an increase in the fraction of the anisotropic phase causes a drop in viscosity. The minimum

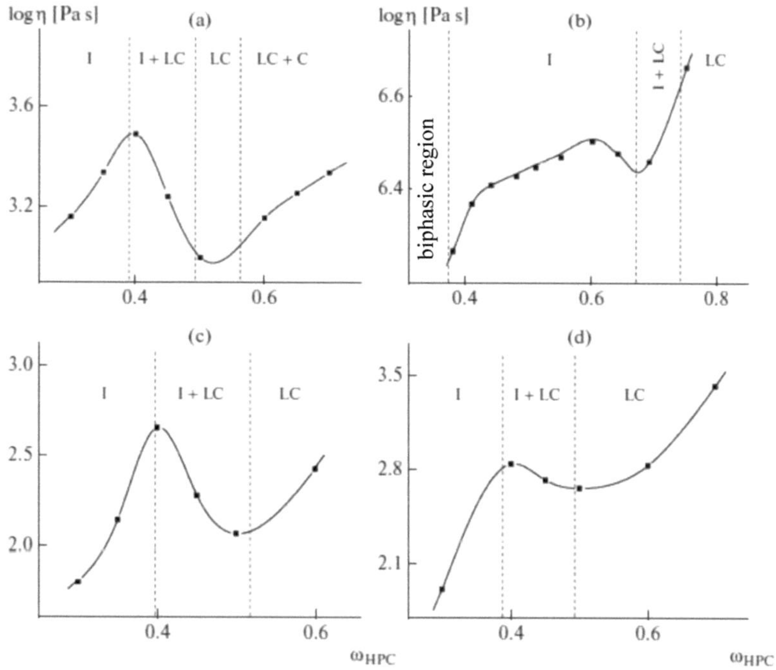

Fig. 33. Concentration dependences of viscosity for solutions of HPC in PG at 23 (a), TEC at 50 (b), water at 23 (c), DMSO at 23°C (d). Dotted lines show phase boundaries according to phase diagrams constructed.

on the curve corresponds to the critical concentration at which the system becomes fully LC. After further increase in concentration, viscosity increases again due to a reduction in the free volume. Thus, an increase in viscosity with concentration is typical for single-phase (isotropic and LC) solutions, whereas in the two-phase corridor, viscosity drops with increase of concentration (the fraction of the LC phase).

It is pertinent to note the specifics inherent in the HPC-TEC system in the concentration range neighboring amorphous phase separation. In this case, the abnormal decrease in viscosity with a decrease in the concentration of HPC is apparently related to formation of nuclei (species) of the amorphous phase highly concentrated with respect to the polymer.

The above dependences of viscosity on concentration vividly reflect the specifics inherent in solutions of rigid-chain polymers, which form the LC phase in a certain temperature-concentration range. Exactly for these systems viscosity is as a parameter suitable to refine the position of boundary lines of phase equilibrium. Very likely this is the key point in a given analysis. The region of amorphous equilibrium is practically unfeasible for the correct study by means of rheological methods since the system experiences phase separation below the binodal.

6. Conclusion

The different kinds of phase equilibria (amorphous, crystalline, liquid-crystalline) are considered. Each of them has definite futures that were analyzed on examples of various polymer-polymer and polymer-solvent systems. All above mentioned results relate to thermodynamic properties for binary systems containing polymers, and all of them were obtained by interferometry method. In addition to thermodynamic data, this method gives information about kinetics of components interactions that allows us to measure simultaneously interdiffusion coefficients. This dualism of optical interferometry method in application to interaction in polymer containing systems is its great advantage. That why its application in scientific practice is very useful and fruitful.

The couple PC - low molecular weight PMMA is a good example of polymer system with amorphous phase equilibrium. Method of interferometry is very useful and fruitful for such kind of systems due to informativity and combination of joint approaches: kinetics of mass-transfer and thermodynamics of phase states. Increase of the mutual solubility with temperature indicates on UCST that could not be measured experimentally but was calculated using evolution of the interaction parameter with concentration.

Comparison of the phase diagram with concentration dependence of viscosity allowed us to observe local minimum of viscosity near the binodal (sooner in meta-stable region between binodal and spinodal where microemulsion appears as the first step of the phase decomposition) and to apply the empiric equation for calculation of the blends viscosity by means of accurate determination of the composition from binodal branches. Joint consideration of diffusion and rheological properties seems to be important for judging about structure, kinetic and thermodynamic behaviors of complex, multiphase polymer systems. This approach will be used further for analysis of systems with LC and crystalline equilibria.

Calculations of the interphase layers thickness depending on dimensions of drops and time, based on knowledge of interdiffusion coefficients can be very useful for correct organization of technological process of mixing and, consequently, development of new materials with three-phase structure and new properties.

Several examples of crystalline equilibrium in couples polyester-polyester are considered a well. The neat polymers are semi-crystalline and phase equilibrium for their pairs is describing by liquidus lines, sometimes with eutectic points. But this situation takes place at T<280-290°C, when chemical interaction can be neglected. The original feature of these data consists in using as one component of pair thermotropic LC polyesters and copolyesters. Up to clearing point of LC component the interdiffusion process is registered using one part of the interference patterns. Such approach was demonstrated first time.

At using LC polymer melt as one component, both interdiffusion process and ester-ester exchange is delayed because of high molecular orientation in LC polymer along interface. This dense layer prevents fast penetration of coil-like macromolecules of isotropic polyester melt. But in the case of PET-PEN pair both melts are isotropic and transparent, that is why interaction process can be visualized easy. The hypothetic mechanism of joint action of physical and chemical interaction is proposed. The first stage is diffusion, but after penetration of macromolecules of one polymer into another the trans-esterification starts and transition between these two processes can be estimated from kinetics of movement of iso-concentration planes.

Method of optical interferometry is not so often used for describing phase equilibria in solutions of stiff-chain polymers forming in solutions LC phase. In this part was shown its application for several solutions the same polymer in different solvents. We were wonder to informativity of interferometry for such systems. The main portion of equilibria lines in solutions were detected by interferometric method. Of course, for LC systems method of polarizing microscopy has to be obligatory supporting method for confirmation of the LC phase existence. In addition, viscometry seemed to be very useful again not only for description of features of these systems at flow, but also for additional detection of LC transitions as well as for formation of CS phase.

As opposed to diagrams obtained for other systems, for which two types of phase equilibrium are located in different concentration regions, in the case of aqueous systems, all transitions, including formation of crystal solvates, take place below the convex part of the binodal, that is, at the same concentrations but at different temperatures. This result indicates that the activity of water with respect to formation of hetero H-bonds tends to increase with a decrease in temperature.

7. Acknowledgement

Authors appreciate the partial financial support from Russian Foundation for Basic Research (grants 08-03-12035, 09-03-00285a, 10-03-00994a, 11-03-12042-ofi-m).

8. References

Allan, F.M., (1990). Barton Handbook of Polymer-Liquid Interaction Parameters and Solubility Parameters. CRC Press LLC, ISBN: 0095-9782, Florida

Avdeev, N.N. (1990). Phase equilibria and diffusion in oligomer-polymer systems. PhD Thesis. Moscow, Institute of Physical Chemistry

Avdeev, N.N., Makarova, V.V., Kotomin, S.B., Kulichikhin, V.G. & Kuleznev, V.N., (2001). Compatibility and viscosity of polycarbonate- poly(methyl methacryalate) blends, Polymer Science, Ser. A, Vol. 43, №2, 112-117, ISSN: 1560 0904

Borovskii, I.; Gurov, K.; Marchukova, I. & Ugaste, Yu. (1973). Processes of interdiffusion in alloys, Nauka, Moscow (in Russian) ISBN:

Butzbach, G. D. & Wendorff, J. (1991). Polycarbonate-poly(methyl methacrylate) blends: the role of molecular interactions on miscibility and antiplasticization, Polymer, Vol. 32, № 7, 1155-1159, ISSN: 0032-3861

Chalykh, A.E. & Gerasimov, V.K. (2004). Phase equilibria and phase structure of polymer blends. Uspekhi khimii, Vol.73, №1, 63- (in Russian), ISSN: 0042-1308

Chalykh, A.E., Gerasimov, V.K. & Mikhailov, Yu.M. (1998). Diagrams of Phase State of Polymer Systems, Yanus-K, Moscow (in Russian), ISBN: 5-8037-0016-9

Chalykh, A.E., Aliyev, A.D. & Rubtsov, (1990). Electronic Probe Microanalysis in Polymer Investigation, Nauka, Moscow (in Russian)

Chiou, J. S., Barlow, J. W., & Paul, D. R. (1987). Miscibility of bisphenol-A polycarbonate with poly(methyl ethacrylate). J. Polym. Sci., Polym. Phys. Ed., Vol. 25, № 7, 1459-1471, ISSN 0887-6266

Collyer, A, (1996). Introduction to Liquid Crystal Polymers. In Rheology and Processing of Liquid Crystal Polymers, Acierno, D., Collyer A.A. (Eds.). Chapman&Hall, ISBN: 0412596407

Crank, J. (1956). The Mathematics of Diffusion. Clarendon Press, ISBN: 0198534116, Oxford

Debier, D., Devaus, J. & Legras, R. (1994). Blends of bisphenol a polycarbonate and acrylic polymers. I. A chemical reaction mechanism. J. Polym. Sci., Polym. Phys. Ed., Vol. 33, № 3, 407-414, ISSN: 0887-624X

Eastmond, G.C. & Kotomin, S.V. (1994). Spectroscopic monitoring of phase separation in polymer blends. Polymer, Vol. 35, № 4, 882-884, ISSN: 0032-3861

Fischer, H., Murray, M., Keller, A. & Odell J.A. (1995). On the Phase Diagram of the System Hydroxypropylcellulose-Water. J. Mater. Sci. Vol. 30, № 10, 4623-4627, ISSN: 0022-2461

Flory, P.J. (1956). Phase diagrams of rod-like polymer solutions, Proc. Roy. Soc., Ser. A., Vol. 234, № 1, 73-81

Gallyamov, M.O., (2009). PhD Thesis, Moscow State University

Geller, B.E., Geller, A.A. & Chirtulov, V.G. (1996). Practicum on Physical Chemistry of Fiber-forming Polymers. Khimiya, ISBN 5-7245-0944-X, Moscow

Guido, S. (1995). Macromolecules, Vol. 28, № 13, 4530-4539, ISSN: 0024-9297

Guo, M. & Brittain, W. J. (1998). Structure and Properties of Naphthalene-Containing Polyesters. 4. New Insight into the Relationship of Transesterification and Miscibility. Macromolecules, Vol. 31, 7166-7172, ISSN: 0024-9297

Ihm, D. W.S., Park, Y., Chang, C. G., et al., (1996). J. Polym.Sci., Part A: Polym. Chem., Vol. 34, 2841- 2849, ISSN: 0887-6266

Iovleva, M.M. & Papkov, S.P. (1982). Polymer Crystal Solvates. Polymer Science, Ser. A, Vol.24, № 2, 233-237, ISSN:1560-0904

Iovleva, M.M., Smirnova, V.N. & Prozorova, G.E. (1989). Upper and Lower Critical Solution Temperatures for Solutions of Stiff-Chain Aromatic Polyamides. Polymer Science, Ser. B., Vol. 31, №11, 861- 865, ISSN: 1560-0904

Kim, C. K. & Paul, D. R. (1992). Effects of polycarbonate molecular structure on the miscibility with other polymers. Macromolecules., Vol. 25, № 12, P. 3097- 3105, ISSN 0024-9297

Klenin, V.I. (1995). Thermodynamics of systems with flexible-chain polymers, Saratov University Publ., ISBN: 5-292-01936-4, Saratov (in Russian)

Koningsveld, R. & Kleintjens, L.A. (1977). Thermodynamics of polymer mixtures. J. Polym. Sci., Polym. Symp. №61, 221-249. ISSN: 0887-6258

Kolotyrkin, Ya.M. (1974). Electrochemistry of Metals in non-Aqueous Solutions, Mir, Moscow

Kotomin, S.V. & Kulichikhin,V.G. (1996). Application of Squeezing Flow Technique for Viscosity Measurements of Polymer Liquids. Polymer Science, Ser.B, Vol.38, №12, 2079- ISSN: 1560-0904

Kuleznev, V.N. (1980). Polymer Blends, Khimiya, Moscow (in Russian), ISBN: 5-7245-0944-X

Kuleznev, V.N. & Kandyrin L.B. (2000). Structure-Rheology Behavior of Binary Polymer Blends in Vicinity of Phase Decomposition, Polymer Science, Ser.B. Vol. 42, №4, 711-714. ISSN:1560-0904

Kuleznev, V.N. (1987). Features of Structure and Properties of Polymer Blends near Decomposition Region, Colloid. J., Vol. 49, №5, 881-890, ISSN 0023-2912

Kulichikhin, V.G. & Golova, L.K., (1985). Liquid Crystal State in Solutions of Cellulose Derivatives. Khimiya Drevesiny (Chemistry of Wood), № 3, 9-17, ISSN: 0201-7474

Kulichikhin, V.G. (1989). Rheology, Phase Equilibria and Processing of Lyotropic Liquid Crystalline Polymers. Mol. Cryst. Liq. Cryst., Vol. 169, 51-81, ISSN 0026-8941

Kulichikhin, V.G., Avdeev, N.N., Semakov, A.V., Plate, N.A. (1994). Interphase Interaction in Heterogeneous Polymer Systems Containing Liquid-Crystal Component. Russian Chemical Bulletin, Vol. 43, №11, 1753-1769, ISSN 0002-3353

Kulichikhin V.G., Makarova V.V., Tolstykh M.Yu., Basil'ev G.B., (2010), Phase Equilibria in Solutions of Cellulose Derivatives and Rheological Properties of Solutions in Various Phase States. Polymer Science, A, V.52, №11, P.2001-2013. ISSN: 1560-0904

Kulichikhin, V.G., Kudryavtsev, G.I. & Papkov S.P. (1982). Rheological Properties of Liquid-Crystalline Polymer Solutions, Int. Polym. Mater, Vol. 9, № 3-4, 239-256, ISSN: 0091-4037

Kulvinder, K., (2000). PhD Thesis, Jai Narain Vyas University, Jodhpur, India

Kyu, T., Ko, C.-C., Lim, D.-S., Smith, S. D. & Noda, I. (1993). Miscibility studies on blends of polycarbonate with syndiotactic polymethyl methacrylate. J. Polym. Sci., Polym. Phys. Ed., Vol. 31, № 11, 1641-1648, ISSN: 0887-6266

Landry, Ch. J. T. & Henrichs, P. M. (1989). The influence of blending on the local motions of polymers. Studies involving polycarbonate, poly(methyl methacrylate), and a polyester. Macromolecules, Vol. 22, № 5, 2157-2166, ISSN: 0024-9297

Larez, V., Crescenzi, V. & Ciferri A. (1995). Phase separation of Rigid Polymers in Poor Solvents. 1. (Hydroxypropyl)cellulose in Water. Macromolecules, Vol. 28, № 15, 5280–5284, ISSN: 0024-9297

Lipatov, Yu.S. (1984). Phase Diagrams of Polymer Solutions and Blends, Naukova Dumka, Kiev (in Russian)

Litmanovich, A. D., Plate, N. A. & Y. V. Kudryavtsev, (2002). Reactions in Polymer Blends: Interchain Effects and Theoretical Problems. Prog. Polym. Sci. Vol. 27, № , 915- ISSN: 0079-6700

Makarova, V.V. (2007). Mass-transfer in vicinity of interfaces in polymer-polymer systems. PhD Thesis. Moscow, A.V.Topchiev Institute of Petrochemical Synthesis

Makarova, V. V., Avdeev, N. N., Strelets, B. Kh., et al., (2005). Approaches to Chemical and Physical Modification of Polyethylenterephthalate. Polymer Science, Ser. A, Vol. 47, №7, 1140-1152 , ISSN:1560-0904

Makarova, V.V., Gerasimov, V.K., Tereshin, A.K., Chalykh, A.E. & Kulichikhin, V.G. (2007). Diffusion and Phase Behavior of Hydroxypropylcellulose-Polyethylenglycole System. Polymer Science, Ser. A, Vol. 49, №4, 663-673, ISSN: 1560-0904

Mali, C.S., Chavan, S.D., Kanse, K.S., Kumbharkhane, A.C. & Mehrotra S.C. (2007). Indian J. Pure Appl. Phys., Vol. 45, № 5, 476-480, ISSN: 0019-5596

Malkin, A.Ya, Chalykh, A.E., Kovriga, V.V., Askadsky, A.A. (1983). Experimental Methods of Polymer Rhysics: Measurement of Mechanical Properties, Viscosity and Diffusion, Prentice Hall, New Jersey, ISBN-10: 0132954850

Malkin, A., & Chalykh, A. (1979). Diffusion and viscosity of polymers. Methods of measurement, Khimiya, Moscow (in Russian) ISBN: 5-7245-0944-X

Nishimoto, M., Keskkula, H. & Paul, D. R. (1991). Role of slow phase separation in assessing the equilibrium phase behaviour of PC PMMA blends. Polymer, Vol. 32, № 2, 272-278, ISSN: 0032-3861

Okamoto, M. & Kotaka, T., (1997). Phase separation and homogenization in poly(ethylene naphthalene-2,6-dicarboxylate)/ poly(ethylene terephthalate) blends. Polymer, Vol. 38, № ,1357-1363, ISSN: 0032-3861

Palatnik, L.C. & Landau, A.I., (1961) Phase Equilibria in Multicomponent Systems, Kharkov University Publ., Kharkov (in Russian)

Papkov, S.P. (1981). Phase equilibrium in polymer-solvent system, Khimiya, Moscow (in Russian), ISBN: 5-7245-0944-X

Papkov, S.P. (1971) Physical-chemical Fundamentals of Polymer Solutions Processing, Khimiya, Moscow (in Russian), ISBN: 5-7245-0944-X

Papkov, S.P., Kulichikhin, V.G., Kalmykova, V.D. & Malkin A.Ya. (1974). Rheological properties of anisotropic poly(para-benzamide) solutions, J. Polym. Sci., Polym. Phys. Ed. Vol. 12. № 9. 1753-1759, ISSN: 0887-6266

Papkov, S.P. Phase Equlibria in Polymer Systems Containing a Liquid-Crystalline Phase. (1992). In: Liquid-Crystal Polymers. Plate, N.A. (Ed.) 39-70. Plenum Press, ISBN: 0-306-44219-1, New York

Papkov, S.P., Iovleva, M.M., Smirnova, V.N. & Volokhina, A. V. (1986). Experimental Analysis of the Phase equilibrium in the System Stiff-Chain Polyamide - Sulfuric Acid. Polymer Science, Ser. B, Vol. 28, № 9, 677-681,ISSN:1560-0904

Papkov, S.P., Kulichikhin, V.G. (1977). Liquid Crystalline State in Polymers, Khimiya, Moscow (in Russian), ISBN: 5-7245-0944-X

Rabeony, M., Hseih, D. T., Garner, R. T. & Peiffer, D. G. (1992). Nearly monodisperse spherical domain morphology in polycarbonate/poly (methylmethacrylate) blends following spinodal decomposition coupled with an interchange reaction. J. Chem. Phys., Vol. 97, № 6, 4505-4512, ISSN: 0021-9606

Scott, R. L. (1949). The Thermodynamics of High Polymer Solutions. V. Phase Equilibria in the Ternary System: Polymer 1-Polymer 2-Solvent. J. Chem. Phys. ,Vol. 17, № 2, 279-285, ISSN: 0021-9606

Shimamura, K., White, J.L. & Fellers J.F. (1981). Hydroxypropylcellulose, a Thermotropic Liquid Crystal: Characteristics and Structure Development in Continuous Extrusion and Melt Spinning. J. Appl.Polym. Sci. Vol. 26, № 7, 2165-2180. ISSN: 0021-8995

Stewart, M. S., Cox, A. J. & Naylor, D. M. (1993). Reactive Processing of Poly(ethylene 2,6-naphthalene dicarboxylate)/Poly(ethylene terephthalate) blends. Polymer, Vol.34, 4060-4067, ISSN: 0032-3861

Tager, A.A. (1978). Physical Chemistry of Polymers, Khimiya, Moscow (in Russian), ISBN: 5-7245-0944-X

Tereshin, A.K., Vasil'eva, O.V., Avdeev, N.N., Bondarenko, G.N. & Kulichikhin, V.G. (2000). Effects of Interphase Interactions on Rheological Properties of PET-LC Polyester Blends. Polymer Science, Ser. A, Vol.42, №6, 1009-1015, ISSN:1560-0904

Tolstykh, M., Makarova, V., Semakov, A. & Kulichikhin, V. (2010). Polymer Science, Ser. A, Vol. 52, № 2, 228-234, ISSN: 1560-0904

Tompa, H. (1956). Polymer Solutions. Academic Press, ASIN: B0006AUTUA, New York

Transreactions in Condensation Polymers, (1999). Ed. By Fakirov, S.,Wiley, ISBN: 9783527297900, Weinheim

Uusi-Penttila, M.S., Richards, R.J., Torgerson, B.A. & Berglund, K.A. (1997). Ind. Eng. Chem. Res., Vol. 36, № 2, 510-512, ISSN: 0888-5885

Vshivkov, C.A., Adamova, L.V., Rusinova, E.V., Safronov A.P., Dreval, V.E. & Galyas A.G., (2007). Thermodynamics of Liquid-Crystalline Solutions of Hydroxypropyl-cellulose in Water and Ethanol. Polymer Science, Ser. A, Vol. 49, № 5, 578-583, ISSN: 1560-0904

Woo, E. M. & Su, C. C. (1996). Phase micro-heterogeneity in bisphenol-A polycarbonate/ poly(methyl methacrylate) blends. Polymer, Vol. 37, № 18, 4111-4118, ISSN: 0032-3861

Woolley, E.M. & George, R.E. (1974). Ionization constants for water and for very weak organic acids in aqueous organic mixtures. J. Solut. Chem., Vol. 3. № 2. 119-126, ISSN: 0095-9782

21

Interferometry in Wireless Sensor Networks

Sandor Szilvasi[1], Peter Volgyesi[1], Janos Sallai[1],
Akos Ledeczi[1] and Miklos Maroti[2]
[1]*Institute for Software Integrated Systems, Vanderbilt University*
[2]*Bolyai Institute, University of Szeged*
[1]*USA*
[2]*Hungary*

1. Introduction

Wireless Sensor Networks (WSN) consist of small, low-cost, resource-constrained embedded computers equipped with low-power radios and various sensors. When deployed, they form an ad-hoc wireless network and sense the environment cooperatively. Application areas of WSNs include environmental monitoring, healthcare, structural monitoring and the military among others. Many WSN applications rely on the location of sensor nodes in order to make consistent spatio-temporal observations. Sensor node localization, therefore, has been an active research area for the past decade.

Consequently, a wide variety of localization methods has been proposed for WSNs. Among the most popular are acoustic and radio signal-based approaches. The limited range of acoustic methods make them impractical for anything but small scale deployments. The best known radio-based method is the Global Positioning System (GPS). For outdoor applications that require absolute localization accuracy in the order of meters, GPS is generally a good solution. However, its price and power requirement restrict its applicability to high-end WSNs. In indoor and urban environments, heavily affected by multipath effects, Ultra Wide band (UWB) methods offer high accuracy by measuring the Time of Flight (TOF) or Time Difference of Arrival (TDOA) of a short pulse-waveform to deduce ranges Tuchler et al. (2005), Fontana et al. (2003). The UWB-based techniques typically achieve sub-meter ranging error where the accuracy depends mainly on the available bandwidth, which in turn is constrained by emission regulations. UWB has not become popular in WSNs primarily due to its high cost.

Techniques based on received radio signal strength (RSS) gained significant popularity in WSNs as they require no specialized hardware; they use the radio chip readily available on WSN nodes. RSS-based ranging measurements are also simple as they typically consist of averaging the value of the radio chip RSS signal. The operating range is limited to the communication range of the radio with a precision of a few meters Taubenheim et al. (2005).

This research was partially supported by ARO MURI grant W911NF-06-1-0076 and the TÁMOP-4.2.2/08/1/2008-0008 program of the Hungarian National Development Agency.

However, the measurement is highly environment dependent. Achieving reasonable accuracy requires extensive calibration prior to ranging. Uncalibrated RSS ranging can exhibit 10-20% errors.

A novel ranging method based on radio interferometry (RI) for static WSN node localization was introduced in Maróti et al. (2005). This RI ranging method measures the phase of the radio signals to obtain information on the relative distance between the sensor nodes. It achieves sub-meter localization accuracy and long range simultaneously. As this RI approach also relies solely on the radio chip readily available on popular, low-cost commercial off-the-shelf (COTS) sensor nodes, it became an alternative to RSS-based methods and, thus, garnered considerable attention in the WSN community. Since its original debut, improvements in several directions have been proposed. Dil & Havinga (2011) drops the requirement that a low frequency interference signal need to be synthesized through tuning and copes with the problem taking a stochastic approach. Contrary, Amundson et al. (2010) assumes a priori knowledge of certain node locations and transforms the range estimation into a bearing estimation problem. RI phase measurement based techniques generally outperform other currently available methods that use no specialized hardware, such as RSS-based methods. However, they also require more sophisticated ranging measurements and computationally more intensive fusion of the ranging data as shown in Figure 1. Furthermore, as multipath propagation heavily impacts the phase measurement accuracy, their performance degrades significantly in indoor environments.

	Signal strength based	**Signal phase based**
Accuracy	Low (beyond a few meters)	High (up to hundreds of meters)
Ranging measurement complexity	Very low	High
Localization complexity	Low, trilateration	High
Additional hardware required	No	No
Bandwidth requirement	Low	Medium

Table 1. Comparison of received signal strength (RSS) and signal phase based localization

While the above techniques measure either the signal strength or phase, radio-interferomentric methods exploiting the Doppler effect deduce information from the signal frequency. The measured frequency change of the radio signal produced by a moving transmitter is a function of its velocity relative to the receiver. In a WSN with a few static nodes with known locations, this relative velocity can be used to track mobile transmitters Kusý et al. (2007). Inverting the roles, the Doppler-shift technique is used for localization Lédeczi et al. (2008): the technique replaces the static receivers with rotating transmitters around a known point with a known angular speed. The receiver nodes, in turn, can calculate their bearings from these anchors. RI is applied in both cases to allow precise measurement of the frequency change on resource constrained WSN nodes. Unfortunately, these methods still suffer from the adverse effect of multipath propagation. Furthermore, the rotation of the receivers calls for special hardware, usually unavailable on WSN nodes.

This chapter describes the baseline approach of radio interferometry based localization and details the many developments that occurred since its introduction. Section 2 lays down the mathematical foundation of the localization method, which is then referenced throughout the rest of the chapter. It also presents the first platform using this approach, called the Radio Interferometric Positioning System (RIPS), along with a discussion of the initial results. Section 3 introduces a method for tracking mobile nodes in WSNs utilizing the phenomena of Doppler effects and radio interferometry. The technique shown in Section 4 exploits the Doppler effect in a different way to determine the location of static nodes in a WSN. It relies on a few distinct nodes with known locations (anchors) to first calculate the bearing to the rest of the nodes and to deduce location information based on the bearing estimates. The technique presented in Section 5 also calculates bearings from anchor nodes, but instead of using Doppler shifts, it uses a special arrangement of RIPS which allows for an enhanced, distributed method of node localization.

2. Radio interferometric positioning

The basic idea behind interferometric localization is to use the phase information of a radio signal to measure distance. In practice, however, this is hard to achieve, because it would require (a) nanosecond-precision timing, (b) precise control over the phase of the transmitted signal, and (c) precise phase detection on the receiver. COTS radio transceivers commonly used in sensor nodes do not offer these features. The novelty of the method presented in this chapter is to use a transmitter pair and a receiver pair to eliminate the unknown initial phase of the respective local oscillators (LO), as well well as to allow for a low-speed ADC to process the incoming signal on the receiver.

The interferometric approach thus utilizes a pair of wireless sensor nodes to generate interfering radio signals by transmitting pure sinusoids at slightly different frequencies. The envelope of this composite signal has a beat frequency equal to the frequency difference of the sinusoids, as illustrated in Figure 2. It is the phase of the envelope signal that is of most interest, as it carries location related information. When measured at two receivers at the same time, the instantaneous phase difference (also referred to as the *phase offset*) of the respective interference signal envelopes is directly related to the carrier wavelength and the relative distance between the transmitters and receivers. Working with the envelope signal instead of the "raw" carrier signal has several advantages. On one hand, the envelope frequency depends on only the frequency difference of the transmitted sinusoids. This means that the received envelope frequency is the same at all receivers, and it is independent of the frequency and initial phase of the receivers' local oscillators. On the other hand, the envelope frequency can be precisely controlled (in short term), thus it is adjustable to meet the available time synchronization accuracy, limited sampling rate and processing capability of the wireless sensor nodes.

The rest of the section discusses the theory of radio interferometric localization in detail.

2.1 Idealistic phase-based localization

Let us consider, hypothetically, that the transmitted signal's phase is controllable, and the received phase can be detected accurately.

These sinusoids we refer to as carrier signals, though they are not modulated.

Here, the measured phase at the receiver as a function of the distance from the receiver is

$$\varphi = 2\pi \left(\frac{d}{\lambda}\right) \mod (2\pi), \tag{1}$$

where φ_i is the measured phase, d is the receiver's distance from the transmitter and $\lambda_i = c/f$ is the wavelength of the RF signal. A representative received phase map is is shown in Figure 1.

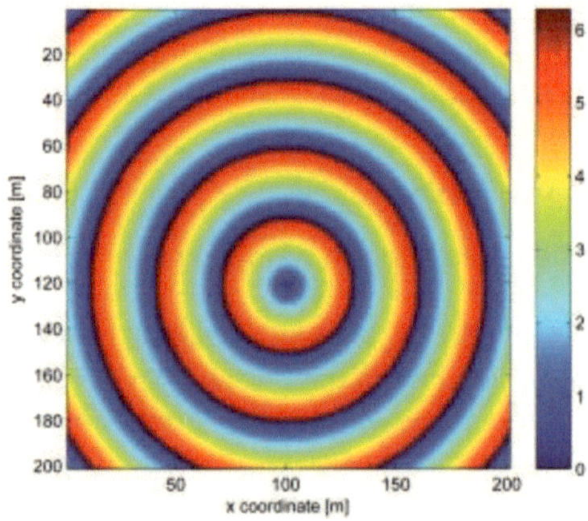

Fig. 1. Phase map of an unmodulated carrier signals assuming controllable transmitter phase and accurate phase detection at the receiver.

From here, we can express the distance as follows.

$$d_i = \frac{\varphi_i}{2\pi}\lambda_i + k \cdot \lambda_i \qquad k \in \mathbb{Z} \tag{2}$$

This means that the receiver is capable of measuring its distance from the transmitter with a mod (λ_i) ambiguity. By repeating the measurement at different wavelengths, it is possible to determine the unambiguous distance from the transmitter within the effective radio range of the transmitter, thereby restricting the possible receiver positions to a single circle around the receiver. Then, this is repeated for several other transmitters to gain multiple circles intersecting at a single point, at the position of the receiver.

2.2 Interferometric localization

In WSNs, however, the nodes have independent LOs and neither their initial phase, nor their relative phase to each other can be assumed to be known. Maróti et al. (2005) suggests the use of transmitter and receiver pairs along with radio interferometry to deduct location information from the phase of propagating radio waves without assumptions on the LO initial phases.

2.2.1 Phase measurement

A ranging measurement starts with two transmitters emitting high-frequency pure sinusoid radio signals with slightly different frequencies f_1 and f_2. Due to variations of the LO crystals, precise tuning of the frequency difference $f_1 - f_2$ is required, which is achieved through a *calibration* process. Once the calibration is done, the received composite signal at a receiver node, see Figure 2, takes the following form:

$$s(t) = a_1 \cos(2\pi f_1 t + \varphi_1) + a_2 \cos(2\pi f_2 t + \varphi_2). \tag{3}$$

Interestingly, the envelope of $s(t)$ preserves important phase information such as the $\varphi_1 - \varphi_2$

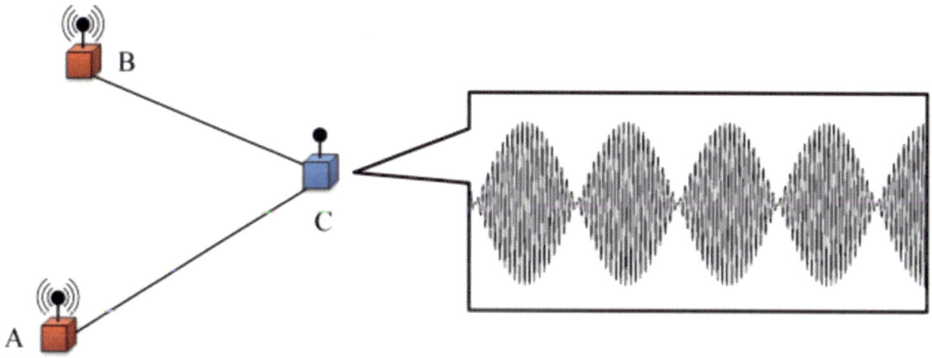

Fig. 2. The radio interferometric signal generated by two independent transmitters received by a third node

phase offset the two carriers. To show this, let the carrier signal be down-mixed to the intermediate frequency f_{IF}. Denoting the mean of the two transmitted frequencies (f_1 and f_2) with $\delta = (f_1 - f_2)/2$ and assuming $f_2 < f_1, \delta \ll f_2$ and that the down-mixing introduces zero phase shift, the intermediate frequency signal can be written as

$$s_{\text{IF}}(t) = a_1 \cos\left(2\pi(f_{\text{IF}} + \delta)t + \varphi_1\right) + a_2 \cos\left(2\pi(f_{\text{IF}} - \delta)t + \varphi_2\right). \tag{4}$$

The envelope of this signal is accessed by first calculating its power:

$$\begin{aligned} s_{\text{IF}}^2(t) &= a_1^2 \cos^2\left(2\pi(f_{\text{IF}} + \delta)t + \varphi_1\right) + a_2^2 \cos^2\left(2\pi(f_{\text{IF}} - \delta)t + \varphi_2\right) \\ &\quad + 2a_1 a_2 \cos\left(2\pi(f_{\text{IF}} + \delta)t + \varphi_1\right) \cos\left(2\pi(f_{\text{IF}} - \delta)t + \varphi_2\right). \end{aligned} \tag{5}$$

Applying the following trigonometric identities

$$\cos^2(x) = \frac{1}{2} + \frac{\cos(2x)}{2} \tag{6}$$

$$\cos(x)\cos(y) = \frac{\cos(x+y)}{2} + \frac{\cos(x-y)}{2}, \tag{7}$$

the square of $s_{\text{IF}}(t)$ can be rewritten as

$$\begin{aligned} s_{\text{IF}}^2(t) &= \frac{1}{2}(a_1^2 + a_2^2) + \frac{a_1^2}{2} \cos\left(4\pi(f_{\text{IF}} + \delta)t + 2\varphi_1\right) + \frac{a_2^2}{2} \cos\left(4\pi(f_{\text{IF}} - \delta)t + 2\varphi_2\right) \\ &\quad + a_1 a_2 \cos\left(4\pi f_{\text{IF}} t + \varphi_1 + \varphi_2\right) + a_1 a_2 \cos\left(4\pi \delta t + \varphi_1 - \varphi_2\right), \end{aligned} \tag{8}$$

where the components are either at zero frequency (DC), $2(f_{\text{IF}} \pm \delta)$, $2f_{\text{IF}}$ or 2δ. Removing the DC, and the $2f_{\text{IF}} \pm \delta$ and $2f_{\text{IF}}$ double-IF frequency components by appropriate filtering we obtain the envelope signal

$$r(t) = a_1 a_2 \cos\left(2\pi(2\delta)t + \varphi_1 - \varphi_2\right), \tag{9}$$

where the frequency of $r(t)$ is $2\delta = f_1 - f_2$ and the $\varphi_1 - \varphi_2$ phase corresponds to the phase offset of the two carrier signals. That is, by measuring the phase of the the envelope signal, $r(t)$, the phase difference of the two carriers can be obtained *independently of the receiver LO initial phase.*

2.2.2 Q-range estimation

The carrier phase offset in the form of Equation 9 is unsuitable to deduct location information directly. Interstingly, however, when it is measured at two different nodes in a time synchronized manner, it allows to gain information on the relative distances between the transmitters and receivers.

To show this, consider Figure 3, where transmitters A and B generate the interfering carrier signals and receivers C and D extract the envelope signals $r_C(t)$ and $r_D(t)$, respectively.

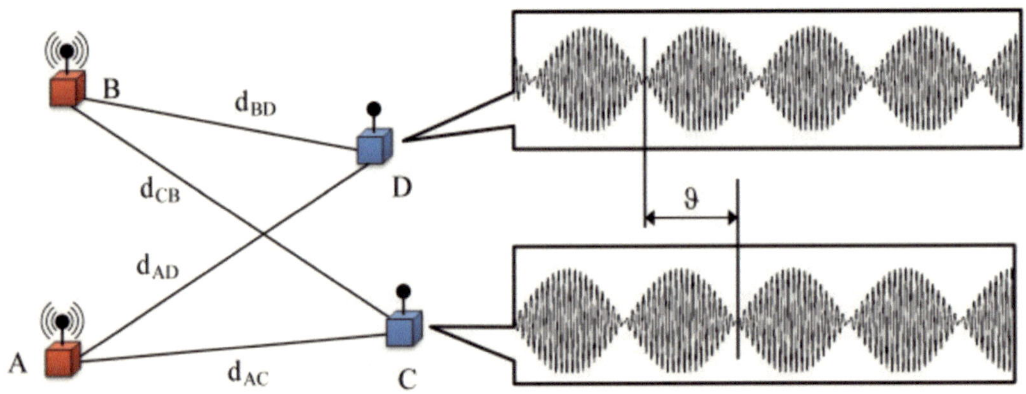

Fig. 3. Radio interferometric ranging performed by a pair of transmitters and a pair of receivers.

Let Y be a receiver node, and let A and B be the two transmitters. Let t_A and t_B be the time when node A and B start to transmit, respectively. Furthermore, let a_{AY} and a_{BY} the amplitude of the attenuated signal as received at node Y from node A and B, respectively. The received composite signal at node Y is

$$\begin{aligned} s_Y(t) &= a_{AY}\cos\left(2\pi f_A(t - t_A - d_{AY}/c)\right) \\ &\quad + a_{BY}\cos\left(2\pi f_B(t - t_B - d_{BY}/c)\right) \\ &= a_{AY}\cos\left(2\pi f_A t - 2\pi f_A(t_A + d_{AY}/c)\right) \\ &\quad + a_{BY}\cos\left(2\pi f_B t - 2\pi f_B(t_B + d_{BY}/c)\right) \end{aligned} \tag{10}$$

after sufficient amount of time, that is when t is greater than $t_A + d_{AY}/c$ and $t_B + d_{BY}/c$. Using Equation 9, the absolute phase offset of the envelope signal $r_Y(t)$ is

$$\vartheta_Y = -2\pi f_A(t_A + d_{AY}/c) + 2\pi f_B(t_B + d_{BY}/c). \tag{11}$$

Now consider the two receivers C and D. The relative phase offset of $r_C(t)$ and $r_D(t)$ is

$$\begin{aligned}\vartheta_C - \vartheta_D &= -2\pi f_A(t_A + d_{AC}/c) + 2\pi f_B(t_B + d_{BC}/c) \\ &\quad + 2\pi f_A(t_A + d_{AD}/c) - 2\pi f_B(t_B + d_{BD}/c) \\ &= 2\pi f_A/c \cdot (d_{AD} - d_{AC}) + 2\pi f_B/c \cdot (d_{BC} - d_{BD}).\end{aligned} \tag{12}$$

From this the relative phase offset of $r_C(t)$ and $r_D(t)$ is

$$2\pi\left(\frac{d_{AD} - d_{AC}}{c/f_A} + \frac{d_{BC} - d_{BD}}{c/f_B}\right) \pmod{2\pi}. \tag{13}$$

Due to the limited range of wireless sensor nodes and their high carrier frequency relative to the envelope frequency, the above formula of the measured relative phase offset can be simplified. Consider Figure 3 again, where nodes A and B transmit pure sine waves at two close frequencies $f_A > f_B$, and two other nodes C and D measure the envelope of the signal. Using the notation $\delta = (f_A - f_B)/2$, the relative phase offset of $r_C(t)$ and $r_D(t)$ is

$$\vartheta_C - \vartheta_D = 2\pi\frac{d_{AD} - d_{AC} + d_{BC} - d_{BD}}{c/f} + 2\pi\frac{d_{AD} - d_{AC} - d_{BC} + d_{BD}}{c/\delta} \pmod{2\pi}.$$

Assuming that $f_A - f_B < 2$ kHz and $d_{AC}, d_{AD}, d_{BC}, d_{BD} \leq 1$ km, it follows that $c/\delta \geq 300$ km and thus the second term can be neglected. Now, if for any four nodes A, B, C and D we define a quantity called the *q-range* as

$$d_{ABCD} = d_{AD} - d_{BD} + d_{BC} - d_{AC}, \tag{14}$$

then for any frequency f the relative phase offset simplifies to

$$\vartheta_{ABCD} = 2\pi\frac{d_{ABCD}}{c/f} \pmod{2\pi}. \tag{15}$$

Thus, if adequately precise time synchronization is available, the receiver can effectively measure ϑ_{ABCD} and calculate d_{ABCD} with a $\lambda = c/f$ wave length ambiguity.

Similarly to Figure 1, the phase ambiguity corresponds to equi-phase-offset curves (or surfaces in three dimension) in Figure 4. Note, that while Figure 1 plots the phase of the signal transmitted by a single transmitter, Figure 4 depicts the phase offset of the interference signal transmitted by a transmitter pair relative to that of a reference receiver. That is, the equi-phase-offset curves with respect to a reference receiver are multiple hyperbolas in Figure 4. If the phase offset measurement is repeated at i different carrier wavelengths λ_i for the same set of nodes A, B, C and D, we get the following set of equations by reorganizing Equation 15:

$$d_{ABCD} = \frac{\vartheta_{ABCD}}{2\pi}\lambda + k\lambda \quad (k \in \mathbb{I}) \tag{16}$$

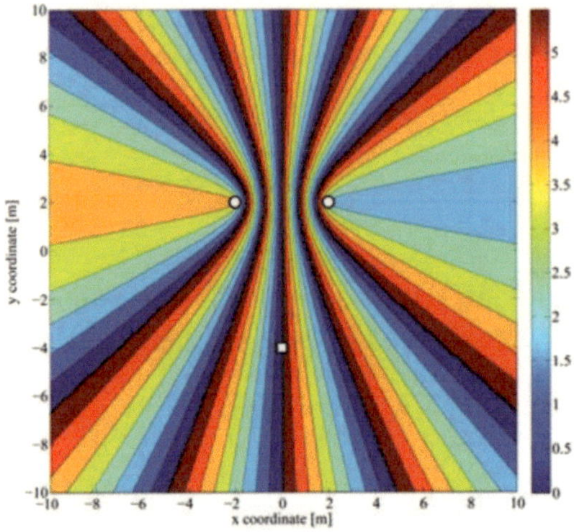

Fig. 4. Map of phase difference with respect to a reference receiver at (0,-4), when the interference signal is created by two independent transmitters located at (-2,0) and (2,0), respectively.

To compute the q-range estimate d_{ABCD}, we have to solve not only for d_{ABCD}, but also for integers k_i. Clearly, no closed form solution exists, given the integer ambiguity in the equations. However, direct search techniques are possible, since in practice the q-range is constrained by the effective radio communications range, thereby restricting the search space. In practice, the measured q-range is always between $-2r$ and $2r$, r being the radio's maximum transmission distance. Kusý et al. (2006) describe a search heuristic that finds the q-range by minimizing the error of the q-range estimate with respect to the available phase offset measurements:

$$\hat{d}_{ABCD} = \underset{\hat{d}_{ABCD} \in [-2r,2r]}{\arg\min} \sum_{i=1}^{N} \left| \frac{\varphi_i}{2\pi}\lambda_i + k_i\lambda_i - \hat{d}_{ABCD} \right| \tag{17}$$

2.2.3 Position estimation

Once the necessary amount of q-ranges is obtained from the radio-interferometric phase measurements, the position of the individual nodes is determined by finding the optimum of a set of constrained non-linear equations. For this optimization problem Maróti et al. (2005) suggests to use least-squares (LS) error fitting of the j q-range measurements:

$$(\hat{x}, \hat{y}) = \underset{\hat{x},\hat{y}}{\arg\min} \sum_{j=1}^{M} (d_{ABCD,j} - \hat{d}_{ABCD,j})^2. \tag{18}$$

Maróti et al. (2005) describe a genetic search algorithm to localize the network. Dil & Havinga (2011) observed that LS approach is highly suspectible to outliers, and propose to use the Least Absolute Deviation (LAD) instead:

$$(\hat{x}, \hat{y}) = \underset{\hat{x},\hat{y}}{\arg\min} \sum_{j=1}^{M} |d_{ABCD,j} - \hat{d}_{ABCD,j}|. \tag{19}$$

2.3 Platform description and results

The first interferometry based localization system, the Radio Interferometric Localization System, was presented in Maróti et al. (2005). The RIPS uses the popular MICA2 wireless sensor nodes equipped with CC1000 radio chip Texas Instruments (2007a). The fine grained LO tuning capability of the CC1000 allows to precisely set the interference frequency to 300-700 Hz in the 433 MHz band. On the downside, the sampling rate of the interference signal is limited to 9 kHz by the ADC found on the MICA2.

Nonetheless, this relatively low sampling rate proves to be sufficient for accurate phase offset measurements when accompanied with precise, microsecond order time synchronization among the independent sensor nodes. In RIPS, two transmitter nodes generate the interference signal and the receiver nodes measure its envelope phase in a time-synchronized manner, at particular time instant. This is repeated on 11 different radio channels in the 400 MHz to 430 MHz range. The measured phase values are sent to a PC using WSN communication infrastructure. (Notice that the CC1000 radio chip is used for both localization and communication.) For a given pair of transmitters, the PC calculates the phase offsets of all possible pairs of receivers. As the number of participating recivers, and thus the phase measurements, is limited only by the communication range, the PC calculates the q-ranges for the same transmitter pair and all possible combination of the receivers with valid phase measurements. The whole process is then repeated with different combination of transmitters to acquire a large number of q-ranges. The q-ranges are then used as input for a genetic algorithm (GA) to optimize the heavily non-linear localization problem expressed by Equation 18. The output of the GA is the relative coordinates of the sensor nodes, assuming no a priori knowledge on any of the node locations.

The 16-node experiment on a 18×18 m outdoors area showed that RIPS was able to localize the nodes with an average error less than 5 cm. For that, 240 different transmitter combinations were used and the overall localization process took approximately 80 minutes.

3. Doppler shift-based tracking

Tracking of mobile sensor nodes in a WSN has been another active area of research in the past years. It is typically more challenging that localizing static nodes as the latency needs to be low enough to keep up with the mobility. On the other hand, estimating the current location of a node is aided by the already obtained estimate of the previous location. Similarly to static localization, a number of different methods have been proposed in the literature. When the most natural solution, GPS is not available due to cost or power constraints, interferometry offers an alternative.

The original idea of the Doppler shift-based tracking presented in Kusý et al. (2007) assumes a set of stationary infrastructure nodes, *anchor nodes*, and another set of *mobile nodes* that need to be tracked – both with known locations initially. The tracking system aims for power-efficiency by remaining inactive as long as the tracked nodes stay still. Tracked nodes are responsible for maintaining their location, detecting their movement and invoking the tracking service only when movement is detected. Once a tracked node reports movement, the tracking system becomes active and assists the mobile node to determine its location and velocity.

On the fundamental level, the provided tracking service is based on the phenomenon of Doppler effect or Doppler shift. The movement of a radio signal transmitter relative to a stationary receiver introduces a change in the observed signal frequency as illustrated in Figure 5.

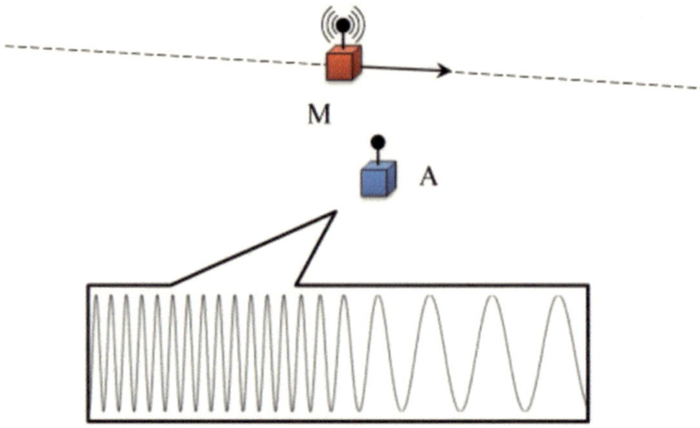

Fig. 5. Illustration of the Doppler frequency shift in the signal introduced by the relative velocity of the moving transmitter to the stationary receiver.

The velocity of the mobile transmitter is directly related to the change in the received signal frequency, thus, by precisely measuring the frequency deviation, the velocity can be estimated. In order to improve measurement accuracy by canceling the receiver side LO uncertainty, an interference signal similar to the one introduced in Section 2 is utilized. When the tracking service is invoked, the mobile node and *one* anchor node transmit their unmodulated carrier at close frequencies. The rest of the anchor nodes measure the frequency change in the interference signal envelope and report the estimates to a central unit. The central unit fuses frequency change information to calculate the location and velocity of the mobile node simultaneously. The derived location and velocity estimates are then fed to a Kalman filter to keep the state information of the nodes up-to-date.

The remainder of the section starts by detailing the interferometry based Doppler shift measurement. That is followed by the description of the location and velocity estimation formulated as a non-linear optimization problem, and the Kalman filter used for tracking. Finally, a prototype implementation on the CC1000 radio chip based MICA2 platform is shown.

3.1 Interferometric tracking

Radio interferometric tracking based on doppler shifts is naturally divided into the *measurement* of doppler shifts and *tracking* phases, that is, the fusion of the doppler shift measurements. Throughout the discussion, nodes are classified either as *anchor nodes* with known positions, or as *mobile nodes* that need to be localized. The interferometry based Doppler shift measurement utilizes an interference signal similar to the one described in Section 2. However, while interferometric ranging estimates the phase of the interference signal, interferometric tracking exploits the frequency shift of the same signal induced by the movement of a transmitting mobile node. The radio interferometric Doppler shift measurement begins with an anchor node and the mobile node to be localized transmitting pure sinusoid waveforms at slightly different frequencies. Consider Figure 6, where A denotes an anchor node and M the mobile node, both acting as transmitters, and the rest as receiver anchor nodes. Let M and A transmit an unmodulated carrier with frequencies f_M, and F_A respectively, such that $f_M > f_A$. The two carriers interfere with each other and create a signal with an envelope frequency of $f_M - f_A$. This interference signal is measured by a number of anchor nodes A_i. As M moves among the anchor nodes A_i, they observe f_M to be Doppler shifted by Δf_M^i, where the value of Δf_M^i depends on the relative speed of M to A_i.

Transmitter A is stationary and its singal is not affected by the doppler shift, thus the measured envelope frequency at receiver node A_i becomes

$$f_i = f_M - f_A + \Delta f_M^i. \tag{20}$$

This allows for the calculation of the Doppler shift measured at node A_i and, consequently, the relative speed of the tracked node.

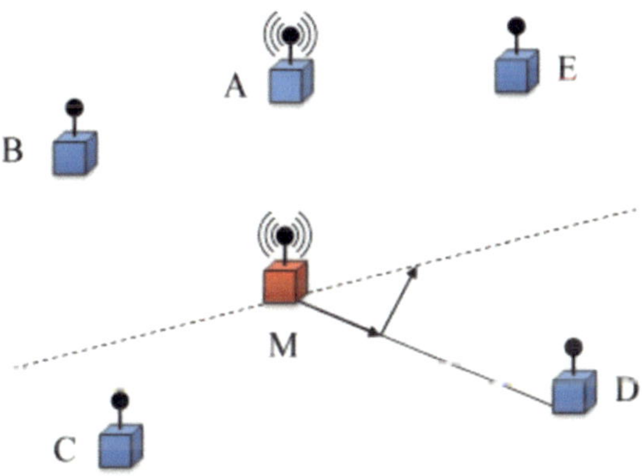

Fig. 6. Doppler shift-based tracking of a mobile node (M) using several stationary nodes (A-E).

Let $\vec{v}$ denote the velocity of the tracked mobile node M and $\vec{u}_i = \overrightarrow{A_iM}/\|A_iM\|$ the unit length vector pointing from anchor node A_i to M. The relative speed of A_i and M can then be defined as the following dot product

$$v_i = \vec{u}_i \cdot \vec{v}, \tag{21}$$

where v_i is a scalar value with positive sign if $\vec{v}$ points away from A_i and negative sign otherwise.

The Doppler equation states that if f is the frequency of the transmitted radio signal, c is the speed of light, and $v \ll c$ is the speed of the source with respect to the observer (with negative sign of v if the source was going towards the observer), then the observed frequency is $f' = (1 - v/c)f$. Therefore,

$$\Delta f = f' - f = -vf/c. \tag{22}$$

Using $\widehat{f} = f_M - f_A$ and $\lambda_M = c/f_M$, the node A_i observes the interference signal with frequency

$$f_i = \widehat{f} - v_i/\lambda_M \tag{23}$$

which can be measured at node A_i. Consequently, if the $\widehat{f}$ difference of the two transmitted frequencies is known, the relative speed of the tracked node M and the anchor node A_i can be calculated.

Note that estimating the frequency difference $\widehat{f}$ with sufficient accuracy becomes a problem when using low-cost radio transceivers due to LO inaccuracies. The use of radio interferometry instead of direct carrier signals partially alleviates this issue by removing the receiver side LO uncertainty. However, as the transmitter side LO may still drift, $\widehat{f}$ is treated as an unknown parameter in the tracking phase.

In the tracking phase the Doppler shifts measured by multiple infrastructure nodes are utilized to calculate the location and the velocity of a tracked node. The tracking problem is modeled as a non-linear optimization problem where the location coordinates (x, y) and the velocity vector $\vec{v} = (v_x, v_y)$ of the tracked node need to be determined. Due to the uncertainty in the transmitted signal frequencies f_M and f_A, the interference frequency $\widehat{f} = f_M - f_A$ is also treated as a parameter to be estimated. Consequently, the full parameter vector $\mathbf{x}$ of the optimization problem is

$$\mathbf{x} = \left(x, y, v_x, v_y, \widehat{f}\right)^{\top}. \tag{24}$$

Assuming that n anchor nodes measure the Doppler shifted radio signal, there are n frequency observations f_i. Therefore, the observation vector $\mathbf{c}$ is defined as

$$\mathbf{c} = (f_1, f_2, \ldots, f_n)^{\top}. \tag{25}$$

The relation of the parameter vector $\mathbf{x}$ and the observation vector $\mathbf{c}$ can be formalized as a function $H : \mathcal{R}^5 \to \mathcal{R}^n$, such that

$$\mathbf{c} = H(\mathbf{x}). \tag{26}$$

The function H is a vector-vector function consisting of n functions $H_i : \mathcal{R}^5 \to \mathcal{R}$, each of them calculating the Doppler shifted interference frequency f_i measured at an infrastructure node A_i. From Equation 23, $H_i(\mathbf{x})$ is defined as

$$H_i(\mathbf{x}) = \widehat{f} - v_i/\lambda_M. \tag{27}$$

The relative speed v_i of the mobile node M to an anchor node A_i can be calculated from the location of the two nodes and the velocity $\vec{v}$ of the tracked node M, as shown in Figure 7. Using the distributive property of the dot product and the identities $\vec{v} = \vec{v}_x + \vec{v}_y$ and $|\vec{u}_i| = 1$ the relative velocity becomes

$$v_i = \vec{u}_i \cdot \vec{v} = \vec{u}_i \cdot \vec{v}_x + \vec{u}_i \cdot \vec{v}_y = |\vec{v}_x| \cos\alpha + |\vec{v}_y| \sin\alpha, \tag{28}$$

where $\sin\alpha$ and $\cos\alpha$ can be calculated from coordinates (x, y) and (x_i, y_i) of nodes M and A_i, respectively. This allows for the calculation of the expected measurements $f_i = H_i(\mathbf{x})$ from

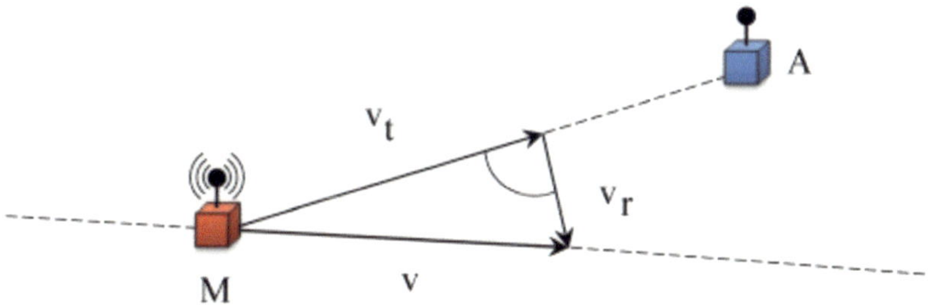

Fig. 7. Vector decomposition of the mobile node velocity $\vec{v} = \vec{v}_t + \vec{v}_r$ as observed by the receiver anchor node A.

the parameter vector $\mathbf{x}$ and the known quantities λ_M, x_i, y_i:

$$H_i(\mathbf{x}) = \widehat{f} - \frac{1}{\lambda_M} \frac{v_x(x_i - x) + v_y(y_i - y)}{\sqrt{(x_i - x)^2 + (y_i - y)^2}} \tag{29}$$

As due to measurement errors, there may exist no $\mathbf{x}$ such that $H(\mathbf{x}) = \mathbf{c}$, the parameters are estimated by finding $\mathbf{x} \in \mathcal{R}^5$ such that $\|H(\mathbf{x}) - \mathbf{c}\|$ is minimized. Note that components of the objective function H are non-linear functions, requiring the use of non-linear optimization methods.

3.2 Platform description and results

The Doppler shift-based tracking uses radio interferometry to enable frequency shift measurements on resource constrained WSN platforms, such as the CC1000 equipped MICA2. The latter platform is capable of synthesizing a 300-700 Hz interference frequency in the 433 MHz band on the transmitter side, and to accurately measure frequency shifts in the envelope of the interference signal despite its limited, 9 kHz, sampling rate. Therefore, identical sensor nodes can be used both as anchor nodes and as mobile nodes.

Kusý et al. (2007) suggests to find the initial locations at deployment time with an accurate and possibly computationally more expensive algorithm, such Maróti et al. (2005), and switch to Doppler shift-based tracking afterwards. Once the initial positions are available, the tracking system starts in listening mode and operates on a per-request basis. A mobile target node can invoke the tracking service of the anchor nodes by sending a request message and starting to transmit an unmodulated carrier signal. In response, one of the anchor nodes start transmitting at a close frequency, while the rest measure frequency changes in the interference signal caused by the movement of the mobile node. The relative speed is calculated from the frequency shift by Equation 23 at every anchor node, which is then fed into an extended Kalman filter (EKF) running on a PC to keep an up-to-date model of the node location.

Measurements with eight anchor nodes and a single mobile node in a 50×30 m field indicated that the Doppler shift-based tracking of a mobile node with a speed between 1 and 3 m/s can be performed with an average location error of 1.5 m, speed error of 0.14 m/s and heading error of 7.2°. The update interval of the EKF is the sum of the 0.3 s coordination phase, 0.4 s measurement time and 1.0 s to route the measurement results to the PC, resulting in 1.7 s latency.

4. Doppler shift-based localization

The methods presented in Section 2 and Section 3 for localization and tracking both have the adventage that they use WSN nodes equipped with the same hardware throughout the entire system. The distinction between any two nodes is their actual role (e.g. *transmitter/receiver, mobile/anchor*) and the a priori knowledge of their location. Also, both methods call for computationally expensive non-linear optimization at the sensor fusion phase. However, by specializing certain nodes in the system the "workload" of the fusion phase can be significantly decreased.

Lédeczi et al. (2008) and Chang et al. (2008) proposed the idea to use physically *rotating anchor* nodes in WSNs for node self-localization. Their underlying approach is similar to the one used for mobile node tracking, described in Section 3, in many aspects. Anchor nodes represent key elements of the localization service and measurements rely on the Doppler effect, which in turn utilize radio interferometry. However, the Doppler effect is used in a different way. Instead of having the tracked node and an anchor node transmitting the interfering carrier signals it is two anchor nodes that transmit. One of the anchor nodes is equipped with no specialized hardware, while the other one has a "spinning" antenna. This spinning either refers to the physical rotation of a single antenna or its imitation using an antenna array. In both cases, the target nodes observe a Doppler shift in the received signal due to the actual or virtual rotational movement of the transmitter antenna. The measured Doppler shift is used by the target nodes to determine their bearing from the anchor nodes. The bearing estimates from multiple anchor node pairs are then used to determine the actual location of the target node.

The remainder of this section details the theory behind the Doppler-shift based localization methods using rotating-antenna and antenna-array anchor nodes, their first WSN implementation, and experimental results.

4.1 Bearing estimation with rotating-antenna anchor nodes

The Doppler shift-based radio interferometric localization with rotating-antenna anchor nodes is divided into two distinct phases, *bearing estimation* and *localization*. In the bearing estimation phase, the nodes to be localized, the *target nodes*, measure the Doppler shift in the carrier signal introduced by the rotational movement of the anchor node antenna and deduct bearing information. The localization phase takes the bearing estimates and the anchor node locations to determine the target node location via triangulation.

The bearing estimation starts with two anchor nodes, a rotating and a fixed one, transmitting unmodulated sinusoid carriers at close frequencies, where the fixed node can be an arbitrarily selected transmitter node with known location. Receiver nodes extract the envelope of the interference signal, see Equation 9, and measure its frequency.

Note that radio interferometry is utilized to make frequency estimation feasible on resource constrained sensor nodes as any frequency change in the envelope signal equals to that of the original carrier. Thus, assuming the additional transmitter node has ideal accuracy, its presence and the details of radio interferometry will be temporarily ignored in the following discussion.

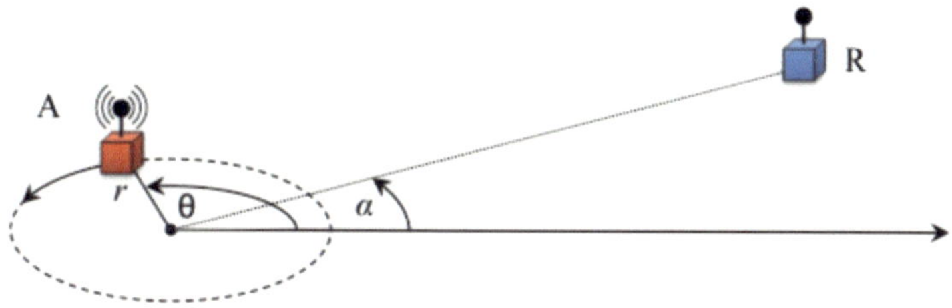

Fig. 8. Doppler shift generated by the rotating-antenna anchor node.

The measured frequency change is related to the movement of the transmitter through the Doppler effect, $\Delta f = vf/c$, where v is the tangential component of the transmitter velocity relative to the receiver, f is the carrier frequency and c is the speed of light. To discuss the case where the transmitter antenna is rotating, consider Figure 8. The antenna of the anchor node, A_{rot}, is rotating counterclockwise around the origin with a radius r and a constant angular velocity ω. Let $\vec{v}(t)$ denote the velocity of the rotating antenna A_{rot} and $\vec{u}_r = \overrightarrow{RA}_{rot}/\|RA_{rot}\|$ the unit length vector pointing from the receiver node R to A_{rot}. The relative speed of A_{rot} and R can then be defined as the following dot product

$$v_p(t) = \vec{u}_r \cdot \vec{v}(t), \tag{30}$$

where

$$\vec{v}(t) = (-\omega r \sin(\omega t + \varphi), \omega r \cos(\omega t + \varphi)) . \tag{31}$$

Therefore, the receiver node R observes the projected velocity component $v_p(t)$:

$$v_p(t) = \vec{u}_r \cdot \vec{v}(t) = \frac{-\omega r d \sin(\omega t + \varphi)}{\sqrt{d^2 + r^2 - 2dr\cos(\omega t + \varphi)}} \tag{32}$$

Consequently, the observed frequency change at R is

$$\begin{aligned} \Delta f = \frac{v_p(t)}{c} f = \frac{f}{c} \frac{-\omega r d \sin(\omega t + \varphi)}{\sqrt{d^2 + r^2 - 2dr\cos(\omega t + \varphi - \alpha)}} \\ = \frac{-\omega r f}{c} \frac{\sin(\omega t + \varphi - \alpha)}{\sqrt{1 + (r/d)^2 - 2(r/d)\cos(\omega t + \varphi - \alpha)}}. \end{aligned} \tag{33}$$

Assuming that the rotational radius is significantly smaller than the node distance, $r \ll d$, the expression of Δf simplifies to

$$\lim_{r/d \to 0} \Delta f = \frac{-\omega r f}{c} \sin(\omega t + \varphi). \tag{34}$$

The argument of the expression in Equation 34 then becomes

$$\omega t + \varphi - \alpha = \sin^{-1}\left(\frac{-c\Delta f}{\omega r f}\right). \tag{35}$$

Thus, instead of estimating the φ initial phase of the rotating anchor, a second receiver, a *reference* node can be used to eliminate the $(\omega t + \varphi)$ term. Let R_2 denote the reference node as in Figure 9. The α angle between the two nodes is obtained by subtracting the two arguments:

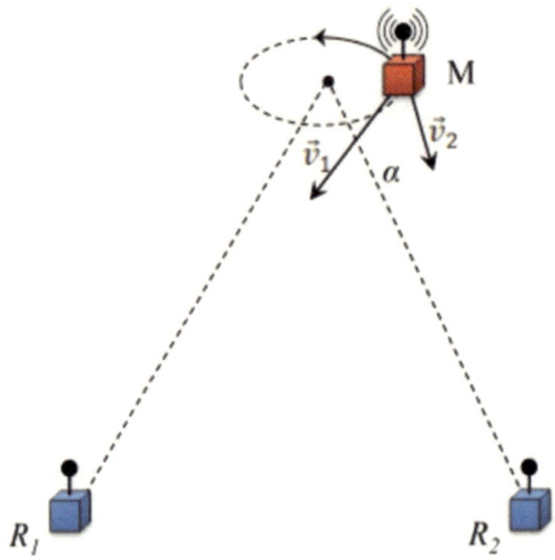

Fig. 9. Estimation of the angle between two receivers and a rotating anchor node.

$$\alpha = \sin^{-1}\left(\frac{c\Delta f_{R_1}}{\omega r f}\right) - \sin^{-1}\left(\frac{c\Delta f_{R_2}}{\omega r f}\right) \tag{36}$$

4.2 Bearing estimation with antenna-array anchor nodes

Similarly to the rotating-antenna case, bearing estimation starts with two anchor nodes transmitting unmodulated sinusoids at close frequencies to construct the interference signal. One of these anchor nodes is equipped with multiple circularly arranged antennas, while the other one can be any node of the WSN with a priori known location. The antenna array mimics the physical rotation of a single antenna by switching between the antennas in sequence - enabling only one at a time. The receivers process the envelope of the interference signal to obtain bearing information through the quasi Doppler shift caused by the imitated movement of the antenna. Note, that in case of ideal transmitters, any change in the carrier frequency introduces the same amount of frequency shift in the detected envelope signal. Therefore, the presence of the interferometry is temporarily disregarded in the following discussion, despite its heavy use in real WSN measurements. Consider an N element uniform circular array (UCA) as depicted in Figure 10. The virtual rotation of the antenna over the N locations results in a constant frequency signal with sudden phase jumps. As a corner case, when $N \to \infty$ the magnitude of phase changes converge to the Doppler shift frequency, essentially giving the same results as the physically rotating antenna discussed in Section 4.1. Thus, by measuring the N discrete phase changes over a full turn of the virtual antenna, the frequency of the mimicked rotating antenna can be calculated at N different locations along the UCA. Assume that the radius of an N-element UCA is negligible compared to receiver to transmitter distance and let the angular speed of the virtually rotating antenna be ω. The phase change between time instants t and Δt then becomes:

$$\Delta\phi(t, t+\Delta t) = \frac{\int_t^{t+\Delta t} sin(\omega t)dt}{\Delta t} = sin\left(\omega t + \frac{\Delta t}{2}\right)\frac{2sin(\frac{\Delta t}{2})}{\omega \Delta t}, \tag{37}$$

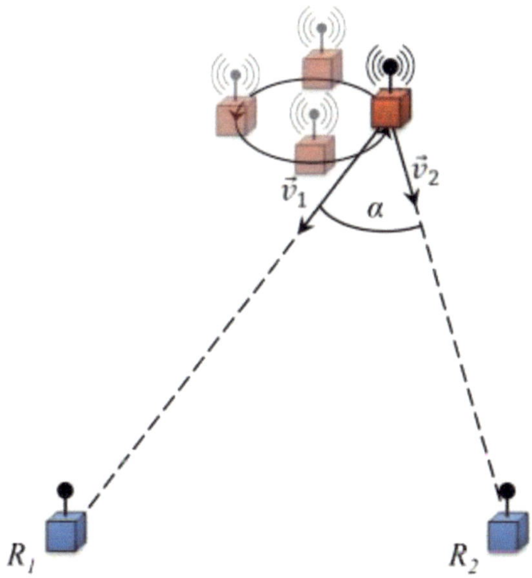

Fig. 10. Quasi doppler bearing estimation using only four discrete (N=4) steps instead of continuous rotation.

where $\omega = 1/(N\Delta t)$. Thus, by measuring the N phase jumps throughout a full turn of the antenna, the $\sin(\omega t + \varphi)$ term can be estimated. The phase of this signal is then used to calculate the bearing to the anchor nodes in a similar fashion as described in Section 4.1.

4.3 Localizaton based on bearing estimates

The localization phase takes multiple angle estimates, α_i, between the target node, rotating-antenna or antenna-array node i and the reference node, as depicted in Figure 11, and deduces the location of node Q. In the optimal case, the bearing estimates intersect at a single point. Due to measurement errors, however, bearing estimates $\alpha_i (i > 2)$ to target node Q may have multiple intersections. In this case, an optimization algorithm can be used to estimate the most probable target node location.

4.4 Platform description and results

The Doppler shift-based SpinLoc localization system presented in Chang et al. (2008) uses MICA2 type CC1000 equipped sensor nodes both for anchor nodes and target nodes, where the rotating-antenna transmitter anchors are physically rotated on a 50 cm radius arm, at 133 revolutions per minute (RPM) by a servo motor. The measurements use the 900 MHz band to generate the 600 Hz interference signal and measure its frequency change.

The measurement round starts with time synchronization initiated by the base station. The rotating anchor node and a static node starts transmitting the unmodulated carriers, which is received and processed by a reference node and the target nodes. The latter two calculate the signal frequencies and send it to the base station. A PC attached to the base station then calculates the orientation of the rotating anchors, the relative bearings and the position of the target nodes.

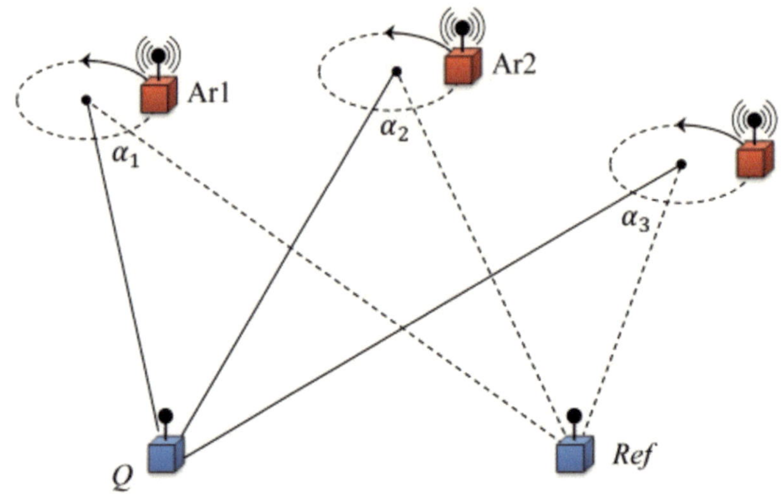

Fig. 11. Localization based on bearing estimates towards rotating transmitter nodes.

SpinLoc experimental measurements in an 8 × 10 m indoor garage area showed an average bearing estimation accuracy of 3° leading to a localization accuracy of approximately 40 cm. The data collection time is reported by Chang et al. (2008) to be less than 1.5 s, while the per target node localization to be approximately 0.5 s.

The quasi-Doppler localization system proposed in Sallai et al. (2009) generates the interference signal with two anchor nodes. One of these anchor nodes switch the continuous, unmodulated carrier between the antennas of the array. A third anchor node and a target node act as receiver and measure the phase jumps caused by the virtual re-location of the transmitter antenna and reconstruct the quasi Doppler-shift of the signal. The Doppler-shift estimates are then sent to the a PC to estimate the bearings and calculate the target node position. Sallai et al. (2009) present experiments that verify the feasibility of the approach, but no thorough evaluation of the expected accuracy is available.

5. Radio interferometric bearing estimation

The first techniques for bearing estimation in WSNs relied both on radio interferometry and a secondary phenomenon, the Doppler-effect. While these techniques provided bearing estimates of adequate precision for node localization and significantly simplified the sensor fusion algorithm, they called for specialized hardware on a subset of the WSN nodes. To mitigate this unwelcome requirement, Amundson et al. (2010) presents another approach that uses the same uniform, simple WSN nodes as the original RIPS discussed in Section 2, yet it is capable of accurately estimating the bearing from anchor nodes. Note, however, that even though specialized hardware is not required, the method assumes a specific arrangement of certain nodes.

The baseline approach of the method presented in Amundson et al. (2010) is to group together three of the four nodes participating in a typical radio interferometric measurement described in 2. The three nodes, two transmitters and a receiver, placed together in a known and fixed geometry, form a composite anchor node. During the interferometric measurement, the estimated phase difference between the receiver of the composite anchor node and the target node constraints the location of the latter to a hyperbola. The bearing of the target node to the anchor node is estimated by the asymptote of the hyperbola.

The rest of the section details the theoretical background of the bearing estimation approach, the prototype system used for experimental measurements and the reported results.

5.1 Bearing estimation using 3-node anchors

The bearing estimation phase starts with a regular interferometric measurement where the location of three closely placed nodes is known a priori. These three nodes, M, A_1 and A_2, form a composite anchor from which a receiver node, R, with unknown location estimates its bearing, see Figure 12 (a). Two nodes of the anchor, A_1 and M, transmit unmodulated sinusoid signals at close frequencies, generating an interference signal with a low-frequency envelope. The receiver nodes, A_2 and R measure the phase of the envelope signal according to Equation 9. The measured $\vartheta_{MA_1A_2R} = \vartheta_R - \vartheta_{A_2}$ phase difference is related to the linear combination of the distances between the transmitters according to Equation 15. Thus,

$$\vartheta_{MA_1A_2R} = 2\pi\frac{d_{MR} - d_{A_1R} + d_{A_1A_2} - d_{MA_2}}{c/f} \pmod{2\pi}, \tag{38}$$

where $\lambda = c/f$ is the wavelength of the carrier signal and d is the distance between the nodes denoted in the index. Assuming that the pairwise distance between the three nodes in the anchor is the same, $d_{A_1A_2} - d_{MA_2} = 0$, the expression of the q-range gets simpler, and Equation 38 becomes

$$\vartheta_{MA_1A_2R} = 2\pi\frac{d_{MR} - d_{A_1R}}{c/f} \pmod{2\pi}. \tag{39}$$

The distance difference $d_{MR} - d_{A_1R}$ is of high importance, thus it is shortened as d_{A_1MR}, and referred to as *t-range*. The d_{A_1MR} t-range takes its maximum when R is colinear with the two transmitters A_1 and M. According to Equation 39 the measured distance difference is 2π-ambiguous if the distance between the two transmitters is larger than $\lambda/2$. Thus, with the assumption that the antennas of the anchor nodes are less than half wavelength apart, the modulo 2π is removed and t-range is simplified to

$$d_{A_1MR} = \frac{\vartheta_{MA_1A_2R}}{2\pi}\cdot\lambda. \tag{40}$$

According to Equation 40, the distances between the receiver node and the two transmitter nodes have a constant difference. Thus, d_{A_1MR} in Equation 40 defines the arm of a hyperbola that intersects with the position of node R as shown in Figure 12 (b).

The possible location coordinates (x, y) of R can be expressed with the general equation of the origin centered hyperbola:

$$\frac{x^2}{a^2} - \frac{y^2}{b^2} = 1. \tag{41}$$

Assuming $d_{A_1R}, d_{MR} \ll d_{A_1M}$, the bearing of the R can estimated with the asymptote of the hyperbola. The bearing α of R relative to the origin can then be written using the two

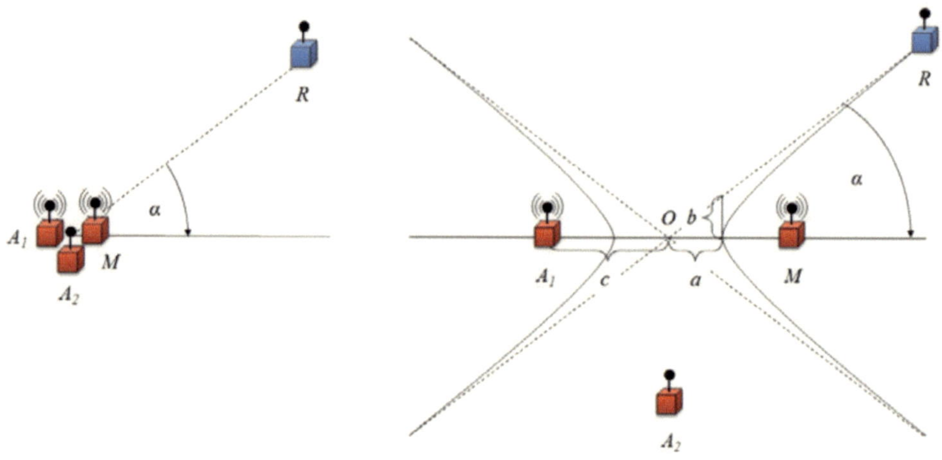

(a) Relative bearing from the 3-node anchor

(b) The 3-node anchor exaggerated to show distances

Fig. 12. Bearing estimation using 3-node anchors

parameters of the hyperbola, a and b:

$$\alpha = \tan^{-1}\left(\frac{b}{a}\right). \tag{42}$$

To find the value of a and b consider the point where the hyperbola arm intersects the line connecting the two focus points. As the distance differences between the foci and *all* points of the hyperbola are the same constant, this point is no exception. Therefore, denoting the distance between the origin and the foci as c and using the t-range definition:

$$d_{A_1R} - d_{MR} = (c + a) - (c - a) = 2a, \tag{43}$$

where c is also known as the linear eccentricity of the hyperbola, for which $c = \sqrt{a^2 + b^2}$ and $c = d_{A_1M}$. Thus, $b = \sqrt{c^2 - a^2}$ and in terms of distances, the value of a and b and the bearing estimate becomes

$$a = (d_{A_1R} - d_{MR})/2 \tag{44}$$

$$b = \sqrt{(d_{A_1M}/2)^2 - ((d_{A_1R} - d_{MR})/2)^2} \tag{45}$$

$$\alpha = \tan^{-1}\left(\frac{\sqrt{(d_{A_1M}/2)^2 - ((d_{A_1R} - d_{MR})/2)^2}}{(d_{A_1R} - d_{MR})/2}\right). \tag{46}$$

The bearing estimate α is, however, ambiguous. First, the hyperbola has two arms. The sign of $d_{A_1R} - d_{MR}$, and thus the sign of the phase difference $\vartheta_R - \vartheta_{A_2}$ selects the arm of the hyperbola. In case $\vartheta_R < \vartheta_{A_2}$, R lies on the left arm, otherwise it lies on the right arm. Second, the $\pm\alpha$ values, relative to the transverse axis of the hyperbola, are both solutions of Equation 46. A second measurement with swapped anchor node roles is used to resolve this ambiguity. Node A_1 and A_2 switch transmitter-receiver roles at the anchor, defining a new transverse axis and $\pm\alpha$ values relative to it. When the $\pm\alpha$ values from the two measurements are compared to the same reference orientation, two of them coincide, yielding the final bearing estimate. Once the bearing estimates are available from a target node to a set of anchor nodes, simple triangulation can be used for localization.

5.2 Platform description and results

The first experimental implementation of the radio interferometry based bearing estimation was named TripLoc presented in Amundson et al. (2010). The nodes in TripLoc are equipped with fine tunable CC1000 radio chips operating in the 433 MHz band and processing it with a microcontroller at a 9 kHz rate.

The basic TripLoc measurement is a series of radio interferometric phase measurements with four participants, three of which form an anchor node with a priori known location. The measurement starts with two nodes of the anchor transmitting unmodulated carriers at close frequencies. The third node of the anchor and another node with unknown location then synchronize their clocks and measure the phase of the envelope signal. The phase measured by the anchor-receiver is transmitted to the receiver to be localized, which in turn calculates the phase offset. The same measurement is then repeated with the anchor-receiver switching role with one of the transmitters. Using the two phase offsets the receiver node is able to calculate its t-range, see Equation 40, and estimate its bearing to the anchor node according to Equation 46. After repeating the bearing estimation relative to another anchor node, TripLoc uses triangulation to determine the non-anchor receiver node location. As the non-anchor node always acts as a "passive receiver", multiple receivers can participate in the same measurement and calculate their own location simultaneously.

TripLoc experiments using a 16 node setup on a 20×20 m low-multipath environment, with four anchor nodes at the corners, showed that receiver nodes are capable of bearing estimation with an average error of approximately 3°. After triangulation, the average overall position error was 78 cm, while the whole localization took less then 1 second.

6. Stochastic radio interferometric positioning

The localization algorithms presented in Sections 2, 4 and 5 are all based on radio interferometry, where two independent transmitter nodes generate an interference signal by transmitting unmodulated carriers at close frequencies. The radio interferometric localization discussed in Section 2 and the radio interferometric bearing estimation in Section 5 measure the phase of this interference signal, while the Doppler-shift based bearing the frequency of that. However, regardless of the way this interference signals is processed at further stages, all of these methods assume that the interference signal has a low frequency. The fine grained control over the transmit frequencies available on the 433 MHz/950 MHz ISM band CC1000 radio chip based platforms is not present on the increasingly popular 2.4 GHz radio chips, which renders this assumption invalid. The recently presented radio interferometric localization method introduced in Dil & Havinga (2011), therefore, drops this assumption and places most of its processing steps on stochastic bases. The rest of the section presents the theory behind the stochastic radio interferometric positioning by highlighting the key differences with the original "deterministic" approach, shows a 2.4 GHz band CC2430 radio chip based implementation and the corresponding localization experimentation results.

6.1 Stochastic interferometric localization

Similarly to the original radio interferometric positioning method, the stochastic approach starts with two nodes transmitting unmodulated carriers at close frequencies. As fine-grained control over the transmit frequencies is not assumed, the calibration phase is omitted as shown in Figure 13. As a result, the frequency value of the two transmitted signals becomes unknown and the difference of the two has a significantly higher variance than in case of the original method after calibration. The high variance of the interference frequency, in turn, manifests in numerous measurements with interference frequency too low or too high for precise phase estimation. This ultimately generates a significantly higher number of inaccurate phase measurements, hence, the phase estimate is treated by the upcoming stochastic processing stages as a random variable. Despite the fact that a significant number of the phase measurements are likely to be corrupted, the phase measurement process is identical to that of the one presented in Section 2. That is, with carrier frequencies f_A and f_B of transmitters A and B, where $f_A > f_B$ is assumed:

$$\Delta\varphi_i = 2\pi\left(\frac{d_{AD} - d_{AC}}{\lambda_A} - \frac{d_{BD} - d_{BC}}{\lambda_B}\right) \mod (2\pi) \tag{47}$$

$$\approx 2\pi\left(\frac{d_{ABCD}}{\lambda_i}\right) \mod (2\pi), \tag{48}$$

where $\Delta\varphi_i$ is, again, the relative phase offset of the envelope frequencies at the two receivers, d_{XY} is the distance of the corresponding nodes X and Y, and λ_i is defined as $\lambda_i = 2c/(f_A + f_B)$.

Again, Equation 48 provides the q-range value with a 2π ambiguity, which is removed in the method in Section 2 by repeating the same measurement over a set of different carrier frequencies to obtain the q-range. The stochastic method takes a different approach here, as shown in Figure 13. Rather than calculating the q-range error distribution first, then estimating the q-range values and then the node positions, it calculates only the q-range error distributions and uses them to estimate the positions directly:

$$(\hat{x}, \hat{y}) = \arg\min_{\hat{x},\hat{y}} \sum_{j=1}^{M}\sum_{i=1}^{N}\left(\Delta\varphi_{i,j} - \Delta\hat{\varphi}_{i,j}\right)^2 \tag{49}$$

$$= \arg\min_{\hat{x},\hat{y}} \sum_{j=1}^{M}\sum_{i=1}^{N}\left(d_{i,j} - \hat{q}_j\right)^2 \tag{50}$$

$$= \arg\min_{\hat{x},\hat{y}} \sum_{j=1}^{M} \text{error}(\hat{q}_j) \tag{51}$$

That is, the stochastic method minimizes Equation 51, where $\Delta\hat{\varphi}_{i,j}$ is the estimated phase offset calculated by the location estimate, $d_{i,j}$ is defined by Equation 17, $\hat{q}_j$ is the q-range estimate and $\text{error}(\hat{q}_j)$ is the q-range error distribution. As the latter one oscillates rapidly, it is not

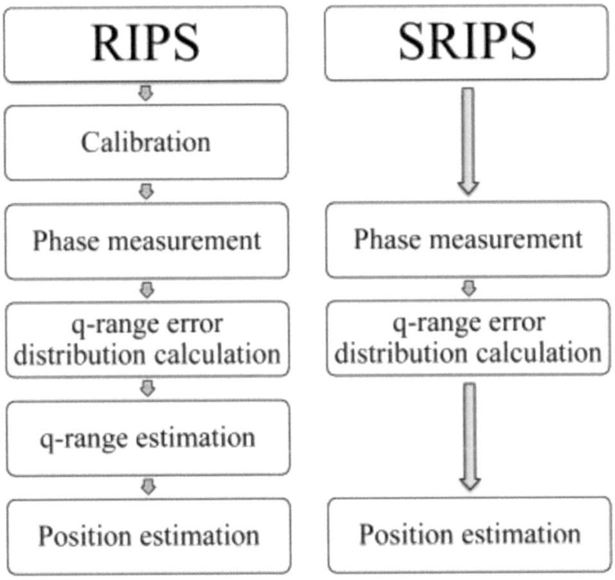

Fig. 13. Comparison of the original (RIPS) and the stochastic interferometric localization algorithms

minimized directly. Rather, it is smoothed first by taking its envelope and then, the minimum of the envelope is searched.

6.2 Platform description and results

The key motivation behind the Stochastic Radio Interferometric Positioning System (SRIPS), detailed in Dil & Havinga (2011), was to make inteferometric localization available on WSN platforms using radio chips operating in the 2.4 GHz band. Popular 2.4 GHz radio chips, such as the CC2430 Texas Instruments (2007b), offer no mechanism to fine tune the transmit frequency. However, they support much higher sampling rates than the CC1000 radio. As a result, the synthesized interference signal envelope frequency has a larger variance, with high confidence in the 200 Hz to 14 kHz range, but it is sampled at a rate of 62.5 kHz.

The SRIPS phase measurement starts with two nodes transmitting unmodulated carriers at nominally same frequencies – without making attempts to fine tuning. The receiver nodes measure the envelope phase and transmit it to a PC. The same phase measurement is carried out over a set of frequencies and is repeated for different transmitter pair combinations. Due to the inaccurately set interference signals, several of the phase measurements are invalid. To cope with the large number of inaccurate measurements, SRIPS skips the q-range calculation step and optimizes Equation 51 on the PC using the phase offsets directly.

SRIPS experiments on a 20×20 m field resulted in an average error of 50 cm.

7. Summary

The algorithms presented in this chapter all rely on radio interferometry in order to localize or track nodes in WSNs. In each case, the interference signal is generated the same way, by two independent transmitters, but the different algorithms exploit features of this signal in different ways. Table 2 compares the algorithms with an emphasis on the experimental implementation and key results. All the prototype systems expect for SRIPS rely on the CC1000 radio chipset, which is designed operate in the 433 MHz and 900 MHz bands. They are characterized by a precisely tuned interference frequency which allows for processing at the limited 9 kHz sampling rate. Clearly, it is the unique fine-tuning capability of the CC1000 LO that made this radio chip - along with the MICA2 platform - the most favorable choice. SRIPS, on the other hand, aimed to make interferometric localization available on platforms that lack a fine-tunable LO. Driven by the reduced control over the interference frequency in the CC2430, SRIPS decided to drop tuning altogether. The available higher 62.5 kHz sampling rate made the processing of the highly varying frequency interference signal possible. This came at the cost of significantly increased number of inaccurate phase measurements, which SRIPS deals with in later processing stages.

	RIPS	SRIPS	RIPS bearing	Doppler tracking	Doppler bearing
Radio chip	CC1000	CC2430	CC1000	CC1000	CC1000
Operating band	433 MHz	2400 MHz	433 MHz	433 MHz	900 MHz
Sampling frequency	9 kHz	62.5 kHz	9 kHz	9 kHz	9 kHz
Interference frequency	300-700 Hz	200-12000 Hz	300-700 Hz	300-700 Hz	300-700 Hz
Special hardware required	No	No	No	No	Yes
Anchor nodes required	No	No	Yes	Yes	Yes
Localization time	Several minutes	Several minutes	Few seconds	Few seconds	Few seconds
Average error	< 5 cm	50 cm	80 cm	1.5 m	40 cm

Table 2. Comparison of RI localization systems

An important aspect in several applications is whether the localization or tracking service requires additional hardware. RIPS, SRIPS, RIPS bearing and Doppler tracking are especially favorable as they rely on no extra hardware. RIPS bearing, however, uses a special arrangement of certain nodes, which might be unfavorable in certain scenarios. Doppler bearing approaches utilize either a physically rotating antenna, or a relatively complex antenna array—both of which are impractical in low-cost WSNs.

RIPS and SRIPS are unique in the sense that they are able to calculate the relative node locations without anchor nodes, that is, without prior knowledge of locations of a subset of nodes. In this mode, they generally require orders of magnitude more time to do the self-localization. The presence of anchor nodes may potentially speed up the localization process both for RIPS and SRIPS, while it is essential for the other systems. In general,

RIPS bearing, Doppler tracking and Doppler bearing leverage the prior knowledge of the anchor node locations and allow for distributed calculation of the locations, e.g. the nodes can estimate their own positions.

The operating range of the presented radio interferometric platforms is generally in the order of tens of meters up to a few hundred meters. In multipath free environments, the original RIPS proves to be the most accurate, which is, again, mainly due to the favorable properties of the CC1000. This is followed by the Doppler bearing, SRIPS and RIPS bearing, each of which achieves sub-meter accuracy. The average error of Doppler tracking is difficult to compare as it assumes the initial positions to be known and because the error is calculated over several iterations.

8. Nomenclature

Anchor node	Node with a priori known location
COTS	Commercial Off-The Shelf
IF	Intermediate Frequency
LO	Local Oscillator
PC	Personal Computer
RIPS	Radio Interferometric Positioning System
RSS	Received Signal Strength
SRIPS	Stochastic Radio Interferometric Positioning System
WSN	Wireless Sensor Network

Table 3. List of acronyms

9. References

Amundson, I., Sallai, J., Koutsoukos, X. & Lédeczi, A. (2010). Radio Interferometric Angle of Arrival Estimation, *7th European Conference on Wireless Sensor Networks*, Springer, Coimbra, Portugal.
URL: *http://www.isis.vanderbilt.edu/node/4139*

Chang, H.-l., Tian, J.-b., Lai, T.-T., Chu, H.-H. & Huang, P. (2008). *Spinning beacons for precise indoor localization*, ACM Press, New York, New York, USA.
URL: *http://portal.acm.org/citation.cfm?id=1460412.1460426*

Dil, B. J. & Havinga, P. J. M. (2011). Stochastic Radio Interferometric Positioning in the 2.4 GHz Range, *Proceedings of the 9th ACM conference on Embedded network sensor systems - SenSys'11*, ACM Press, Seattle, WA, USA.

Fontana, R. J., Richley, E. & Barney, J. (2003). Commercialization of an ultra wideband precision asset location system, *IEEE Conference on Ultra Wideband Systems and Technologies 2003*, number November, IEEE, IEEE, pp. 369–373.
URL: *http://ieeexplore.ieee.org/lpdocs/epic03/wrapper.htm?arnumber=1267866*

Kusý, B., Balogh, G., Völgyesi, P., Sallai, J., Nádas, A., Maróti, M., Meertens, L. & Lédeczi, A. (2006). Node-density independent localization, *Information Processing in Sensor Networks (IPSN 06) SPOTS Track.*

Kusý, B., Lédeczi, A. & Koutsoukos, X. (2007). *Tracking mobile nodes using RF Doppler shifts*, ACM Press, New York, New York, USA.
URL: *http://portal.acm.org/citation.cfm?id=1322263.1322267*

Lédeczi, A., Völgyesi, P., Sallai, J. & Thibodeaux, R. (2008). *A novel RF ranging method*, IEEE. URL: *http://ieeexplore.ieee.org/xpl/freeabs_all.jsp?arnumber=4623303*

Maróti, M., Völgyesi, P., Dóra, S., Kusý, B., Nádas, A., Lédeczi, A., Balogh, G. & Molnár, K. (2005). *Radio interferometric geolocation*, ACM Press, New York, New York, USA. URL: *http://portal.acm.org/citation.cfm?id=1098918.1098920*

Sallai, J., Völgyesi, P. & Lédeczi, A. (2009). Radio interferometric Quasi Doppler bearing estimation, *IPSN '09 Proceedings of the 2009 International Conference on Information Processing in Sensor Networks*, IEEE Computer Society, Washington, DC, USA, pp. 325–336. URL: *http://portal.acm.org/citation.cfm?id=1602165.1602195*

Taubenheim, D., Kyperountas, S. & Correal, N. (2005). Distributed radiolocation hardware core for ieee 802.15.4, *Technical report*, Motorola Labs, Plantation, Florida.

Texas Instruments (2007a). Single-chip very low power RF transceiver, *Technical report*, Texas Instruments. URL: *http://www.ti.com/lit/ds/symlink/cc1000.pdf*

Texas Instruments (2007b). A true system-on-chip solution for 2.4 GHz IEEE 802.15.4 / zigbee(tm), *Technical report*, Texas Instruments. URL: *http://www.ti.com/lit/ds/symlink/cc2430.pdf*

Tuchler, M., Schwarz, V. & Huber, A. (2005). Location accuracy of an UWB localization system in a multi-path environment, *2005 IEEE International Conference on UltraWideband* (September): 414–419. URL: *http://ieeexplore.ieee.org/lpdocs/epic03/wrapper.htm?arnumber=1570023*